An Integrated Approach to Genomics

An Integrated Approach to Genomics

Editor: Joel White

Callisto Reference
www.callistoreference.com

Callisto Reference,
118-35 Queens Blvd., Suite 400,
Forest Hills, NY 11375, USA

Visit us on the World Wide Web at:
www.callistoreference.com

ISBN: 978-1-64116-545-7 (Hardback)

Cataloging-in-Publication Data

An integrated approach to genomics / edited by Joel White.
p. cm.
Includes bibliographical references and index.
ISBN 978-1-64116-545-7
1. Genomics. 2. Genomes. 3. Molecular genetics. I. White, Joel.
QH447 .I58 2022
572.86--dc23

Table of Contents

Preface

Genes are the functional units of heredity that direct the production of proteins. The field of genomics is concerned with the collective quantification and characterization of genes, as well as the analysis and sequencing of genomes using bioinformatics and high throughput DNA. It also studies intragenomic processes like heterosis, epistasis, pleiotropy, etc. Some of the research areas in genomics are the study of gene translation, transcription and protein-protein interactions, 3D structures of the protein encoded by a genome, epigenetic modifications on a cell's DNA, and bacterial and viral genomes. Advances in genomics have furthered the understanding of complex biological processes. It has also revolutionized the sectors of medicine and biotechnology. Genomics is an upcoming field of science that has undergone rapid development over the past few decades. It elucidates new techniques in a multidisciplinary manner. This book includes contributions of experts and scientists which will provide innovative insights into this field.

This book is a comprehensive compilation of works of different researchers from varied parts of the world. It includes valuable experiences of the researchers with the sole objective of providing the readers (learners) with a proper knowledge of the concerned field. This book will be beneficial in evoking inspiration and enhancing the knowledge of the interested readers.

In the end, I would like to extend my heartiest thanks to the authors who worked with great determination on their chapters. I also appreciate the publisher's support in the course of the book. I would also like to deeply acknowledge my family who stood by me as a source of inspiration during the project.

Editor

1

New Genomic Approaches to Enhance Biomass Degradation by the Industrial Fungus *Trichoderma reesei*

Renato Graciano de Paula, Amanda Cristina Campos Antoniêto, Liliane Fraga Costa Ribeiro, Cláudia Batista Carraro, Karoline Maria Vieira Nogueira, Douglas Christian Borges Lopes, Alinne Costa Silva, Mariana Taíse Zerbini, Wellington Ramos Pedersoli, Mariana do Nascimento Costa, and Roberto Nascimento Silva

Molecular Biotechnology Laboratory, Department of Biochemistry and Immunology, Ribeirao Preto Medical School (FMRP), University of Sao Paulo, Ribeirao Preto, SP, Brazil

Correspondence should be addressed to Renato Graciano de Paula; renatogpaula@usp.br

Academic Editor: Raul A. Platero

The filamentous fungi *Trichoderma reesei* is one of the most well-studied cellulolytic microorganisms. It is the most important fungus for the industrial production of enzymes to biomass deconstruction being widely used in the biotechnology industry, mainly in the production of biofuels. Here, we performed an analytic review of the holocellulolytic system presented by *T. reesei* as well as the transcriptional and signaling mechanisms involved with holocellulase expression in this fungus. We also discuss new perspectives about control of secretion and cellulase expression based on RNA-seq and functional characterization data of *T. reesei* growth in different carbon sources, which comprise glucose, cellulose, sophorose, and sugarcane bagasse.

1. *Trichoderma reesei*: Environmental and Lignocellulosic Biomass Degrader

Trichoderma species are ubiquitous and cosmopolitan. They are very efficient colonizers of a variety of habitats and can be found from the tundra to the tropics [1], especially in lignocellulosic material and plant rhizospheres, and this effectiveness is translated by the ability of competently degrading the available substrate and of secreting different enzymes and metabolites used in the process [2–5]. The capability of growing in such a variety of carbon sources is also due to the high and fast capacity of responding to diverse environmental signals, being able to adapt according to that current background and regulate its growth, conidiation, and the production of enzymes and secondary metabolites. These signals may vary from different nutrients found in the milieu to the absence and presence of light, and adjusting to them is crucial for the survival of the microorganism [2, 4, 6–8].

As a result of this versatility, *Trichoderma* species are very useful in many aspects that range from plant biocontrol [9] to various sorts of industries [10–13], especially for the cellulolytic enzymes produced by them. Among all species from this genus and which are industrially used, *Trichoderma reesei* is the most studied one regarding lignocellulosic biomass degradation, since it is the main producer of cellulolytic and xylanolytic enzymes [14–17]. The ability of growing in a wide range of carbon sources allows great variability in the production of cellulases, since the gene expression and secretion of enzymes are directly dependent on the different chemical signals produced from the diverse substrates. Considering that the plant biomass, one of the most important and complex substrates used by *Trichoderma*, is composed of mono-, di-, and polysaccharides, the different sugars may have different levels of induction or repression of cellulase genes. Some of the cellulase inducers are cellulose, β-glucan, xylan, lactose, cellobiose, and sophorose, while glucose is the main repressor carbon source [18]. When *T. reesei* degrades

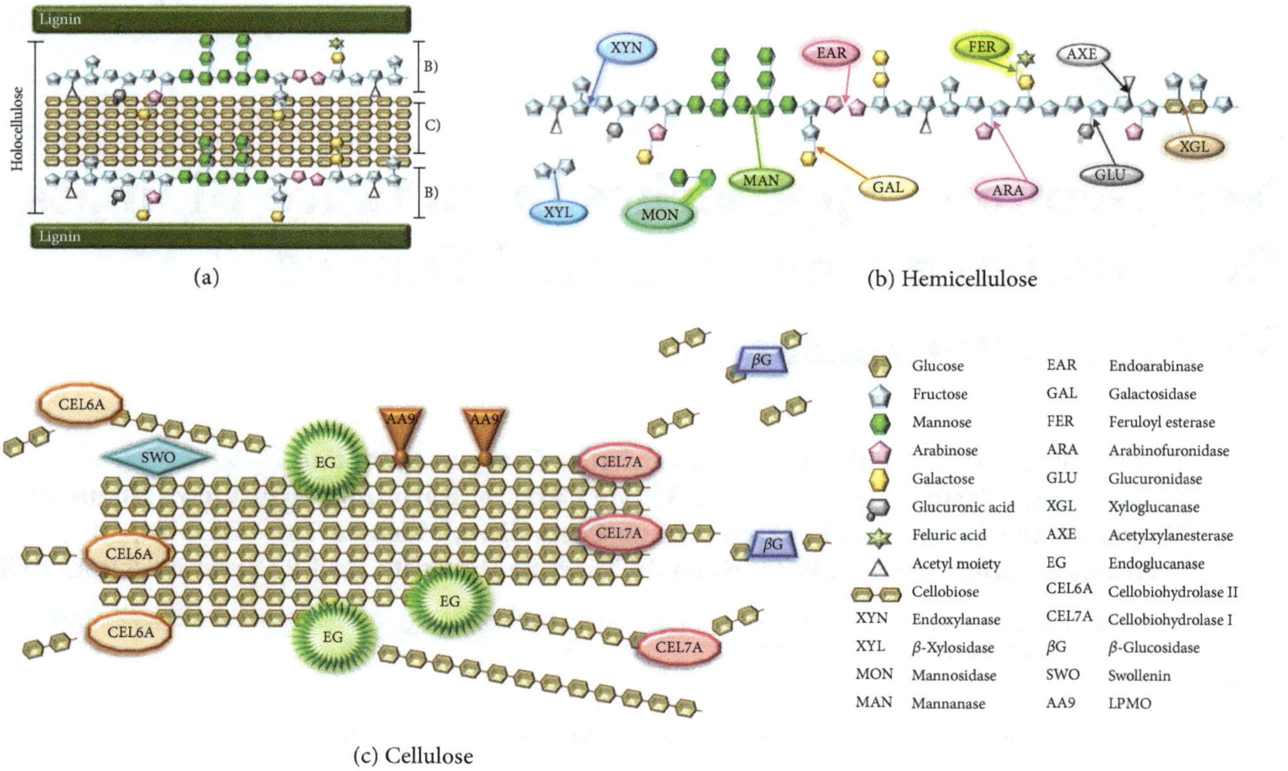

Figure 1: Global regulation of holocellulase expression in *T. reesei*. (a) The schematic structure of the lignocellulosic biomass, which is constituted by lignin and holocellulose, composed of hemicellulose and cellulose. All chains are drawn from the reducing (left) to the nonreducing end (right). (b) Enzymes that attack hemicellulose act in synergy in order to efficiently hydrolyze it and promote a more accessible surface area on cellulose, to enhance cellulase activities. (c) The enzymatic degradation of cellulose: EG acts by cleaving in amorphous regions of the chain, while CEL6A and CEL7A cleave at the nonreducing and reducing ends, respectively. The resulting oligosaccharides from this cleavage are then broken into monosaccharides by β-glucosidase, so they become capable of being directly metabolized by the organism. SWO expands the cellulolytic chain to improve cellulase accessibility to it, and AA9 works through bivalent-metallic-ion-dependent oxidative metabolism (based, among others, on [39, 144–146]).

the lignocellulosic biomass, cellobiose may be converted into sophorose by a transglycosylation activity of a β-glucosidase [19, 20]. The comparison of the genomes of *Trichoderma* species, including *T. reesei*, suggests they have a mycoparasitic common ancestral, probably from fungi that degrade lignocellulosic material. Considering this, *T. reesei* may have maintained the mycoparasitic characteristic, which allows it to have advantages over other species when competing for substrate, through the conversion of cellobiose into sophorose by transglycosylation to be metabolized [21, 22]. Differently from other fungi, in *T. reesei*, sophorose acts as a very potent cellulase inducer in very low concentrations, being able to induce the expression of some xylanases as well [23, 24].

With the increasing concern about the environmental disadvantages that fossil fuels and nuclear energy represent nowadays, there has been considerable pursuit for new ways of generating large-scale renewable green energy. One of the currently most promising possibilities is the usage of lignocellulosic material from agricultural and industrial waste [25, 26]. This material is mainly composed of cellulose and hemicellulose, which suffer enzymatic hydrolysis and are converted into simple carbohydrate monomers and finally into second-generation bioethanol [13, 27, 28]. However, considering that the plant biomass is highly recalcitrant, large amounts of enzymes are needed during the hydrolysis process, which make the biofuel production economically unfeasible [29–31]. In this context, *Trichoderma reesei* may play an important role in decreasing costs for bioethanol production, whereas it is the filamentous fungus with the greatest capacity of degrading the lignocellulosic biomass [2, 15, 32].

2. The Global Analysis of the Enzymatic Repertoire of *Trichoderma reesei*

The filamentous fungus *T. reesei* obtains energy through the degradation of the lignocellulosic biomass, composed specially of the cellulose polymer [2, 33, 34]. The cellulolytic enzymes synergistically act to degrade the cellulose polymer. Regarding the site of action, these enzymes are classified in at least three large groups: the endoglucanases, which cleave the internal bonds of the cellulose fiber; the exoglucanases, which act in the external region of the cellulose chain; and β-glucosidases, which hydrolyze soluble oligosaccharides into glucose molecules [2, 35] (Figure 1). Recently, LPMOs (lytic polysaccharide monooxygenases) have been suggested to make the cellulose polymer more accessible to the action of traditional cellulases through an oxidative mechanism [36]. In addition to the LPMOs, a protein known as swollenin also participates in the deconstruction of the plant cell wall by breaking the hydrogen bonds between the cellulose microfibrils without hydrolyzing them, thus increasing the biomass degradation efficiency [37] (Figure 1).

In 2012, Häkkinen and coworkers performed a reannotation of all genes encoding enzymes belonging to CAZy (Carbohydrate-Active Enzymes—http://www.cazy.org) in *T. reesei* and identified 201 genes of glycosyl hydrolases, 22 carbohydrate esterases, and 5 genes of polysaccharide lyases [38]. The cellulases produced by *T. reesei* belong to six GH families: endo-β-1,4-D-glucanases are found in the GH5, GH7, GH12, and GH45 families, the exoglucanases in the GH6 and GH7 families, and the β-glucosidases in the family GH3 [39]. The Cel7a is the dominant enzyme of the cellulolytic complex, comprising about 60% of the total proteins secreted by *T. reesei*, followed by Cel6a (20%), and then the endoglucanases, mostly Cel7b (10%). In fewer amounts, the β-glucosidases represent only 1% of the total proteins secreted by this species [40–42].

In 2014, dos Santos Castro and coworkers showed that when *T. reesei* is grown in cellulase inducing carbon sources, such as cellulose and sophorose, CAZyme-encoding genes are highly transcriptionally induced. An opposite profile is observed during growth in glucose, in which most of the genes encoding GH families had low expression. In this study, 61 genes encoding CAZymes were highly expressed in the presence of cellulose in comparison to glucose [21]. Among these, the main gene upregulated was a copper-dependent polysaccharide monooxygenase *cel61b* (ID 120961), almost 3 thousand-fold more expressed in cellulose than in glucose. The endoglucanase *cel12a* (ID 123232) was the second most upregulated gene in this condition, followed by a β-mannanase. In sophorose, 58 genes were transcriptionally induced when compared to glucose. The most expressed gene in this condition was the endoglucanase *cel12a* (ID 123232, as well as observed in cellulose). The two cellobiohydrolases of *T. reesei cel7a* (ID 123989) and *cel6a* (ID 72567) ranked the second and third among the most expressed genes in sophorose [21] (Figure 2). In a study by de Paula [43], 98 CAZyme-encoding genes were upregulated during growth in sugarcane bagasse compared to glycerol in the *T. reesei* QM6a strain. The endoglucanase *cel12a* was the third most expressed gene in this condition, similar to that found by dos Santos Castro and coworkers [21] in the presence of cellulose. The two CAZyme genes most expressed in sugarcane bagasse encode a mannanase from the GH76 family (ID 122495) and the hemicellulase *xyn3* (ID 120229). Similar to that found by dos Santos Castro and coworkers in sophorose [21], the cellobiohydrolase *cel7a* was also highly expressed in sugarcane bagasse, achieving 326-fold higher expression in sugarcane bagasse than in glycerol (ID 122495) (Figure 2).

CAZyme gene expression is regulated at the transcriptional level by transcription factors such as XYR1 and CRE1 (for more details, see Transcriptional Regulation of

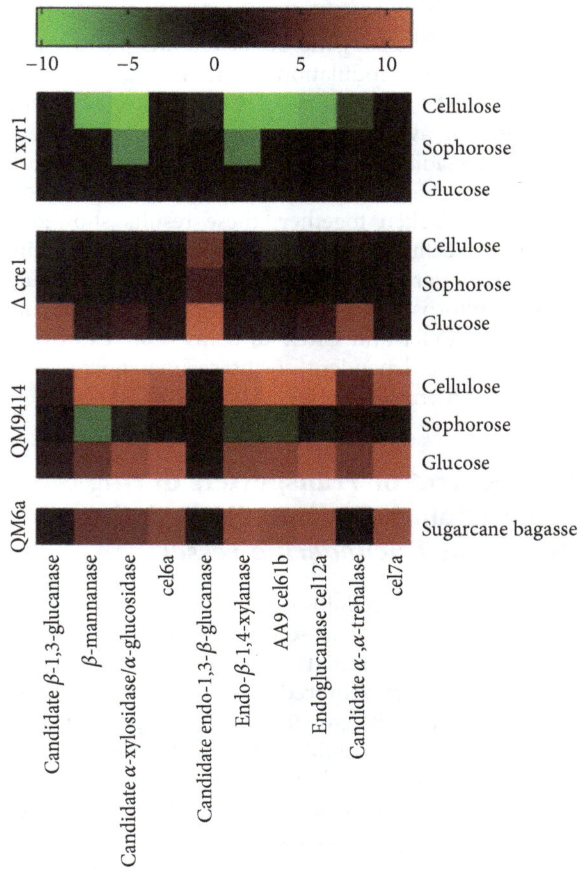

Figure 2: Heatmap expression of Top CAZy-encoding genes differentially expressed in *T. reesei* strains grown in cellulose, sophorose, glucose, and sugarcane bagasse. The results of gene expressions were transformed in Log2FoldChange values and employed to heat map construction using the GraphPad Prism version 7 program (https://www.graphpad.com/).

Biomass Degradation). In 2016, dos Santos Castro and coworkers showed that the XYR1 regulator positively regulates expression of at least 61 CAZyme genes in cellulose and 46 genes in sophorose [35]. The genes encoding a candidate α-xylosidase/α-glucosidase (ID 69944), endo-β-1,4-xylanase (ID 120229), and a copper-dependent monooxygenase polysaccharide candidate AA9 (ID 120961) were the major XYR1 target in the presence of cellulose. All of them were more than 500-fold less expressed in the *Δxyr1* mutant compared to the parental QM9414, and the first two genes were also modulated by XYR1 in the presence of sophorose. However, none of these genes are under CRE1-mediated carbon catabolic repression (CCR) [44, 45] (Figure 2). CRE1-mediated CCR was also evaluated regarding the CAZymes of *T. reesei* by Antoniêto and coworkers [44, 45]. In these studies, the most evident CCR occurred during growth of *T. reesei* in glucose [45]. In this condition, several genes encoding CAZymes were repressed by CRE1. A gene encoding a candidate endo-1,3-β-glucanase (ID 73256) was the most repressed by CRE1, being more than 2 thousand-fold more expressed in the mutant *Δcre1* compared to the QM9414, followed by another candidate β-1,3-glucanase

(ID 56418) and a candidate α,α-trehalase (ID 123456) (Figure 2). The trehalase gene was also under positive regulation of XYR1 [35]. Modulation of the trehalase gene by the XYR1 and CRE1 regulators may be a strategy adopted by *T. reesei* to avoid the unnecessary use of the energy stock when a readily available carbon source is in the culture medium, since this disaccharide acts as an energetic reserve in fungi [44]. Taken together, these results showed that, although the cellulase gene expression profile is similar during *T. reesei* growth in inducing carbon sources such as cellulose, sophorose, and sugarcane bagasse, the mechanisms employed by XYR1 and CRE1 to control the expression of specific genes are dependent on the carbon source available in the environment (Figure 2).

3. Involvement of Transporters during the Lignocellulosic Biomass Degradation Process in *Trichoderma reesei*

The process of lignocellulosic biomass utilization involves the capacity of *T. reesei* to sense the insoluble cellulose in the environment and initiates the rapid production of the enzymatic machinery required to breakdown cellulose and offer the nutrients to its growth [46–48]. Appropriated sense of the extracellular insoluble cellulose is key to initiating the rapid synthesis of cellulases by this fungus, and the uptake of soluble sugars released from biomass hydrolysis denote a potential point of control in the induced cascade [48]. In this context, transporter proteins have an important role during the biomass degradation process. Transport systems act in the sensing and uptaking of essential nutrients and ions, allowing excretion of end products of metabolism and toxic substances. Also, these transporters are involved in communication between cells and the environment [49, 50]. It is possible that organisms sense cellulose through recognition of sugars by a transporter in the membrane. In fact, two MFS sugar transporters, *Stp1* and *Crt1*, were implicated in cellulose sensing and cellulase induction in *T. reesei* [48]. Transporter proteins can carry different small molecule inducers from the extracellular environment into the fungus influencing the expression of CAZyme-encoding genes [51–53]. Approximately, 5% (459 genes) of the *T. reesei* genome comprises genes that encode proteins involved in transport [21]. Among these, the largest group of identified transporters belongs to the major facilitator superfamily (MFS), class of sugar transporters, followed by ABC (ATP binding cassette) transporters. These two families of transporter proteins have been most intensively studied among the fungal transporters [34, 54].

The genomes of the filamentous fungus encode large numbers of MFS transporters [55, 56]. These proteins can transport a broad variety of substrates and are divided into 17 distinct families among which three families (1, 5, and 7) are involved in sugar transport into the cell. These transporters carried out the transport of carbon sources, including hexose and pentose sugars, and small soluble molecules in response to ion gradients [50, 57, 58]. In the filamentous fungus, sugar transporters, which belong to the MFS permease family, have a characteristic to recognize and carry more than one type of sugar into the fungal cell. For example, the *T. reesei* STP1 transporter is involved in glucose and cellobiose uptake [48], as well as *Aspergillus nidulans* transporter XtrD was shown to be able to transport xylose, glucose, and several other monosaccharides [59].

Despite advances in studies about the involvement of MFS sugar transporters during biomass degradation, a very few sugar transporters have been functionally characterized in *T. reesei* [48, 60, 61]. A deep transcriptomic and proteomic study investigating the molecular basis for lignocellulose-degrading enzyme production in *T. reesei* during growth in cellulose, sophorose, and glucose revealed new components involved in cellulose degradation, including transporters, while the MFS permeases family was the most present in the different carbon sources evaluated [21]. In this study, MFS permeases differentially expressed in the three carbon sources were analyzed. The gene encoding MFS permease (ID 69957) and MFS permease (ID 76800), targets of light signaling in *Trichoderma*, were highly expressed in cellulose when compared to glucose and sophorose. In sophorose, a gene encoding MFS maltose permease (ID 48444) was highly induced by this carbon source, while the MFS permease ID 76641 was expressed at a higher level in glucose than in sophorose or cellulose [21]. Conversely, cultures with cellulose or sophorose promoted the expression of MFS permeases at similar expression levels, including *crt1* (ID 3405), required for cellulase induction by cellulose and lactose, besides mediating the cellulose sensing process in *T. reesei* [48]; *Hxt1*, a glucose permease [60]; *Str1* (ID 50894), a xylose transporter [61]; and MFS ID 79202, which is critical for cellulase production in lactose cultures, although it is related to a sucrose transporter. In addition, the stp1 exhibited a higher level of expression in sophorose than in cellulose, although it is involved in cellobiose and glucose transport [21].

In addition, in *T. reesei*, the global induction of genes in response to exposure to wheat straw showed that MFS transporters were highly transcribed in straw and repressed in glucose-rich conditions. The genes encoding MFS ID 3405 (*crt1*), ID 50894 (*str1*), and ID 69957 were upregulated in the presence of straw and downregulated in the presence of glucose [62]. Besides, during cultivation of *T. reesei* RUT C-30 strain in sugarcane bagasse, the genes encoding the MFS permeases *ctr1* and *str1* were strongly induced by this carbon source at all the time points analyzed (6, 12, and 24 hours) [63]. The induction of MFS transporters may be controlled by endogenous regulatory systems over a range of in vivo conditions. In *T. reesei*, the modulation of transporter expression was shown to be carbon source-dependent [21, 35, 44]. A global transcriptome analysis of the *Δxyr1 T. reesei* mutant strain and the parental strain QM9414 grown in the presence of cellulose, sophorose, and glucose as sole carbon sources showed that genes encoding to proteins belonging to MFS permease superfamily are the most modulated genes by XYR1 (Figure 3). The genes that encode a MFS permease ID 69957, ID 44175, ID 54632, ID 79202, ID 50894, and the ctr1 showed to be strongly repressed by XYR1 in the presence of cellulose [35].

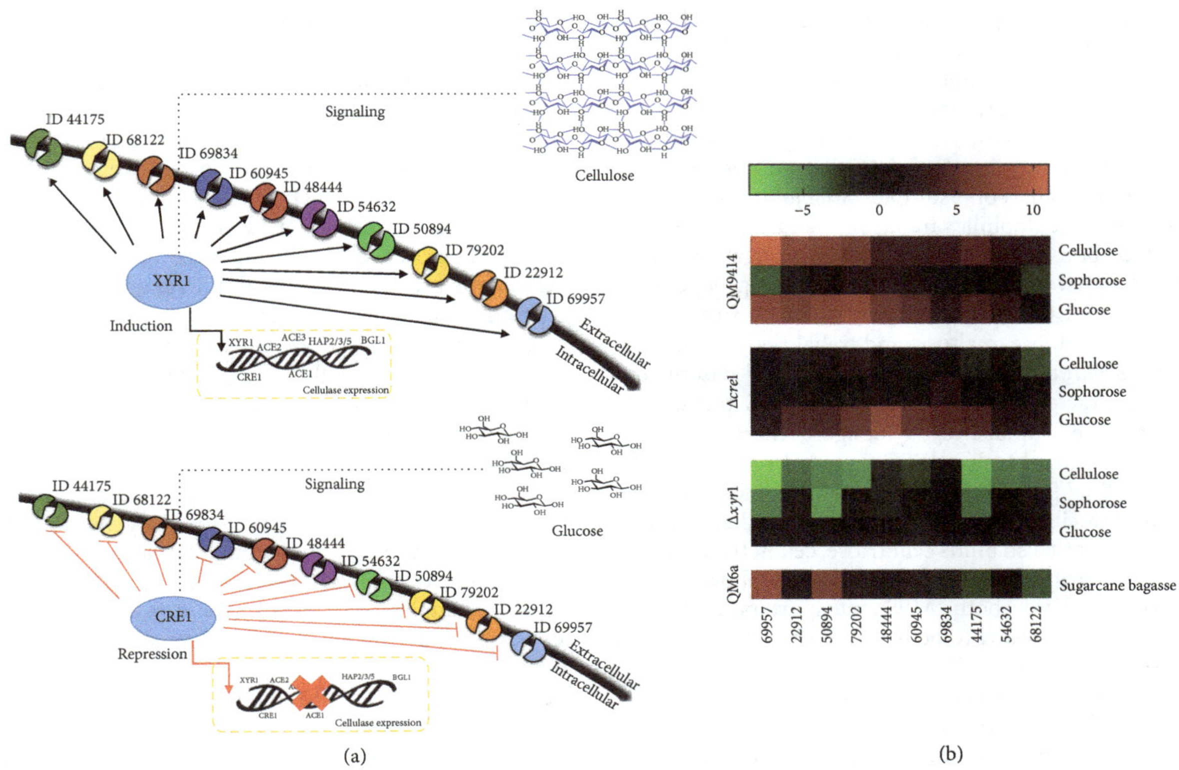

FIGURE 3: Sugar transporters potentially associated with the degradation of biomass in *T. reesei*. (a) The predicted model of XYR1 and CRE1-mediated MFS permease regulation in *T. reesei* under cellulose inducing condition and glucose repressed condition. (b) Heatmap expression of Top MFS transporter-encoding genes differentially expressed in *T. reesei* strains grown in cellulose, sophorose, glucose, and sugarcane bagasse. The results of gene expressions were transformed in Log2FoldChange values and employed to heat map construction using the GraphPad Prism version 7 program (https://www.graphpad.com/).

Furthermore, it is possible that the protein transporters are involved in carbon catabolic repression (CCR) in *T. reesei*. During analysis between *Δcre1 T. reesei* mutant strain and parental strain QM9414, grown in cellulose, glucose, and sophorose, the main genes marked by CRE1-mediated CCR encode to proteins belonging to MFS permeases superfamily. Most genes of MFS permeases were upregulated by CRE1 in the presence of glucose. Among these, the expression of the maltose permease gene (ID 48444) and three other MFS permease genes (ID 60945, ID 79202, and ID 44174) was higher in the *Δcre1* mutant than in the parental strain [45] (Figure 3). In sophorose, the gene encoding a xylose transporter gene (ID 104072) was highly repressed by CRE1. During the transcriptional analysis of *T. reesei* grown in wheat straw and glucose, this gene was strongly transcribed in wheat straw and repressed in glucose [62].

Furthermore, a large number of ABC transporters are encoded in *T. reesei* [34]. Generally, ABC proteins are integral to the membrane, acting as ATP-driven transporters for several substrates, including lipids, sugar, drugs, heavy metals, and auxin [64]. In fungi, ABC transporters are involved with the secretion of secondary metabolites, resistance to toxic compounds, and cell signaling [63]. Although the ABC transporters have been shown to be important for different processes in fungi, the role of the transporters belonging to the ABC family is still unclear in *T. reesei*. In this fungus, the ABC transporters showed carbon source-dependent transcriptional regulation, which were upregulated in cellulose and sophorose [35]. In cellulose, the ABC transporter ID 76682 was highly upregulated, while the gene encoding the ABC transporter ID 80028 was upregulated in sophorose [21]. Additionally, in *T. reesei*, it was demonstrated that the transport of molecules by ABC transporters is highly induced by wheat straw and lactose than glucose [53, 65].

The ABC transporters showed to be modulated by transcription factors such as XYR1 and CRE1. In *Δxyr1 T. reesei* mutant strain, the genes encoding ABC transporters ID 55814, ID 60116 (MRP-type ABC transporter), and ID 120114 were downregulated in the presence of cellulose, sophorose, and glucose, respectively. Oppositely, the gene encoding ABC transporter ID 58899 (MDR-type ABC transporter) was upregulated in the presence of cellulose [35]. In *Δcre1* mutant strain, the gene encoding ABC transporter ID 76682 (PDR-type ABC transporter) was upregulated in the presence of cellulose [45] while in sophorose, the major CRE1 repressed gene was ID 76682. However, four other ABC transporter genes (ID 82105, 123293, 73924, and 58899) were repressed by CRE1 in the presence of this carbon source [44]. Among the identified ABC transporters, most of them are correlated with multidrug resistance (MDR), pleiotropic drug resistance (PDR), and multidrug resistance-related protein (MRP) ABC protein subfamilies [66]. Although the ABC transporters contribute to multidrug resistance in microbial pathogens and tumor cells, in fungi, these transporters have been scarcely studied [34]. In other fungi, the ABC transporters are also related with mycoparasitic interaction and antifungal resistance [67, 68], and interestingly, several studies have shown the involvement of ABC transporters with the sugar transport [69]. For example, *Sulfolobus solfataricus* (extreme thermoacidophilic archaeon) uses several sugars as the sole carbon and energy source. This sugar transport is mediated by two families of protein binding dependent ABC transporters that may transport different sugars such as arabinose, cellobiose, maltose, and trehalose [70]. In *Clorynebacterium glutamicum*, Watanabe et al. [71] described a functional characterization of a xyloside ABC transporter and an enhanced uptake of xylooligosaccharides in the presence of a functional xylEFG-encoded xyloside ABC transporter. In *Pyrococcus furiosus*, cellobiose uptake involves an inducible ABC transporter system that not only binds cellobiose but also binds cellotriose, cellotetraose, cellopentaose, laminaribiose, laminaritriose, and sophorose [72].

As exposed here, both MFS transporters and ABC transporters are suggested to be involved during the process of biomass degradation in *T. reesei*. The novel transporters identified offer new perspectives for studies about the involvement of protein transporters in the expression and secretion of cellulase for degradation of lignocellulosic biomass. With this, new proteins might be revealed as involved in the process to sense and transduce signals related to biomass deconstruction, providing future strain improvement for cellulase production. The characterization of these transporters may allow the construction of more efficient strains in the degradation of plant biomass and contribute to the implementation of *T. reesei* in the bioethanol industry.

4. Transcriptional Regulation of Biomass Degradation

The regulation of holocellulose degradation by the fungus *T. reesei* is a highly coordinated phenomenon dependent on the carbon sources available in the medium. In the presence of readily metabolizable carbon sources, such as glucose, the fungus represses the genes accountable for the expression of cellulolytic enzymes as a way of saving energy, while in the presence of inductive carbon sources, such as cellobiose and sophorose, the fungus activates the cellulase production. At the transcriptional level, transcription factors are the key proteins for regulating the expression of genes that act on the hydrolysis of the cellulose polymer [32, 73–75]. In *T. reesei*, ten transcription factors involved in the regulation of this process have been identified so far. They are the positive regulators XYR1, ACEII, ACEIII, LAE1, VEL1, BglR, and the HAP2/3/5 complex, as well the repressors ACEI, RCE1, and CRE1 (Figure 4).

In *T. reesei*, the XYR1 is the master positive regulator of cellulase production, and deletion of this transcription factor totally eliminates the cellulase gene expression and also impairs the induction of hemicellulolytic genes involved with the degradation of xylans and arabinans [73, 76–78]. In 2016, dos Santos Castro and coworkers showed that XYR1 mainly regulates genes belonging to the CAZymes family, in a carbon source-dependent manner [35]. In this work, two xylosidases were the main downregulated genes in the *Δxyr1* mutant in the presence of cellulose (ID 69944; 1260-fold) and sophorose (ID 121127; 209-fold). In addition to CAZymes, transcription factors and transporters were also regulated by XYR1. Besides XYR1, the deletion of the *aceII* transcription factor causes a decrease in the transcription levels of most cellulases and significantly reduces the cellulolytic activity when the fungus is grown in a cellulose-containing medium [79]. Furthermore, the ACEIII transcription factor was more recently discovered, and its deletion was detrimental to the production of cellulases and xylanases in *T. reesei* [80]. It is believed that the HAP2/3/5 complex promotes the formation of an open chromatin structure, necessary for the activation of the transcription process [81]. In addition, the BglR transcription factor acts as a positive regulator of β-glycosidases-specific genes [82]. Deletion of the *lae1* reduces the production of cellulases, xylanases, and the auxiliary factors CIP1 and swollenin [83]. Finally, deletion of *vel1* completely decreases the expression of cellulases, xylanases, and *xyr1* genes in the presence of lactose. Interestingly, studies have shown that a combined action occurs between the LAE1/VEL1 complex for regulation of genes involved in biomass degradation [83, 84].

Among the negative regulators of cellulase gene transcription, the main transcription factor is the CRE1 catabolic repressor [75]. In 2014, Antoniêto et al. showed that, in addition to traditional cellulases, CRE1 deletion affects the expression of genes involved in nutrient transport, other transcription factors, and oxidative metabolism. In the presence of cellulose, this regulator represses genes encoding copper transporters and ferric reductase enzymes and, consequently, inhibits the access of the traditional cellulases to the cellulose polymer. In the presence of glucose, CRE1 acts suppressing the expression of genes related to the entry of the inducers into the cell [45]. It was also shown that the transcription factor *xyr1* (ID 122208) is the main target regulator of CRE1 in the presence of glucose, being almost 40 times more expressed in the mutant *Δcre1* when compared to the parental QM9414 [45]. In sophorose, CRE1 mainly modulates CAZYmes and membrane permeases, including maltose permeases that possibly act transporting the disaccharide sophorose [44]. Regarding the other negative regulators of cellulose degradation, studies have shown that ACEI represses the expression of the major cellulolytic genes (*cbh1*, *cbh2*, *egl1*, and *egl2*) and xylanases (*xyn1* and *xyn2*), under the cellulose- and sophorose-inducing conditions [85]. In 2017, Cao and coworkers identified a new transcription factor, Rce1, which acts as a repressor of cellulase gene expression, antagonizing XYR1 by binding to the *cbh1* promoter [86].

Despite the numerous studies about the transcription factors already identified in *T. reesei*, several regulatory proteins involved in the regulation of the lignocellulosic biomass deconstruction have not been characterized yet (Figure 4). In a study of dos Santos Castro and coworkers [21], the expression of 7 genes encoding transcription factors was affected during growth in cellulose. Likewise, 18 genes

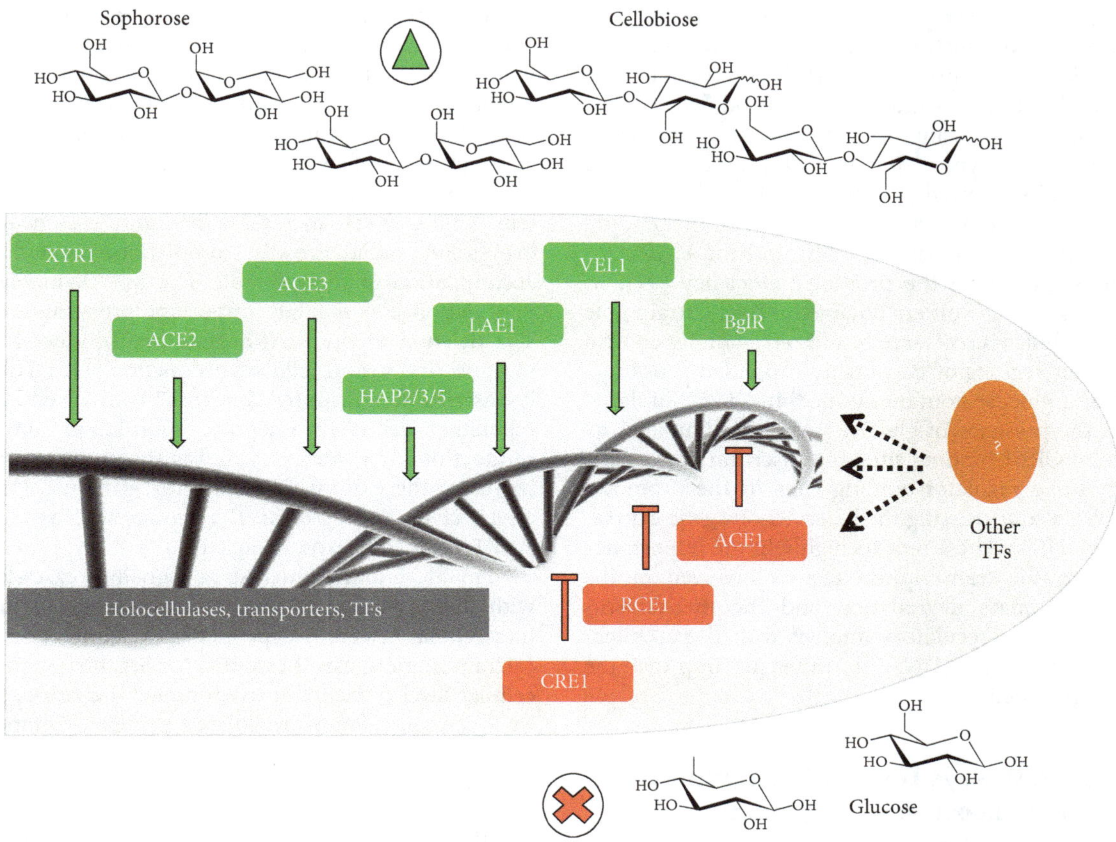

Figure 4: Overview of the transcriptional regulation of biomass degradation. The regulation of cellulose deconstruction involves at least 10 transcription factors: the positive regulators XYR1, ACEII, ACEIII, LAE1, VEL1, BglR, and the HAP2/3/5 complex, as well the repressors ACEI, RCE1, and CRE1. However, several transcription factors still not identified are potentially involved in this mechanism.

were also modulated in the presence of sophorose and glucose. In sugarcane bagasse, 88 transcription factors were modulated in the QM6a strain compared to the growth in glycerol [43]. Expression of several genes encoding transcription factors was also under regulation of XYR1 and CRE1. During cultivation in cellulose and glucose, 14 transcription factors were targeted by the CRE1-mediated mechanism of carbon catabolic repression (CCR) [45]. In sophorose, 8 regulators were also regulated by CRE1 [44]. The main positive regulator of cellulase production, XYR1, regulates the expression of other 31 transcription factors in the presence of cellulose, 17 in sophorose, and 7 in glucose [35]. In all these studies, most of the transcription factors modulated have not been characterized yet, which highlights the importance of exploring more about the regulatory proteins involved in the degradation of biomass.

5. Epigenetic Regulation of Holocellulase Expression

Another important mechanism of gene transcriptional regulation involves the chromatin modification, a phenomenon that includes the modification of the histones which are the proteins responsible for the DNA packaging. In *T. reesei*, the evidence of nucleosome rearrangement in the promoter region of *cel6a* and *cel7a* suggests that chromatin remodeling is necessary to promote cellulolytic enzyme expression [81, 87]. Interestingly, deleting the methyltransferase LAE1 impairs cellulase expression, while overexpressing this protein increases cellulase expression and changes in the H3K4 methylation pattern in the promoter region of cel5b [83]. This mechanism of regulation was also demonstrated in *M. oryzae*, in which deleting the methyltransferase MoSET1 decreased cellulase induction [88]. Likewise, acetyltransferases are important enzymes involved with chromatin remodeling. In *T. reesei*, acetyltransferases belonging to the GCN5 family are crucial for cellulase expression [89]. During growth of the *T. reesei* QM6a in sugarcane bagasse, de Paula [43] showed that at least 9 genes related to chromatin remodeling were upregulated and 8 downregulated in comparison to glycerol. Of these genes, the acetyltransferase SidF (ID 82628) was almost 40 times more expressed in sugarcane bagasse, reaching the top of the list of genes modulated by this carbon source [43]. Also, dos Santos Castro and coworkers showed that two genes encoding a SWI-SNF chromatin-remodeling complex protein (ID 123327 and ID 122943) were also highly expressed in the presence of cellulose and sophorose [21]. This last gene was a target of CCR by CRE1 under cellulose condition [45]. All these findings reinforce the evidence of the involvement of the chromatin modifications in the mechanism of biomass degradation by the filamentous fungus *T. reesei*.

The genetic engineering has become an important way for improving the production of holocellulolytic enzymes. The modification of gene promoters has also been extensively studied. Recently, Hirasawa and coworkers modified the promoter region of *xyn3* by using the *xyn1* cis-acting region and obtained improved enzyme expression [90]. A study of Zou and coworkers also showed that the CRE1 binding motifs in the promoter region of *cbh1* were replaced by the binding motifs of the positive regulators ACEII and the HAP2/3/5 complex, thus improving the promoter efficiency [91]. In addition, Uzbas and coworkers obtained a *Δxyr1* strain able to secrete the cellulases *cel7a*, *cel7b*, and *cel12a* under control of the promoter regions of two highly expressed genes *tef1* and *cdna1* in a glucose-containing medium [92]. Similarly, the CCR in the presence of glucose was also eliminated by the deletion of CRE1 binding sites and insertion of multiple copies of positive regulator-binding sites in the promoter region of *cbh1*, also increasing the heterologous gene expression in *T. reesei* [93]. These reports highlight the importance of investigating the transcription factors involved in the regulation of biomass degradation and the mechanisms through which these regulators interact with the architectural framework of the DNA in order to improve the holocellulase production.

6. Signaling Pathways Involved in Cellulose, Sophorose, Glucose, and Sugarcane Bagasse Recognition

The sensing of the changing environment is an important event for both survival and fungi competition. The fungi have appreciated mechanisms involved in sending signals and reacting to them [34, 94, 95]. Right after the perception of environmental signals, a complex network can be initiated, which is responsible for integrating all signals and promoting a suitable gene response to desirably react to the environment conditions [34]. In *T. reesei*, different signaling pathways have been described to be involved with fungal development and environment sensing. Among these, the signaling pathways dependent on G protein, cAMP, Ras-GTPases, protein kinases, phosphatases, calcium, and MAPK are the mostly characterized intracellular pathways although some aspects remain unclear [95–104]. Several studies showed the involvement of Gα proteins in the regulation of cellulase gene transcription by light [105, 106]. A functional characterization of a GPCR-encoding gene of *Trichoderma atrovoviride* showed its role in vegetative growth and conidiation [107] and expression of chitinase-encoding genes [108]. Recently, it was demonstrated that two G proteins Gβ and Gγs act along with a class I phosducin protein controlling the expression of glycoside hydrolase genes [109]. The cAMP pathway is a highly conserved signaling cascade in which cAMP acts as secondary messenger promoting the integration of different pathways [99]. In *T. reesei*, cAMP levels control cellulase gene expression [110] and according to Nogueira et al. [111], this regulation is in a carbon source-dependent manner.

The cellulase gene expression can also be regulated by the dynamics of protein phosphorylation and dephosphorylation involving protein kinases and phosphatases [34], respectively. Schuster et al. [112] demonstrated the role of protein kinase A (PKA) in the regulation of cellulase expression in the presence of light. Additionally, the deletion of a *T. reesei* casein kinase II promotes the decrease of chitinase expression and repression of sporulation and glucose metabolism [113]. Similarly, He et al. [114] showed that the deletion of protein kinase EKi1 increases the chitinase-encoding gene expression, radial growth, conidiation, and ethanolamine accumulation in the cell wall. The MAPK-mediated phosphorylation can regulate important processes in *T. reesei*. The deletion of the MAPK gene *tmk3* induced a decrease in transcription and cellulase production [115]. Furthermore, it was also demonstrated that *tmk2* is involved in processes regulating cell wall integrity, sporulation and, cellulase production [116]. Recently, it was shown that MAPK IME2 represses the expression of the three main cellulase genes (*cbh1*, *cbh2*, and *egl1*) in *T. reesei* as well as activates the XYR1 and CRE1 expressions [117].

Finally, another interesting signaling cascade involved with the regulation of important processes in fungi is the intracellular calcium-dependent signaling pathway. The calcium homeostasis is essential for organisms, and its intracellular level reflects the environmental changes [118, 119]. In *T. reesei*, calcium-modulated protein (calmodulin) has an important role in xylanase expression [120]. Equally, it was demonstrated that NCS (neuronal calcium sensor-like) interacts with calcineurin and phosphodiesterase, regulating the light-dependent signaling pathway involved with the control of cellulase expression [121]. Recent reports have shown the differential expression of genes encoding calcium transporters and other calcium signaling pathway members during cultivation of *T. reesei* QM9414, *Δxyr1* e *Δcre1* strains in the presence of cellulose, sophorose, and glucose [21, 35, 44, 45]. Similarly, the cultivation of *T. reesei* QM6a strain in sugarcane bagasse, glucose, and glycerol induced the expression of genes encoding Ca^{2+}-ATPases and other calcium signaling transporters [43]. These results suggest the involvement of different signaling pathways controlling cellulase expression in the filamentous fungus *T. reesei*.

A deep analysis of gene expression in two functional mutant strains of main transcription factors (XYR1 and CRE1) [35, 44, 45] and two MAPK-encoding genes showed that different signaling pathways and regulatory mechanisms are involved with the regulation of cellulase expression [43]. The main finding of these studies was that the enzyme regulation is carbon source-dependent. The growth of the QM9414 parental strain in the presence of cellulose and sophorose revealed a distinct expression pattern of genes encoding proteins involved with intracellular signaling. Additionally, 50 and 28 genes related to signal transduction were differentially expressed in the presence of cellulose and sophorose, respectively [35]. In cellulose, the main upregulated gene encoded to a conidiospore surface protein cmp1 (ID 72379), which was 21-fold more expressed in the parental strain. Oppositely, the most upregulated gene in sophorose encoded to an unknown protein (ID 73119), which was 8-fold more in the parental strain. Regarding

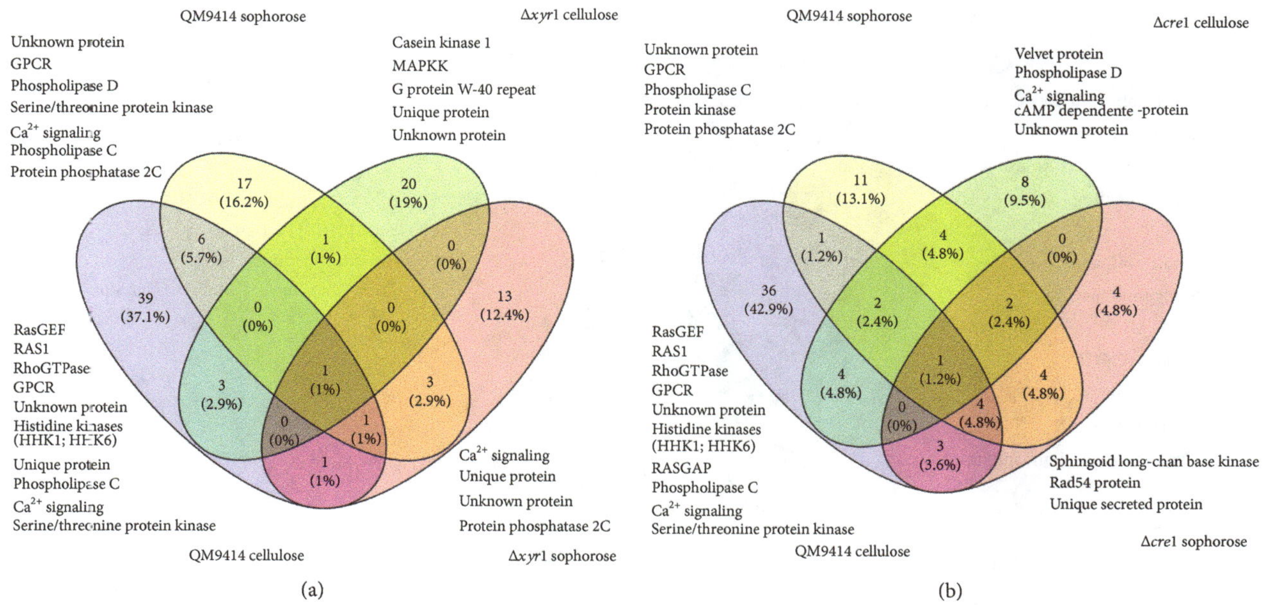

FIGURE 5: Expression pattern of differentially expressed signaling pathway genes in *T. reesei* during cultivation in cellulose and sophorose. (a) Comparative Venn diagram of expressed genes between the QM9414 and *Δxyr1* mutant strains in the presence of cellulose and sophorose. Venn diagram clustering was designed using Venny 2.1 tools. (b) Comparative Venn diagram of expressed genes between the QM9414 and *Δcre1* mutant strains in the presence of cellulose and sophorose. Venn diagram clustering was designed using Venny 2.1.

downregulated genes, an unknown protein (ID 65522, 12-fold) was the most repressed in the presence of cellulose. Moreover, the most repressed gene in the presence of sophorose encoded the conidiospore surface protein cmp1 (ID 72379) being 69-fold less expressed in this condition [35]. Curiously, this gene was the most upregulated one when the QM9414 mutant strain was grown in the presence of cellulose. This result suggests that this protein may have an important role in cellulose recognition although the aspects about the regulatory mechanisms involved with the signal transduction remain unclear.

Here, we compared the expression patterns of signaling pathway-encoding genes in two functional mutant strains of the XYR1 positive and CRE1 negative regulators of cellulase expression in *T. reesei*. The reported studies have shown there is a specific pattern of signaling pathways for the recognition of different carbon sources (Figure 5). In the QM9414 parental strain, the RAS-GTPAses (RAS-EGF, RAS1, and RhoGTPase) and histidine kinases (HHK1 and HHK6) are the main activated proteins (Figure 5(a)). Moreover, *ras1* gene (ID 67275) and *hhk6* gene (ID 62751) were 8-fold and 2-fold more expressed in the presence of cellulose, respectively [35]. Zhang et al. [97] showed that *ras1* plays important roles in some cellular processes such as polarized apical growth, hyphal branch formation, sporulation, and cAMP level adjustment. Additionally, the deletion of GTPase *ras2* modulates the expression of major cellulase genes and transcription factors in cellulose. The growth and protein secretion of *T. reesei* in cellulose cultures were decreased in *Δrho3* GTPase mutant strain, suggesting *rho3* is involved with secretion processes in this fungus [100].

In the presence of sophorose, we observe distinct expression patterns, being GPCR, phospholipases D and C, and Ca^{2+}- signaling the main pathways involved with carbon source recognition (Figure 5(a)). The *Δxyr1* mutant strain demonstrated a distinct pattern of regulation, and, in the presence of cellulose, we observed the activation of casein kinase 1 and MAPKK signaling pathways (Figure 5(a)). The gene encoding to casein kinase 1 (ID 55049) was 2.5-fold more expressed in this condition, and the MAPKK (ID 57513) was increased about 2-fold in cellulose as the carbon source [35]. Wang et al. [113] showed that the casein kinase pathway governs chitinase expression, and beyond that, casein kinase-dependent phosphorylation has been suggested to be an important mechanism of regulation of DNA-binding zinc finger proteins, such as CRE1 [122]. These results suggest that XYR1 is a negative regulator of some genes encoding components of intracellular signaling pathways, such as MAPK and casein kinase cascades, involved with cellulose recognition being this regulation an additional mechanism of transcriptional regulation of cellulase expression.

The functional *Δcre1* mutant strain exhibited a specific expression pattern related to signaling pathways. In the presence of cellulose, the main signaling pathways involved with cellulose sensing were Ca^{2+}, phospholipase, cAMP, and velvet-dependent signaling pathways (Figure 5(b)). The Ca^{2+}-dependent protein kinase (ID 62181, 5.7-fold) was the most upregulated gene in this condition. Moreover, a cAMP-dependent protein (ID 119614) was 2-fold more expressed in *Δcre1* grown in cellulose [45]. The phospholipase D gene (ID 22331) was 3.8-fold more expressed in *Δcre1* in the presence of cellulose [45]. Finally, the deletion of *cre1* promoted a decrease of 2.3-fold in the expression

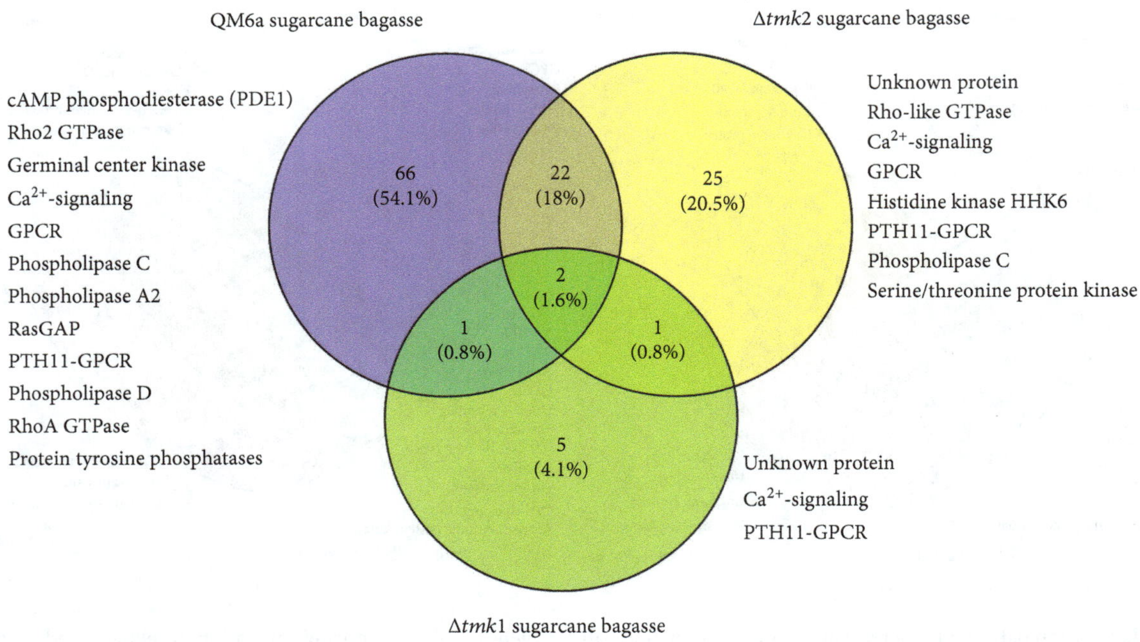

FIGURE 6: Expression pattern of differentially expressed signaling pathway genes in QM6a parental and two functional MAPK-mutant genes in the presence of sugarcane bagasse. (a) Comparative Venn diagram of expressed genes between the QM6a, *Δtmk1*, and *Δtmk2* mutant strains in the presence of sugarcane bagasse. Venn diagram clustering was designed using Venny 2.1.

of regulatory protein velvet 1 (ID 122284). Karimi-Aghcheh et al. [84] showed the deletion of *vel1* completely blocked the expression of xylanases, cellulases, and the regulator XYR1 in the presence of lactose. Since *vel1* was downregulated with *cre1* deletion, this result suggests that CRE1 might indirectly regulate the XYR1 expression in a *vel1*-dependent mechanism, and this may be an unclear mechanism of CCR regulation in *T. reesei*.

The growth of *T. reesei* QM6a in sugarcane bagasse demonstrated a specific expression profile of signaling-encoding genes. In this condition, the expression of genes belonging to cAMP, GTPase, germinal center kinase, calcium, phospholipases, GPCR, protein phosphatases, and PTH11 signaling pathways was observed (Figure 6). Among these, PTH11-encoding genes were an overrepresented category, where eight genes were differentially expressed in this condition. The most upregulated PTH11 (ID 69500) was 14.5-fold more expressed in the presence of sugarcane bagasse [43]. In the filamentous fungus *Neurospora crassa*, PTH11 GPCR seems to be a yet-unclear key role in cellulose recognition and in the control of holocellulytic enzyme expression [123]. De Paula [43] showed in a global transcriptional analysis of two functional mutants for the MAPK signaling pathway that there is a cross talk among different signaling pathways in the presence of sugarcane bagasse. In the *Δtmk2* mutant strain, the main signaling pathways involved with the sugarcane recognition are Rho GTPase, GPCR, histidine kinase, and PTH11 GPCR (Figure 6). Interestingly, PTH11-encoding genes were also overrepresented in the *tmk2* deletion when compared to the parental strain [43]. The deletion of *tmk1* alters the expression of few genes belonging to signaling pathways. Only 4.1% of all genes related with intracellular signaling were involved with sugarcane bagasse recognition in *Δtmk1* mutant strain. Curiously, Vercoe et al. [124] demonstrated in *Ruminococcus flavefaciens*, a cellulolytic bacterium, that the phosphorylation and dephosphorylation dynamics are important for carbon source recognition and regulation of carbon metabolism. Together, our results suggest that in *T. reesei*, there may be a mechanism for carbon source recognition involving PTH11 GPCR, and that this process might be regulated by posttranslational events.

The understanding of how signal transduction pathways modulate cellulase gene expression has acquired attention, such as tools for strain improvement in *T. reesei* [125]. The adequate environmental perception is a crucial step for regulation of gene expression. Thereby, changing transmission of signals aiming at the adjustment of enzyme secretion in response to environmental conditions is an advantageous alternative to enhance the activity efficiency of genes of interest [18]. Different approaches can be employed to improve the production of enzymes by *T. reesei*. One of them is the manipulation of the expression of holocellulolytic-encoding genes by increasing gene expression of activators and/or decreasing expression of repressors that control the expression of the cellulose- and hemicellulose-degrading enzymes. Thus, to carry out this task, a deeper knowledge of holocellulase regulation is crucial [23, 35, 45, 76, 126–128].

In order to accomplish improvement in cellulase expression, different signal transduction mechanisms can be exploited. The holocellulytic gene expression may be regulated by different factors such as light, carbon and nitrogen sources, pH, temperature, inorganic compounds, transcription factors, and epigenetic events [21, 35, 44, 45, 81, 87, 126, 129–133]. These factors and their effects in regulating intracellular signaling pathways have been well studied in *Trichoderma*. As an example, in *T. reesei*, light and photoreceptors BLR1 and BLR2, which belong to light signaling, are known to regulate expression of cellulase genes [134, 135]. Regarding *cbh1* and *cbh2*, the two main cellobiohydrolases, it has been demonstrated that growth of *T. reesei* in the presence of light promotes an increase of 2-fold in gene expression in the presence of cellulose [134, 135]. The signal integration of light and nutrient signaling was an important discovery in the understanding of cellulase gene expression [105, 106]. The G-protein alpha subunits GNA1 and GNA3 were described to be involved in light-dependent regulation of cellulase gene expression. Additionally, transcriptional analysis of the effect of deletion of GNB1 and GNG1 as well as the phosducin-like protein PHLP1 showed that glycosyl hydrolases are the major targets of light-dependent signaling by heterotrimeric G-proteins [109]. The intracellular level of cAMP directly affects cellulase expression [110, 136]. Nogueira et al. [111] showed that intracellular levels of cAMP were 4-fold in the presence of sophorose, and cAMP may regulate secretion of cellulolytic enzymes in *T. reesei* in the presence of this sugar. Interestingly, de Paula [43] showed that deletion of MAPK gene *tmk1* promoted the downregulation of the main holocellulolytic genes of *T. reesei*. These results suggest that TMK1 is a positive regulator of cellulase gene expression and point this gene as a potential target for improvement engineering approaches to enhance cellulase expression in this fungus. Finally, all the knowledge about intracellular signaling pathways and its effects in the cellulase expression in *T. reesei* will provide important insights for metabolic engineering for strain improvement to be used in biotechnological industries.

7. The New Players Potentially Involved in *T. reesei* Lignocellulosic Biomass Degradation

In order to survive in different environments, filamentous fungi must be able to sense the surroundings and respond to it accordingly. Once the fungus senses the carbon source (whether it is sugarcane bagasse, cellulose, or sophorose—cellulase inducers), a signaling cascade is activated, and it ends with increasing/decreasing the transcription level of certain genes involved in the degradation of each carbon source. However, all this sensing and responding processes are still not completely characterized. Attempts to solve this puzzle have been made by analyzing transcriptome results, and several studies have shown that *T. reesei* presents genes that code for proteins of unknown function [137].

A differential analysis of expressed genes when *T. reesei* was cultivated in cellulose versus glucose, in sophorose versus cellulose, and in sophorose versus glucose revealed that about 35 to 46% of the genes presented unknown function [21]. Also, when analyzing the top 10 upregulated genes in cellulose, five of them corresponded to protein of unknown function [21]. Among the top 10 upregulated genes in sophorose, there were three of unknown function. These results suggested these genes (were upregulated in inducing conditions) play an important and so far a neglected role in biomass degradation.

Some transporters have been reported to be involved with sensing external carbon sources and thus with the induction of CAZymes [48, 61]. Although the importance of these proteins has been known, specially sugar transporters, for the cellular response to the environment, there are putative transporters that have been differentially expressed [21, 35, 44, 45] and have not yet been characterized. Figure 7(a) summarizes a few genes that have been expressed in very different levels depending upon the carbon source. Among them, four copper transporters (ID 52315, ID 62716, ID 71029, and ID 108749) had transcriptional levels decreased in the presence of sophorose [21]. Supporting this result, Bak showed in 2015 that a copper transporter was downregulated in *Phanerochaete chrysosporium* cultivated in rice straw compared to no carbon source [138]. In 2016, dos Santos Castro showed that the deletion of *xyr1* caused the transcriptional levels of these genes to be upregulated in the knockout strain compared to the wild type in sophorose [35]. Deletion of the major transcription factor involved in CCR, *cre1*, did not result in changes in these four genes [44, 45]. Therefore, these copper transporters may be involved in biomass degradation since copper is a cofactor required for LPMO activity [139], and the results obtained so far suggest they are regulated (directly or indirectly) by XYR1 but not by CRE1.

Another class of proteins related to biomass degradation regulation and that is still poorly characterized is transcription factor (TF). A lot of effort have been put into discovering new TF that would be important for the regulation of expression of genes that is important for an efficient biomass degradation [73, 80, 82, 140]. Antoniêto et al. [44, 45] and dos Santos Castro et al. [21, 35] studied differential expression in *T. reesei* cultivated in different carbon sources. Also, they have studied two transcription factors, CRE1 and XYR1, and how they affect the transcriptional level of genes related to biomass degradation. Among the transcription factors that showed differences in expression, a few unknown TF presented particularly high differences of expression Figure 7(b). The high expression of these TFs in sugarcane bagasse, cellulose, and sophorose suggests that they can play a role on the regulation of biomass degradation and should be considered targets for further studies [137]. Particularly, gene 107641, classified as a transcription factor by KOG, was repressed in sugarcane bagasse and induced in sophorose. Also, the deletion of *xyr1* caused this gene (ID 107641) to be highly expressed when *T. reesei* was cultivated in sophorose, a cellulase inductor. Moreover, this TF (ID 107641) was also highly expressed in the knockout of *cre1* when cultivated in cellulose Figure 7(b). All these results combined suggest that this TF (ID 107641) is involved in the regulation of transcription of enzymes related to biomass degradation and that XYR1 and CRE1 supposedly act as repressors for the expression of this TF. Several studies have identified genes encoding putative fungal C6 zinc finger-type transcription factors enriched among the differential

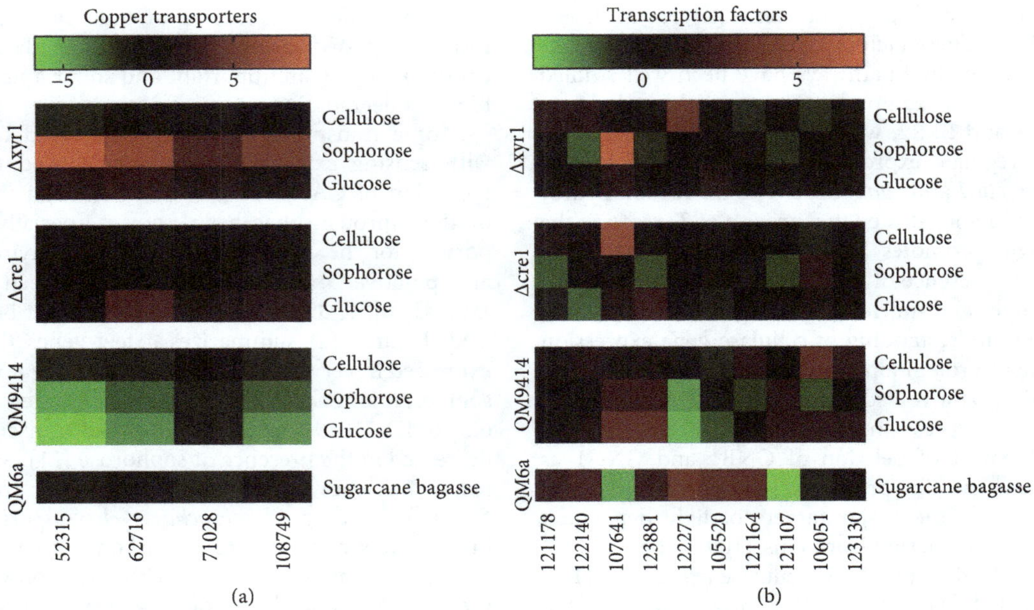

FIGURE 7: Heatmap expression of new players potentially involved with *T. reesei* biomass degradation. (a) Expression profile of copper transporter-encoding gene differentially expressed in *T. reesei* strains grown in cellulose, sophorose, glucose, and sugarcane bagasse. (b) Expression profile of transcription factor-encoding gene differentially expressed in *T. reesei* strains grown in cellulose, sophorose, glucose, and sugarcane bagasse. The results of gene expressions were transformed in Log2FoldChange values and employed to heat map construction using the GraphPad Prism version 7 program (https://www.graphpad.com/).

analysis [80, 141–143]. However, characterizing them according to the binding site and which gene expression it regulates is still necessary in order to have a greater knowledge of biomass degradation regulation. Therefore, the identification of novel transcription factors which transcription levels are affected by different carbon sources may be the first step for understanding more deeply how the regulation of transcription occurs. Therefore, these new findings could be used for engineering *T. reesei* in order to have a new hyperproducer strain for the industry to use in biomass degradation with higher efficiency.

8. Conclusions

The adoption of the filamentous fungus *T. reesei* as the most important holocellulase producer in biotechnology industries has been stabilized over the years. However, some aspects about holocellulase expression remain unclear. Thus, global transcriptional analyses are an excellent approach to understanding gene expression and promoting the selection of candidate genes for construction of strains producing high levels of holocellulases for plant cell wall degradation. This way, knowledge about the mechanisms involved in the recognition of environmental signals, sugar transport, and transcriptional regulatory events involved with fungal adaptation for different conditions consists in an extremely important step in the fungal biology comprehension. The integration of all these data might contribute to a better understanding of the regulatory mechanisms during lignocellulosic biomass degradation by *T. reesei* facilitating its application in fungal biotechnology.

Conflicts of Interest

The authors declare that there is no conflict of interest regarding the publication of this paper.

Authors' Contributions

Renato Graciano de Paula and Amanda Cristina Campos Antoniêto contributed equally to this work.

Acknowledgments

This work was supported by The State of São Paulo Research Foundation (FAPESP) (Processes 2016/20358-5, 2016/23233-9, and 2017/04206-3).

References

[1] G. J. Samuels, "Trichoderma: a review of biology and systematics of the genus," *Mycological Research*, vol. 100, no. 8, pp. 923–935, 1996.

[2] V. K. Gupta, A. S. Steindorff, R. G. de Paula et al., "The post-genomic era of Trichoderma reesei: what's next?," *Trends in Biotechnology*, vol. 34, no. 12, pp. 970–982, 2016.

[3] G. E. Harman, A. H. Herrera-Estrella, B. A. Horwitz, and M. Lorito, "Special issue: *Trichoderma*—from basic biology to biotechnology," *Microbiology*, vol. 158, no. 1, pp. 1-2, 2012.

[4] A. Schuster and M. Schmoll, "Biology and biotechnology of *Trichoderma*," *Applied Microbiology and Biotechnology*, vol. 87, no. 3, pp. 787–799, 2010.

[5] C. P. Kubicek, M. Komon-Zelazowska, and I. S. Druzhinina, "Fungal genus *Hypocrea/Trichoderma*: from barcodes to biodiversity," *Journal of Zhejiang University Science B*, vol. 9, no. 10, pp. 753–763, 2008.

[6] D. Tisch and M. Schmoll, "Targets of light signalling in Trichoderma reesei," *BMC Genomics*, vol. 14, no. 1, p. 657, 2013.

[7] A. Schuster, C. P. Kubicek, M. A. Friedl, I. S. Druzhinina, and M. Schmoll, "Impact of light on Hypocrea jecorina and the multiple cellular roles of ENVOY in this process," *BMC Genomics*, vol. 8, no. 1, p. 449, 2007.

[8] I. Druzhinina and C. P. Kubicek, "Species concepts and biodiversity in *Trichoderma* and *Hypocrea*: from aggregate species to species clusters?," *Journal of Zhejiang University Science B*, vol. 6, no. 2, pp. 100–112, 2005.

[9] G. E. Harman, "Overview of mechanisms and uses of *Trichoderma* spp," *Phytopathology*, vol. 96, no. 2, pp. 190–194, 2006.

[10] Y. Galante, A. De Conti, and R. Monteverdi, *Application of Trichoderma Enzymes in the Food and Feed Industries*, G. E. Harman and C. P. Kubicek, Eds., Taylor and Francis, London, 1998.

[11] Y. Galante, A. De Conti, and R. Monteverdi, *Application of Trichoderma Enzymes in the Textile Industry*, G. E. Harman and C. P. Kubicek, Eds., Taylor and Francis, London, 1998.

[12] J. Buchert, T. Oksanen, J. Pere, M. Siika-Aho, A. Suurnäkki, and L. Viikari, *Applications of Trichoderma reesei Enzymes in the Pulp and Paper Industry*, G. E. Harman and C. P. Kubicek, Eds., Taylor and Francis, London, 1998.

[13] S. Pereira, L. Maehara, C. Machado, and C. Farinas, "2G ethanol from the whole sugarcane lignocellulosic biomass," *Biotechnology for Biofuels*, vol. 8, no. 1, p. 44, 2015.

[14] R. H. Bischof, J. Ramoni, and B. Seiboth, "Cellulases and beyond: the first 70 years of the enzyme producer *Trichoderma reesei*," *Microbial Cell Factories*, vol. 15, no. 1, p. 106, 2016.

[15] Y. Li, C. Liu, F. Bai, and X. Zhao, "Overproduction of cellulase by *Trichoderma reesei* RUT C30 through batch-feeding of synthesized low-cost sugar mixture," *Bioresource Technology*, vol. 216, pp. 503–510, 2016.

[16] J. Huang, D. Chen, Y. Wei et al., "Direct ethanol production from lignocellulosic sugars and sugarcane bagasse by a recombinant *Trichoderma reesei* strain HJ48," *The Scientific World Journal*, vol. 2014, Article ID 798683, 8 pages, 2014.

[17] S. S. Adav and S. K. Sze, "Trichoderma secretome: an overview," in *Biotechnology and Biology of Trichoderma*, V. K. Gupta, M. Schmoll, A. Herrera-Estrella, R. S. Upadhyay, I. Druzhinina, and M. G. Tuohy, Eds., pp. 103–114, Elsevier, Waltham, 1st edition, 2014.

[18] H. Bazafkan, D. Tisch, and M. Schmoll, "Regulation of glycoside hydrolase expression in Trichoderma," in *Biotechnology and Biology of Trichoderma*, V. K. Gupta, M. Schmoll, A. Herrera-Estrella, R. Upadhyay, I. Druzhinina, and M. G. Tuohy, Eds., p. 527, Elsevier, 2014.

[19] J. Gao, Y. Qian, Y. Wang, Y. Qu, and Y. Zhong, "Production of the versatile cellulase for cellulose bioconversion and cellulase inducer synthesis by genetic improvement of *Trichoderma reesei*," *Biotechnology for Biofuels*, vol. 10, no. 1, p. 272, 2017.

[20] Q. Zhou, J. Xu, Y. Kou et al., "Differential involvement of β-glucosidases from Hypocrea jecorina in rapid induction of cellulase genes by cellulose and cellobiose," *Eukaryotic Cell*, vol. 11, no. 11, pp. 1371–1381, 2012.

[21] L. dos Santos Castro, W. Pedersoli, A. C. Antoniêto et al., "Comparative metabolism of cellulose, sophorose and glucose in *Trichoderma reesei* using high-throughput genomic and proteomic analyses," *Biotechnology for Biofuels*, vol. 7, no. 1, p. 41, 2014.

[22] C. P. Kubicek, A. Herrera-Estrella, V. Seidl-Seiboth et al., "Comparative genome sequence analysis underscores mycoparasitism as the ancestral life style of *Trichoderma*," *Genome Biology*, vol. 12, no. 4, article R40, 2011.

[23] L. dos Santos Castro, A. C. C. Antoniêto, W. R. Pedersoli, R. Silva-Rocha, G. F. Persinoti, and R. N. Silva, "Expression pattern of cellulolytic and xylanolytic genes regulated by transcriptional factors XYR1 and CRE1 are affected by carbon source in *Trichoderma reesei*," *Gene Expression Patterns*, vol. 14, no. 2, pp. 88–95, 2014.

[24] E. A. Znameroski, S. T. Coradetti, C. M. Roche et al., "Induction of lignocellulose-degrading enzymes in *Neurospora crassa* by cellodextrins," *Proceedings of the National Academy of Sciences of the United States of America*, vol. 109, no. 16, pp. 6012–6017, 2012.

[25] Z. Anwar, M. Gulfraz, and M. Irshad, "Agro-industrial lignocellulosic biomass a key to unlock the future bio-energy: a brief review," *Journal of Radiation Research and Applied Sciences*, vol. 7, no. 2, pp. 163–173, 2014.

[26] V. Vandenbossche, J. Brault, G. Vilarem et al., "A new lignocellulosic biomass deconstruction process combining thermo-mechano chemical action and bio-catalytic enzymatic hydrolysis in a twin-screw extruder," *Industrial Crops and Products*, vol. 55, pp. 258–266, 2014.

[27] J. K. Saini, R. Saini, and L. Tewari, "Lignocellulosic agriculture wastes as biomass feedstocks for second-generation bioethanol production: concepts and recent developments," *3 Biotech*, vol. 5, no. 4, pp. 337–353, 2015.

[28] C. A. Cardona, J. A. Quintero, and I. C. Paz, "Production of bioethanol from sugarcane bagasse: status and perspectives," *Bioresource Technology*, vol. 101, no. 13, pp. 4754–4766, 2010.

[29] A. V. Gusakov, "Alternatives to *Trichoderma reesei* in biofuel production," *Trends in Biotechnology*, vol. 29, no. 9, pp. 419–425, 2011.

[30] L. R. Lynd, W. H. van Zyl, J. E. Mcbride, and M. Laser, "Consolidated bioprocessing of cellulosic biomass: an update," *Current Opinion in Biotechnology*, vol. 16, no. 5, pp. 577–583, 2005.

[31] L. R. Lynd, P. J. Weimer, W. H. van Zyl, and I. S. Pretorius, "Microbial cellulose utilization: fundamentals and biotechnology," *Microbiology and Molecular Biology Reviews*, vol. 66, no. 3, pp. 506–577, 2002.

[32] Y. Shida, T. Furukawa, and W. Ogasawara, "Deciphering the molecular mechanisms behind cellulase production in *Trichoderma reesei*, the hyper-cellulolytic filamentous fungus," *Bioscience, Biotechnology, and Biochemistry*, vol. 80, no. 9, pp. 1712–1729, 2016.

[33] D. Martinez, R. M. Berka, B. Henrissat et al., "Genome sequencing and analysis of the biomass-degrading fungus *Trichoderma reesei* (syn. *Hypocrea jecorina*)," *Nature Biotechnology*, vol. 26, no. 5, pp. 553–560, 2008.

[34] M. Schmoll, C. Dattenböck, N. Carreras-Villaseñor et al., "The genomes of three uneven siblings: footprints of the lifestyles of three Trichoderma species," *Microbiology and Molecular Biology Reviews*, vol. 80, no. 1, pp. 205–327, 2016.

[35] L. dos Santos Castro, R. G. de Paula, A. C. C. Antoniêto, G. F. Persinoti, R. Silva-Rocha, and R. N. Silva, "Understanding the role of the master regulator XYR1 in Trichoderma reesei by global transcriptional analysis," *Frontiers in Microbiology*, vol. 7, p. 175, 2016.

[36] R. J. Quinlan, M. D. Sweeney, L. Lo Leggio et al., "Insights into the oxidative degradation of cellulose by a copper metalloenzyme that exploits biomass components," *Proceedings of the National Academy of Sciences of the United States of America*, vol. 108, no. 37, pp. 15079–15084, 2011.

[37] M. Saloheimo, M. Paloheimo, S. Hakola et al., "Swollenin, a *Trichoderma reesei* protein with sequence similarity to the plant expansins, exhibits disruption activity on cellulosic materials," *European Journal of Biochemistry*, vol. 269, no. 17, pp. 4202–4211, 2002.

[38] M. Häkkinen, M. Arvas, M. Oja et al., "Re-annotation of the CAZy genes of *Trichoderma reesei* and transcription in the presence of lignocellulosic substrates," *Microbial Cell Factories*, vol. 11, no. 1, p. 134, 2012.

[39] I. S. Druzhinina and C. P. Kubicek, "Genetic engineering of Trichoderma reesei cellulases and their production," *Microbial Biotechnology*, vol. 10, no. 6, pp. 1485–1499, 2017.

[40] S. K. Brady, S. Sreelatha, Y. Feng, S. P. S. Chundawat, and M. J. Lang, "Cellobiohydrolase 1 from *Trichoderma reesei* degrades cellulose in single cellobiose steps," *Nature Communications*, vol. 6, no. 1, p. 10149, 2015.

[41] V. K. Gupta, *Biotechnology and Biology of Trichoderma*, Elsevier, 2014.

[42] J. Strakowska, L. Błaszczyk, and J. Chełkowski, "The significance of cellulolytic enzymes produced by *Trichoderma* in opportunistic lifestyle of this fungus," *Journal of Basic Microbiology*, vol. 54, no. S1, pp. S2–S13, 2014.

[43] R. G. De Paula, *Characterization of the MAPK-dependent signaling pathway in the regulation of cellulase expression by the fungus Trichoderma reesei (Hypocrea jecorina)*, University of Sao Paulo, 2017.

[44] A. C. Campos Antoniêto, R. Graciano de Paula, L. d. Santos Castro, R. Silva-Rocha, G. Felix Persinoti, and R. Nascimento Silva, "Trichoderma reesei CRE1-mediated carbon catabolite repression in response to sophorose through RNA sequencing analysis," *Current Genomics*, vol. 17, no. 2, pp. 119–131, 2016.

[45] A. C. C. Antoniêto, L. dos Santos Castro, R. Silva-Rocha, G. F. Persinoti, and R. N. Silva, "Defining the genome-wide role of CRE1 during carbon catabolite repression in *Trichoderma reesei* using RNA-Seq analysis," *Fungal Genetics and Biology*, vol. 73, pp. 93–103, 2014.

[46] C. P. Kubicek, R. Messner, F. Gruber, M. Mandels, and E. M. Kubicek-Pranz, "Triggering of cellulase biosynthesis by cellulose in *Trichoderma reesei*: involvement of a constitutive, sophorose-inducible, glucose-inhibited β-diglucoside permease," *Journal of Biological Chemistry*, vol. 268, no. 26, pp. 19364–19368, 1993.

[47] C. P. Kubicek, R. Messner, F. Gruber, R. L. Mach, and E. M. Kubicek-Pranz, "The *Trichoderma* cellulase regulatory puzzle: from the interior life of a secretory fungus," *Enzyme and Microbial Technology*, vol. 15, no. 2, pp. 90–99, 1993.

[48] W. Zhang, Y. Kou, J. Xu et al., "Two major facilitator superfamily sugar transporters from *Trichoderma reesei* and their roles in induction of cellulase biosynthesis," *Journal of Biological Chemistry*, vol. 288, no. 46, pp. 32861–32872, 2013.

[49] E. M. Quistgaard, C. Löw, F. Guettou, and P. Nordlund, "Understanding transport by the major facilitator superfamily (MFS): structures pave the way," *Nature Reviews Molecular Cell Biology*, vol. 17, no. 2, pp. 123–132, 2016.

[50] S. S. Pao, I. T. Paulsen, and M. H. Saier Jr., "Major facilitator superfamily," *Microbiology and Molecular Biology Reviews*, vol. 62, no. 1, pp. 1–34, 1998.

[51] P. Cai, R. Gu, B. Wang et al., "Evidence of a critical role for cellodextrin transporte 2 (CDT-2) in both cellulose and hemicellulose degradation and utilization in *Neurospora crassa*," *PLoS One*, vol. 9, no. 2, article e89330, 2014.

[52] J. d. O. Porciuncula, T. Furukawa, Y. Shida et al., "Identification of major facilitator transporters involved in cellulase production during lactose culture of *Trichoderma reesei* PC-3-7," *Bioscience, Biotechnology, and Biochemistry*, vol. 77, no. 5, pp. 1014–1022, 2014.

[53] C. Ivanova, J. A. Bååth, B. Seiboth, and C. P. Kubicek, "Systems analysis of lactose metabolism in *Trichoderma reesei* identifies a lactose permease that is essential for cellulase induction," *PLoS One*, vol. 8, no. 5, article e62631, 2013.

[54] N. Chaudhary, I. Kumari, P. Sandhu, M. Ahmed, and Y. Akhter, "Proteome scale census of major facilitator superfamily transporters in *Trichoderma reesei* using protein sequence and structure based classification enhanced ranking," *Gene*, vol. 585, no. 1, pp. 166–176, 2016.

[55] W. Vongsangnak, M. Salazar, K. Hansen, and J. Nielsen, "Genome-wide analysis of maltose utilization and regulation in aspergilli," *Microbiology*, vol. 155, no. 12, pp. 3893–3902, 2009.

[56] J. E. Galagan, S. E. Calvo, K. A. Borkovich et al., "The genome sequence of the filamentous fungus *Neurospora crassa*," *Nature*, vol. 422, no. 6934, pp. 859–868, 2003.

[57] C. J. Law, P. C. Maloney, and D.-N. Wang, "Ins and outs of major facilitator superfamily antiporters," *Annual Review of Microbiology*, vol. 62, no. 1, pp. 289–305, 2008.

[58] N. Yan, "Structural advances for the major facilitator superfamily (MFS) transporters," *Trends in Biochemical Sciences*, vol. 38, no. 3, pp. 151–159, 2013.

[59] A. C. Colabardini, L. N. Ries, N. Brown et al., "Functional characterization of a xylose transporter in *Aspergillus nidulans*," *Biotechnology for Biofuels*, vol. 7, no. 1, p. 46, 2014.

[60] W. Zhang, Y. Cao, J. Gong, X. Bao, G. Chen, and W. Liu, "Identification of residues important for substrate uptake in a glucose transporter from the filamentous fungus *Trichoderma reesei*," *Scientific Reports*, vol. 5, no. 1, article 13829, 2015.

[61] Z. B. Huang, X. Z. Chen, L. N. Qin, H. Q. Wu, X. Y. Su, and Z. Y. Dong, "A novel major facilitator transporter TrSTR1 is essential for pentose utilization and involved in xylanase induction in *Trichoderma reesei*," *Biochemical and Biophysical Research Communications*, vol. 460, no. 3, pp. 663–669, 2015.

[62] L. Ries, S. T. Pullan, S. Delmas, S. Malla, M. J. Blythe, and D. B. Archer, "Genome-wide transcriptional response of *Trichoderma reesei* to lignocellulose using RNA sequencing and comparison with *Aspergillus niger*," *BMC Genomics*, vol. 14, no. 1, p. 541, 2013.

[63] L. Atanasova, S. L. Crom, S. Gruber et al., "Comparative transcriptomics reveals different strategies of *Trichoderma* mycoparasitism," *BMC Genomics*, vol. 14, no. 1, p. 121, 2013.

[64] J. Kang, J.-U. Hwang, M. Lee et al., "PDR-type ABC transporter mediates cellular uptake of the phytohormone abscisic acid," *Proceedings of the National Academy of Sciences of the United States of America*, vol. 107, no. 5, pp. 2355–2360, 2010.

[65] R. Bischof, L. Fourtis, A. Limbeck, C. Gamauf, B. Seiboth, and C. P. Kubicek, "Comparative analysis of the *Trichoderma reesei* transcriptome during growth on the cellulase inducing substrates wheat straw and lactose," *Biotechnology for Biofuels*, vol. 6, no. 1, p. 127, 2013.

[66] E. Lamping, P. V. Baret, A. R. Holmes, B. C. Monk, A. Goffeau, and R. D. Cannon, "Fungal PDR transporters: phylogeny, topology, motifs and function," *Fungal Genetics and Biology*, vol. 47, no. 2, pp. 127–142, 2010.

[67] M. Ruocco, S. Lanzuise, F. Vinale et al., "Identification of a new biocontrol gene in Trichoderma atroviride: the role of an ABC transporter membrane pump in the interaction with different plant-pathogenic fungi," *Molecular Plant-Microbe Interactions*, vol. 22, no. 3, pp. 291–301, 2009.

[68] M. Karlsson, M. B. Durling, J. Choi et al., "Insights on the evolution of mycoparasitism from the genome of *Clonostachys rosea*," *Genome Biology and Evolution*, vol. 7, no. 2, pp. 465–480, 2015.

[69] S. Wilkens, "Structure and mechanism of ABC transporters," *F1000Prime Reports*, vol. 7, p. 14, 2015.

[70] M. G. L. Elferink, S. V. Albers, W. N. Konings, and A. J. M. Driessen, "Sugar transport in *Sulfolobus solfataricus* is mediated by two families of binding protein-dependent ABC transporters," *Molecular Microbiology*, vol. 39, no. 6, pp. 1494–1503, 2001.

[71] A. Watanabe, K. Hiraga, M. Suda, H. Yukawa, and M. Inui, "Functional characterization of *Corynebacterium alkanolyticum* β-xylosidase and xyloside ABC transporter in *Corynebacterium glutamicum*," *Applied and Environmental Microbiology*, vol. 81, no. 12, pp. 4173–4183, 2015.

[72] S. M. Koning, M. G. L. Elferink, W. N. Konings, and A. J. M. Driessen, "Cellobiose uptake in the hyperthermophilic archaeon *Pyrococcus furiosus* is mediated by an inducible, high-affinity ABC transporter," *Journal of Bacteriology*, vol. 183, no. 17, pp. 4979–4984, 2001.

[73] T. Furukawa, Y. Shida, N. Kitagami et al., "Identification of specific binding sites for XYR1, a transcriptional activator of cellulolytic and xylanolytic genes in Trichoderma reesei," *Fungal Genetics and Biology*, vol. 46, no. 8, pp. 564–574, 2009.

[74] T. Nakari-Setala, M. Paloheimo, J. Kallio, J. Vehmaanpera, M. Penttila, and M. Saloheimo, "Genetic modification of carbon catabolite repression in *Trichoderma reesei* for improved protein production," *Applied and Environmental Microbiology*, vol. 75, no. 14, pp. 4853–4860, 2009.

[75] T. Portnoy, A. Margeot, R. Linke et al., "The CRE1 carbon catabolite repressor of the fungus *Trichoderma reesei*: a master regulator of carbon assimilation," *BMC Genomics*, vol. 12, no. 1, p. 269, 2011.

[76] T. Portnoy, A. Margeot, V. Seidl-Seiboth et al., "Differential regulation of the cellulase transcription factors XYR1, ACE2, and ACE1 in *Trichoderma reesei* strains producing high and low levels of cellulase," *Eukaryotic Cell*, vol. 10, no. 2, pp. 262–271, 2011.

[77] E. Akel, B. Metz, B. Seiboth, and C. P. Kubicek, "Molecular regulation of arabinan and L-arabinose metabolism in *Hypocrea jecorina* (*Trichoderma reesei*)," *Eukaryotic Cell*, vol. 8, no. 12, pp. 1837–1844, 2009.

[78] A. R. Stricker, K. Grosstessner-Hain, E. Wurleitner, and R. L. Mach, "Xyr1 (xylanase regulator 1) regulates both the hydrolytic enzyme system and D-xylose metabolism in Hypocrea jecorina," *Eukaryotic Cell*, vol. 5, no. 12, pp. 2128–2137, 2006.

[79] N. Aro, A. Saloheimo, M. Ilmén, and M. Penttilä, "ACEII, a novel transcriptional activator involved in regulation of cellulase and xylanase genes of *Trichoderma reesei*," *The Journal of Biological Chemistry*, vol. 276, no. 26, pp. 24309–24314, 2001.

[80] M. Häkkinen, M. J. Valkonen, A. Westerholm-Parvinen et al., "Screening of candidate regulators for cellulase and hemicellulase production in *Trichoderma reesei* and identification of a factor essential for cellulase production," *Biotechnology for Biofuels*, vol. 7, no. 1, p. 14, 2014.

[81] S. Zeilinger, M. Schmoll, M. Pail, R. L. Mach, and C. P. Kubicek, "Nucleosome transactions on the *Hypocrea jecorina* (*Trichoderma reesei*) cellulase promoter *cbh2* associated with cellulase induction," *Molecular Genetics and Genomics*, vol. 270, no. 1, pp. 46–55, 2003.

[82] M. Nitta, T. Furukawa, Y. Shida et al., "A new $Zn(II)_2Cys_6$-type transcription factor BglR regulates *β-glucosidase* expression in *Trichoderma reesei*," *Fungal Genetics and Biology*, vol. 49, no. 5, pp. 388–397, 2012.

[83] B. Seiboth, R. A. Karimi, P. A. Phatale et al., "The putative protein methyltransferase LAE1 controls cellulase gene expression in *Trichoderma reesei*," *Molecular Microbiology*, vol. 84, no. 6, pp. 1150–1164, 2012.

[84] R. Karimi Aghcheh, Z. Németh, L. Atanasova et al., "The VELVET A orthologue VEL1 of *Trichoderma reesei* regulates fungal development and is essential for cellulase gene expression," *PLoS One*, vol. 9, no. 11, article e112799, 2014.

[85] N. Aro, M. Ilmen, A. Saloheimo, and M. Penttila, "ACEI of *Trichoderma reesei* is a repressor of cellulase and xylanase expression," *Applied and Environmental Microbiology*, vol. 69, no. 1, pp. 56–65, 2003.

[86] Y. Cao, F. Zheng, L. Wang et al., "Rce1, a novel transcriptional repressor, regulates cellulase gene expression by antagonizing the transactivator Xyr1 in Trichoderma reesei," *Molecular Microbiology*, vol. 105, no. 1, pp. 65–83, 2017.

[87] L. Ries, N. J. Belshaw, M. Ilmén, M. E. Penttilä, M. Alapuranen, and D. B. Archer, "The role of CRE1 in nucleosome positioning within the cbh1 promoter and coding regions of Trichoderma reesei," *Applied Microbiology and Biotechnology*, vol. 98, no. 2, pp. 749–762, 2014.

[88] B. Van Vu, K. T. M. Pham, and H. Nakayashiki, "Substrate-induced transcriptional activation of the MoCel7C cellulase gene is associated with methylation of histone H3 at lysine 4 in the rice blast fungus *Magnaporthe oryzae*," *Applied and Environmental Microbiology*, vol. 79, no. 21, pp. 6823–6832, 2013.

[89] Q. Xin, Y. Gong, X. Lv, G. Chen, and W. Liu, "*Trichoderma reesei* histone acetyltransferase Gcn5 regulates fungal growth, conidiation, and cellulase gene expression," *Current Microbiology*, vol. 67, no. 5, pp. 580–589, 2013.

[90] H. Hirasawa, K. Shioya, T. Furukawa et al., "Engineering of the Trichoderma reesei xylanase3 promoter for efficient enzyme expression," *Applied Microbiology and Biotechnology*, vol. 102, no. 6, pp. 2737–2752, 2018.

[91] G. Zou, S. Shi, Y. Jiang et al., "Construction of a cellulase hyper-expression system in *Trichoderma reesei* by promoter and enzyme engineering," *Microbial Cell Factories*, vol. 11, no. 1, p. 21, 2012.

[92] F. Uzbas, U. Sezerman, L. Hartl, C. P. Kubicek, and B. Seiboth, "A homologous production system for *Trichoderma reesei* secreted proteins in a cellulase-free background," *Applied Microbiology and Biotechnology*, vol. 93, no. 4, pp. 1601–1608, 2012.

[93] T. Liu, T. Wang, X. Li, and X. Liu, "Improved heterologous gene expression in *Trichoderma reesei* by cellobiohydrolase I gene (*cbh1*) promoter optimization," *Acta Biochimica et Biophysica Sinica*, vol. 40, no. 2, pp. 158–165, 2008.

[94] Y.-S. Bahn, C. Xue, A. Idnurm, J. C. Rutherford, J. Heitman, and M. E. Cardenas, "Sensing the environment: lessons from fungi," *Nature Reviews Microbiology*, vol. 5, no. 1, pp. 57–69, 2007.

[95] M. Schmoll, "The information highways of a biotechnological workhorse-signal transduction in Hypocrea jecorina," *BMC Genomics*, vol. 9, no. 1, p. 430, 2008.

[96] M. Saloheimo and T. M. Pakula, "The cargo and the transport system: secreted proteins and protein secretion in *Trichoderma reesei* (*Hypocrea jecorina*)," *Microbiology*, vol. 158, no. 1, pp. 46–57, 2012.

[97] J. Zhang, Y. Zhang, Y. Zhong, Y. Qu, and T. Wang, "Ras GTPases modulate morphogenesis, sporulation and cellulase gene expression in the cellulolytic fungus *Trichoderma reesei*," *PLoS One*, vol. 7, no. 11, article e48786, 2012.

[98] S. Zeilinger, B. Reithner, V. Scala, I. Peissl, M. Lorito, and R. L. Mach, "Signal transduction by Tga3, a novel G protein alpha subunit of Trichoderma atroviride," *Applied and Environmental Microbiology*, vol. 71, no. 3, pp. 1591–1597, 2005.

[99] C. A. D'Souza and J. Heitman, "Conserved cAMP signaling cascades regulate fungal development and virulence," *FEMS Microbiology Reviews*, vol. 25, no. 3, pp. 349–364, 2001.

[100] T. Vasara, M. Saloheimo, S. Keränen, and M. Penttilä, "*Trichoderma reesei rho3*, a homologue of yeast *RHO3*, suppresses the growth defect of yeast *sec15-1* mutation," *Current Genetics*, vol. 40, no. 2, pp. 119–127, 2001.

[101] R. Morawetz, H. Mischak, J. Goodnight, T. Lendenfeld, J. F. Mushinsky, and C. P. Kubicek, "A protein kinase-encoding gene, *pktl*, from *Trichoderma reesei*, homologous to the yeast *YPK1* and *YPK2* (*YKR2*) genes," *Gene*, vol. 146, no. 2, pp. 309-310, 1994.

[102] R. Morawetz, T. Lendenfeld, H. Mischak et al., "Cloning and characterisation of genes (*pkc1* and *pkcA*) encoding protein kinase C homologues from *Trichoderma reesei* and *Aspergillus niger*," *Molecular and General Genetics MGG*, vol. 250, no. 1, pp. 17–28, 1996.

[103] T. Lendenfeld and P. C. Kubicek, "Characterization and properties of protein kinase C from the filamentous fungus Trichoderma reesei," *Biochemical Journal*, vol. 330, no. 2, pp. 689–694, 1998.

[104] M. Wang, M. Zhang, L. Li et al., "Role of *Trichoderma reesei* mitogen-activated protein kinases (MAPKs) in cellulase formation," *Biotechnology for Biofuels*, vol. 10, no. 1, p. 99, 2017.

[105] C. Seibel, G. Gremel, R. do Nascimento Silva, A. Schuster, C. P. Kubicek, and M. Schmoll, "Light-dependent roles of the G-protein α subunit GNA1 of *Hypocrea jecorina* (anamorph *Trichoderma reesei*)," *BMC Biology*, vol. 7, no. 1, p. 58, 2009.

[106] M. Schmoll, A. Schuster, R. N. Silva, and C. P. Kubicek, "The G-alpha protein GNA3 of *Hypocrea jecorina* (Anamorph *Trichoderma reesei*) regulates cellulase gene expression in the presence of light," *Eukaryotic Cell*, vol. 8, no. 3, pp. 410–420, 2009.

[107] K. Brunner, M. Omann, M. E. Pucher et al., "Trichoderma G protein-coupled receptors: functional characterisation of a cAMP receptor-like protein from Trichoderma atroviride," *Current Genetics*, vol. 54, no. 6, pp. 283–299, 2008.

[108] M. R. Omann, S. Lehner, C. Escobar Rodriguez, K. Brunner, and S. Zeilinger, "The seven-transmembrane receptor Gpr1 governs processes relevant for the antagonistic interaction of *Trichoderma atroviride* with its host," *Microbiology*, vol. 158, no. 1, pp. 107–118, 2012.

[109] D. Tisch, C. P. Kubicek, and M. Schmoll, "The phosducin-like protein PhLP1 impacts regulation of glycoside hydrolases and light response in Trichoderma reesei," *BMC Genomics*, vol. 12, no. 1, p. 613, 2011.

[110] S. Sestak and V. Farkas, "Metabolic regulation of endoglucanase synthesis in *Trichoderma reesei*: participation of cyclic AMP and glucose-6-phosphate," *Canadian Journal of Microbiology*, vol. 39, no. 3, pp. 342–347, 1993.

[111] K. M. V. Nogueira, M. d. N. Costa, R. G. de Paula, F. C. Mendonça-Natividade, R. Ricci-Azevedo, and R. N. Silva, "Evidence of cAMP involvement in cellobiohydrolase expression and secretion by *Trichoderma reesei* in presence of the inducer sophorose," *BMC Microbiology*, vol. 15, no. 1, p. 195, 2015.

[112] A. Schuster, D. Tisch, V. Seidl-Seiboth, C. P. Kubicek, and M. Schmoll, "Roles of protein kinase A and adenylate cyclase in light-modulated cellulase regulation in *Trichoderma reesei*," *Applied and Environmental Microbiology*, vol. 78, no. 7, pp. 2168–2178, 2012.

[113] M. Wang, H. Yang, M. Zhang et al., "Functional analysis of Trichoderma reesei CKIIα2, a catalytic subunit of casein kinase II," *Applied Microbiology and Biotechnology*, vol. 99, no. 14, pp. 5929–5938, 2015.

[114] R. He, W. Guo, and D. Zhang, "An ethanolamine kinase Eki1 affects radial growth and cell wall integrity in *Trichoderma reesei*," *FEMS Microbiology Letters*, vol. 362, no. 17, article fnv133, 2015.

[115] M. Wang, Q. Zhao, J. Yang et al., "A mitogen-activated protein kinase Tmk3 participates in high osmolarity resistance, cell wall integrity maintenance and cellulase production regulation in *Trichoderma reesei*," *PLoS One*, vol. 8, no. 8, article e72189, 2013.

[116] M. Wang, Y. Dong, Q. Zhao et al., "Identification of the role of a MAP kinase Tmk2 in *Hypocrea jecorina* (*Trichoderma reesei*)," *Scientific Reports*, vol. 4, no. 1, article 6732, 2015.

[117] F. Chen, X. Z. Chen, X. Y. Su et al., "An Ime2-like mitogen-activated protein kinase is involved in cellulase expression in the filamentous fungus *Trichoderma reesei*," *Biotechnology Letters*, vol. 37, no. 10, pp. 2055–2062, 2015.

[118] D. E. Clapham, "Calcium signaling," *Cell*, vol. 131, no. 6, pp. 1047–1058, 2007.

[119] L. Navazio, B. Baldan, R. Moscatiello et al., "Calcium-mediated perception and defense responses activated in plant cells by metabolite mixtures secreted by the biocontrol fungus *Trichoderma atroviride*," *BMC Plant Biology*, vol. 7, no. 1, p. 41, 2007.

[120] R. L. Mach, S. Zeilinger, D. Kristufek, and C. P. Kubicek, "Ca^{2+}-calmodulin antagonists interfere with xylanase formation and secretion in *Trichoderma reesei*," *Biochimica et*

Biophysica Acta (BBA) - Molecular Cell Research, vol. 1403, no. 3, pp. 281–289, 1998.

[121] D. Tisch, C. P. Kubicek, and M. Schmoll, "New insights into the mechanism of light modulated signaling by heterotrimeric G-proteins: ENVOY acts on gna1 and gna3 and adjusts cAMP levels in Trichoderma reesei (Hypocrea jecorina)," *Fungal Genetics and Biology*, vol. 48, no. 6, pp. 631–640, 2011.

[122] A. Cziferszky, R. L. Mach, and C. P. Kubicek, "Phosphorylation positively regulates DNA binding of the carbon catabolite repressor Cre1 of Hypocrea jecorina (Trichoderma reesei)," *The Journal of Biological Chemistry*, vol. 277, no. 17, pp. 14688–14694, 2002.

[123] I. E. Cabrera, I. V. Pacentine, A. Lim et al., "Global analysis of predicted G protein–coupled receptor genes in the filamentous fungus, *Neurospora crassa*," *G3: Genes, Genomes, Genetics*, vol. 5, no. 12, pp. 2729–2743, 2015.

[124] P. E. Vercoe, S. A. Kocherginskaya, and B. A. White, "Differential protein phosphorylation–dephosphorylation in response to carbon source in *Ruminococcus flavefaciens* FD-1," *Journal of Applied Microbiology*, vol. 94, no. 6, pp. 974–980, 2003.

[125] D. Tisch and M. Schmoll, "Novel approaches to improve cellulase biosynthesis for biofuel production—adjusting signal transduction pathways in the biotechnological workhorse *Trichoderma reesei*," in *Biofuel Production—-Recent Developments and Prospects*, M. A. Santos Bernardes, Ed., pp. 199–224, Intech, Rijeka, Croatia, 2011.

[126] T. M. Mello-de-Sousa, A. Rassinger, M. E. Pucher et al., "The impact of chromatin remodelling on cellulase expression in *Trichoderma reesei*," *BMC Genomics*, vol. 16, no. 1, p. 588, 2015.

[127] G. P. Borin, C. C. Sanchez, A. P. de Souza et al., "Comparative secretome analysis of *Trichoderma reesei* and *Aspergillus Niger* during growth on sugarcane biomass," *PLoS One*, vol. 10, no. 6, article e0129275, 2015.

[128] R. Peterson and H. Nevalainen, "*Trichoderma reesei* RUT-C30—thirty years of strain improvement," *Microbiology*, vol. 158, no. 1, pp. 58–68, 2012.

[129] R. K. Aghcheh and C. P. Kubicek, "Epigenetics as an emerging tool for improvement of fungal strains used in biotechnology," *Applied Microbiology and Biotechnology*, vol. 99, no. 15, pp. 6167–6181, 2015.

[130] A. B. Sanz, R. García, J. M. Rodríguez-Peña, C. Nombela, and J. Arroyo, "Cooperation between SAGA and SWI/SNF complexes is required for efficient transcriptional responses regulated by the yeast MAPK Slt2," *Nucleic Acids Research*, vol. 44, pp. 7159–7172, 2016.

[131] R. M. Duran, S. Gregersen, T. D. Smith et al., "The role of *Aspergillus flavus veA* in the production of extracellular proteins during growth on starch substrates," *Applied Microbiology and Biotechnology*, vol. 98, no. 11, pp. 5081–5094, 2014.

[132] J. Schumacher, A. Simon, K. C. Cohrs et al., "The VELVET complex in the gray mold fungus *Botrytis cinerea*: impact of BcLAE1 on differentiation, secondary metabolism, and virulence," *Molecular Plant-Microbe Interactions*, vol. 28, no. 6, pp. 659–674, 2015.

[133] N. Trushina, M. Levin, P. K. Mukherjee, and B. A. Horwitz, "PacC and pH-dependent transcriptome of the mycotrophic fungus Trichoderma virens," *BMC Genomics*, vol. 14, no. 1, p. 138, 2013.

[134] F. Castellanos, M. Schmoll, P. Martínez et al., "Crucial factors of the light perception machinery and their impact on growth and cellulase gene transcription in *Trichoderma reesei*," *Fungal Genetics and Biology*, vol. 47, no. 5, pp. 468–476, 2010.

[135] M. Schmoll, L. Franchi, and C. P. Kubicek, "Envoy, a PAS/LOV domain protein of Hypocrea jecorina (Anamorph Trichoderma reesei), modulates cellulase gene transcription in response to light," *Eukaryotic Cell*, vol. 4, no. 12, pp. 1998–2007, 2005.

[136] W. Dong, Q. Yinbo, and G. Peiji, "Regulation of cellulase synthesis in mycelial fungi: participation of ATP and cyclic AMP," *Biotechnology Letters*, vol. 17, no. 6, pp. 593–598, 1995.

[137] G. P. Borin, C. C. Sanchez, E. S. de Santana et al., "Comparative transcriptome analysis reveals different strategies for degradation of steam-exploded sugarcane bagasse by *Aspergillus niger* and *Trichoderma reesei*," *BMC Genomics*, vol. 18, no. 1, p. 501, 2017.

[138] J. S. Bak, "Lignocellulose depolymerization occurs via an environmentally adapted metabolic cascades in the wood-rotting basidiomycete Phanerochaete chrysosporium," *Microbiology*, vol. 4, no. 1, pp. 151–166, 2015.

[139] G. R. Hemsworth, E. M. Johnston, G. J. Davies, and P. H. Walton, "Lytic polysaccharide monooxygenases in biomass conversion," *Trends in Biotechnology*, vol. 33, no. 12, pp. 747–761, 2015.

[140] W. C. Kim, J. Y. Kim, J. H. Ko, H. Kang, and K. H. Han, "Identification of direct targets of transcription factor MYB46 provides insights into the transcriptional regulation of secondary wall biosynthesis," *Plant Molecular Biology*, vol. 85, no. 6, pp. 589–599, 2014.

[141] Y. Xiong, V. W. Wu, A. Lubbe et al., "A fungal transcription factor essential for starch degradation affects integration of carbon and nitrogen metabolism," *PLoS Genetics*, vol. 13, no. 5, article e1006737, 2017.

[142] S. T. Coradetti, J. P. Craig, Y. Xiong, T. Shock, C. Tian, and N. L. Glass, "Conserved and essential transcription factors for cellulase gene expression in ascomycete fungi," *Proceedings of the National Academy of Sciences*, vol. 109, no. 19, pp. 7397–7402, 2012.

[143] P. K. Foreman, D. Brown, L. Dankmeyer et al., "Transcriptional regulation of biomass-degrading enzymes in the filamentous fungus Trichoderma reesei," *The Journal of Biological Chemistry*, vol. 278, no. 34, pp. 31988–31997, 2003.

[144] F. Segato, A. R. Damásio, R. C. de Lucas, F. M. Squina, and R. A. Prade, "Genomics review of holocellulose deconstruction by aspergilli," *Microbiology and Molecular Biology Reviews*, vol. 78, no. 4, pp. 588–613, 2014.

[145] C. M. Payne, B. C. Knott, H. B. Mayes et al., "Fungal cellulases," *Chemical Reviews*, vol. 115, no. 3, pp. 1308–1448, 2015.

[146] A. Berlin, "No barriers to cellulose breakdown," *Science*, vol. 342, no. 6165, pp. 1454–1456, 2013.

The Effect of Canertinib on Sensitivity of Cytotoxic Drugs in Tamoxifen-Resistant Breast Cancer Cells In Vitro

Hesham A. M. Gomaa,[1,2] Asmaa T. Ali,[3] M. Abdel Gabbar,[4] and M. A. Kandeil[5]

[1]Pharmacology Department, College of Pharmacy, Jouf University, Sakakah, Saudi Arabia
[2]Assistant Professor at Biochemistry Department, Faculty of Pharmacy, Nahda University, Beni-Suef, Egypt
[3]Assistant Lecturer at Biochemistry Department, Faculty of Pharmacy, Nahda University, Beni-Suef, Egypt
[4]Lecturer at Biochemistry Division, Chemistry Department, Beni-Suef University, Ph.D. in Biochemistry, Beni-Suef, Egypt
[5]Professor and Head of Biochemistry Department Faculty of Veterinary Medicine, Beni-Suef University, Beni-Suef, Egypt

Correspondence should be addressed to Asmaa T. Ali; asmaa.tarek@nub.edu.eg

Academic Editor: Atsushi Kurabayashi

Aims and Objectives. To investigate and examine the reversal effects of canertinib on the activity of EGFR and tamoxifen resistance in drug-resistant human breast carcinoma cells (MCF-7/TamR). *Materials and Methods.* The antiproliferative activity of canertinib alone or in combination with a conventional EGFR-targeting chemotherapies cytotoxic drugs differing in the mechanism(s) of action, such as paclitaxel, carboplatin, etoposide, vinorelbine, and daunorubicin as well as resistance mechanisms of EGFR targeting, have been investigated. *Results.* With an elevated dosage of canertinib, a significant decrease in proliferation and increase in apoptosis was observed. The treatment with higher doses of canertinib resulted in a 2-3-fold increase in apoptosis. In the combined treatment, it had been noticed a significant developed apoptotic cell death rather induced by single agent treatment. A significant downregulation of the antiapoptotic protein bcl-2 was exposed by immunocytochemistry investigation. Sensitivity to paclitaxel was also measured and was found to inversely correlate to bcl-2 status. *Conclusion.* Proliferation inhibition and apoptosis in MCF-7/TAM-R cells increase with increasing dosage of canertinib. This suggests that canertinib can reverse tamoxifen resistance in breast cancer cells. The antitumor effect of this EGFR-irreversible tyrosine kinase inhibitor provides a rationale for its clinical evaluation in combination with other cytotoxic drugs.

1. Introduction

Epidermal growth factor (EGF) was one of the first growth factors discovered which was first isolated by Cohen in 1962 [1]. Moreover, EGF induces cell proliferation or differentiation in mammalian cells through binding to the epidermal growth factor receptor (EGFR) [2]. Further studies elucidated more precise structural details such as the exact amino acid composition [3], the primary structure [4], and the position of the three disulfide bridges shown in Figure 1 [5]. Canertinib was designed as an inhibitor for pan-ErbB tyrosine kinase; it inhibits all four ErbB receptor family members [6] Figure 2.

2. Materials and Methods

2.1. Materials. Clinical grade canertinib (CI-1033) and cytotoxic drugs (paclitaxel, etoposide, vinorelbine, and daunorubicin) were purchased from Sigma Chemical Co. (Sigma-Aldrich, Egypt). Carboplatin was provided by Sanofi Italia.

2.1.1. Cell Lines and Growth Conditions. The human breast cancer cell line MCF-7 and T47D cells were purchased from VACSERA (Cairo, Egypt); growth conditions of these cells are full RMPl medium supplemented with 5% (*v*/*v*) foetal calf serum, 10 μg/ml penicillin-streptomycin, and 2.5 μg/ml fungizone. Hereafter referred to as RPMl + 5%. To minimize

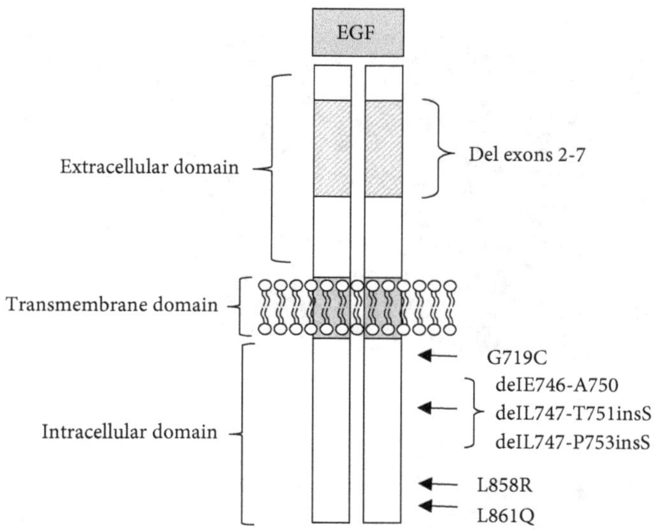

FIGURE 1: Schematic structure of EGFR with the extra- and intracellular domains.

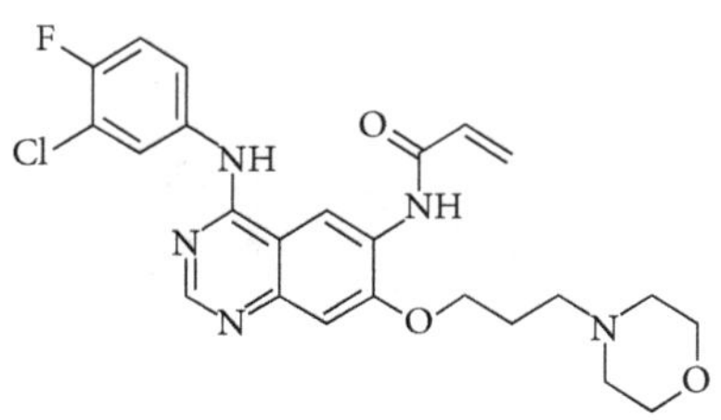

FIGURE 2: Chemical structure of canertinib (EGFR-irreversible tyrosine kinase inhibitor).

the unwanted oestrogenic effects of phenol red and foetal calf serum, cells were transferred to alternative media prior to the experimentation, containing RMPl medium (w/o phenol red and L-glutamine) supplemented with 5% (*v*/*v*) charcoal stripped foetal calf serum, 4 mM l-glutamine, and 10 μg/ml penicillin-streptomycin.

Establishment of MCF-7(TamR) and T47d (TamR) cells: W + 5% media supplemented with 10^{-7} M 4-hydroxytamoxifen (Sigma-Aldrich, Egypt).

2.2. Methods

2.2.1. Western Blot Analysis. Expression and activation levels of several proteins were assayed by Western blot analysis, which was carried out according to the following method (all reagents from Sigma-Aldrich).

Cells were lysed on ice with a lysis solution (50 mM TRIS base, 5 mM EGTA, and 150 mM NaCl, (1% Trion)) containing a cocktail of protease inhibitors (2 mM NaVO4, 50 mM NF, 1 mM PMSF, 20 μM phenylalanine, 10 Mm sodium molybdate, 10 μg/ml leupeptin, and 8 μg/ml aprotinin), the latter two reagents both being added on day of use. The lysates were then centrifuged at 15.300g for 15 minutes at 4°C. The level of EGFR and phosphorylated and unphosphorylated Rb protein was determined by Western blot transfer analysis followed by immunoprobing with respective antibodies and was detected by the enhanced chemiluminescence system.

2.2.2. Cell Viability. Cell viability and sensitivity of the cells to different compounds were determined by 3-(4,5-dimethylthiazol-2-yl)-2,5-diphenyl tetrazolium bromide assay. Cells were plated in duplicate in 96-well cloning tissue culture dishes in specified media. After 48 h, the medium was changed, and Go6976 was added at the desired concentration every 2 days along with fresh medium. After the specified days of growth, the viability of cells was measured and was expressed as a percentage of the untreated control cells grown in the presence of the same concentration of the solvent (DMSO).

2.3. Detection of Cellular Apoptosis by Flow Cytometry. Cells were seeded at a density of 40.000 cells/cm^2 in a 24-well plate and left overnight before being treated with a medium containing 5 mg/ml FITC-dextran (Sigma-Aldrich, Egypt) for 4 hrs. Cells were then washed in situ three times with PBS (37°C) before being trypsinised for 5 minutes in 200 μl trypsin/EDTA (Sigma-Aldrich, Egypt). Following trypsinisation, cells were transferred to centrifuge tubes, pelleted by centrifugation, and resuspended in PBS.

2.3.1. Statistics. All measured data were presented as mean ± standard deviation (± SD) and analyzed using the SPSS program for Windows (SPSS Inc., Chicago, US). ANOVA was performed on each set of data, and a Levene statistic calculated to establish whether variances were equal in sets of data were to be analyzed. Where the Levene statistic was equal to or more than 0.05 and therefore variances were measured to be equal, Dunnett's *t*-test was used to establish whether each treatment had an effect relative to the untreated control group.

3. Results

The tamoxifen-resistant cells developed by long-term culture in 10^{-7} M tamoxifen were examined to ensure that they were able to proliferate in the presence of tamoxifen and that they displayed behavior consistent with previous cell culture models of tamoxifen resistance. A number of

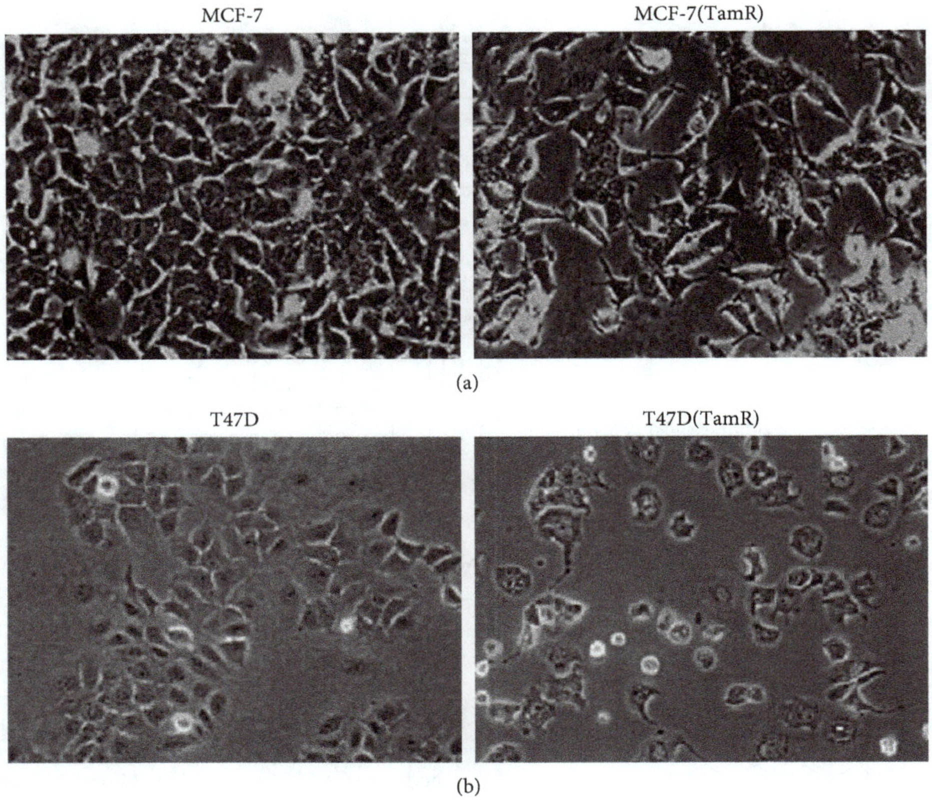

FIGURE 3: Morphological changes of both cell lines: (a) MCF-7 and its tamoxifen-resistant derivative MCF-7(TamR) cells, (b) T47D and its tamoxifen resistant derivative T47D (TamR).

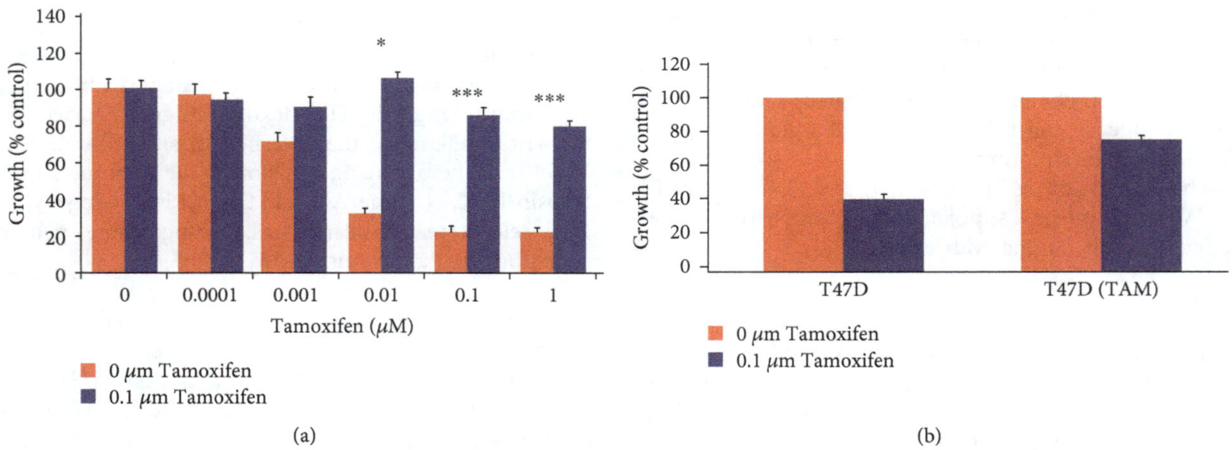

FIGURE 4: (a) Effect of tamoxifen on the growth of both cell lines MCF-7 (red), and its tamoxifen-resistant derivative MCF-7(TamR) (blue) cells of a 7-day treatment with tamoxifen, added to culture medium and refreshed on day 4. (b) Effect of a 14-day tamoxifen treatment (0.1 μM tamoxifen added to normal growth media) on T47D and its tamoxifen-resistant derivative T47D (TamR) cell line. $N = 6$. P values calculated from a paired t-test comparing growth inhibition between cell lines at a given concentration of tamoxifen, $*$ indicates $p < 0.05$, $**$ indicates $p < 0.01$, $***$ indicates $p < 0.001$. Error bars indicate standard deviation.

changes were observed in both MCF-7 and T47D cells following the development of tamoxifen resistance as the morphology of the cells was significantly altered, with the tamoxifen-resistant MCF-(TamR) and T47D (TamR) cells, respectively, growing in a more scattered fashion and had a more angular appearance which displayed the longer processes previously observed in culture compared to its corresponding wild-type which generally grew in

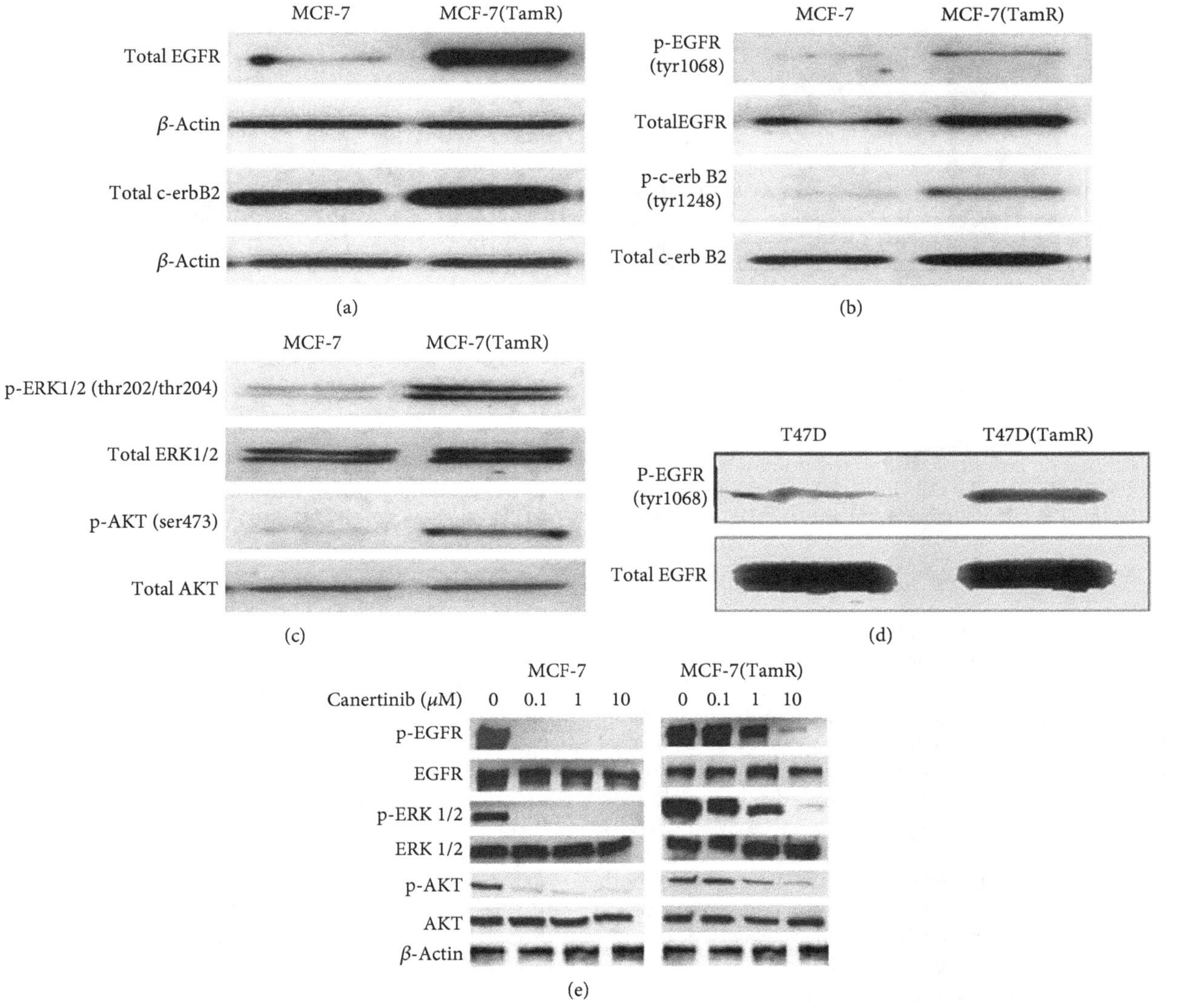

FIGURE 5: Measuring of EGFR and c-erbB2 expression; activation of ERK1/2 and AKT in MCF-7, T47D, and their tamoxifen-resistant derivative cells MCF-7(TamR) and T47D (TamR) cells, respectively. (a) Changes in levels of total EGFR and c-erbB2 expression by Western blot. (b) Changes in levels of p-EGFR and p-c-erbB2 expression by Western blot. (c) Changes in levels of expression and activation of ERK1/2 and AKT by Western blot. (d) Changes in levels of EGFR expression and activation in t47d cells before and after the development of tamoxifen resistance by Western blot. (e) The inhibitory effects of canertinib on the activation of EGFR, Akt, and ERK1/2 and their encoded protein in both MCF-7 and MCF-7(TamR) cells by Western blot.

more clearly defined colonies (Figure 3). In addition to their routine growth in 10^{-7} M tamoxifen, MCF-7(TamR) cells also demonstrated the ability to proliferate in a range of concentrations of tamoxifen which were shown to inhibit the growth of wild-type MCF-7 cells (Figure 4), with only a 25% reduction in growth of MCF-7(TamR) cells being observed in a 7-day treatment with 10^{-7} M tamoxifen, compared to 85% reduction in growth of MCF-7 cells. During the development of tamoxifen resistance, levels of expression and activation of several key signaling molecules such as EGFR and AKT are increased, given that increased levels of these proteins, in particularly, EGFR are generally associated with tumors that have increased resistance to cytotoxic drugs. In addition to increased expression of EGFR and c-erbB2, increased activation of these proteins was also observed, with clearly increased levels of phosphorylation of both molecules detected by Western blotting (Figure 5). ERK1/2 activation revealed a clear increased in ERK1/2 phosphorylation, with MCF-7(TamR) cells exhibiting both an increase in staining density and an increase in the number of densely staining cells. Examination of AKT activation also displayed a similar pattern, with a clear increase in staining density following the development of tamoxifen resistance. Western blot analysis confirmed the increase in activation of both ERK1/2 and AKT compared to their total expression levels. The total expression of ERK1/2 seems to increase with the development of tamoxifen resistance, while levels of total AKT seem to remain

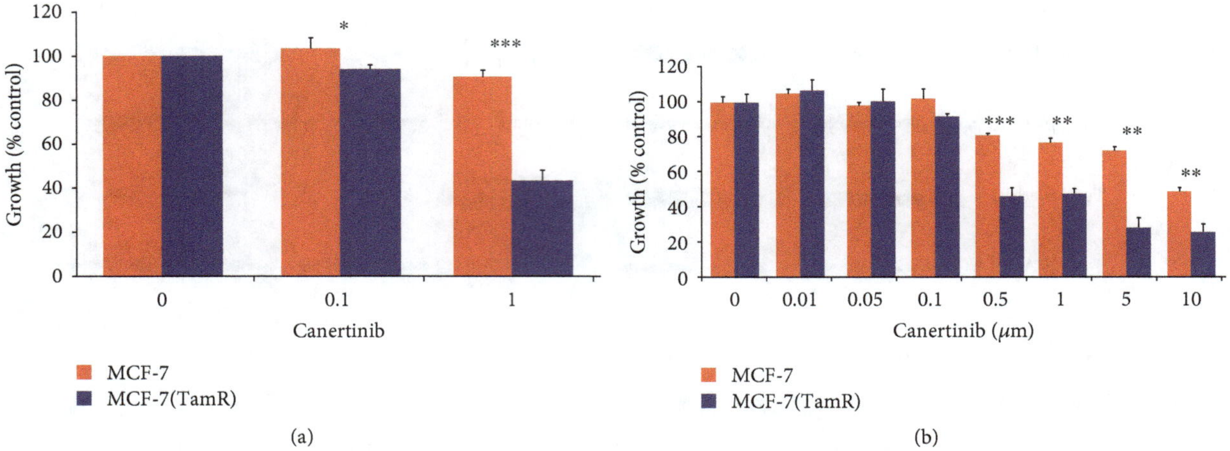

FIGURE 6: Effect on growth of MCF-7 (red) and MCF-7(TamR) (blue) cells of a 9-day treatment of the EGFR inhibitors (a) canertinib and (b) canertinib added to the cell culture medium and refreshed on day 5 (test comparing growth inhibition between cell lines at a given concentration of EGFR inhibitor). $*$ indicates $p < 0.05$, $**$ indicates $p < 0.01$, $***$ indicates $p < 0.001$. Error bars indicate standard deviation.

comparable. MCF-7(TamR) cells became much more sensitive to the growth inhibitory effects of EGFR inhibitors, highlighting their reliance on EGFR signaling for the transmission of proliferative signals. The small molecule inhibitor canertinib, delivered at a concentration of $1\,\mu$M over a 48 hours (Figure 6) and 9 days (Figure 7), inhibited the growth of MCF-7(TamR) cells by around 65%, while having no detectable effect on the growth of MCF-7 cells. Much higher concentrations of canertinib (around $10\,\mu$M) were able to significantly reduce the growth of both cell lines (Figure 5(e)); however, at all concentrations of canertinib, the growth inhibitory effect was significantly higher in MCF-7(TamR) cells than MCF-7. This is consistent with the observations made in several other studies using various inhibitors of EGFR including gefitinib, lapatinib, erlotinib, LY294002, and Herceptin (Figure 8). Once the expected changes in the activity of signaling molecules in tamoxifen-resistant cells had been confirmed, the effect of these changes on cytotoxic sensitivity was examined. The sensitivity of wild-type, low EGFR expressing MCF-7 cells and tamoxifen resistant, high EGFR expressing MCF-7(TamR) cells to a panel of cytotoxic agents was compared. When treated with paclitaxel there was a marked difference in sensitivity observed between MCF-7 and MCF-7(TamR) cells. Paclitaxel reduced the growth of both cell lines but had a much larger effect on the growth of MCF-7(TamR) cells. Over a 6-day treatment, $0.05\,\mu$g/ml paclitaxel reduced the growth rate of MCF-7(TamR) by around 50% while having no significant effect on the growth of MCF-7 cells (Figure 9(a)). Higher concentrations of paclitaxel over 6 days reduced the growth rate of both cell lines but had a much larger effect on MCF-7(TamR) compared to MCF-7 (Figure 9(a)). Treatment with carboplatin also caused a reduction in growth rate of both cell lines, but unlike paclitaxel that had a comparable effect across a wide concentration range. Growth was reduced to a similar extent in both cell lines over the course of a 6-day (Figure 9(b)) treatment period, with MCF-7(TamR) displaying a slightly increased sensitivity at the highest concentrations of carboplatin in both cases. Daunorubicin reduced the growth rate of both MCF-7 and MCF-7(TamR) with a slightly increased effect on MCF-7(TamR) at the highest concentrations (Figure 9(c)). Aside from paclitaxel, the second largest difference in response between MCF-7 and MCF-7(TamR) cells was seen following a 6-day treatment with etoposide with the drug having a significantly larger effect on MCF-7(TamR) cells over a range of concentrations (Figure 9(d)). MCF-7 and MCF-7(TamR) cells displayed broadly similar sensitivity to vinorelbine (Figure 9(e)). The highest concentration of vinorelbine did however have a greater growth reduction effect on MCF-7(TamR) cells in both treatment regimens.

4. Discussion

Canertinib is an irreversible inhibitor that binds covalently to specific cysteine residues in the ATP-binding pocket such as cysteine 773 of EGFR, cysteine 784 of ErbB2, and cysteine 778 of ErbB4 thereby blocking the ATP-binding site in the kinase domain of ErbB proteins, preventing their kinase activity and downstream signaling, and additionally, it also prevents transmodulation of ErbB2 [7]. The covalent binding of canertinib results in prolonged suppression of ErbB activity [8]. Since canertinib blocks signaling through all members of the ErbB receptor family, it is more efficient than inhibitors that only prevent signaling from one of the ErbB receptors [9, 10]. Canertinib has been shown to inhibit growth and induce apoptosis in several cancer cell lines [11–13]. In clinical studies, it has been shown to have acceptable side effects. However, in phase II studies canertinib was only able to show modest effects on breast cancer and NSCLC patients [14, 15]. Such drug is able to dephosphorylate p70S6-kinase T389 in a dose-dependent manner as well as to block

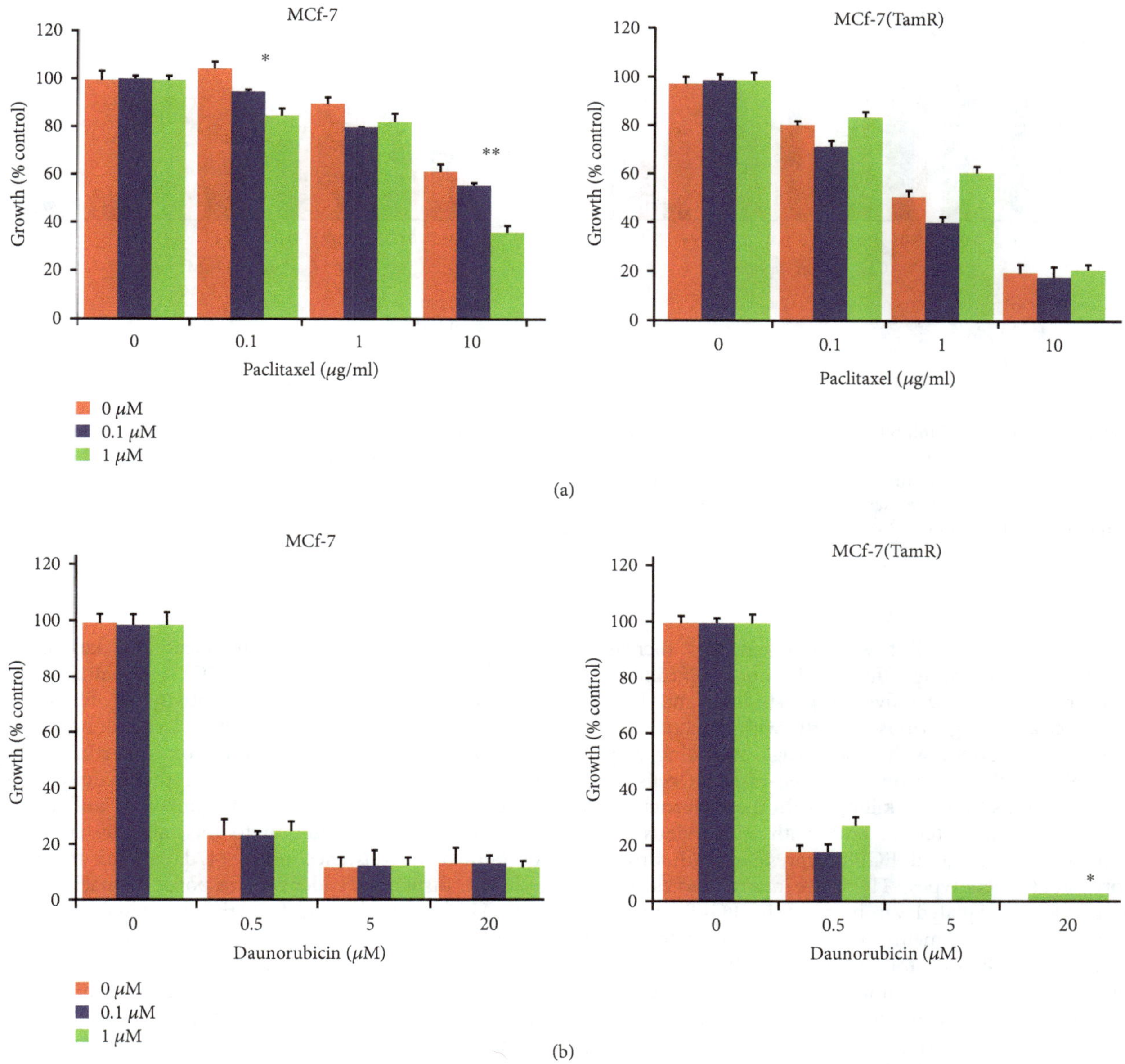

Figure 7: Effect of EGFR inhibition with canertinib treatment on the sensitivity of MCF-7 and MCF-7(TamR) cells to a 48-hour treatment with (a) paclitaxel and (b) daunorubicin. Cells were subjected to a 48-hour cytotoxic treatment with the addition of (red) 0 μM, (blue) 0.1 μM, or (green) 1 μM canertinib before 4 days recovery in the same medium minus the cytotoxic. P values calculated from a paired t-test comparing growth inhibition between cells untreated with canertinib, both of which were treated with a given concentration of daunorubicin or paclitaxel. $*$ indicates $p < 0.05$ of 0 μM canertinib. Error bars indicate standard deviation.

downstream signaling molecules in all cell lines. Additionally, it induces the apoptosis through an increased expression of the proapoptotic protein BIM (Bcl2-interacting mediator), caspase-3 cleavage, and inhibits proliferation of BCR/ABL-cells resistant to TKIs [16]. It is able to bind not only to the ErbB receptor family but also to intracellular proteins. For instance, the Src kinase family consists of eight members, five of which are mainly expressed in hematopoietic cells, Blk, Hck, Lck, Fyn, and Lyn, where the Lck protein seems to have a stronger binding to canertinib as shown in a protein-binding assay [17].

Initial observation of the MCF-7/MCF-7(TamR) cell culture models revealed that tamoxifen resistance had developed and that the characteristics of interest previously reported by others (Jones et al., 2004; EL-Zarruk and Van den Berg, 1999; Gee et al., 2001; Knowlden et al., 2003) were present [18–21]. These included an increased rate of growth, increased EGFR activity of growth, and increased activation of downstream signaling molecules linked to proliferation and survival. The ability of the increased EGFR signaling to protect MCF-7(TamR) cells from the effects of a panel of cytotoxic drugs was

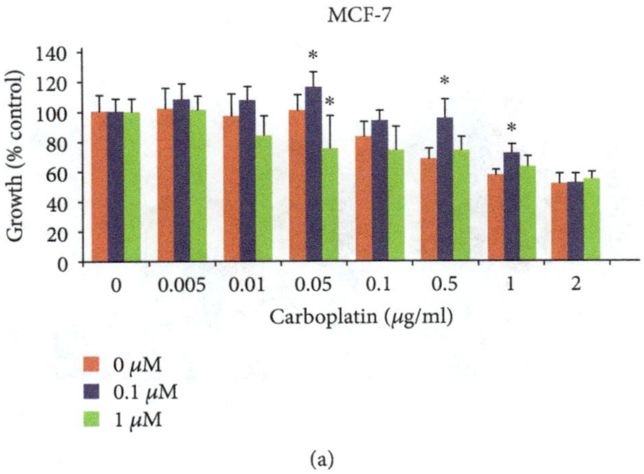

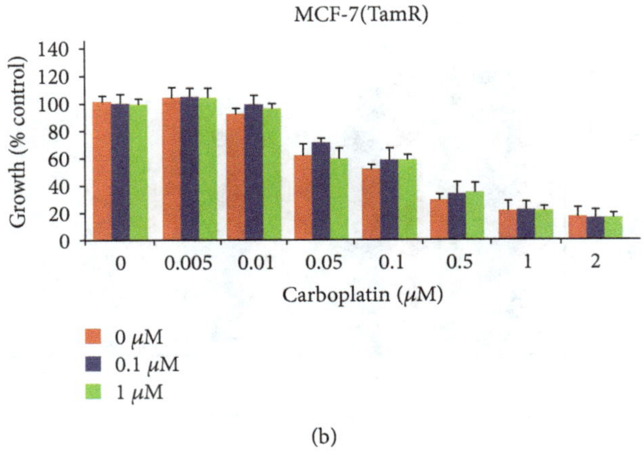

FIGURE 8: Effect of EGFR inhibition with canertinib treatment on the sensitivity of (a) MCF-7 and (b) MCF-7(TamR) cells to carboplatin. Cells were subjected to a 9-day treatment with carboplatin with the addition of (red) 0 μM, (blue) 0.1 μM, or (green) 1 μM canertinib to the culture medium for the full length of the treatment, refreshed on day 5. *P* values calculated from a paired *t*-test comparing growth inhibition between cells treated with canertinib in addition to a given concentration of paclitaxel. $*$ indicates $p < 0.05$, $**$ indicates $p < 0.01$, $***$ indicates $p < 0.001$ of 0 μM canertinib. Error bars indicate standard deviation.

carefully examined, and it was observed that increased levels of EGFR signaling offered cells no significant protection from any of the five drugs studied, and in the case of certain drugs (most clearly with paclitaxel, but also to a lesser extent with other drugs, including etoposide), high EGFR expressing tamoxifen-resistant cells were significantly sensitized to killing with these agents. This observation is completely at odds with all previous studies, which link increased EGFR expression with drug or radioresistant phenotypes. There are many studies which have previously equated an increase in EGFR signaling during tumor development with a resistance to treatment with ionising radiation, for example EGFR signaling was found to be a strong determinant of tumor radioresponse in a panel of murine carcinomas [22] and in human glioblastomas [23], with strong EGFR expression correlating to poor response to radiotherapy in all cases. Consistent with its well-established role in the modulation of cell survival and apoptosis, increased AKT activity has also been linked with cytotoxic resistance. An inducible form of AKT has been shown to protect human myeloid cells from the cytotoxic effects of etoposide and AraC (1-β-arabinofuranosylcytosine) [24]. This peculiar sensitivity of tamoxifen-resistant MCF-7 cells in spite of their apparent increase in proliferative and prosurvival signaling was entirely the opposite of what was expected and raises a number of questions concerning the changes that occur in breast cancer cells following the development of tamoxifen resistance. The change in paclitaxel sensitivity looks likely to be caused by a change in signaling molecules or apoptotic machinery lying downstream of the damage event. Bcl-2 looks to be a good candidate, being downregulated in tamoxifen-resistant cells and having a key role in the prevention of apoptosis but attempts to alter levels of bcl-2 and examine the effect on paclitaxel cytotoxicity did not show any clear correlation. It was established that the extent of EGFR signaling appeared to have little effect on the sensitivity of either MCF-7 or MCF-7(TamR) cells to paclitaxel or daunorubicin, as the addition of the EGFR inhibitor, canertinib to cytotoxic treatments had no effect on the toxicity of these drugs to either cell line over a number of different treatment regimens. In addition to this, an alternative model of tamoxifen resistance using t47d cells as the parental cell line displays an identical response to paclitaxel as in the MCF-7-based model, with the tamoxifen-resistant derivative of t47d, (t47d (TamR)) displaying a greatly increased sensitivity, a probable explanation for the inability of increased EGFR levels in tamoxifen-resistant breast cancer to protect against the cytotoxic effects of paclitaxel is that in addition to an increase in EGFR signaling, many other changes in cellular signaling pathways occur during the development of tamoxifen resistance, and that at least one of these changes is responsible for the marked increase in sensitivity to paclitaxel; there are a number of possible candidate pathways that may become altered following the development of tamoxifen resistance and have an effect of paclitaxel sensitivity, including changes in the ability of tamoxifen-resistant cells to take-up or expel drug molecules, differences in the ability of the cells to deal with the damage to DNA caused by cytotoxic drugs, or changes in the apoptotic machinery which may render cells particularly sensitive to attack with certain cytotoxic [25].

5. Conclusion

EGFR is an important target for therapeutic intervention in patients with cancer. The present study focused on the inhibitory and antitumor activity of canertinib and

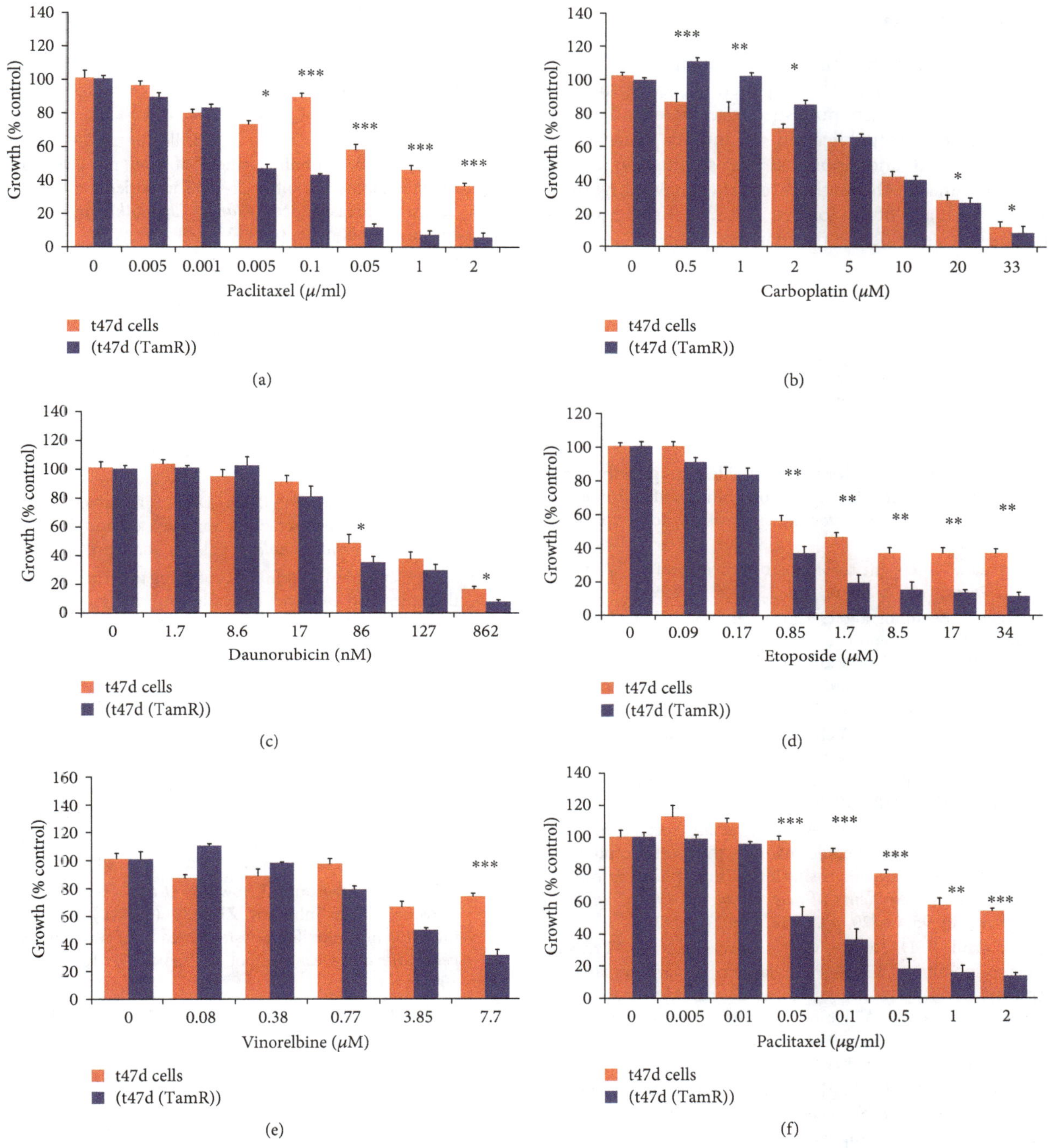

FIGURE 9: Effect of tamoxifen-resistant breast cancer cells to different cytotoxic agents; cells cultured in media containing paclitaxel for 6 days. (a) Effect on paclitaxel sensitivity, (b) effect on carboplatin sensitivity, (c) effect on daunorubicin sensitivity, (d) effect on etoposide sensitivity, (e) effect on vinorelbine sensitivity, (f) sensitivity of wild-type t47d cells (red) and tamoxifen-resistant t47d cells (t47d (TamR)) (blue) to a 9-day treatment with paclitaxel. Paclitaxel was added directly to the cell culture medium and refreshed on day 5. $N = 6$. P values calculated from a paired t-test comparing growth inhibition between MCF-7 and MCF-7(TamR) cells caused by a given concentration of cytotoxic agent of interest. $*$ indicates $p < 0.05$, $**$ indicates $p < 0.01$, $***$ indicates $p < 0.001$. Error bars indicate standard deviation.

provide a rationale for the evaluation of the anticancer activity of this EGFR-specific tyrosine kinase irreversible inhibitor alone and in combination with cytotoxic drugs in different breast cancer cell lines that express functional EGFR, our preliminary results have been reported that the combination of canertinib which is still under phase II clinical studies with different cytotoxic agents significantly inhibits the overexpression of EGFR and markedly increase the sensitivity of different cytotoxic drugs in tamoxifen-resistant breast cancer cells in vitro.

Conflicts of Interest

The authors declare that there is no conflict of interest regarding the publication of this paper. I declare that the manuscript has not been published elsewhere and is not being considered for publication in another journal. All authors have read and approved the submission of the manuscript. Finally, the authors did not have any prior discussions with any of the Journal Editorial Board Member about the work described in the manuscript.

References

[1] S. Cohen, H. Ushiro, C. Stoscheck, and M. Chinkers, "A native 170,000 epidermal growth factor receptor-kinase complex from shed plasma membrane vesicles," *The Journal of Biological Chemistry*, vol. 257, no. 3, pp. 1523–1531, 1982.

[2] Y. Yarden and J. Schlessinger, "Epidermal growth factor induces rapid, reversible aggregation of the purified epidermal growth factor receptor," *Biochemistry*, vol. 26, no. 5, pp. 1443–1451, 1987.

[3] J. J. Hsuan, "Oncogene regulation by growth factors," *Anticancer Research*, vol. 13, no. 6B, pp. 2521–2532, 1993.

[4] C. Soler, L. Beguinot, and G. Carpenter, "Individual epidermal growth factor receptor autophosphorylation sites do not stringently define association motifs for several SH-2 containing proteins," *The Journal of Biological Chemistry*, vol. 269, no. 16, pp. 12320–12324, 1994.

[5] B. Margolis, N. Li, A. Koch et al., "The tyrosine phosphorylated carboxyterminus of the EGF receptor is a binding site for GAP and PLC-gamma," *The EMBO Journal*, vol. 9, no. 13, pp. 4375–4380, 1990.

[6] D. W. Fry, A. J. Bridges, W. A. Denny et al., "Specific, irreversible inactivation of the epidermal growth factor receptor and erbB2, by a new class of tyrosine kinase inhibitor," *Proceedings of the National Academy of Sciences of the United States of America*, vol. 95, no. 20, pp. 12022–12027, 1998.

[7] J. B. Smaill, B. D. Palmer, G. W. Rewcastle et al., "Tyrosine kinase inhibitors. 15. 4-(Phenylamino)quinazoline and 4-(phenylamino)pyrido[*d*]pyrimidine acrylamides as irreversible inhibitors of the ATP binding site of the epidermal growth factor receptor," *Journal of Medicinal Chemistry*, vol. 42, no. 10, pp. 1803–1815, 1999.

[8] E. A. Djerf, C. Trinks, A. Abdiu, L. K. Thunell, A. L. Hallbeck, and T. M. Walz, "ErbB receptor tyrosine kinases contribute to proliferation of malignant melanoma cells: inhibition by gefitinib (ZD1839)," *Melanoma Research*, vol. 19, no. 3, pp. 156–166, 2009.

[9] E. A. Djerf Severinsson, C. Trinks, H. Gréen et al., "The pan-ErbB receptor tyrosine kinase inhibitor canertinib promotes apoptosis of malignant melanoma *in vitro* and displays anti-tumor activity *in vivo*," *Biochemical and Biophysical Research Communications*, vol. 414, no. 3, pp. 563–568, 2011.

[10] E. Ako, Y. Yamashita, M. Ohira et al., "The pan-erbB tyrosine kinase inhibitor CI-1033 inhibits human esophageal cancer cells in vitro and in vivo," *Oncology Reports*, vol. 17, no. 4, pp. 887–893, 2007.

[11] M. K. Nyati, D. Maheshwari, S. Hanasoge et al., "Radiosensitization by pan ErbB inhibitor CI-1033 *in vitro* and *in vivo*," *Clinical Cancer Research*, vol. 10, no. 2, pp. 691–700, 2004.

[12] W. J. Slichenmyer, W. L. Elliott, and D. W. Fry, "CI-1033, a pan-erbB tyrosine kinase inhibitor," *Seminars in Oncology*, vol. 28, Supplement 16, pp. 80–85, 2001.

[13] O. Rixe, S. X. Franco, D. A. Yardley et al., "A randomized, phase II, dose-finding study of the pan-ErbB receptor tyrosine-kinase inhibitor CI-1033 in patients with pretreated metastatic breast cancer," *Cancer Chemotherapy and Pharmacology*, vol. 64, no. 6, pp. 1139–1148, 2009.

[14] P. A. Jänne, J. von Pawel, R. B. Cohen et al., "Multicenter, randomized, phase II trial of CI-1033, an irreversible pan-ERBB inhibitor, for previously treated advanced non–small-cell lung cancer," *Journal of Clinical Oncology*, vol. 25, no. 25, pp. 3936–3944, 2007.

[15] M. E. Irwin, L. D. Nelson, J. M. Santiago-O'Farrill et al., "Small molecule ErbB inhibitors decrease proliferative signaling and promote apoptosis in Philadelphia chromosome–positive acute lymphoblastic leukemia," *PLoS One*, vol. 8, no. 8, article e70608, 2013.

[16] M. A. Fabian, W. H. Biggs, D. K. Treiber et al., "A small molecule–kinase interaction map for clinical kinase inhibitors," *Nature Biotechnology*, vol. 23, no. 3, pp. 329–336, 2005.

[17] K. Dorsey and M. Agulnik, "Promising new molecular targeted therapies in head and neck cancer," *Drugs*, vol. 73, no. 4, pp. 315–325, 2013.

[18] S. L. Jones, I. D. Hickson, A. L. Harris, and P. R. Harnett, "Repair of cisplatin-DNA adducts by protein extracts from human ovarian carcinoma," *International Journal of Cancer*, vol. 59, no. 3, pp. 388–393, 1994.

[19] A. A. El-Zarruk and H. W. van den Berg, "The anti-proliferative effects of tyrosine kinase inhibitors towards tamoxifen-sensitive and tamoxifen-resistant human breast cancer cell lines in relation to the expression of epidermal growth factor receptors (EGF-R) and the inhibition of EGF-R tyrosine kinase," *Cancer Letters*, vol. 142, no. 2, pp. 185–193, 1999.

[20] J. Gee, I. Hutcheson, J. Knowlden et al., "The EGFR-selective tyrosine kinase inhibitor ZD1839 ('Iressa') is an effective inhibitor of tamoxifen-resistant breast cancer growth," *European Journal of Cancer*, vol. 37, article S261, Supplement 6, 2001.

[21] J. M. Knowlden, I. R. Hutcheson, H. E. Jones et al., "Elevated levels of epidermal growth factor receptor/c-erbB2 heterodimers mediate an autocrine growth regulatory pathway in tamoxifen-resistant MCF-7 cells," *Endocrinology*, vol. 144, no. 3, pp. 1032–1044, 2003.

[22] T. Akimoto, N. R. Hunter, L. Buchmiller, M. Kathy, K. Kian Ang, and L. Milas, "Inverse relationship between epidermal growth factor receptor expression and radiocurability of murine carcinomas," *Clinical Cancer Resarch*, vol. 5, no. 10, pp. 2884–2890, 1999.

[23] F. G. Barker II, M. L. Simmons, S. M. Chang et al., "EGFR overexpression and radiation response in glioblastoma multiforme," *International Journal of Radiation Oncology, Biology, Physics*, vol. 51, no. 2, pp. 410–418, 2001.

[24] V. L. Grandage, R. E. Gale, D. C. Linch, and A. Khwaja, "PI3-kinase/AKt is constitutively active in primary acute myeloid leukaemia cells and regulates survival and chemoresistance via NF-KappaB, Mapkinase and p53 pathways," *Leukemia*, vol. 19, no. 4, pp. 586–594, 2005.

3

Transcriptome Analysis of Rainbow Trout (*Oncorhynchus mykiss*) Eggs Subjected to the High Hydrostatic Pressure Treatment

Artur Gurgul,[1] Klaudia Pawlina-Tyszko,[1] Monika Bugno-Poniewierska,[2] Tomasz Szmatoła,[1] Igor Jasielczuk,[1] Stefan Dobosz,[3] and Konrad Ocalewicz[4]

[1]*Laboratory of Genomics, Department of Animal Genetics and Molecular Biology, National Research Institute of Animal Production, Krakowska 1, 32-083 Balice, Poland*
[2]*Institute of Veterinary Sciences, University of Agriculture in Krakow, Mickiewicza 24/28 Av, 30-059 Krakow, Poland*
[3]*Department of Salmonid Research, Inland Fisheries Institute in Olsztyn, Rutki, 83-330 Żukowo, Poland*
[4]*Department of Marine Biology and Ecology, Institute of Oceanography, Faculty of Oceanography and Geography, University of Gdansk, M. Piłsudskiego 46 Av, 81-378 Gdynia, Poland*

Correspondence should be addressed to Konrad Ocalewicz; konrad.ocalewicz@ug.edu.pl

Academic Editor: Elena Pasyukova

High hydrostatic pressure (HHP) causes depolymerization of the spindle microtubules. HHP applied to fish eggs prevents extrusion of the second polar body and inhibits the first cell cleavage, and it is used to produce triploids and diploid gynogenetic and androgenetic individuals. HHP has been also found to affect biomolecules including nucleic acids, and it may be presumed that HHP administered to the rainbow trout (*Oncorhynchus mykiss*) eggs disturbs cytoplasmic maternal RNA indispensable for the early embryogenesis. To verify this assumption, quality and quantity of RNA extracted from the rainbow trout eggs subjected to the high hydrostatic pressure shock were analyzed. Provided results exhibited that maternal transcriptome was resistant to a three-minute exposure to 65.5 MPa of HHP treatment. Some trend showing increase of the RNA integrity was observed in the HHP-treated eggs; however, the difference was not statistically significant. Alterations in the expression profiles in the rainbow trout eggs subjected to HHP were also negligible. Greater differences in the maternal gene expression were observed between eggs from different clutches than between HHP-treated and untreated eggs from the same clutch. It may be assumed that exposure to HHP shock was too short to modify significantly maternal transcripts in the rainbow trout eggs.

1. Introduction

High hydrostatic pressure (HHP) treatment may be lethal for the prokaryotic and eukaryotic cells. HHP about 200 MPa causes apoptotic death of the mammalian cells while treatment with HHP higher than 300 MPa results in the cell death by necrosis [1]. HHP inactivates many pathogens and keeps unchanged functional and nutritional characteristics of many products. Hence, it has been widely used in the food industry to inactivate microorganisms [2, 3]. Moreover, HHP might be applied for sterilization and disinfection of biomaterials, modulation of enzymatic activities, or genetic transformations, among others [1]. It has been observed that mammalian cells subjected to HHP of 40–60 MPa are more resistant to the cryopreservation process and thawed oocytes, spermatozoa, and embryos show increased stress tolerance and the postthaw survival [4, 5].

HHP causes depolymerization of the spindle microtubules in the fish eggs what impairs chromosome movement and disturbs meiotic and mitotic divisions. HHP shock implemented to fish eggs shortly after insemination with normal or UV-irradiated spermatozoa results in abortion of the 2nd polar body extrusion and enables development of the triploid or diploid meiogynogenetic individuals, respectively [6]. In the gamma-irradiated and inseminated eggs or eggs activated by the UV-irradiated spermatozoa, HHP applied at the time of prophase of the 1st zygotic division inhibits the cell cleavage which leads to duplication of the haploid set of chromosomes and production of the androgenetic and mitogynogenetic doubled haploids (DHs), respectively [7]. Apart from the cellular organelles [8, 9], HHP has been also found to affect DNA-protein complexes [10] and transcriptomes [11–13]. In fish, maternal mRNA deposited in the oocyte cytoplasm during oogenesis controls early embryonic development before activation of the zygotic genome which takes part at cell cycle 10 [14]. Thus, hydrostatic pressure shock administered to the rainbow trout eggs has been assumed to affect cytoplasmic maternal RNA indispensable for the early embryogenesis. To examine such assumption, quality and quantity of RNA isolated from rainbow trout (*Oncorhynchus mykiss*) eggs subjected to a three-minute exposure to 65.5 MPa of HHP were analyzed. Obtained results exhibited that rainbow trout maternal transcriptome was resistant to the HHP shock. Analysis of the transcriptome integrity did not reveal any statistically significant differences between HHP-treated and nontreated eggs. Alterations in the expression profiles of genes related to the fish development and growth, response to the DNA damage, actin filament polymerization, and function of the spindle microtubules observed in the HHP-treated eggs were not substantial. Interestingly, differences in the maternal gene expression were greater between eggs from different clutches than between HHP-treated and untreated eggs originated from an individual female.

2. Material and Methods

2.1. Gamete Collection and Egg Treatment. An experiment performed in 18 November 2015 was approved by the Animal Experiments Local Committee in Gdansk, Poland (no. 28/2015). Eggs originated from the winter spawning rainbow trout (*Oncorhynchus mykiss* Walbaum 1792) from the broodstock raised in the Department of Salmonid Research, Inland Fisheries Institute in Olsztyn, Rutki, Poland. In the case of rainbow trout from the studied broodstock, a three-minute HHP shock of 65.5 MPa was applied to the eggs 35 min after activation (early shock) or 350 min after activation (late shock) efficiently inhibits release of the 2nd polar body or prevents the 1st cell cleavage, respectively [15]. Before handling, the fish were anesthetized with Propiscin (etomidate, IRŚ, Poland) at a dose of 0.5 $ml \cdot l^{-1}$ of water. Three rainbow trout females were stripped and eggs were collected into the separate plastic bowls. In the present research, portions of 100 eggs from each female were shocked with 65.5 MPa for 3 min using TRC-APV electric/hydraulic apparatus (TRC Hydraulics Inc., Dieppe, Canada). Both HHP-treated and untreated (control) eggs were stored for 15 min at +4°C and then divided into two batches and either frozen on the dry ice or placed in RNA stabilization solution for tissues (RNAlater®) (Sigma-Aldrich), incubated overnight at +4°C, and finally stored at −80°C for further use.

2.2. RNA Purification and Degradation Analysis. RNA was isolated from the untreated eggs and eggs exposed to HHP originated from three females in five technical replicates including different methods of egg conservation and homogenization. For RNA extraction, eggs were thawed on ice and homogenized in TRIzol reagent (Thermo Fisher Scientific) using a manual and two homogenizer methods, Bullet Blender (bead mill) homogenizer (Next Advance) and TissueRuptor (Qiagen), in order to find the most efficient approach to provide high-quality RNA and to estimate the level of method-specific homogenization-induced RNA degradation. Two eggs were used for each replicate. Extracted RNA was then purified using a modified TRIzol procedure established at Igor Babiak Laboratory (University of Nordland, Bodø, Norway). The purified RNA was quantified using the NanoDrop2000 spectrophotometer (Thermo Fisher Scientific) and assessed for the quality using the Agilent 2200 TapeStation system (RNA screen tapes). The Agilent RIN (RNA integrity number) algorithm was used for the comparisons of RNA integrity between HHP-treated and untreated eggs. The algorithm analyzed not only the 28S and 18S rRNA ratios but also a whole electrophoretic trace of RNA samples, which includes the presence and absence of the degradation products [16]. RIN was standardized to be used for the comparative studies [17].

2.3. A Whole Transcriptome Sequencing and Data Analysis. Before sequencing library construction, a total RNA was additionally purified using Agencourt RNAClean XP beads (Beckman Coulter) according to the manufacturer protocol. A total of 800 ng purified RNA was used as an input for TruSeq RNA Sample Prep v2 kit (Illumina). Standard library construction steps (mRNA selection, fragmentation, cDNA synthesis, end repair, adenylation, indexed adapter ligation, and amplification) were followed by a qualitative (Agilent TapeStation 2200) and quantitative (Qubit, Thermo Fisher Scientific) evaluation. Validated and normalized libraries were eventually sequenced in a single 50 bp run (1×50 bp) on the HiScanSQ system using TruSeq SBSv3 Sequencing kit (Illumina) to obtain approx. 25 million reads per sample.

The obtained demultiplexed raw reads were filtered using the Flexbar software [18] to trim accidental sequences of adapters and remove the low-quality reads. The resulting read set was mapped against the newest available and supplemented rainbow trout transcriptome (encompassing 44,990 transcripts; downloaded from http://www.animalgenome.org/repository/pub/MTSU2014.1218/) [19] using Bowtie aligner [20], permitting for unlimited multimappings (–a). The obtained sample files were analyzed using eXpress (http://bio.math.berkeley.edu/eXpress/overview.html) software which can be used to estimate transcript abundances in the multi-isoform genes, and it is also able to resolve multimappings of reads across the gene families and does not require a reference genome so that it can be used even in the conjunction with de novo transcriptome assemblies. The underlying model is based on the previously described probabilistic models developed for RNA-seq [21] but it is also applicable to other settings where target sequences are sampled and includes parameters for the fragment length distributions, errors in reads, and sequence-specific fragment bias [22]. The obtained estimated rounded effective counts for separate transcripts and samples were put into a DESeq2

TABLE 1: RNA integrity and concentration for all RNA purification replicates in the HHP-treated and control eggs.

Replicate	1		2		3		4		5			
Homogenization/conservation	Bead mill/ RNAlater		Manual/ RNAlater		Bead mill/ RNAlater		Bead mill/ RNAlater		TissueRuptor/ dry ice		All	
Group	Control	HHP	Control	HHP	Control	HHP	Control	HHP	Control	HHP	Control	HHP
Mean RIN	8.90	8.70	7.83	8.03	7.97	8.90	8.80	8.87	8.73	8.93	8.45	8.69
RIN SD	0.14	0.20	0.62	0.45	0.50	0.16	0.14	0.42	0.05	0.12	0.58	0.47
Mean concentration (ng/μl)	224.50	318.43	134.80	122.60	175.87	163.13	178.93	187.17	152.37	164.57	173.29	191.18
Concentration SD (ng/μl)	19.77	72.54	11.21	20.66	19.99	76.60	11.42	96.53	18.89	11.43	34.59	93.18
Min RIN	8.70	8.50	7.30	7.50	7.30	8.70	8.70	8.30	8.70	8.80	7.30	7.50
Max RIN	9.00	8.90	8.70	8.60	8.50	9.10	9.00	9.30	8.80	9.10	9.00	9.30
Min conc.	197.40	220.10	119.60	103.10	148.50	59.30	163.70	55.60	126.90	149.70	119.60	55.60
Max conc.	244.00	392.90	146.30	151.20	195.70	241.80	191.20	284.50	172.10	177.50	244.00	392.90

[23] software, which is a count-based differential expression analysis tool.

3. Results

The assigned RIN was independent of the sample concentration. No clear evidence for the RNA degradation was detected in the rainbow trout eggs exposed to 65.5 MPa of HHP shock that lasted 3 minutes. The mean observed RIN for all 15 replicates in control egg RNA was 8.44 ($\pm$0.58) and was similar to that observed for the eggs subjected to HHP (8.69 ± 0.47) (Table 1, Supplementary Figure S1). The paired t-test for group means exhibited some trend ($p = 0.072$), unexpectedly showing a slightly higher RNA integrity in the HHP-treated eggs. No significant differences ($p = 0.380$) have been found in the quantity of provided RNA when mean concentrations of RNA extracted from the eggs exposed to HHP and untreated eggs were similar (191.2 ± 93.2 versus 173.3 ± 34.6 ng/μl, resp.). The detailed statistics on RNA integrity and concentrations are presented in Supplementary Table S1.

In total, 162.2 million sequencing reads were obtained and about 60% of them were successfully mapped to the reference transcriptome (Supplementary Table S1). eXpress software was used to estimate transcript abundance in each sample. In the control eggs 16,346 expressed transcripts were detected with at least one RPKM (reads per kilobase of exon per million reads mapped). A similar number of transcripts (16,243) were detected in the HHP-treated eggs. The obtained effective read counts analyzed with DESeq2 showed no significantly altered transcript levels (after FDR correction from p values using the Benjamini-Hochberg procedure) (Supplementary File Data). Examination of the expression patterns with the principal component analysis (PCA) showed expression profile clustering according to the individual from which eggs were taken but not according to experimental conditions (Figure 1). PCA also showed some interindividual variation with first two PCA components explaining 42 and 37% of variance, respectively. Pearson's correlation between expression profiles of untreated eggs in the pairwise comparisons for the separate individuals was high and ranged from 0.977 to 0.997.

The top 10 transcripts with the lowest pointwise p value ($p < 0.01$) for the difference in the expression level between HHP-treated and untreated eggs encompassed transport protein *Sec31A*-like *Oreochromis niloticus* homolog, insulin-like growth factor 1b receptor *Oncorhynchus kisutch* homolog, telomere-associated protein *RIF1 Oreochromis niloticus* homolog, *IGF-I* receptor subtype A, *FH1/FH2* domain-containing protein 3 (*Danio rerio* homolog), domain-containing glycophosphatidylinositol anchor protein, centromere protein F (*Oreochromic niloticus* homolog), *FH1/FH2* domain-containing protein 3 (*Oreochromis niloticus* homolog), histone-lysine N-methyltransferase *SETD2*-like (*Oreochromis niloticus* homolog), and male-specific lethal 3-like 1 (*Salmo salar* homolog) (Table 2, Supplementary File Data). However, RNA-seq results presented here should be confirmed using quantitative RT-PCR approach to verify differences between HHP-treated and control eggs.

4. Discussion

Induced development of triploids and diploid gynogenetic and androgenetic fish includes 1 to 10 min exposure of activated eggs to HHP of 48.3 to 79.3 MPa in order to damage spindle microtubules and to disrupt extrusion of the 2nd polar body or to inhibit 1st cell cleavage [6, 7]. However, short HHP treatment of fish eggs does not have to affect cytoplasmic maternal mRNA that is indispensable for the early embryonic development in fish. The three-minute exposure to 65.5 MPa of HHP that has been usually applied to recover diploid state in the rainbow trout eggs during gynogenesis [15] did not degrade maternal transcripts. It is not excluded that longer HHP treatment might impair stability of the maternal mRNA as it has been observed in the pig oocytes after one-hour exposure to 20 MPa of hydrostatic pressure [24]. However, such long HHP treatment might be damaging for the fish eggs.

High integrity of RNA in rainbow trout eggs subjected to HHP shock suggests such treatment has not decreased developmental competence of the eggs. HHP treatment prior cryopreservation increased stress tolerance and improved survival rates, fertilizing ability, and development competence of the vitrified porcine oocytes [4, 5]. It has been found that increased stress tolerance in the mammalian oocytes

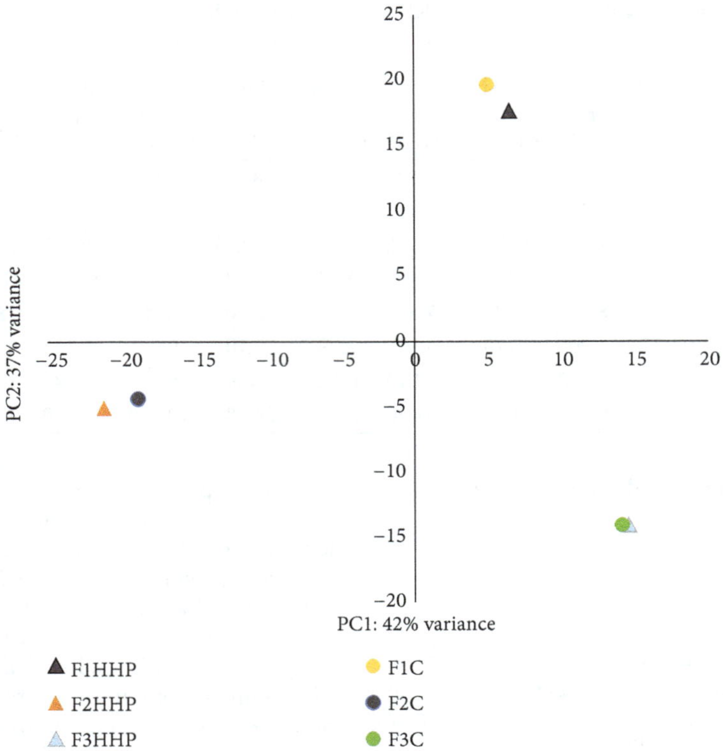

FIGURE 1: Principal component analysis (PCA). RNA-seq expression data from the rainbow trout eggs treated with high hydrostatic pressure (HHP) and untreated control eggs (C). Female (F1–F3) eggs that were treated with HHP are marked with triangles, and their untreated eggs (C) are marked with dots.

pretreated with the sublethal HHP results from the induction of the posttranscriptional activation of the shock proteins [4]. Bovine embryos treated with 40 and 60 Mpa of HHP exhibited increased embryo competence through downregulation of genes connected to cell death and apoptosis process and upregulation of genes responsible for the RNA processing, cellular growth, and proliferation [12]. In turn, higher dose of HHP (80 MPa) decreased embryo competence that showed downregulated cell cycle-related genes and upregulated genes engaged in the apoptosis mechanism [12].

Many studies performed so far using RNA-seq method are based on the three replicates per treatment that is accepted as minimal number of samples needed for analysis. Although three biological replicates is not many and the obtained statistical power may be low, the expression changes with high fold change and high uniformity of change across replicates should be reliably identified. In our study, profiles of expression transcripts in the HHP-treated rainbow trout eggs were altered when compared to those in the untreated eggs; however, the differences were not significant at the genome-wide level. This may be explained by the large expression differences between eggs from the different females. Indeed, PCA of expression profiles in the rainbow trout eggs exhibited stronger similarity between gametes from the individual female rather than from the experimental conditions (Figure 1). This observation suggests that inter-individual differences between rainbow trout females from the winter broodstock had larger influence on the egg transcriptome and developmental ability than HHP treatment of the eggs. Similar has been observed in the zebrafish (*Danio rerio*) eggs where differences in the expression of maternal genes in eggs from different clutches reflected differences between mothers [25].

Genes with transcripts affected on a pointwise level, because of their biological functions, are promising candidates for the further studies on HHP. Although, our statistical analysis cannot fully confirm their involvement in the process of cellular response to HHP, observed minor differences in their expression are suggestive and may give more information and clues for further research than classical typing of the candidate genes. To clarify provided results, expression of the genes whose transcripts differed the most in the control and HHP-treated eggs (Table 2) should be compared using qRT-PCR. These genes are involved in the regulation of development and somatic growth (*1a IGF1RB* and *1b IGF1RB*, *IGF-I* receptor subtype A) [26], response to the DNA damage and regulation of the DNA mismatch repair (*SETD2*, *RIF1*) [27, 28], actin filament polymerization (*FH1/FH2* domain-containing protein 3, *(FHOD3)*), transcriptional regulation (male-specific lethal 3-like 1 *(MS3L1)*), and formation of centromere/kinetochore complex (centromere protein F *(CENPF)*) [29] (Table 2, Supplementary File Data). The *CENPF* gene plays a role in attaching the kinetochore and spindle microtubules [29]. Thus, alterations of the CENPF transcript made by HHP may be a part of the molecular mechanism leading to a damage or dysfunction of

TABLE 2: The top 10 transcripts with the lowest pointwise p value ($p < 0.01$) for the differences in the expression level between HHP-treated and untreated rainbow trout eggs. Downregulated transcripts are written in bold.

Transcript	Gene description	Base mean	Log2 fold change	lfcSE	Stat	p value	p adj	Treated1	Treated2	Treated3	Untreated1	Untreated2	Untreated3
C29972_c2_seq1_	**Predicted: protein transport protein Sec31A-like *(Oreochromis niloticus)***	**807.49**	**−2.19**	**0.76**	**−2.90**	**0.0037**	**1.0**	**1411.93**	**665.51**	**818.05**	**1073.05**	**793.53**	**82.86**
C169673_c0_seq1_	Insulin-like growth factor 1b receptor *(Oncorhynchus kisutch)*	66.73	2.17	0.78	2.80	0.0051	1.0	7.43	48.60	60.68	49.72	35.99	197.95
C63960_c0_seq1_	Predicted: telomere-associated protein RIF1 *(Oreochromis niloticus)*	383.98	2.05	0.73	2.79	0.0053	1.0	95.54	414.58	360.60	230.52	352.88	849.79
C153473_c0_seq1_	IGF-I receptor subtype a *(Cyprinus carpio)*	123.58	2.13	0.77	2.78	0.0054	1.0	22.29	78.35	144.71	91.30	80.76	324.08
C9983_c0_seq1_	Predicted: FH1/FH2 domain-containing protein 3 *(Danio rerio)*	270.05	2.06	0.74	2.78	0.0055	1.0	50.96	325.32	240.40	163.62	313.37	526.63
C93043_c1_seq1_	MAM domain-containing glycosylphosphatidylinositol anchor protein 1 *(Dicentrarchus labrax)*	98.40	2.09	0.76	2.76	0.0058	1.0	11.68	104.14	114.36	89.50	93.05	177.69
C96308_c0_seq1_	Predicted: centromere protein F *(Oreochromis niloticus)*	1329.72	1.85	0.67	2.75	0.0059	1.0	406.59	1473.85	1256.84	1318.94	1195.56	2326.56
C35448_c1_seq1_	Predicted: FH1/FH2 domain-containing protein 3 *(Oreochromis niloticus)*	362.46	1.99	0.75	2.68	0.0075	1.0	77.50	404.66	374.60	158.20	419.59	740.23
C78059_c0_seq1_	Predicted: histone-lysine N-methyltransferase SETD2-like *(Oreochromis niloticus)*	606.50	2.04	0.76	2.67	0.0075	1.0	121.02	847.01	212.39	479.12	565.30	1414.17
C39441_c0_seq1_	**Male-specific lethal 3-like 1 *(Salmo salar)* gb**	**217.77**	**−1.92**	**0.72**	**−2.66**	**0.0077**	**1.0**	**350.33**	**259.86**	**179.72**	**280.24**	**191.36**	**45.11**

the spindle microtubules during cell meiosis and mitosis. It is not excluded that altered transcript levels of *FHOD3* may also be involved in the molecular reaction for HHP leading to the inhibition of cell cleavages in fish. HHP treatment of porcine oocytes resulted in alteration of 44 transcripts [24]. Gene ontology analysis exhibited that altered transcripts played some role during the embryonic development; however, we did not observe any overlap between expression profiles in the HHP-treated porcine and rainbow trout female gametes. Most of the porcine altered transcripts usually showed lower expression levels in the HHP-treated oocytes. Thus, it was suggested that 1 h exposure to HHP promotes precocious degradation of maternal transcriptome [24].

Survivability of salmonid androgenetic and gynogenetic early embryos is lower in eggs that were subjected to the HHP shock [30, 31]. It is not excluded that increased mortality was related to the very discrete changes in the egg cytoplasmic maternal mRNA, though it is more likely that decreased survivability of fish developing in pressurized eggs may be associated with the HHP-induced damages of the cellular organelles and protein structures. Short exposure of fish eggs to HHP depolymerizes microtubules that also play some roles in the transportation of the cytoplasmic particles and factors necessary for the early cellular differentiation of blastomeres [32]. Changes in the microtubule structure caused by HHP may thus disturb fish embryonic development. In crucian carp (*Carassius auratus* Linnaeus 1758), HHP shock administered to inseminated eggs disrupted the proper formation of blastodiscs, impaired the development of cytoplasm, and triggered the delay of epiboly and suppression of the dorsoventral differentiation [9]. Triploid rainbow trout produced using HHP usually exhibits higher incidences of deformities than control diploids [33]; however, it is hard to evaluate which malformations result from the HHP exposure.

5. Conclusions

Three-minute exposure to HHP (65.5 MPa) was presumably too short to degrade cytoplasmic RNA and/or significantly alter expression of the maternal transcripts in the rainbow trout eggs. On the other hand, some changes of the maternal transcripts responsible for the embryo growth and development, DNA synthesis and repair, cell divisions, and function of the spindle microtubules, although statistically insignificant, may impair mechanisms crucial for the proper development of the early fish embryos before activation of the zygotic genome.

Disclosure

Some of the results described in this paper were already presented during the World Aquaculture 2017 conference held in Cape Town (South Africa).

Conflicts of Interest

The authors declare that there is no conflict of interest regarding the publication of this article.

Acknowledgments

The authors thank Tomasz Zalewski from the Department of Salmonid Research, Inland Fisheries Institute in Olsztyn, Rutki, Poland, for the technical assistance and providing fish for the described research. This study was supported by the National Science Centre (NCN, Poland) Project no. 2014/15/BNZ9/00510.

Supplementary Materials

Supplementary 1. Table S1: sequencing reads and mapping statistics of eggs originating from three females (F1–F3) treated with HHP and untreated (control, C). Figure S1: exemplary TapeStation trace for RNA provided from the control (A) and HHP-treated (B) eggs. Electropherogram shows the separation profile of the selected mRNA samples showing from the left: lower marker (50 bp, preset in the sample buffer) and 18S and 28S rRNA peaks.

Supplementary 2. File Data: results of expression analysis of rainbow trout eggs subjected to the high hydrostatic pressure (HHP) shock and untreated eggs.

References

[1] N. Rivalain, J. Roquain, and G. Demazeau, "Development of high hydrostatic pressure in biosciences: pressure effect on biological structures and potential applications in biotechnologies," *Biotechnology Advances*, vol. 28, no. 6, pp. 659–672, 2010.

[2] D. Knorr, "Effects of high-hydrostatic-pressure processes on food safety and quality. Use of hydrostatic pressure in food processing overview, outstanding symposia in food science and technology," *Food Technology*, vol. 47, pp. 156–161, 1993.

[3] M. Yaldagard, S. A. Mortazavi, and F. Tabatabaie, "The principles of ultra-high pressure technology and its application in food processing/preservation: a review of microbiological and quality aspects," *African Journal of Biotechnology*, vol. 7, no. 16, pp. 2739–2767, 2008.

[4] Y. Du, C. S. Pribenszky, M. Molnar et al., "High hydrostatic pressure: a new way to improve *in vitro* developmental competence of porcine matured oocytes after vitrification," *Reproduction*, vol. 135, no. 1, pp. 13–17, 2008.

[5] C. Pribenszky, Y. Du, M. Molnar, A. Harnos, and G. Vajta, "*Increased stress tolerance of matured pig oocytes after high hydrostatic pressure treatment*," *Animal Reproduction Science*, vol. 106, no. 1-2, pp. 200–207, 2008.

[6] T. J. Pandian and R. Koteeswaran, "Ploidy induction and sex control in fish," *Hydrobiologia*, vol. 384, no. 1/3, pp. 167–243, 1998.

[7] H. Komen and G. A. Thorgaard, "Androgenesis, gynogenesis and the production of clones in fishes: a review," *Aquaculture*, vol. 269, no. 1–4, pp. 150–173, 2007.

[8] E. D. Salmon, "Pressure-induced depolymerization of brain microtubules in vitro," *Science*, vol. 189, no. 4206, pp. 884–886, 1975.

[9] E. Yamaha, S. Otani, A. Minami, and K. Arai, "Dorso-ventral axis perturbation in goldfish embryos caused by heat- and pressure-shock treatments for chromosomes set manipulation," *Fisheries Science*, vol. 68, no. 2, pp. 313–319, 2002.

[10] T. W. Lynch and S. G. Sligar, "Experimental and theoretical high pressure strategies for investigating protein–nucleic acid assemblies," *Biochimica et Biophysica Acta (BBA) - Protein Structure and Molecular Enzymology*, vol. 1595, no. 1-2, pp. 277–282, 2002.

[11] Z. Jiang, P. Harrington, M. Zhang et al., "Effects of high hydrostatic pressure on expression profiles of *in vitro* produced vitrified bovine blastocysts," *Scientific Reports*, vol. 6, no. 1, article 21215, 2016.

[12] P. M. Fernandes, T. Domitrovic, C. M. Kao, and E. Kurtenbach, "Genomic expression pattern in *Saccharomyces cerevisiae* cells in response to high hydrostatic pressure," *FEBS Letters*, vol. 556, no. 1–3, pp. 153–160, 2004.

[13] T. Domitrovic, C. M. Fernandes, E. Boy-Marcotte, and E. Kurtenbach, "High hydrostatic pressure activates gene expression through Msn 2/4 stress transcription factors which are involved in the acquired tolerance by mild pressure precondition in *Saccharomyces cerevisiae*," *FEBS Letters*, vol. 580, no. 26, pp. 6033–6038, 2006.

[14] D. A. Kane and C. B. Kimmel, "The zebrafish midblastula transition," *Development*, vol. 119, no. 2, pp. 447–456, 1993.

[15] M. Polonis, T. Fujimoto, S. Dobosz, T. Zalewski, and K. Ocalewicz, "Genome incompatibility between rainbow trout (*Oncorhynchus mykiss*) and sea trout (*Salmo trutta*) and induction of the interspecies gynogenesis," *Journal of Applied Genetics*, vol. 59, no. 1, pp. 91–97, 2018.

[16] S. Heumüller-Klug, C. Sticht, K. Kaiser et al., "Degradation of intestinal mRNA: a matter of treatment," *World Journal of Gastroenterology*, vol. 21, no. 12, pp. 3499–3508, 2015.

[17] A. Schroeder, O. Mueller, S. Stocker et al., "The RIN: an RNA integrity number for assigning integrity values to RNA measurements," *BMC Molecular Biology*, vol. 7, no. 1, p. 3, 2006.

[18] M. Dodt, J. T. Roehr, R. Ahmed, and C. Dieterich, "FLEXBAR—flexible barcode and adapter processing for next-generation sequencing platforms," *Biology*, vol. 1, no. 3, pp. 895–905, 2012.

[19] M. Salem, B. Paneru, R. Al-Tobasei et al., "Transcriptome assembly, gene annotation and tissue gene expression atlas of the rainbow trout," *PLoS One*, vol. 10, no. 3, article e0121778, 2015.

[20] B. Langmead and S. Salzberg, "Fast gapped-read alignment with Bowtie 2," *Nature Methods*, vol. 9, no. 4, pp. 357–359, 2012.

[21] L. Pachter, "Models for transcript quantification from RNA-Seq," 2011, http://arxiv.org/abs/1104.3889.

[22] A. Roberts, C. Trapnell, J. Donaghey, J. L. Rinn, and L. Pachter, "Improving RNA-Seq expression estimates by correcting for fragment bias," *Genome Biology*, vol. 12, no. 3, article R22, 2011.

[23] M. I. Love, W. Huber, and S. Anders, "Moderated estimation of fold change and dispersion for RNA-seq data with DESeq2," *Genome Biology*, vol. 15, no. 12, p. 550, 2014.

[24] L. Lin, Y. Luo, P. Sørensen et al., "Effects of high hydrostatic pressure on genomic expression profiling of porcine parthenogenetic activated and cloned embryos," *Reproduction, Fertility and Development*, vol. 26, no. 3, pp. 469–484, 2014.

[25] H. Rauwerda, P. Wackers, J. F. B. Pagano et al., "Mother-specific signature in the maternal transcriptome composition of mature, unfertilized *zebrafish* eggs," *PLoS One*, vol. 11, no. 1, article e0147151, 2016.

[26] S. Moriyama, F. G. Ayson, and H. Kawauchi, "Growth regulation by insulin-like growth factor-I in fish," *Bioscience, Biotechnology, and Biochemistry*, vol. 64, no. 8, pp. 1553–1562, 2000.

[27] F. Li, G. Mao, D. Tong et al., "The histone mark H3K36me3 regulates human DNA mismatch repair through its interaction with MutSα," *Cell*, vol. 153, no. 3, pp. 590–600, 2013.

[28] J. Silverman, H. Takai, S. B. Buonomo, F. Eisenhaber, and T. de Lange, "Human Rif 1, ortholog of a yeast telomeric protein, is regulated by ATM and 53BP1 and functions in the S-phase checkpoint," *Genes Development*, vol. 18, no. 17, pp. 2108–2119, 2004.

[29] V. A. Volkov, P. M. Grissom, V. K. Arzhanik et al., "Centromere protein F includes two sites that couple efficiently to depolymerizing microtubules," *Journal of Cell Biology*, vol. 209, no. 6, pp. 813–828, 2015.

[30] K. Ocalewicz, S. Dobosz, H. Kuzminski, J. Nowosad, and K. Goryczko, "Chromosome rearrangements and survival of androgenetic rainbow trout (*Oncorhynchus mykiss*)," *Journal of Applied Genetics*, vol. 51, no. 3, pp. 309–317, 2010.

[31] O. Michalik, S. Dobosz, T. Zalewski, M. Sapota, and K. Ocalewicz, "Induction of gynogenetic and androgenetic haploid and doubled haploid development in the brown trout (*Salmo trutta* Linnaeus 1758)," *Reproduction in Domestic Animals*, vol. 50, no. 2, pp. 256–262, 2015.

[32] T. A. Webb, W. J. Kowalski, and R. A. Fluck, "Microtubule-based movements during ooplasmic segregation in the medaka fish egg (*Oryzias latipes*)," *The Biological Bulletin*, vol. 188, no. 2, pp. 146–156, 1995.

[33] P. Haffray, J. Aubin, V. Houis, L. Labbe, and B. Jalabert, "Comparison of pressure or thermal treatments on triploid yields and malformations up to swim up stage in rainbow trout (*Oncorhynchus mykiss*)," *Aquaculture*, vol. 272, article S265, 2007.

4

Comparative Transcriptome Analysis of Temperature-Induced Green Discoloration in Garlic

Ningyang Li,[1] Zhichang Qiu,[1] Xiaoming Lu,[1] Bingchao Shi,[1] Xiudong Sun,[2] Xiaozhen Tang,[1] and Xuguang Qiao[1]

[1]*Key Laboratory of Food Processing Technology and Quality Control in Shandong Province, College of Food Science and Engineering, Shandong Agricultural University, No. 61, Daizong Road, Tai'an, Shandong Province 271018, China*
[2]*State Key Laboratory of Crop Biology, College of Horticulture Science and Engineering, Shandong Agricultural University, No. 61, Daizong Road, Tai'an, Shandong Province 271018, China*

Correspondence should be addressed to Xiaozhen Tang; txz@sdau.edu.cn and Xuguang Qiao; xgqiao@sdau.edu.cn

Academic Editor: Wilfred van IJcken

Green discoloration is one of the most important problems that cause low quality of product in the processing of garlic, which can be induced by low-temperature stress. But the mechanism of low temperature-induced green discoloration is poorly understood. In the present study, the control garlic and three low temperature-treated garlic samples (stored at 4°C with 10, 15, and 40 days, respectively) were used for genome-wide transcriptome profiling analysis. A total of 49280 garlic unigenes with an average length of 1337 bp were de novo assembled, 20231 of which were achieved for functional annotation. When being suffered from 10, 15, and 40 days of low-temperature treatment, an increased degree of discoloration was observed, and a total of 4757, 4401, and 2034 unigenes showed a differential expression, respectively. Finally, 5923 differentially expressed genes (DEGs) were found to respond to the low-temperature stress, of which 3921 were identified in at least two treatments. Among these stress-responsive unigenes, there were large numbers of enzyme-encoding genes, which significantly enriched the pathway "proteasome," many genes of which are potentially involved in the garlic discoloration, such as 7 alliinase-encoding genes, 5 γ-glutamyltranspeptidase-encoding genes, and 1 δ-aminolevulinic acid dehydratase-encoding gene. These stress-responsive enzyme-encoding genes are possibly responsible for the low-temperature-induced garlic discoloration. The identification of large numbers of DEGs provides a basis for further elucidating the mechanism of low-temperature-induced green discoloration in garlic.

1. Introduction

Garlic (*Allium sativum*) is one of the most widely used food ingredients worldwide and also possesses excellent medicinal benefits. To provide conveniences for consumers, the bulbs of garlic are processed as products with various forms, such as powder, granules, puree, minced paste, and oleoresin. However, the garlic processing frequently causes a green discoloration, resulting in the loss of visual appeal and commercial value of garlic products [1]. Mechanistic studies indicated that green discoloration of garlic is a complex, multistep process beginning with the alliinase-catalyzed conversion of isoalliin and other S-substituted cysteine derivatives (alliin and methiin) to the corresponding sulfenic acids [2–4]. Subsequent reactions of the sulfenic acids result in the generation of 1-propenyl-containing thiosulfinates (including cepathiolanes), acting as a color developer to react with certain amino acids to produce color precursors of 3,4-dimethylpyrrole. The color precursors further react with allicin and various (thio) carbonyls to produce color compounds that are the hallmarks of garlic discoloration [4, 5]. In addition, recent studies have suggested that the membrane permeability of garlic cells also plays a key role in the greening process [6]. This is consistent with the finding that alliinase and isoalliin are physically separated from each other under normal circumstances but would only come into contact upon the injury of garlic tissues [7, 8]. Therefore, there are a large number of metabolites and enzymes, including isoalliin, alliinase, and unknown compounds, involved in the formation of green discoloration.

Garlic discoloration can be induced by some environmental stresses, such as low temperature, acetic acid compound, and various amino compounds [3, 9–11]. However, garlic has a giant genome (about 16 Gb), leading to the fact that little genetic and genomic information can be obtained in this crop. Therefore, the exact mechanism about these environmental factors inducing garlic greening is poorly understood.

In recent years, the next-generation sequencing (NGS) technology has emerged to offer a cost-effective and powerful tool for sequence determination [12, 13]. Transcriptome analysis by NGS is rapid, inexpensive, and unlimited by genomic complexity and has been widely used as a primary tool in many researching areas, including gene discovery, SSR marker development, and investigation of the domesticated patterns of crops, particularly in the characterization of the gene expression profiling [14–16]. The garlic transcriptome has been de novo assembled by several previous studies, and large numbers of expressed genes have been discovered [17, 18]. These works extended our knowledge in understanding of expression profiles, regulation, and networks of important traits of garlic, and thus they offered a new sight for garlic study.

In this study, to detect the potential mechanism of garlic discoloration caused by low-temperature stress, three low-temperature-treated garlic samples with different times which resulted in different degrees in discoloration and control garlic were exploited for genome-wide transcriptome profiling analysis. The characterization of expression profiling and identification of differentially expressed genes (DEGs) would provide a basis for further elucidating the formation mechanism of green discoloration in garlic.

2. Materials and Methods

2.1. Materials. Freshly harvested garlic cloves were used for the green discoloration and transcriptome analysis. Three treated groups (DS1, DS2, and DS3) and one control group (DS0) were set up. Each group contained four replications, and each replication included three cloves. Four replications of DS0, DS1, DS2, and DS3 were stored at 4°C for 0, 10, 15, and 40 days, respectively. After treatment, the cloves of three replications of each group were individually frozen in liquid nitrogen and then stored at −80°C until use for sequencing. The residual one replication of each group was used for the green discoloration.

To measure the extent of discoloration, 20 g of garlic cloves were used and the green sprouts were removed to minimize potential spectrometric interference. The sprout-free garlic samples were minced, mixed thoroughly with 5 mL of 2% citric acid solution, and incubated at 80°C for 30 min. The resultant mixture was cooled to room temperature, and then 95% ethanol was added to a final volume of 50 mL, followed by incubation at 4°C for 24 h. The garlic extract was then centrifuged at 12000 ×g for 5 min, and the absorbance of the supernatant was measured using an ultraviolet-visible spectrometer (Beckman, Michigan, USA) at 440 nm and 590 nm, respectively. The extent of green discoloration was represented by multiplying the absorbance at 440 or 590 nm by 10.

2.2. cDNA Library Construction and RNA Sequencing. Twelve samples from the four groups were used to extract total RNA using the RNeasy Plant Mini Kit (Qiagen, Germany) according to the manufacturer's instructions. The obtained RNA was quantified by NanoDrop 2000 (Thermo Fisher Scientific, USA) and Qubit 2.0 RNA Broad Range Assay Kit (Invitrogen, USA). The RNA integrity was assessed using an Agilent Bioanalyzer 2100 (Agilent Technologies, USA). An RNA integrity number (RIN) greater than or equal to 8 was considered to be useful. Subsequently, the library of each sample was constructed using the TruSeq RNA Sample Preparation Kit (Illumina, USA) and SuperScript II Reverse Transcriptase (Invitrogen, USA) according to the manufacturers' recommended protocols. Paired-end sequencing was then performed using the Illumina sequencing platform (HiSeq™ 4000) according to the manufacturer's instructions (Illumina, San Diego, CA). The quality of all sequences (in FASTQ format) was assessed by FastQC [19]. Adaptors and low-quality bases were trimmed using Trimmomatic [20]. Reads with a Phred quality score over 30 were used for the following transcriptome assembly.

2.3. Transcriptomic Analysis. The clean reads were further assembled using Trinity (version 2.4.0) [21] with min_kmer_cov set to 5 and other parameters to default values. All possible open reading frames within the assembled transcripts were extracted using TransDecoder (https://github.com/TransDecoder). Transcripts missing a likely CDS were discarded, and all predicted CDS sequences were translated into protein sequences and clustered by CD-HIT (version 4.6.6) [22] with 95% global sequence identity. Only the unique transcript with CDS sequence was designated as a unigene. The longest sequence in each cluster was transferred to the final data set. The translated protein sequences of all ORFs were annotated by performing BLAST searches against the NCBI nonredundant protein database (NR), Swiss-Prot, Gene Ontology (GO), Kyoto Encyclopedia of Unigenes and Genomes (KEGG), eggNOG, and protein family (PFAM) databases with e-value threshold set to $1e^{-5}$.

2.4. Identification and Quantization of Gene Expression Levels. Transcript abundances were separately estimated by RSEM (version 1.3.0) [23] for each sample. Differential gene expression analysis was performed using DEGseq2, an R package [24]. A unigene was considered differentially expressed between two garlic groups if the criteria of the Q value < 0.05 and fold change > 2 were met. GO and KEGG enrichment analyses were performed using the topGO package [25] and KOBAS (version 3.0) [26], respectively.

2.5. Quantitative Reverse-Transcription Polymerase Chain Reaction (qRT-PCR) Validation. Twenty candidates were chosen from the top 100 most differentially expressed unigenes for qRT-PCR analysis. PCR primers were designed using Primer Premier 5.0 software and summarized in Table S1. Total RNA was reverse-transcribed into single-stranded cDNA using Takara PrimeScript RT Reagent Kit with gDNA Eraser (Clontech laboratories, USA). qRT-PCR was set up using SYBR Premix Ex Taq polymerase (Clontech laboratories, USA) following the manufacturer's recommended protocol and conducted on an Applied Biosystems 7500 Fast Real-Time PCR System (Thermo Fisher Scientific, USA). All PCR reactions were performed in triplicate, and the relative expression of each unigene was calculated using the $2^{-\Delta\Delta CT}$ method with actin as the reference [27].

3. Results

3.1. Investigation of Green Discoloration in Four Groups. As shown in Figure 1(a), with the extension of the treatment time, the samples DS1, DS2, and DS3 displayed an increase in the extent of greening, whereas the control DS0 served as the negative control in which discoloration had not been induced. The progression of greening from DS0 to DS3 was also confirmed by the detection of an increasing amount of green pigments as evidenced by colorimetry (Figure 1(b)).

3.2. Transcriptome Sequencing, Assembly, and Annotation. Three repeats of DS0, DS1, DS2, and DS4 were individually performed for Illumina sequencing. After quality filtering, we obtained an average of 70.48, 74.97, 72.05, and 69.74 million clean reads from DS0, DS1, DS2, and DS3 samples, respectively, representing 129 Gb of sequencing data in total (Table S2). De novo assembly of the reads by Trinity generated 49280 unigenes with an average length of 1337 bp. Among these 49280 unigenes, 14374 (29.17%), 8647 (17.55%), 4405 (8.94%), 20210 (41.01%), 3527 (7.16%), and 9127 (18.52%) showed significant similarity to known proteins in the eggNOG, GO, KEGG, NR, PFAM, and SwissProt databases, respectively. In total, there were 20231 (41.05%) unigenes that were achieved for functional annotation in at least one database (Table S3).

3.3. Identification of DEGs. When garlic was stored at 4°C for 10 (DS1), 15 (DS2), and 40 days (DS3), there were 4757, 4401, and 2034 unigenes that showed differential expression, respectively (Figure 2, Table S4). Among these DEGs, 3921 unigenes were identified in at least two treatments, including 2051 in D1 and D2, 38 in D1 and D3, 222 in D2 and D3, and 1610 in all three treatments, respectively (Figure 2, Table S4). In total, we identified 5923 DEGs, including 7 allinase-encoding genes, 5 γ-glutamyltranspeptidase-encoding genes, and 1 δ-aminolevulinic acid dehydratase gene (Table S4). We also analyzed the expression trends of these 5923 DEGs, and the result showed that they could be classified into six distinct clusters (Table S4), and each group showed a similar expression pattern (Figure 3). DEGs in three clusters (clusters 1, 4, and 5) exhibited a trend of upregulated expression, whereas the expression trends of DEGs in the other clusters (clusters 2, 3, and 6) were downregulated. Therefore, under our experimental conditions, although the expressed difference of the 4313 out of 5923 DEGs was only significant in one or two of three treatments, the trends of continuously up- or down regulated expression in all three treatments could be observed (Figure 3).

3.4. Validation of the Differential Expression by qRT-PCR. To validate the expression profiling by Illumina sequencing, the expression levels of twenty DEGs (11 upregulated and 9 downregulated) involved with beta-glucosidase, pectate lyase, aquaporin, vacuolar processing, heat shock protein, chaperone, and glutamate-cysteine ligase were selected for qRT-PCR analysis (Table S1). The result indicated that the trends of the expression change of all unigenes analyzed by qRT-PCR were identical to those by Illumina sequencing, except for c224948_g1 (Figure 4). However, the change folds of the expression levels of these DEGs checked by qRT-PCR have slight difference with those by Illumina sequencing. These results confirmed that our RNA sequencing and quantitation results were accurate and could be used for subsequent functional analysis to identify candidates mechanistically implicated in garlic greening.

3.5. Enrichment Analysis of DEGs. To determine the function categories that DEGs were involved in, the GO categories enriched by DEGs were analyzed. The functional enrichment analysis indicated that upregulated DEGs and downregulated DEGs in garlic greening samples were significantly enriched in different GO terms (Table S5). It was found that coupregulated DEGs are highly related to cellular component functions and metabolic and catabolic functions, such as organelle part/lumen, intracellular, membrane, organelle envelope, protein-containing complex, proteasome core complex, small molecule metabolic process, and oxidation-reduction process. Other functions such as lyase activity, oxidoreductase activity and catalytic activity were also enriched. Surprisingly, when the codownregulated DEGs were analyzed, the functions of binding, cellular metabolic process, and response-related functions were significantly enriched, such as carbohydrate derivative binding, small molecule binding, heterocyclic/organic compound binding, response to stress, and biotic stimulus (Figure 5). The GO enrichment analysis indicated that the genes coupregulated in greening process are involved in divergent biology processes and cellular component functions, especially in the structural maintenance of organelle and intracellular but not restricted to stress-related.

The biological pathways influenced by DEGs were determined by KEGG enrichment analysis. Our result showed that one pathway, proteasome (ID: ko03050) was dramatically enriched by DEGs of DS1 (Q value = 0.015), and another pathway, glycolysis/gluconeogenesis (ID: ko00010) was markedly enriched by DEGs of DS3 (Q value = 0.014). None of pathways showed significant enrichment by DEGs of DS2.

4. Discussion

Comparative transcriptomics analysis has demonstrated enormous utilities in helping researchers understand complex biological processes in plants. Villacorta-Martin et al. explored the transcriptomic changes that occurred in the meristems of lily bulbs at several time points during cold exposure and identified 9872 dysregulated unigenes from 42430 unigenes during the first two weeks of cold exposure [28]. Similarly, Chaturvedi et al. found 3307 upregulated

(a)

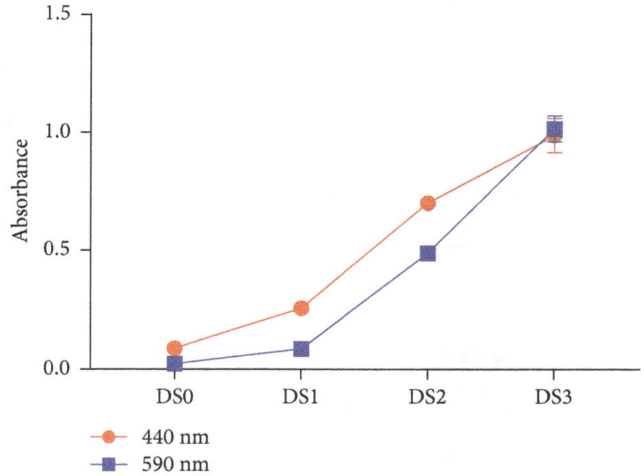

(b)

FIGURE 1: Assessment of greening in different experiment groups. (a) Appearance of pigment extracts prepared from DS0 to DS3. (b) Colorimetric measurement of pigment content at 440 nm and 590 nm.

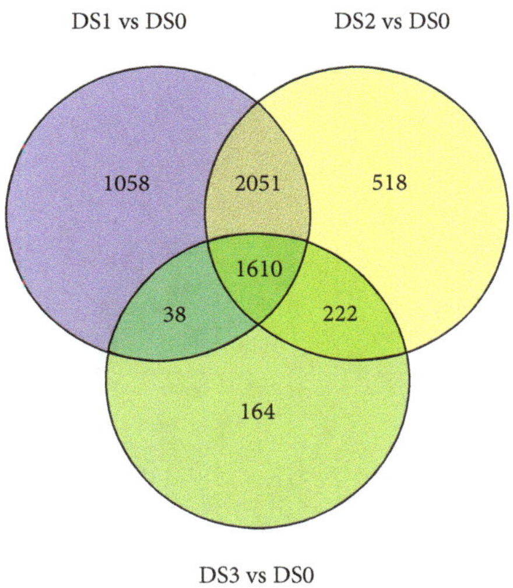

FIGURE 2: Venn diagram of the DEGs identified in three treatments.

and 4996 downregulated unigenes in the internal bud (IB) of garlic cloves at 33°C compared to 2°C treatment. They also found 5703 upregulated and 8444 downregulated unigenes in storage leaf (SL) at 33°C compared to 2°C treatment [29]. Our current study aimed at generating mechanistic insight into garlic greening via systematic transcriptomic profiling. We successfully identified 49280 unigenes from a total of 87570 unique transcripts, from which 5923 DEGs were found to be differentially expressed in garlic samples undergoing discoloration. Among them, the most upregulated unigenes in the greening garlic cloves were compared to the control including β-glucosidases, pectatelyases, NRT1/PTR family proteins, aquaporins, and various proteases, whereas the most downregulated candidates were shown to consist primarily of chaperones, heat shock proteins, pectins, and metabolic enzymes. Subsequent functional analysis suggested that the transcriptomic alterations observed in the garlic specimens undergoing discoloration could signify changes in various metabolic, stress response, and cellular component functions.

Some of the differentially expressed unigenes in our transcriptomic study unearthed have been validated in previous studies. For example, our results indicated that γ-glutamyl-transpeptidase (GGT, c187783_g1, c199606_g1, c222769_g1, and c222769_g2) was among one of the most upregulated unigenes in greening garlic cloves. Previously, Li et al. reported that GGT activity was significantly augmented in garlic bulbs stored at 4°C, which could be reversed by incubation at 35°C [30]. The striking similarity between the trend of GGT activity and that of total thiosulfinate level prompted further investigation into the putative role of the enzyme in the biosynthesis of color developers responsible for garlic greening. Indeed, GGT was subsequently shown by Yoshimoto and coworkers to catalyze the deglutamylation of γ-glutamyl-S-(1-propenyl)-L-cysteine, a hypothetical isoalliin precursor [31]. Therefore, the upregulation of GGT could be an early indicator of activated isoalliin production and garlic greening.

It has been proposed that certain stress conditions such as low temperature could cause injury to vacuolar membranes in plant cells [32, 33]. This allows alliinase to be released from the vacuoles and come into contact with its substrate isoalliin present in the cytosol [7, 8]. Consistent with these findings, Bai et al. reported that treatment of garlic cloves with 5% (*v*/*v*) acetic acid resulted in significant tonoplast damage and a concomitant increase in the concentration of thiosulfinates [6]. The differential proteomic signatures between the greening groups and the control group in our current study also provided several clues to the possible involvement of vacuole disruption. For example, c199231_g1 and c221124_g1, annotated as a possible aquaporin TIP2-2 homolog, was shown to be consistently among the top 10 most upregulated unigenes. Aquaporins form the water channels on cellular and vacuolar membranes and are mechanistically implicated in the regulation of osmotic stress signals [34]. Not surprisingly, expression

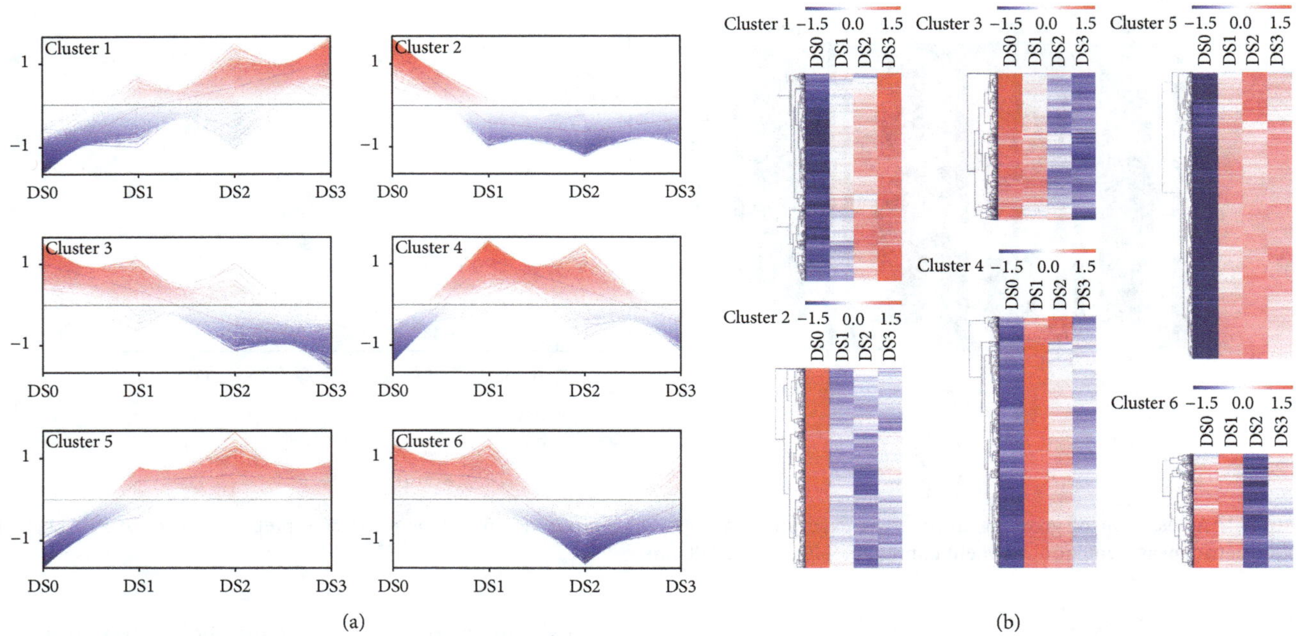

Figure 3: Clustering of DEGs based on their expression pattern control and three treatments. (a) The expression trend of six clusters. (b) The hot map of gene expression level in each cluster.

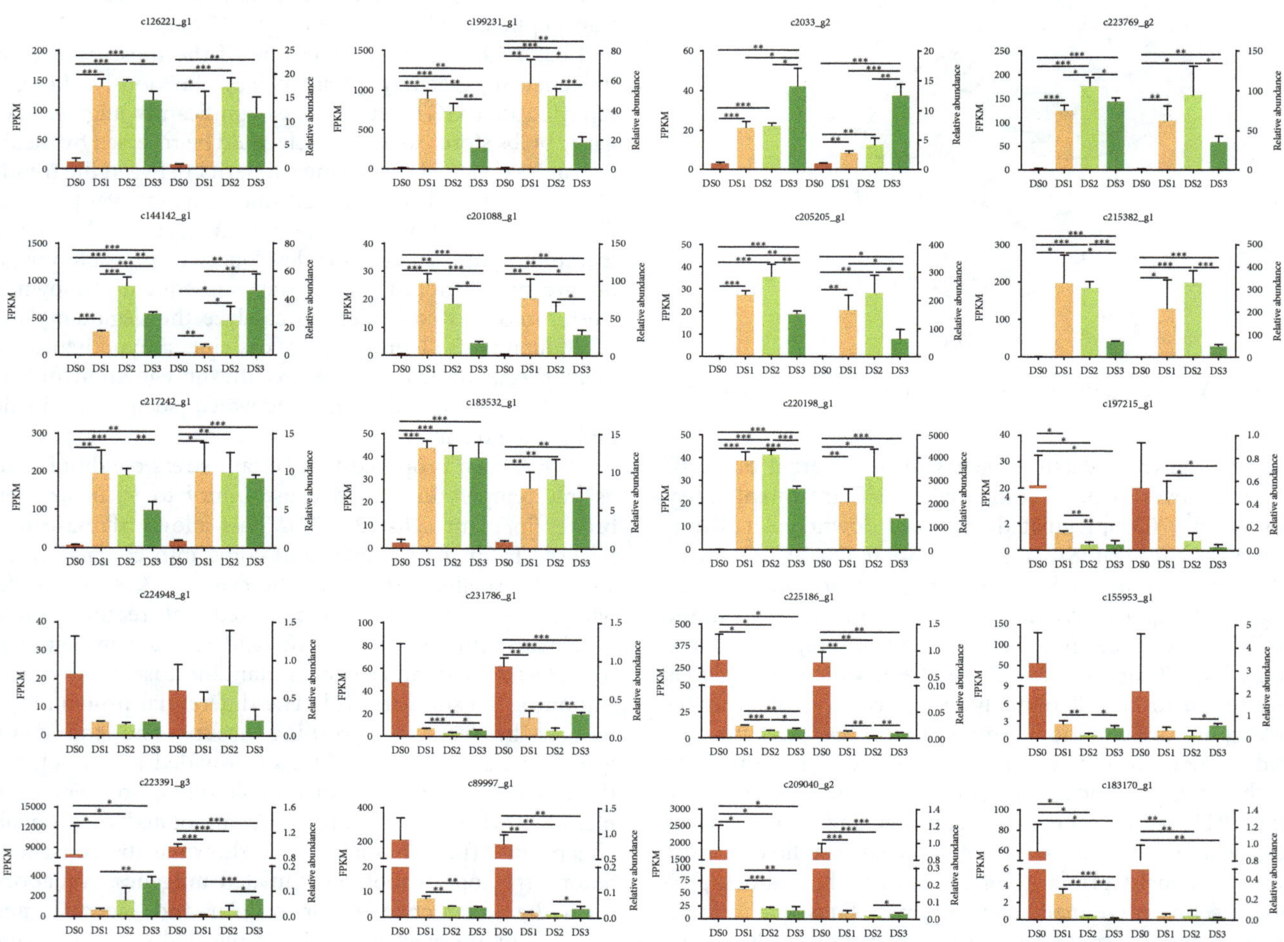

Figure 4: qRT-PCR validation of DEGs. All experiments were performed in triplicate. $^{*}P < 0.05$; $^{**}P < 0.01$; $^{***}P < 0.001$.

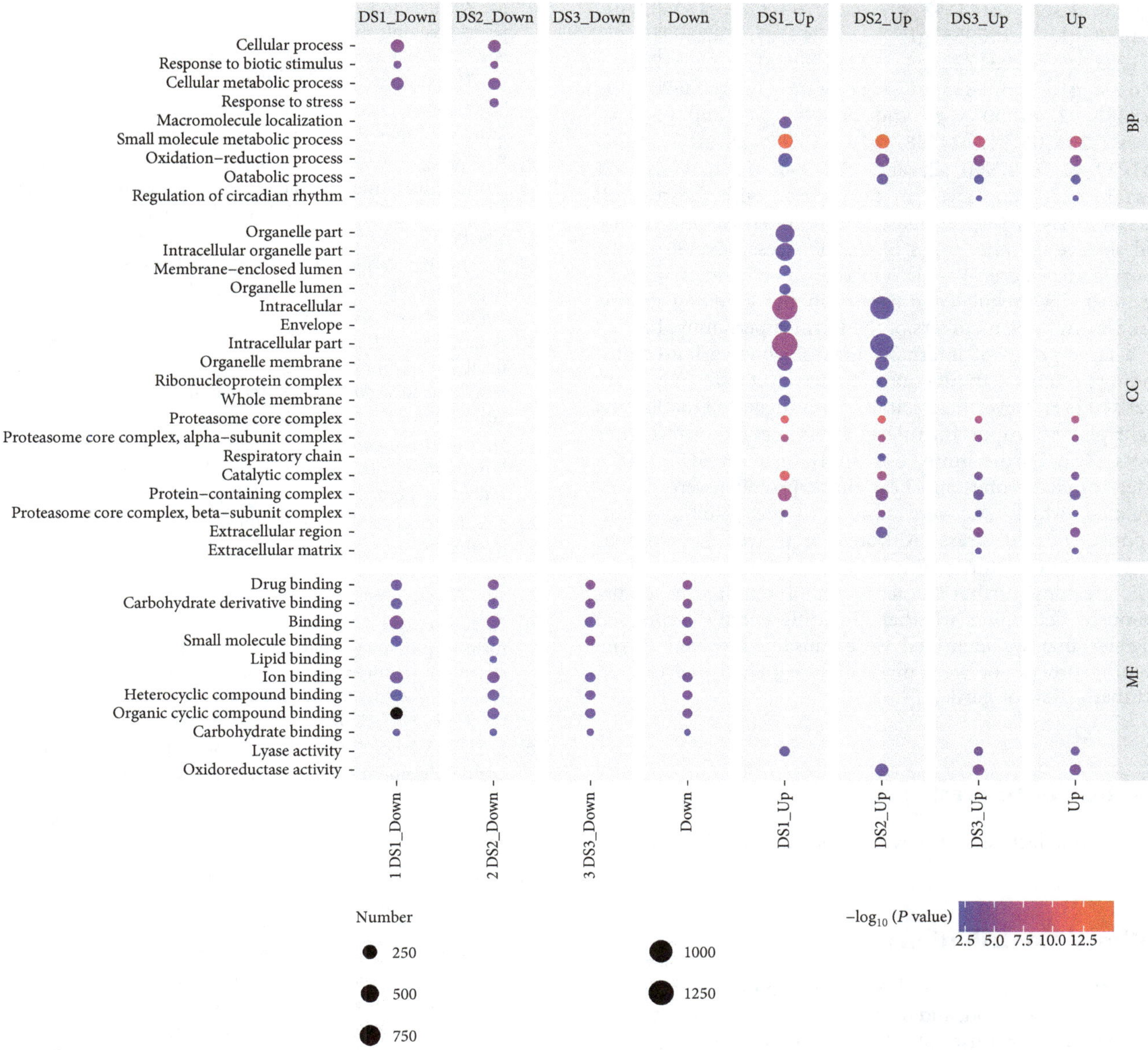

Figure 5: GO enrichment analysis of up- and downregulated DEGs in garlic greening samples.

of aquaporins is often enhanced by low temperature as the change in water density and solubility can result in altered osmotic pressure [35]. Increased expression of an aquaporin SlTIP2-2 was found to enhance the osmotic water permeability of *Arabidopsis* mesophyll protoplasts, suggesting underlying structural changes in the vacuolar membranes [36]. Additional evidence implied that possible vacuolar damage involved in the increased expression of various proteases, particularly several vacuolar-processing enzyme (VPE) homologs (c217242_g1, c176154_g2, c176154_g3, and c200027_g3). Upregulation of VPEs has been shown to play a crucial role in the stress response of *Arabidopsis* plants by promoting the destruction of vacuolar structures, leading to the release of diverse hydrolytic enzymes into the cytosol [36–38]. Therefore, it seemed plausible that cold stress-induced osmotic change and activation of processing proteases could be a key contributing factor to the leakage of alliinase from vacuole to cytosol. Of course, further examination would be needed to determine the exact localization patterns and functions of the upregulated proteolytic enzymes that we identified.

Many of the differentially expressed unigenes uncovered in our current study could have been the result of low-temperature storage. NRT1/PTR family proteins (c220198_g1, c225222_g1, c208304_g1, c169637_g1, c212989_g1, c228990_g1, c204357_g1, c210835_g1, c229656_g3, c75070_g1, c207568_g1, and c214187_g1) are involved in nitrogen transport [39] and have been reported to undergo upregulation in response to cold stress in *Saussurea involucrata* [40]. Cold stress has also been known to modulate sugar metabolism and alter the composition of plant cell wall [41]. This was consistent with our identification of several putative

β-glucosidases (c144142_g1, c223769_g2, c183532_g1, c201088_g1, c155953_g1, c197248_g1, c230818_g2, c231722_g1, c231722_g3, c218711_g1, c155121_g2, c186294_g1, c320855_g1, c176603_g1, c228932_g1, c218383_g2, c195036_g1 and c178982_g1) and pectatelyases (c205205_g1, c215382_g1, c215382_g2, c193411_g2, c211032_g1, c207990_g2, and c207990_g1) that showed aberrant expression levels in comparison with the control. Altered levels of β-glucosidase and pectatelyase induced by cold storage or other types of abiotic stress have also been reported elsewhere [42–44]. On the other hand, several of the most downregulated unigenes in the greening groups were shown to be members of the lectin superfamily. Lectins are a large group of internally heterologous carbohydrate-binding proteins with functions that have yet to be fully characterized [45]. Nevertheless, it has been suggested that lectins might play an important role in stress regulation in higher plants [46]. Furthermore, our study also revealed that a variety of chaperones and heat shock proteins were downregulated, which was very similar to the result of plant response to cold stress. Although it is well known that low-temperature storage was an effective method to induce garlic greening, further studies would be required to unambiguously determine whether the differentially expressed unigenes that we identified were causally involved in the greening process or were merely dysregulated due to cold acclimatization of garlic.

Conflicts of Interest

The authors declare that they have no conflict of interest.

Authors' Contributions

NYL, ZCQ, XML, BCS, and XDS contributed to the study design, data collection, and data analysis; wrote the first draft; and revised the manuscript. XZT and XGQ were the supervisors of the project and contributed to the study design, data analysis, and manuscript revision. All authors reviewed and accepted the content of the final manuscript. Ningyang Li and Zhichang Qiu contributed equally to this work.

Acknowledgments

This work was supported by the National Key R&D Program of China (2016YFD040024), Natural Science Foundation of Shandong Province (ZR2016CM33), and Shandong "Double Tops" Program (SYT2017XTTD04).

References

[1] C. Mou, X. Hao, Z. Xu, and X. Qiao, "Effect of porphobilinogen on the formation of garlic green pigments," *Journal of the Science of Food and Agriculture*, vol. 93, no. 10, pp. 2454–2457, 2013.

[2] M. Aguilar and F. Rincon, "Improving knowledge of garlic paste greening through the design of an experimental strategy," *Journal of Agricultural and Food Chemistry*, vol. 55, no. 25, pp. 10266–10274, 2007.

[3] T. M. Lukes, "Factors governing the greening of garlic puree," *Journal of Food Science*, vol. 51, no. 6, pp. 1577–1582, 1986.

[4] R. Kubec, P. Curko, P. Urajova, J. Rubert, and J. Hajslova, "Allium discoloration: color compounds formed during greening of processed garlic," *Journal of Agricultural and Food Chemistry*, vol. 65, no. 48, pp. 10615–10620, 2017.

[5] J. Zang, D. Wang, and G. Zhao, "Mechanism of discoloration in processed garlic and onion," *Trends in Food Science & Technology*, vol. 30, no. 2, pp. 162–173, 2013.

[6] B. Bai, L. Li, X. Hu, Z. Wang, and G. Zhao, "Increase in the permeability of tonoplast of garlic (Allium sativum) by monocarboxylic acids," *Journal of Agricultural and Food Chemistry*, vol. 54, no. 21, pp. 8103–8107, 2006.

[7] G. S. Ellmore and R. S. Feldberg, "Alliin lyase localization in bundle sheaths of the garlic clove (*Allium sativum*)," *American Journal of Botany*, vol. 81, no. 1, pp. 89–94, 1994.

[8] J. E. Lancaster and H. A. Collin, "Presence of alliinase in isolated vacuoles and of alkyl cysteine sulphoxides in the cytoplasm of bulbs of onion (Allium cepa)," *Plant Science Letters*, vol. 22, no. 2, pp. 169–176, 1981.

[9] R. Kubec and J. Velisek, "Allium discoloration: the color-forming potential of individual thiosulfinates and amino acids: structural requirements for the color-developing precursors," *Journal of Agricultural and Food Chemistry*, vol. 55, no. 9, pp. 3491–3497, 2007.

[10] L. Li, D. Wang, X. Li, Y. Wang, and X. Ju, "Elucidation of colour development and microstructural characteristics of *Allium sativum* fumigated with acetic acid," *International Journal of Food Science & Technology*, vol. 50, no. 5, pp. 1083–1088, 2015.

[11] C. Y. Liang, J. Xiong, and J. Ye, "Study on storage conditions of greening of garlic purees," *Advanced Materials Research*, vol. 1052, pp. 290–293, 2014.

[12] S. Goodwin, J. D. McPherson, and W. R. McCombie, "Coming of age: ten years of next-generation sequencing technologies," *Nature Reviews Genetics*, vol. 17, no. 6, pp. 333–351, 2016.

[13] E. L. van Dijk, H. Auger, Y. Jaszczyszyn, and C. Thermes, "Ten years of next-generation sequencing technology," *Trends in Genetics*, vol. 30, no. 9, pp. 418–426, 2014.

[14] P. A. McGettigan, "Transcriptomics in the RNA-seq era," *Current Opinion in Chemical Biology*, vol. 17, no. 1, pp. 4–11, 2013.

[15] Z. Wang, M. Gerstein, and M. Snyder, "RNA-Seq: a revolutionary tool for transcriptomics," *Nature Reviews Genetics*, vol. 10, no. 1, pp. 57–63, 2009.

[16] Y. X. Qi, Y. B. Liu, and W. H. Rong, "RNA-Seq and its applications: a new technology for transcriptomics," *Hereditas*, vol. 33, no. 11, pp. 1191–1202, 2011.

[17] R. Kamenetsky, A. Faigenboim, E. Shemesh Mayer et al., "Integrated transcriptome catalogue and organ-specific profiling of gene expression in fertile garlic (Allium sativum L.)," *BMC Genomics*, vol. 16, no. 1, pp. 12–12, 2015.

[18] X. Sun, S. Zhou, F. Meng, and S. Liu, "De novo assembly and characterization of the garlic (Allium sativum) bud transcriptome by Illumina sequencing," *Plant Cell Reports*, vol. 31, no. 10, pp. 1823–1828, 2012.

[19] S. Andrews, "FastQC a quality control tool for high throughput sequence data," 2013, http://www.bioinformatics.babra

ham.ac.uk/projects/fastqc/ http://www.bioinformatics.babraham.ac.uk/projects/.

[20] A. M. Bolger, M. Lohse, and B. Usadel, "Trimmomatic: a flexible trimmer for Illumina sequence data," *Bioinformatics*, vol. 30, no. 15, pp. 2114–2120, 2014.

[21] M. G. Grabherr, B. J. Haas, M. Yassour et al., "Full-length transcriptome assembly from RNA-Seq data without a reference genome," *Nature Biotechnology*, vol. 29, no. 7, pp. 644–652, 2011.

[22] L. Fu, B. Niu, Z. Zhu, S. Wu, and W. Li, "CD-HIT: accelerated for clustering the next-generation sequencing data," *Bioinformatics*, vol. 28, no. 23, pp. 3150–3152, 2012.

[23] B. Li and C. N. Dewey, "RSEM: accurate transcript quantification from RNA-Seq data with or without a reference genome," *BMC Bioinformatics*, vol. 12, no. 1, pp. 323–323, 2011.

[24] L. Wang, Z. Feng, X. Wang, X. Wang, and X. Zhang, "DEGseq: an R package for identifying differentially expressed genes from RNA-seq data," *Bioinformatics*, vol. 26, no. 1, pp. 136–138, 2010.

[25] A. Alexa and J. Rahnenführer, "Gene set enrichment analysis with topGO," 2018, (http://www.bioconductor.org).

[26] C. Xie, X. Mao, J. Huang et al., "KOBAS 2.0: a web server for annotation and identification of enriched pathways and diseases," *Nucleic Acids Research*, vol. 39, Supplement 2, pp. W316–W322, 2011.

[27] K. J. Livak and T. D. Schmittgen, "Analysis of relative gene expression data using real-time quantitative PCR and the $2^{-\Delta\Delta C}{}_{T}$ method," *Methods*, vol. 25, no. 4, pp. 402–408, 2001.

[28] C. Villacorta-Martin, F. F. Núñez de Cáceres González, J. de Haan et al., "Whole transcriptome profiling of the vernalization process in *Lilium longiflorum* (cultivar White Heaven) bulbs," *BMC Genomics*, vol. 16, no. 1, p. 550, 2015.

[29] A. K. Chaturvedi, S. R. Shalom, A. Faigenboim-Doron et al., "Differential carbohydrate gene expression during preplanting temperature treatments controls meristem termination and bulbing in garlic," *Environmental and Experimental Botany*, vol. 150, pp. 280–291, 2018.

[30] L. Li, D. Hu, Y. Jiang, F. Chen, X. Hu, and G. Zhao, "Relationship between γ-glutamyl transpeptidase activity and garlic greening, as controlled by temperature," *Journal of Agricultural and Food Chemistry*, vol. 56, no. 3, pp. 941–945, 2008.

[31] N. Yoshimoto, A. Yabe, Y. Sugino et al., "Garlic γ-glutamyl transpeptidases that catalyze deglutamylation of biosynthetic intermediate of alliin," *Frontiers in Plant Science*, vol. 5, p. 758, 2015.

[32] M. Murai and S. Yoshida, "Vacuolar membrane lesions induced by a freeze-thaw cycle in protoplasts isolated from deacclimated tubers of Jerusalem artichoke (Helianthus tuberosus L.)," *Plant and Cell Physiology*, vol. 39, no. 1, pp. 87–96, 1998.

[33] R. P. Willing and A. C. Leopold, "Cellular expansion at low temperature as a cause of membrane lesions," *Plant Physiology*, vol. 71, no. 1, pp. 118–121, 1983.

[34] I. Johansson, M. Karlsson, U. Johanson, C. Larsson, and P. Kjellbom, "The role of aquaporins in cellular and whole plant water balance," *Biochimica et Biophysica Acta*, vol. 1465, no. 1-2, pp. 324–342, 2000.

[35] C. Maurel, M. J. Chrispeels, C. Lurin et al., "Function and regulation of seed aquaporins," *Journal of Experimental Botany*, vol. 48, Special, pp. 421–430, 1997.

[36] N. Sade, B. J. Vinocur, A. Diber et al., "Improving plant stress tolerance and yield production: is the tonoplast aquaporin SlTIP2;2 a key to isohydric to anisohydric conversion?," *New Phytologist*, vol. 181, no. 3, pp. 651–661, 2009.

[37] I. Hara-Nishimura and N. Hatsugai, "The role of vacuole in plant cell death," *Cell Death & Differentiation*, vol. 18, no. 8, pp. 1298–1304, 2011.

[38] N. Hatsugai, M. Kuroyanagi, K. Yamada et al., "A plant vacuolar protease, VPE, mediates virus-induced hypersensitive cell death," *Science*, vol. 305, no. 5685, pp. 855–858, 2004.

[39] S. Léran, K. Varala, J. C. Boyer et al., "A unified nomenclature of nitrate transporter 1/peptide transporter family members in plants," *Trends in Plant Science*, vol. 19, no. 1, pp. 5–9, 2014.

[40] J. Li, H. Liu, W. Xia et al., "De novo transcriptome sequencing and the hypothetical cold response mode of *Saussurea involucrata* in extreme cold environments," *International Journal of Molecular Sciences*, vol. 18, no. 6, p. 1155, 2017.

[41] J. Krasensky and C. Jonak, "Drought, salt, and temperature stress-induced metabolic rearrangements and regulatory networks," *Journal of Experimental Botany*, vol. 63, no. 4, pp. 1593–1608, 2012.

[42] H. Le Gall, F. Philippe, J.-M. Domon, F. Gillet, J. Pelloux, and C. Rayon, "Cell wall metabolism in response to abiotic stress," *Plants*, vol. 4, no. 1, pp. 112–166, 2015.

[43] Y. Oono, M. Seki, M. Satou et al., "Monitoring expression profiles of Arabidopsis genes during cold acclimation and deacclimation using DNA microarrays," *Functional & Integrative Genomics*, vol. 6, no. 3, pp. 212–234, 2006.

[44] X. Peng, L. Teng, X. Yan, M. Zhao, and S. Shen, "The cold responsive mechanism of the paper mulberry: decreased photosynthesis capacity and increased starch accumulation," *BMC Genomics*, vol. 16, no. 1, p. 898, 2015.

[45] H. Lis and N. Sharon, "Lectins: carbohydrate-specific proteins that mediate cellular recognition," *Chemical Reviews*, vol. 98, no. 2, pp. 637–674, 1998.

[46] S.-Y. Jiang, Z. Ma, and S. Ramachandran, "Evolutionary history and stress regulation of the lectin superfamily in higher plants," *BMC Evolutionary Biology*, vol. 10, no. 1, pp. 79–79, 2010.

5

The *YABBY* Genes of Leaf and Leaf-Like Organ Polarity in Leafless Plant *Monotropa hypopitys*

Anna V. Shchennikova,[1] Marya A. Slugina,[1,2] Alexey V. Beletsky,[1] Mikhail A. Filyushin,[1] Andrey A. Mardanov,[1] Olga A. Shulga,[1] Elena Z. Kochieva,[1,2] Nikolay V. Ravin,[1] and Konstantin G. Skryabin[1,2]

[1]*Federal State Institution "Federal Research Centre "Fundamentals of Biotechnology" of the Russian Academy of Sciences", Moscow 119071, Russia*
[2]*Lomonosov Moscow State University, Moscow 119991, Russia*

Correspondence should be addressed to Anna V. Shchennikova; shchennikova@yandex.ru

Academic Editor: Ferenc Olasz

Monotropa hypopitys is a mycoheterotrophic, nonphotosynthetic plant acquiring nutrients from the roots of autotrophic trees through mycorrhizal symbiosis, and, similar to other extant plants, forming asymmetrical lateral organs during development. The members of the YABBY family of transcription factors are important players in the establishment of leaf and leaf-like organ polarity in plants. This is the first report on the identification of *YABBY* genes in a mycoheterotrophic plant devoid of aboveground vegetative organs. Seven *M. hypopitys YABBY* members were identified and classified into four clades. By structural analysis of putative encoded proteins, we confirmed the presence of YABBY-defining conserved domains and identified novel clade-specific motifs. Transcriptomic and qRT-PCR analyses of different tissues revealed *MhyYABBY* transcriptional patterns, which were similar to those of orthologous *YABBY* genes from other angiosperms. These data should contribute to the understanding of the role of the *YABBY* genes in the regulation of developmental and physiological processes in achlorophyllous leafless plants.

1. Introduction

Monotropa hypopitys (syn. *Hypopitys monotropa*) is a member of the flowering seed plant family Ericaceae, which in turn belongs to the order Ericales splitting from the base of the clade Asterids [1]. This mycoheterotrophic, nonphotosynthetic, achlorophyllous plant acquires carbon from the roots of autotrophic trees through monotropoid mycorrhizal symbiosis [2, 3]. The *M. hypopitys* root system consists of fleshy roots, on which shoot buds develop, and finer mycorrhizal roots [2]. This plant has a typical aboveground structure, although the stem and leaves can be taken for flowering parts—floral axis and sterile bracts [4, 5]. Similar to extant plants, *M. hypopitys* forms asymmetrical lateral organs on the flanks of a shoot or inflorescence apical meristem, with adaxial and abaxial surfaces adjacent to or distant from, respectively, the meristem. Paleobotanic studies indicate that such structural asymmetry first appeared in a true leaf (euphyll) transformed from a radially symmetric stem as a consequence of the need to absorb more sunlight [6–9]. Studies of asymmetry in plants indicate that during plant development, cell fate is determined mostly by positional signals. The correct maintenance of the apical meristem and abaxial-adaxial differentiation of lateral organs requires reciprocal signal interaction between the meristem and derived structures [10–12].

It has been shown that the polarity of leaves and floral organs is defined by the network of genes encoding the Class III Homeodomain Leucine Zipper (HD-ZIPIII), ASYMMETRIC LEAVES (AS1/AS2), KANADI (KAN), AUXIN RESPONSE FACTOR (ARF3/ARF4), and YABBY families of transcription factors [8, 13–15]. Among them, the HD-ZIPIII REVOLUTA (REV) is expressed in the adaxial domain of lateral organs, whereas the GARP-family transcription factors KAN1–4 are involved in abaxial differentiation. Together, REV and KAN1 antagonistically regulate the expression of a number of genes encoding auxin signaling and transport components [16, 17].

The *YABBY* genes originating from the lineage leading to seed plants are identified in various green spermatophyte plant species; they are closely associated with the evolutionary emergence of flat-shaped leaves and are presumably diversified during evolution, which resulted in the appearance of family members with specific functions in the leaf, carpel, and ovule [8, 18–20]. YABBY transcription factors are characterized by their nuclear localization and the presence of the C_2C_2 zinc-finger and DNA-binding YABBY (High Mobility Group- (HMG-) box-like) domains [21, 22]. In gymnosperms, the *YABBY* genes are distributed among the A, B, C, and D clades [23]. In extant angiosperms, five *YABBY* subfamilies, *FILAMENTOUS FLOWER (FIL* or *YAB1)/YABBY3 (YAB3* or *AFO), CRABS CLAW (CRC), INNER NO OUTER (INO* or *YAB4), YAB2,* and *YAB5* [21] are distinguished by conserved functions in the initiation of lamina outgrowth, polarity maintenance, and establishment of the leaf margin [23–25]. Almeida et al. [26] and Morioka et al. [27] have provided evidence for the involvement of the *YAB2* and *YAB5* genes in the evolutionary diversification of style and filament morphology. The branching of *INO* and *CRC* from other *YABBY* genes has most likely occurred in parallel with the evolution of the carpel and outer integument via modification of reproductive leaf-like sporophylls [20, 28].

There are current theories related to the history of the *YABBY* genes in angiosperms. Bartholmes et al. [29] suggested that "vegetative" *YABBYs (FIL/YAB3, YAB2* and *YAB5)* do not form a monophyletic clade and that *CRC* and *FIL* evolved from a common ancestor gene, while the *INO* genes are sisters to that ancestral gene. On the other hand, Finet et al. [23] clustered *INO* together with clades *YAB5* and *YAB2*. In addition, two alternative evolutionary scenarios, that is, monophyly or paraphyly of the gymnosperm *YABBY* family towards angiosperm *YABBY* genes, suggest that all spermatophyte *YABBY* genes were derived from one or two, respectively, *YABBY* genes of the last common ancestor of extant seed plants [20, 23, 29]. The reconstruction of *YABBY* evolution in spermatophytes based on these theories suggests that at least one *YABBY* predecessor has already functioned as a polarity regulator and that the diversification of the gene family occurred in both angiosperms and gymnosperms [23]. Although the presence of the YABBY genes is presumably restricted to seed plants [19, 20], genomics studies conducted on marine picoeukaryotes revealed *YABBY* homologs in Chlorophyta [30]. Phylogenetic analysis of identified sequences suggested, with equal probability that either Chlorophyta *YAB* genes are evolutionarily related to seed plant *YABBYs* or emerged independently from ancestral HMG-box sequences [23].

The expression data available for the angiosperm *YABBY* genes suggested that the *FIL-, YAB2-,* and *YAB5*-like genes retained a more ancestral expression pattern in both vegetative and reproductive tissues, while the expression of the *CRC* and *INO*-like genes is more variable [29]. In Eudicots, *FIL, YAB3, YAB2,* and *YAB5* transcripts were detected in the abaxial side of primordia in all aboveground lateral organs (except ovules) determining the abaxial cell fate [31, 32]. The *CRC* genes are expressed abaxially in the carpel, placenta, and nectaries promoting the development of the gynoecium and abaxial part of the carpel wall, and terminating the floral meristem [33–38]. *INO* mRNA is detected in the abaxial epidermis of the outer integument [20, 28, 39, 40]. The *YABBY* expression pattern differences between cereal monocots and other angiosperms indicate the modification of genetic pathways involving *YABBYs* during the process of angiosperm diversification [21, 27, 41–46]. It is assumed that FIL, together with REVOLUTA (REV), APETALA1 (AP1), and LEAFY (LFY), corrects the spatial activity of the *AGAMOUS (AG), AP3, PISTILLATA (PI),* and *SUPERMAN (SUP)* genes, and, thus, is involved in the initiation of floral organ primordia at the correct position and numbers, defining the fate of appropriate cells [31, 44, 47].

Thus, the bifunctional YABBY transcription factors have an important role in driving the evolution of the leaf and gynoecium, as well as in the initiation, growth, and structural organization of almost all aboveground lateral organs, and in the control of shoot apical meristem organization and activity.

In the present study, we identified and phylogenetically classified seven *YABBY* members from *M. hypopitys* and characterized their expression profiles in various tissues during flowering. The structural features and composition of conserved motifs belonging to the predicted MhyYABBY proteins were also analyzed. Our data should further the understanding of possible links between polarity determination and the physiology of achlorophyllous mycoheterotrophic plants.

2. Materials and Methods

2.1. Plants and Transcriptomes. The previous study divided *M. hypopitys* specimens into a North American cluster and two Eurasian (excluding Russia) sister lineages (Swedish and pan-Eurasian) [48]. Analysis of *M. hypopitys* from the European part of Russia revealed two types, A and B, which showed 99 and 100% homology with specimens from Japan, Finland, and Great Britain, and with Swedish specimens, respectively [49–51]. The study of *H. monotropa* specimens from Northern Ireland showed that they occur in small, highly fragmented populations, and exhibit a relatively high level of within-population genetic diversity and a low level of clonality [52].

In the present study, two *M. hypopitys* plants of type B from one clone of the same genet were used. Flowering plants were collected in a coniferous forest, Kaluga region, Russia, in August, 2015. The individual plant was a 15 cm reproductive axis with bracts and raceme of 10–12 flowers (each of 4 sepals, 4 petals, 8 stamens, and 4 fused carpels), and root system comprised mycorrhizal and fleshy roots with adventitious buds. The annual floral axes arise from adventitious buds on the perennial roots and carry the laminar appendages (there are no flowers in their axils, but they are above the soil level), which are termed sterile bracts (in our study, bracts, for simplicity) [4].

Plants were dissected into flowers, bracts, fleshy roots with adventitious buds, and predominantly haustoria-enriched roots, immediately frozen and homogenized in liquid nitrogen, and stored at −80°C. Total RNA was isolated from tissue of each *M. hypopitys* bracts (two individual plants), flowers (two individual plants), roots containing buds (individual plant), and haustoria-enriched roots (individual plant) and used for mRNA library preparation, which was sequenced on the Illumina HiSeq2500 platform (Illumina Inc., San Diego, CA, USA). The *M. hypopitys* RNA-seq data for each of six transcriptomes were assembled into the 98,350 unigenes with a length of 201–12,993 bp [51, 53]. Individual reads were mapped on contigs using Bowtie 2 [54], and protein-coding genes in contigs were identified using TransDecoder (https://transdecoder.github.io/).

2.2. Identification and Bioinformatics Characterization of M. hypopitys YABBY-Coding Sequences. To identify the *M. hypopitys* genes homologous to the known organ polarity genes, we searched unique transcripts revealed by the RNA-seq against the NCBI database (http://blast.ncbi.nlm.nih.gov/). To predict *YABBY* transcripts in *M. hypopitys*, we additionally searched the assembled transcriptomes with the known *YABBY*-related sequences coding for conserved zinc-finger and HMG-like domains extracted from the NCBI database. Selected *YABBY* candidates were examined for open reading frames (ORFs), translated using Clone Manager v.7.11 (http://clone-manager-professional.software.informer.com/), and the conserved domains of putative MhyYABBY proteins were identified using the NCBI-CDD analyzer (http://www.ncbi.nlm.nih.gov/Structure/cdd/wrpsb.cgi) and specified according to [21].

To evaluate the overlap between transcriptomes, Venn diagrams were generated using the online program Venny [55]. To illustrate the transcriptome-based gene expression pattern, the data were clustered with the Average linkage method and Spearman rank correlation (distance measurement method), and visualized as a heat map (http://www2.heatmapper.ca/) [56].

Conserved MhyYABBY amino acid motifs were identified using the MEME (Multiple Expectation Maximization for Motif Elicitation) 4.11.2 online analysis (http://meme-suite.org/tools/meme) [57] and used to construct a schematic diagram. To search for motifs, the "Normal" motif discovery mode, the default width range of 6–50 amino acids (aa), and the motif site distribution "zero or one per sequence" were used. The identified motifs were manually compared with previously suggested specific motifs. Initial search was performed on a set of 34 complete sequences, including identified MhyYABBYs, independently of YABBY clade affiliation. In addition, variable regions between conserved domains within the same group of proteins were searched. Since most of the identified motifs were clade-specific, we further analyzed individual YABBY clades.

To investigate the evolutionary relationship of the *MhyYABBY* genes, the MhyYABBY proteins and YABBY homologs from other species available in NCBI were aligned using ClustalX [58]. For analysis, full-size amino acid sequences, as well as conserved regions consisting of the zinc-finger and HMG-like domains were used. Evolutionary divergence between the genes and proteins was estimated using the maximum composite likelihood and equal input models, respectively, in MEGA7 [59–62]. The phylogenetic tree topology was estimated using the maximal likelihood method based on the JTT matrix-based model in MEGA7 [62].

2.3. Analysis of Tissue-Specific Gene Expression. The *MhyYABBY* gene expression was calculated in each transcriptome. Transcript quantification based on RNA-seq data was performed without a reference genome using the RSEM [63] and Bowtie 2 [54] programs, including normalization of transcripts per kilobase of exon per million fragments mapped (FPKM) values and between samples.

To perform quantitative real-time PCR (qRT-PCR), the first strand cDNA was synthesized from 1 μg of each mixture of two-root, two-bract, and two-flower RNA preparations using the Reverse Transcription System (Promega, Madison, WI, USA) and an oligo-dT primer, and quantified using the Qubit® Fluorometer.

Based on the identified *YABBY*-like transcripts and corresponding draft genomic sequences (our unpublished data), gene-specific primers separated by at least one big intron were designed to amplify parts of gene-coding sequences (Supplementary Table 1). The qRT-PCR was performed in three technical replicates using 2.5 ng of cDNA and an SYBR Green and ROX RT-PCR mixture (Syntol, Moscow, Russia) at the following cycling conditions: initial denaturation at 95°C for 5 min, 40 cycles of denaturation at 95°C for 15 sec, and annealing/synthesis at 60°C for 40 sec. Obtained PCR-fragments were additionally purified and sequenced to confirm certain gene specificities. Gene expression levels were normalized to those of the reference pinesap *Actin5*, *Actin3*, and *SAND* genes (primers are provided in Supplementary Table 1), which transcripts were evenly represented in six transcriptomes [51]. Normalized expression data were statistically evaluated using GraphPad Prism version 7.02 (San Diego, CA, USA; https://www.graphpad.com/scientific-software/prism/). Three values (3 technical replicates) were used for SD calculation. The error bars were generated based on mean with SD calculation. Significance of the qRT-PCR data within the same tissue between species was estimated by unequal variance Welch's t-test and additionally treated with Bonferroni's correction: if any of the t-tests in the list had $p \leq 0.05$/number of t-tests in the list, then the null hypothesis was rejected; that is, the difference between samples was recognized as significant.

3. Results

3.1. M. hypopitys Organ-Polarity Genes. The obtained six transcriptomes [53] showed significant overlap between the whole set of reads of paired libraries (74–77%), and in three libraries for each of the two plants (>80%) (Figure 1), and reflected a number of the known plant organ polarity genes that are being expressed in different tissues of flowering pinesap (Supplementary Table 2). Among them, the genes of the KAN and REV transcription factors, responsible for the abaxial and adaxial cell identity in lateral organs, respectively,

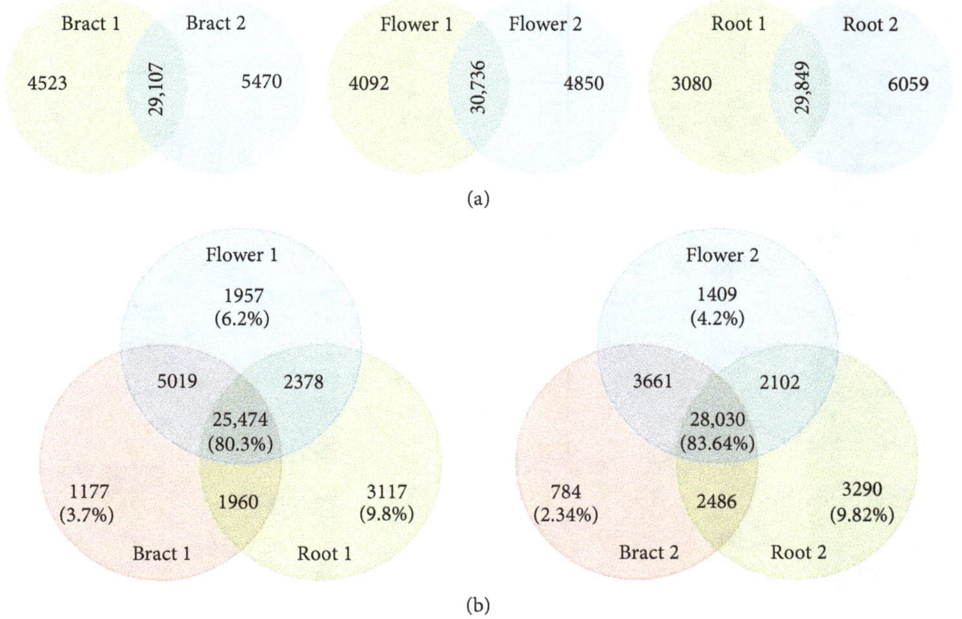

FIGURE 1: Venn diagrams of the overlap between the whole set of reads in paired transcriptomes (a) and in three transcriptomes for each of the two plants analyzed (b).

and their common targets (*ARF*, *SAUR*, *Aux/IAA*, *PID*, *PIN*, *NPY*, *GH3*, etc.) involved in auxin biology [17] were identified. The genes encoding SEUSS (SEU) and LEUNIG (LUG) involved in petal polarity determination along the adaxial/abaxial axis [64], AP2-like transcription factor AINTEGUMENTA (ANT) which contributes to organ polarity [65], the ADP ribosylation factor guanine nucleotide exchange factor GNOM essential for basal polarity establishment in *A. thaliana* [66], ULTRAPETALA1 (ULT1) which acts antagonistically with KAN1 to pattern the adaxial-abaxial polarity axis but jointly to pattern the apical-basal axis restricting the expression domain of the *SPATULA* gene [67], and HD-ZIPIII transcription factors HAT and ATHB positively regulated by REV [68] were also found (Supplementary Table 2). Heat map-based clustering of transcriptomic reads associated with organ polarity genes revealed similarities between pair libraries (according to the column dendrogram in Figure 2(a)). For most genes, the expression levels in flowers were higher than those in bracts (Figure 2(b)).

All six transcriptomes of bracts, flowers and roots contained seven unique *YABBY*-like transcripts, which were considered as putative *MhyYABBYs (MhyYAB1–MhyYAB7)*. The size of putative *MhyYABBY* ORFs varied from 502 to 673 bp, and the length of predicted proteins was from 166 to 224 aa. One transcript, *MhyYAB4*, had a partial 3′-truncated coding sequence (CDS), while the remaining six mRNAs contained complete CDSs. Putative MhyYABBY proteins included both a conserved N-terminal 37-aa C_2C_2 zinc finger-like domain and a C-terminal 48-aa helix-loop-helix domain resembling a part of an HMG box (Figure 3) [21, 33]. A cluster of amino acids at the beginning of the HMG-like domain could potentially serve as a nuclear localization signal [33, 69].

Among *MhyYABBYs*, we distinguished two groups (comprising *MhyYAB2*, *MhyYAB5*, and *MhyYAB6*; and *MhyYAB3* and *MhyYAB4*) based on high identity outside the conserved domains (Supplementary Figures 1a, 1b). The estimated low evolutionary pairwise divergence in the nucleotide and amino acid sequences between the members of each group compared to that between the members of different groups suggested the presence of two sets of paralogs (Supplementary Tables 4 and 5). *MhyYAB1* showed high pairwise divergence with other *MhyYABBYs*, which indicates the affiliation of *MhyYAB1* to the separate clade.

3.2. Phylogeny of the M. hypopitys YABBY Family. Previous phylogenetic studies considered only conserved YABBY domains; however, it has also been shown that variable protein regions contain clade-specific conserved motifs of potential functional importance [29]. In transcription factors, variable regions are often essential for their activity and/or formation of multimeric protein complexes, as it has been shown for MADS-box transcription factors [70]. We aligned complete amino acid sequences and only conserved domains of MhyYABBYs. For the group comprising MhyYAB2, MhyYAB5, and MhyYAB6 paralogs, two different results within the same clade were obtained. The full-size proteins were orthologous to FIL, while the conserved domains were orthologous to YAB3, another member of the FIL/YAB3 clade. We decided to use complete sequences to increase the sensitivity of phylogenetic analysis.

The generated tree, rooted with the *Micromonas commode* (Chlorophyta) YABBY-like protein, classified MhyYABBY1–7 by comparing them with YABBY-like proteins of *A. thaliana* and other angiosperm species belonging to Asterids and Rosids. As a result, *MhyYABBY* transcripts were distributed into the *FIL*, *INO*, *CRC*, and *YAB5* clades and renamed accordingly as *MhyCRC (MhyYAB1)*, *MhyINO1* and *MhyINO2 (MhyYAB3* and *MhyYAB4*, resp.),

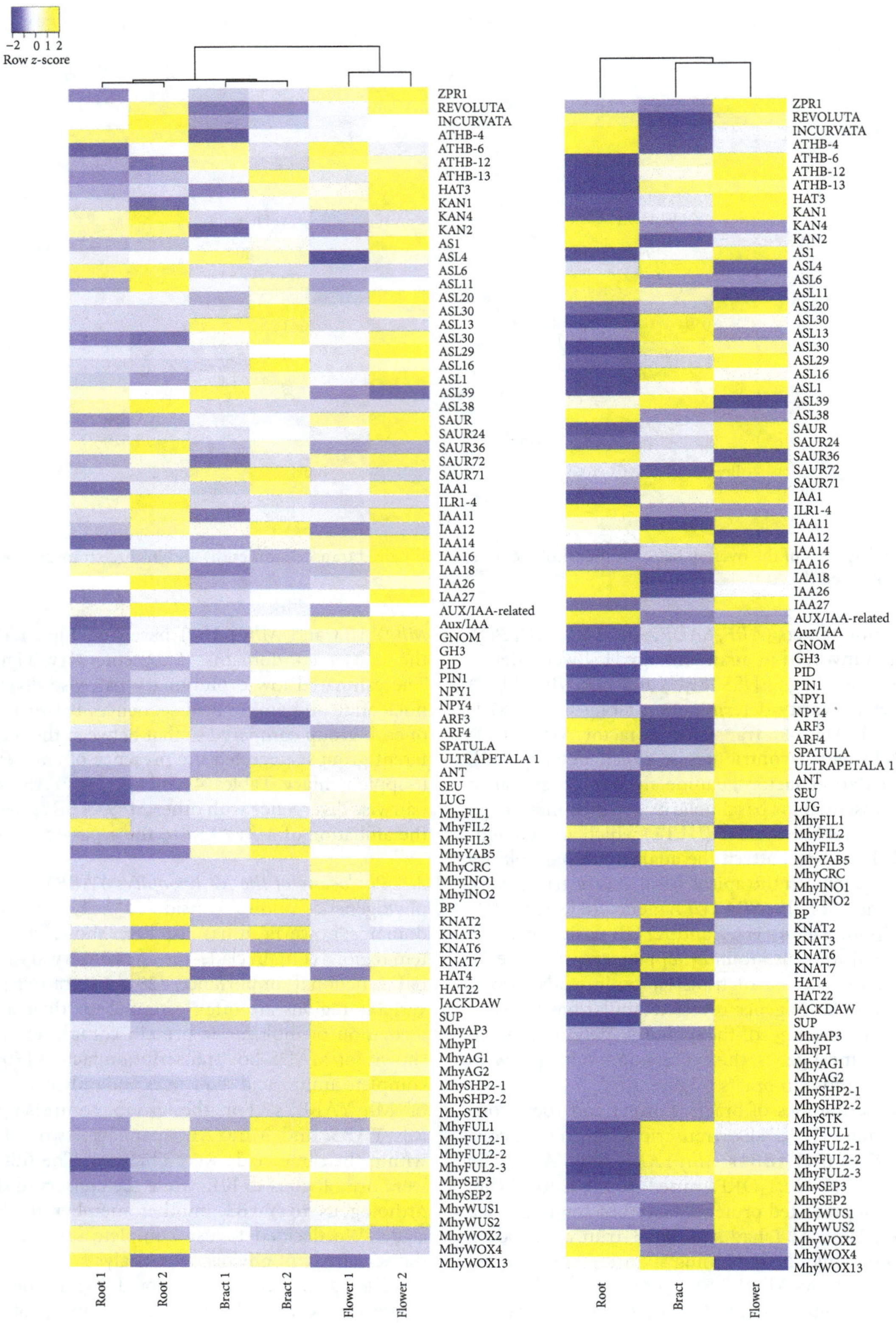

FIGURE 2: Heat maps of the organ polarity gene expression in six transcriptomes (a) and the mean expression of these genes in paired transcriptomes (b).

```
MhyYAB3 EQICYVQCGYCTTILLVGVPCGSLTMVVTVRCGHCTS...RPPEKRQRAPSAYNNFIKEEIRRLKGKHPSMTHKEAFSAAAKNWAHFP
MhyYAB4 EQICYVQCGYCTTILLVGVPCGSLTMVVTVRCGHCTS...RPPEKRQRAPSAYNNFIKEEIRRLKGKHPSMTHKEAFSAAAKNWAHFP
AthINO  GQICHVQCGFCTTILLVSVPFTSLSMVVTVRCGHCTS...KPPEKRQRAPSAYNCFIKEEIRRLKAQNPSMAHKEAFSLAAKNWAHFP
MhyYAB7 EQLCYIPCNFCNIVLAVSVPCSSLFDTVTVRCGHCTN...RPPEKRQRVPSAYNQFIKEEIQRIKANNPDISHREAFSTAAKNWAHFP
AthYAB5 EQLCYIPCNFCNIILAVNVPCSSLFDIVTVRCGHCTN...RPPEKRQRVPSAYNQFIKEEIQRIKANNPDISHREAFSTAAKNWAHFP
AthYAB2 ERVCYVHCSFCTTILAVSVPYASLFTLVTVRCGHCTN...RPPEKRQRVPSAYNRFIKEEIQRIKACNPEISHREAFSTAAKNWAHFP
MhyYAB2 DQLCYVQCNICDTVLAVSVPCSSLFKAVTVRCGHCTS...RPPEKRQRVPSAYNRFIKDEIQRIKAGNPDISHREAFSAAAKNWAHFP
MhyYAB5 DQLCYAHCEFCNTVLVVSVPCSSLFKTATVRCGHCTS...RPPEKRQRVPSAYNRFIKDEIQRIKAGNPDISHREAFSAAAKNWAHFP
MhyYAB6 DQLCYVQCNICDTVLAVSVPCSSLFKTATVRCGHCTS...RPPEKRQRVPSAYNRFIKDEIQRIKAGNPDISHREAFSAAAKNWAHFP
AthFIL  DHLCYVQCNFCQTILAVNVPYTSLFKTVTVRCGCCTN...RPPEKRQRVPSAYNRFIKEEIQRIKAGNPDISHREAFSAAAKNWAHFP
AthYAB3 DQLCYVHCSFCDTVLAVSVPPSSLFKTVTVRCGHCSN...RPPEKRQRVPSAYNRFIKEEIQRIKAGNPDISHREAFSAAAKNWAHFP
MhyYAB1 DNLCYVRCNFCNTVLAVGIPCKRLLDTVTVKCGHCSN...KPPEKKHRLPSAYNRFMKEEIQRIKAANPEIPHREAFSAAAKNWARYI
AthCRC  EHLYYVRCSICNTILAVGIPLKRMLDTVTVKCGHCGN...KPPEKKQRLPSAYNRFMRDEIQRIKSANPEIPHREAFSAAAKNWAKYI
PtYAB   DHLCYVHCDYCNTILAVNVPCNSLYRTVTVRCGHCNN...RQPEKKQRVPSAYNRFIRDEIQRIKAHNPKITHKEAFSAAAKNWAHYP
```

Zinc finger domain | HMG-like YABBY domain

FIGURE 3: Alignment of conserved zinc-finger and HMG-like domains from putative *M. hypopitys* YABBY proteins 1–7, known as *A. thaliana* YABBYs (AthINO, AF195047; AthCRC, AF132606; AthFIL, AF136538; AthYAB3, AF136540; AthYAB5, NP_850081; AthYAB2, AF136539), and *Pinus taeda* PtYAB (DR100835; gymnosperms).

MhyYAB5 (*MhyYAB7*), and *MhyFIL1*, *MhyFIL2*, and *MhyFIL3* (*MhyYAB5*, *MhyYAB6*, and *MhyYAB2*, resp.). It should be noted that no *YABBY2*-like transcripts were found in all six transcriptomes. The sequences have been deposited in GenBank (KX12839–KX12841, KX12843–KX12846; Supplementary Table 3). The maximum likelihood reconstruction of the MhyYABBY family with bootstrap values at tree nodes is shown in Figure 4. Within the individual clades, MhyYABBY sequences are sisters to other YABBY-like sequences from Asterids, and the closest homologs are YABBYs from Ericales.

3.3. MhyYABBY-Specific Motifs Identified outside of the YABBY Domains. To further investigate the structural divergence of pinesap YABBY proteins, we searched for the motifs conserved within individual clades or the whole MhyYABBY group. Comparison of complete sequences of YABBY orthologs (Supplementary Table 6) revealed two major domains, zinc-finger and YABBY, specific to all YABBY proteins. The number of motifs in variable regions was 4 to 9 with the length from 5 to 26 residues. The obtained data were compared with previously defined motifs specific to individual YABBY clades [29].

In MhyINO1/2, the known INO-A motif was found immediately after the zinc-finger domain. Comparison of MhyINO1/2 with other INO-like sequences in the NCBI database revealed novel putative INO-specific motifs. At the N-terminus and between the zinc-finger and YABBY domains, a highly conserved 11-aa INO-B motif and an Eudicot-specific 15-aa INO-C motif were identified. In addition, we found a C-terminal 26-aa INO-D motif specific to MhyINO1/2 and INO-like proteins in Solanaceae. All previously predicted CRC-specific motifs, CRC-A, CRC-B, CRC-C, and CRC-D, were found in MhyCRC. In addition, we identified a putative conserved Eudicot-specific C-terminal 15-aa CRC-E motif. In MhyYAB5, the presence of the known YAB5-specific motifs YAB5-A, GY/YAB2/5-A, YAB5-B, and GY/YAB2/5-B was confirmed. The YAB5-D motif was not detected, while a modified YAB5-C sequence was identified before the YABBY domain as a 12-aa YAB5-Cm motif. In MhyFIL1/2/3, we found all previously predicted motifs, FIL-A, FIL-B, FIL-C, FIL-D, FIL-E, FIL-F, and FIL-G. In addition, we suggested two motifs, 12-aa FIL-I (monocot- and eudicot-specific) and 11-aa FIL-H (eudicot-specific) located between the zinc-finger and YABBY domains, as candidates for conserved FIL-specific motifs.

Thus, we characterized MhyYABBY proteins by sequence conservation on the clade level. The motifs identified in MhyYABBYs were shown in Figure 5 and sequences of the predicted novel motifs are presented in Supplementary Figure 2).

3.4. Expression Pattern of "Vegetative" and Flower-Specific M. hypopitys YABBY Genes. The transcriptome-based data on *MhyYABBY* expression in the roots and buds, haustoria-enriched roots, bracts, and flowers (Figure 6, Supplementary Table 3) showed that the *MhyYABBY* mRNAs (except *MhyFIL2*) were present in the flowers. Except for *INO*- and *CRC*-like *MhyYABBYs*, all other five transcripts were detected in the bracts. The highest bract-specific expression was observed for *MhyYAB5*, while the remaining four genes were transcribed at similarly low levels. Finally, in roots and buds, only *MhyYAB5* and *MhyFIL2* mRNAs were expressed at very low levels, while in the haustoria-enriched roots none of genes were expressed. The *MhyFIL1* and *MhyFIL3* transcripts were increased from the bract to the flower, maintaining the same expression profile. In contrast with this, the number of *MhyYAB5* transcripts was decreased from the bract to the flower.

Quantitative (q) RT-PCR data on *MhyFIL3*, *MhyYAB5*, *MhyINO1*, *MhyINO2*, and *MhyCRC* expression are represented at Figure 6(d) and Supplementary Table 7. The relative expression of *MhyYABBY* genes was estimated in the flowers, bracts, and roots and buds. All analyzed genes were expressed in the flowers with the highest *MhyYAB5* level, and the lowest *MhyINO2* and *MhyFIL3* levels. In the bracts, the *MhyYAB5* gene was also highly expressed, but only traces of the *MhyINO2* and *MhyFIL3* mRNAs were observed. In the roots and buds, the only *MhyYAB5* mRNA was detected at low level. The difference in the *MhyYAB5* gene expression between the pinesap tissues was statistically significant. The flower-specific expression of *MhyCRC* and *MhyINO1* was significantly different from that in bracts and roots. The expression of *MhyINO2* and *MhyFIL3* was similar between tissues (Supplementary Table 7). All the analyzed gene expression modes were the same as it was shown in their

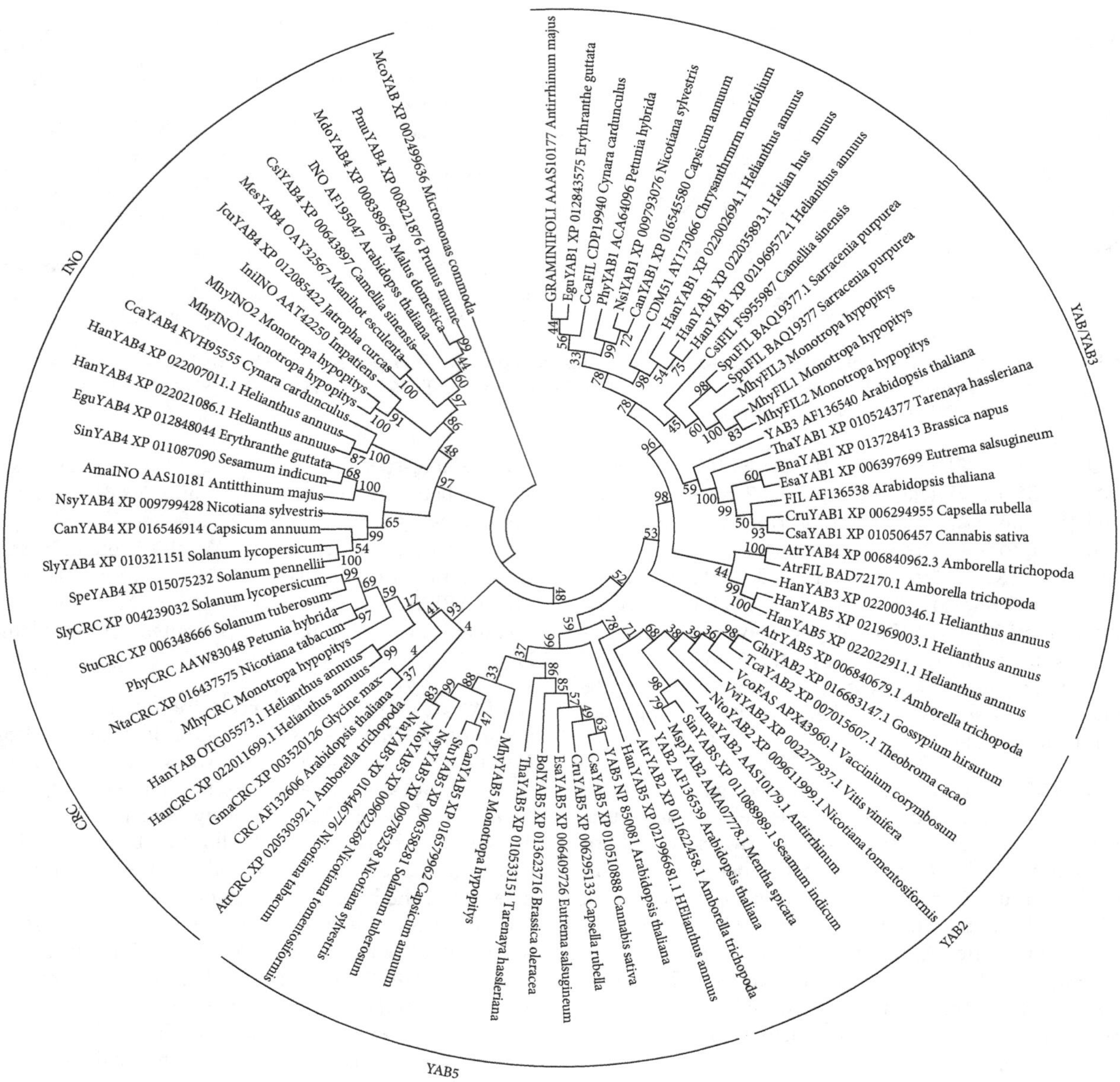

FIGURE 4: A phylogenetic tree of MhyYABBY proteins was generated with 83 full-size YABBY-like amino acid sequences from *M. hypopitys*, *A. thaliana*, and other angiosperm species. The *M. commode* (Chlorophyta) YABBY-like protein was used as outgroup. The *Amborella trichopoda*, Ericales, Rosids, Lamiids (euAsterids), and Campanulids (euAsterids) sequences are colored in dark blue, blue, burgundy, green, and red, respectively. The numbers next to the nodes represent bootstrap values from 1000 replicates.

transcriptome-based patterns, except for *MhyINO2*. In the flower, the measured qRT-PCR expression of this gene was equally low. Given transcriptomic data, *MhyINO2* was absent in flower 2, which may be due to the quality of the libraries or their sequencing. Also, considering the low evolutionary pairwise divergence in the *INO1* and *INO2* sequences, a higher level of *INO1* expression compared to the level of *INO2* may indicate that *INO1* may be more required than *INO2* during plant development.

4. Discussion

The emergence of photosynthesis has become the most significant event in the evolution of plants. The majority of extant plants are autotrophic, except for about 1% of flowering heterotrophic plants. Among the latter, obligate mycoheterotrophs are the results of replicated deevolutionary events of the photosynthetic ability loss, triggering the degradation of both cytoplasmic and nuclear genomes [71]. Full mycoheterotrophs demonstrate a wide range of deevolutionary outcomes such as abrupt morphophysiological changes [2], genome rearrangements, and massive gene loss [71, 72].

In the large and diverse eudicot family Ericaceae with a nearly worldwide distribution, two of nine subfamilies, Pyroloideae and Monotropoideae, contain partial and full mycoheterotrophs, respectively [73]. In Monotropoideae, *M. hypopitys* represents a unique obligate mycoheterotroph. Recent studies on the *M. hypopitys* plastid genome and its

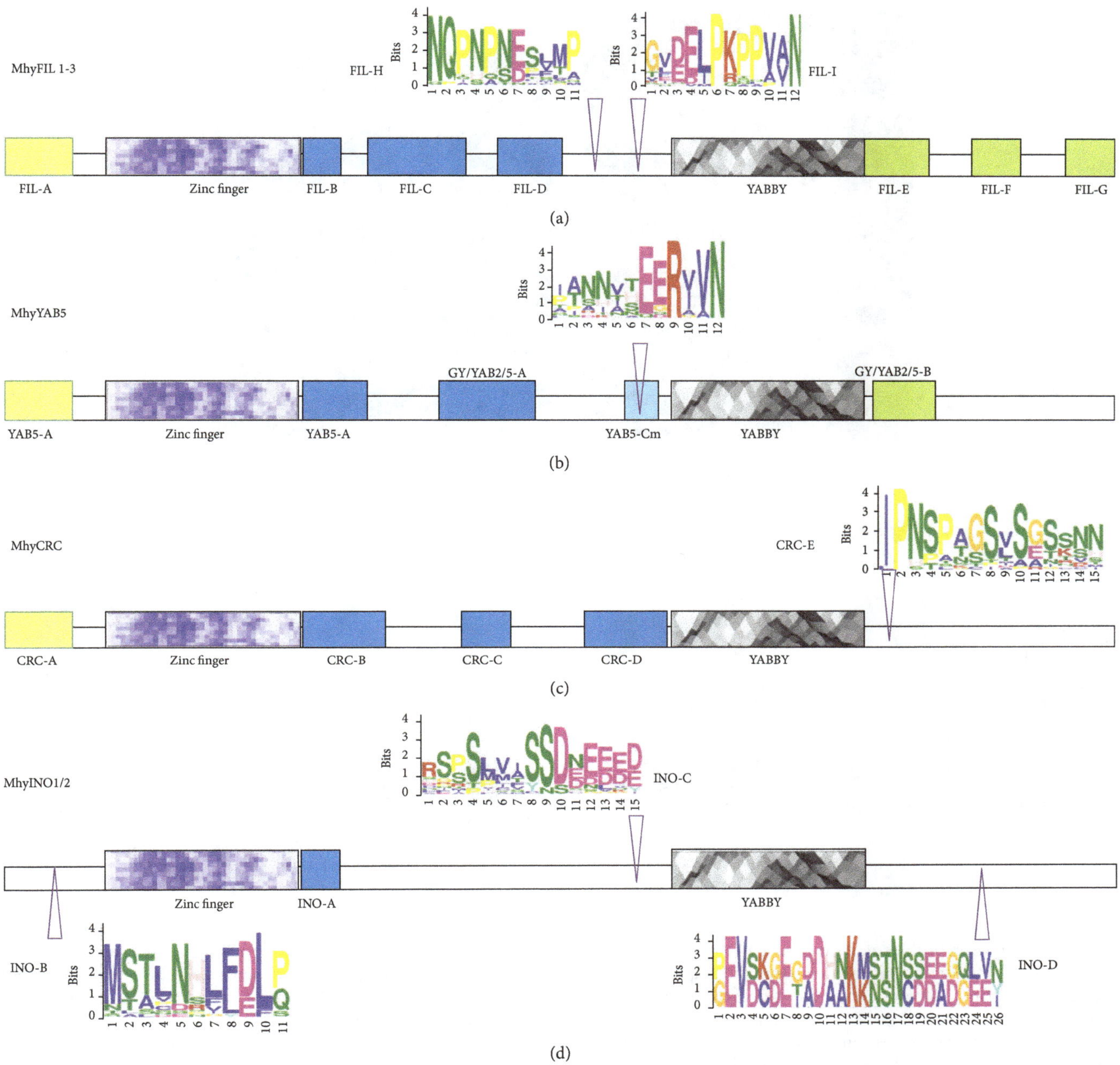

FIGURE 5: Amino acid clade-specific motifs predicted in *M. hypopitys* YABBY proteins. MhyFIL1/2/3 (a); MhyYAB5 (b); MhyCRC (c); MhyINO1/2 (d). Previously suggested clade-specific motifs [29] are shown as boxes. MEME-predicted novel motifs are marked by arrowheads and shown as letter sequences.

comparison with that of photosynthetic relative *Pyrola rotundifolia* indicated that this plant is at the final stages of plastome degradation, which is expressed in highly reduced size and content, dramatic structural rearrangements, and acceleration of nucleotide substitutions in all protein-coding genes [74–76]. Furthermore, the coordinated loss of photosynthesis-related functions in both plastome and nuclear genomes of *M. hypopitys* is a sign of ongoing changes in the nuclear genome of this mycoheterotrophic plant [51].

It is generally accepted that mycoheterotrophic plants have evolved from photosynthetic mycorrhizal lineages, as mycoheterotrophy helps to succeed in the low-light conditions of the forest [77]. It has been established that dark-induced leaf senescence leads to a significant chlorophyll loss and photosynthesis inactivation [78, 79]. During evolution, a *M. hypopitys* ancestor (already with megaphylls) growing in shaded habitats lost the genetic ability to photosynthesize due to symbiosis with fungi, which provided a sufficient amount of carbon to pinesap from the roots of autotrophic trees. It is shown that the loss of photosynthetic ability and full heterotrophy are linked to the degradation and/or modification of vegetative structures [77]. It is believed

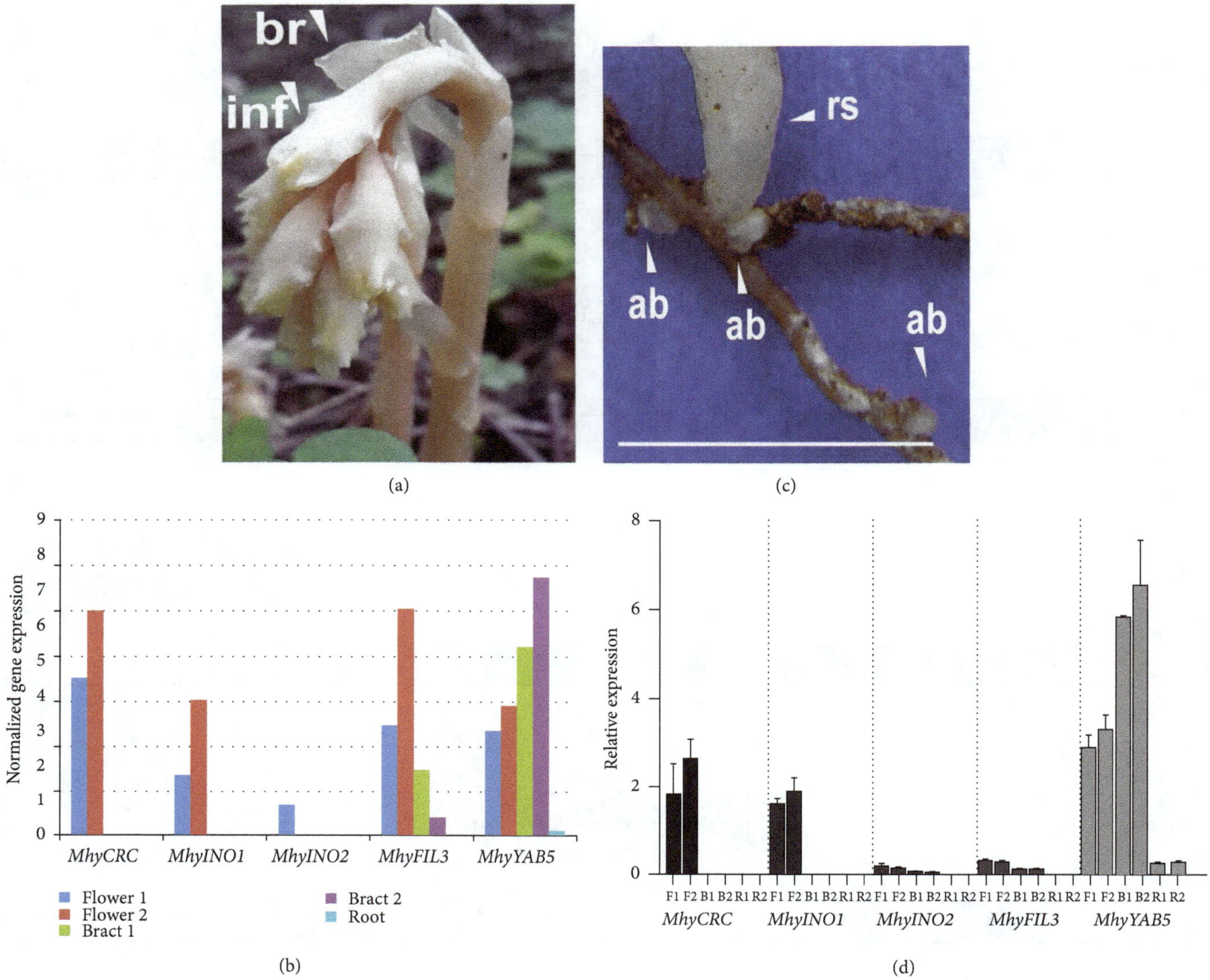

FIGURE 6: Expression profiles of *M. hypopitys YABBY* genes in the bract, flower, root and bud, and haustoria-enriched root tissues. Expression was estimated as the transcript number per million equal to the sum of total transcripts in the tissue. The *M. hypopitys* adult plant; br—bract, fl—flower (a); transcriptome-based expression pattern of the *MhyFIL1*, *MhyFIL2*, *MhyFIL3*, *MhyYAB5*, *MhyCRC*, *MhyINO1*, and *MhyINO2* genes in pinesap tissues (b); the *M. hypopitys* roots with adventitious buds; ab—adventitious bud, rs—growing reproductive stem; scale bar = 1 cm (c); relative expression (qRT-PCR) of the pattern of the *MhyCRC*, *MhyINO1*, *MhyINO2*, *MhyFIL3*, and *MhyYAB5* genes in pinesap tissues, F—flower, B—bract, R—root (d).

that in *M. hypopitys*, an elongated raceme emerges instead of a true stem, developing directly from the adventitious bud on the roots [4].

The photosynthetic ability is closely related to the origin of asymmetrical leaves (providing the absorption of sufficient light energy by seed plants [80, 81]), in particular, due to the *YABBY* genes' evolutionary duplication and diversification [6–9, 25, 82, 83]. Although the role of the *YABBY* genes in plant evolutionary adaptation to light perception is established and they have been systematically studied in model and nonmodel species [27, 84–87], up to now, no *YABBY* genes have been described in mycoheterotrophic plants. It was interesting to figure out, if these genes and, therefore, the conserved mechanism of leaf polarity determination, were exposed to the adaptive deevolution in leafless mycoheterotroph *M. hypopitys*. Therefore, in this study we focused on the diversity and expression profile of the *YABBY* genes in a *M. hypopitys* that may further the understanding of the development and evolution of this plant group.

Peripheral cells of the shoot apical meristem give rise to the leaves that develop along three axes and acquire the adaxial-abaxial and proximal-distal asymmetry and the mediolateral symmetry [88]. The elongated *M. hypopitys* raceme carries bracts below flowers and leaf-like sterile bracts [4]. Bracts are thin, 8–15 mm long, 3–15 mm broad, ovate, and expanding to the top, with irregularly toothed margins [4], rudimentary midvein, and parallel veins of similar thickness (Figures 7(a) and 7(b)). Interestingly, sterile bracts are not

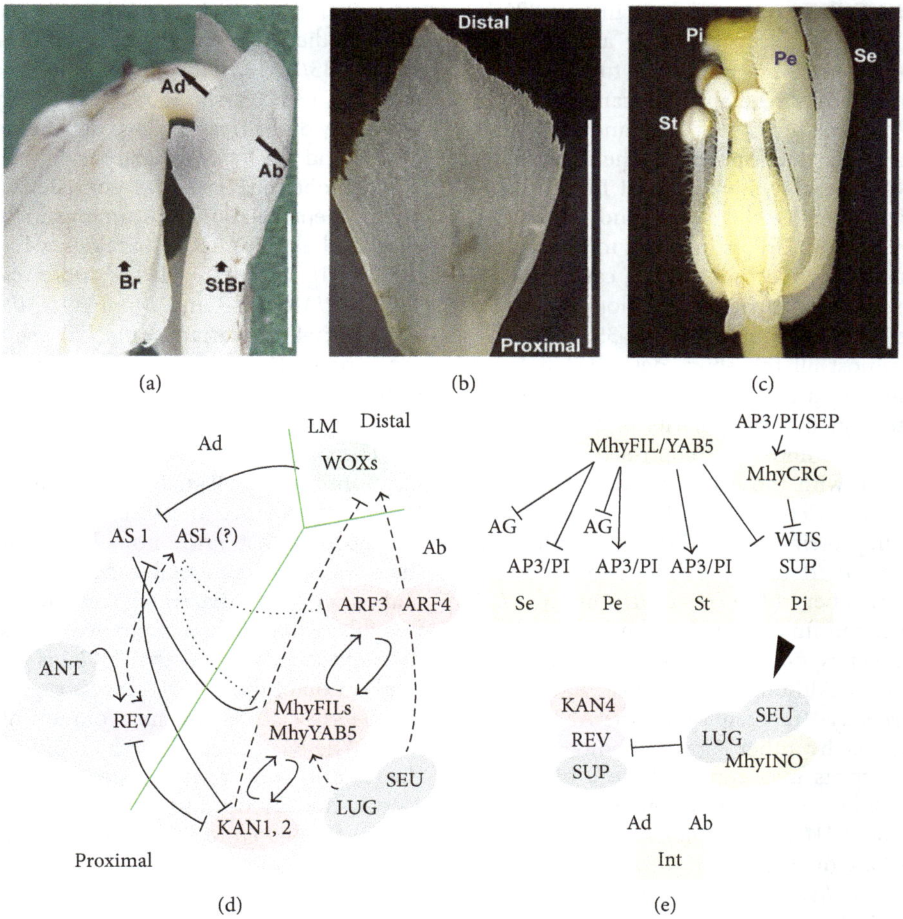

FIGURE 7: A hypothetical scenario for the functioning of the MhyYABBYs during the development of *M. hypopitys* bracts and flowers. (a) *M. hypopitys* raceme. (b) *M. hypopitys* sterile bract. (c) *M. hypopitys* flower with a partially removed perianth. (d) Scheme of possible relations between "abaxial" and "adaxial" factors in the *M. hypopitys* bracts. (e) Scheme of possible interactions of MhyYABBYs in the *M. hypopitys* flowers. Br—bract, StBr—sterile bract, Ad—adaxial side, Ab—abaxial side, LM—leaf margin, Se—sepal, Pe—petal, St—stamen, Pi—pistil, Int—integument. Scale bar = 0.5 cm.

exactly leaves as they have the genetic signatures of reproductive organs, which are manifested in the expression of the floral organ identity MADS-box genes [89] (Figure 2).

The lack of leaves in *M. hypopitys* may correlate with possible changes in the conserved genetic network of lateral organ polarity. In pinesap transcriptomes, we found the number of genes associated with this network, including seven *YABBY*-like sequences encoding proteins, which contain conserved domains and nuclear localization signals characteristic for YABBY transcription factors.

Initially, "adaxial" and "abaxial" genes are expressed throughout the leaf primordium, and, as the leaf develops, their expression becomes restricted to their respective domains due to the mutually exclusive actions of their protein products [90]. *YABBYs* are "abaxial" genes involved in stimulation of the cellular division during lamina outgrowth in all aboveground lateral organs, vegetative or reproductive (e.g., [25, 91–93]). *M. hypopitys* flowers do not show any visible abnormalities compared to those of other eudicots. Thence, *MhyYABBYs* may play common roles in the proper development of the floral meristem into a mature flower. However, the question arises how did the lack of leaves affect the function of the "vegetative" *MhyFIL1–3* and *MhyYAB5* genes.

In *A. thaliana* leaves and sepals, *FIL* and *YAB3* genes are upregulated by KAN1 and ARF4, and, in turn, FIL and YAB3 stimulate the expression of *ARF4*, *KAN1*, and *AS1* [92, 94, 95], and besides, in complex with LUG and SEU promote not only organ polarity, but embryonic shoot apical meristem initiation and maintenance [93]. "Abaxial" KAN represses the "adaxial" *HD-ZIPIII* genes [90, 96, 97]. At the boundary between adaxial and abaxial tissues, the FIL/YAB3 and KAN, respectively, up- and downregulate the *WOX1* and *WOX3* genes that specify redundantly lateral lamina outgrowth and leaf margin cell fate [98]. The *YAB5*, *KAN2*, *ARF3*, and *ARF4* genes are repressed by the "adaxial" AS1 and AS2 implicated in the proper leaf formation along all three axes [99–101].

During flowering, *YABBYs* are required to establish a correctly developed flower primordium through the interaction with *REV*, *KAN4*, *SEU*, *LUG*, *ANT*, *SUP*, *LFY*, and the floral homeotic MADS-box genes [28, 47, 64, 65, 91, 93]. The SEU and LUG are needed to promote and maintain the *FIL*/*YAB3* and *HD-ZIPIII* expression [64]. In turn, FIL in combination with ANT acts to upregulate the "adaxial" gene *PHB* and MADS-box gene *AP3* [65], and together with REV, AP1, and LFY, spatially regulates the transcription of the *SUP* and the MADS-box genes *AG*, *AP3*, and *PI* [31, 44, 47]. To maintain the polar development of the ovule outer integument, the INO interacts with SEU and LUG, but its expression is restricted by KAN4, REV, and SUP [28]. CRC, upregulated by AP3/PI/SEP, is involved in the control of radial and longitudinal gynoecium growth, carpel fusion, and nectary location, and participates in the floral meristem termination through the *WUS* repression [33, 35, 36, 102].

The finding of almost all the above-described polarity genes in the analyzed transcriptomes (Supplementary Table 2; Figure 2) suggests that the polarity of *M. hypopitys* bracts and floral organs is under the control of conserved mechanisms (Figure 7), with the exception of the absence of transcripts *PHB* and *PHV* (HD-ZIPIII), *WOX1* and *WOX3* (homeodomain protein), and *AS2* (LBD domain transcription factor). The *PHB* and *PHV* function may be replaced by another member of the *HD-ZIPIII* family, *REV*, since these three genes can function redundantly [90]. Similarly, the *MhyWOX* genes may perform the functions of *WOX1* and *WOX3* [103]. Although 12 *AS2-like (ASL)* genes were found in *M. hypopitys* transcriptomes, the AS2 cannot be functionally replaced by other family members [104]. The lack of *AS2* transcripts may contribute to a high level of *MhyYAB5* expression compared to other "vegetative" *MhyYABBYs*, since the AS1/AS2 complex suppresses *YAB5* [88]. Moreover, the lack of *AS2* activity may be related to the characteristics of *M. hypopitys* bracts having a plump lamina base, the midvein indistinguishable from parallel veins, and the absence of petioles. This conclusion may be supported by the phenotype of the mutant *as2*, which has a significantly rudimentary leaf midvein (and several parallel veins of very similar thickness), shortened petioles and leaf blades, and a plump and swelled leaf lamina base [105].

Most of the genes that define organ polarity existed before the emergence of a flat leaf in seed plants. Based on the analyses of the families of organ polarity genes, such as *HD-ZIPIII* [106], *ARFs* [23], and *ASLs* [107], it is assumed that after the ferns' divergence, multiple paralogs arose in the seed-plant common ancestor [23]. Unlike other polarity genes, *YABBYs* originated in the lineage leading to seed plants, and it is proposed that they are implicated in the transition of an ancestral shoot-specific network into a leaf-specific one [19, 25]. At least four gene duplication events in the *YABBY* family led to the emergence of at least five *YABBY* genes with both novel and redundant functions in the last common ancestor of extant flowering plants [23, 29].

The structural and phylogenetic analysis based on comparison with the YABBY orthologs revealed that each MhyYABBY belonged to one of the four highly conserved clades in the angiosperm YABBY family [94]. In the dendrogram, it is possible to single out a cluster consisting of the FIL/YAB3, YAB5, and YAB2 clades (Figure 3). CRC- and INO-orthologs have formed separate clusters, which corresponds to the previously proposed origin of the *CRC* and *INO* genes from different ancestors [20, 23]. The tree composition was not completely congruent with the data of other studies with observations of the two clusters CRC/FIL and YAB5/YAB2/INO [23], or the two clusters FIL/CRC/INO and YAB2/YAB5 [20, 29], probably due to the inherent instability of the tree topology depending on the composition of taxa and the mode of analysis.

The obtained tree was consistent with the established phylogenetic relationships among higher plants. The presence of the MhyYABBY paralogs, which are coorthologous to the FIL and INO clades, indicates that the *MhyFIL1*/*MhyFIL2*/*MhyFIL3* and *MhyINO1*/*MhyINO2* groups could represent allelic variants (for *FIL* group), alternative splicing variants (for *INO* group), or may have originated as a result of a recent gene duplication event unique to the Ericales order.

The simplest explanation of the absence of the *YABBY2*-like transcripts in all analyzed pinesap transcriptomes may be the low abundance and insufficient transcriptome size. It is also possible that *YAB2* homologous genes are expressed at the earlier developmental stages during the formation of lateral organ primordia, which were not analyzed. One of three possible evolutionary scenarios explaining the *YABBYs*' diversification suggests that *YAB2* is the result of the earliest duplication of the *YABBY* ancestor gene [20], and therefore, can be associated with the evolution of a leaf stronger than other *YABBYs*. The *YAB2* ortholog is present in photosynthetic Ericaceae relative species *Vaccinum corymbosum*. Hence, the absence of the *YAB2* gene in *M. hypopitys* may be due to the loss of the gene during the adaptive evolution (deevolution of the genome) of the autotrophic ancestor of *M. hypopitys* accompanied by the loss of the leaf. MhyYAB2 functions could be partially complemented through neofunctionalization of MhyYAB5 or MhyFIL paralogs. Studies in *Oryza sativa* and other plants suggest that within the multicomponent regulatory network composed of homo- and heterodimers formed by "vegetative" FIL/YAB3, YAB2, and YAB5 orthologs, protein substitutions and replacements are possible [29, 108]. It was shown that in *O. sativa* the loss of the *YAB5* genes was complemented by the *FIL* and *YAB2* paralogs [108]. Accordingly, *M. hypopitys MhyYAB2* could have been replaced by the *MhyYAB5* or the three *MhyFILs*. It is also known that YAB2 and YAB5 are important for laminar style and filament morphology evolution in angiosperms, when *YAB2* expression over a certain threshold disturbs the balance in the regulatory network, leading to radialization of the laminar structure [26, 27]. Therefore, given the *M. hypopitys* style and filament radial structure and the expression of the *MhyYAB5* gene in flower tissue, the absence of *MhyYAB2* transcripts may also indicate a possible substitution of YAB2 by YAB5 in *M. hypopitys*.

Bioinformatics analysis of the MhyYABBY structural organization revealed the presence of 18 previously known conserved motifs [29] and 7 putative novel candidate sequences for clade-specific motifs (Figure 5), which may be used as markers to identify appropriate genes. Given the shared evolutionary history of YABBYs, the novel motifs could be biologically relevant and involved in subfamily-specific functions, which need further investigation.

The *MhyYABBY* gene orthology data are supported by the *MhyYABBY* expression patterns (Figure 6). In *Arabidopsis*, "vegetative" genes *FIL*, *YAB3*, and *YAB5* are expressed in leaves and leaf-like cotyledons, sepals, petals, stamens, and carpels, whereas expression of *CRC* and *INO* is restricted to specific floral organs that are evolutionarily derived from leaves [20]. *MhyYABBY* transcripts were also found in aboveground tissues (bracts and flowers). In the case of the *MhyYAB5* gene, its atypical expression in roots indicates that it may have some roles in the development of the *M. hypopitys* root system. On the other hand, the perennial plant *M. hypopitys* commonly develops underground adventitious buds on the roots, which presumably contain an embryonic inflorescence [109], and, thus, the *MhyYAB5* gene can be expressed in the buds. Interestingly, in bracts, the expression level of *MhyYAB5* is much higher than that of *MhyFILs*. Given the possible synergy of their action, it can be assumed that the low expression of *MhyFILs* was compensated by an increase in the expression of the *MhyYAB5* gene not only in bracts, but also in roots, or rather in adventitious buds.

The *MhyFIL* expression profiles indicate the possible subfunctionalization of the paralogs. The *MhyFIL3* gene, which according to phylogenetic analysis is at the base of the *M. hypopitys FIL* clade, is expressed approximately at the same level as *MhyFIL1*, while the extremely low number of transcripts of the third paralog *MhyFIL2* is present only in two of the six transcriptomes (Supplementary Table 3). It is possible that *MhyFIL3* and *MhyFIL1* may have redundant functions, and *MhyFIL2* may be a pseudogene. Given trace amounts of *MhyFIL2* transcripts in the root and bud library, similar to *MhyYAB5*, it is likely that *MhyFIL2* may be involved in the development of inflorescence at the early stages after bud dormancy release [110].

The *MhyCRC* expression was detected only in flower tissue, confirming its potentially conserved roles in carpel fusion, style/stigma and nectary development, and in the floral meristem termination as it was shown for *A. thaliana* CRC [33, 38, 102], as well as in vascular development, as indicated by a recent report on the functional role of *Pisum sativum* CRC [111]. Similar to *A. thaliana* INO [40, 112], MhyINO1 and MhyINO2 may redundantly define and promote the outer ovule integument growth in *M. hypopitys*, while MhyFIL1/2/3 and MhyYAB5 may influence the abaxial cell fate in all aboveground lateral organs like their corresponding YAB1/3 and YAB5 orthologs [23].

It has recently been shown that "vegetative" YABBYs act as transcriptional activators of jasmonate-triggered responses. Jasmonate-induced degradation releases YABBYs from complexes with JAZ3 to mediate anthocyanin accumulation and chlorophyll breakdown [113]. The analysis of *M. hypopitys* transcriptome data did not reveal *JAZ3*-like transcripts, which may be consistent with complete chlorophyll loss in *M. hypopitys*. Thus, such mechanism may become evolutionarily obsolete in mycoheterotrophic plants.

The current study is the first to identify the *YABBY* genes in a mycoheterotrophic plant devoid of vegetative leaf-like organs. Seven *MhyYABBY* members were detected and classified in *M. hypopitys*, and putative protein structure, conserved motifs, and phylogenetic relationship were systematically analyzed. *MhyYABBY* transcription profiling in different plant tissues indicated the involvement of MhyYABBY proteins in the regulatory network controlling bract and flower formation. Our findings should further the investigation of YABBY functional roles in the regulation of developmental and physiological processes in achlorophyllous plant species and help to reveal possible differences in generally conserved molecular mechanisms underlying plant development and evolution.

Abbreviations

CRC: CRABS CLAW
FIL: FILAMENTOUS FLOWER
HMG: High Mobility Group
INO: INNER NO OUTER
MhyYABBY: *M. hypopitys* YABBY
YAB: YABBY
qRT-PCR: Quantitative RT-PCR.

Conflicts of Interest

The authors declare no conflict of interest.

Authors' Contributions

Konstantin G. Skryabin and Nikolay V. Ravin conceived and designed the research. Elena Z. Kochieva and Mikhail A. Filyushin provided the plant material. Andrey A. Mardanov, Elena Z. Kochieva, Marya A. Slugina, Mikhail A. Filyushin, Olga A. Shulga, and Anna V. Shchennikova performed the transcriptome sequencing and qRT-PCR experiment. Alexey V. Beletsky and Anna V. Shchennikova analyzed the data. Anna V. Shchennikova wrote the paper. All authors read and approved the manuscript.

Acknowledgments

This work was supported by the Russian Science Foundation (Grant no. 14-24-00175) and was performed using the experimental climate control facility of the Institute of Bioengineering, Research Center of Biotechnology, RAS. The authors would like to thank Dr. Marina Chuenkova for providing language help.

Supplementary Materials

Supplementary 1. Tables 1–7: primers used for qRT-PCR analysis of pinesap gene expression (1); assembled transcripts homologous to the known organ polarity genes in six pinesap transcriptomes (2); MhyYABBY characteristics (3); estimates of evolutionary divergence between nucleotide sequences of the MhyYABBY genes (4); estimates of evolutionary divergence between amino acid sequences of the MhyYABBY proteins (5); set of YABBY orthologues from different plant species used for MEME-mediated identification of clade-specific conserved motifs (6); the significance (p-value) of qRT-PCR results for one gene expression between pinesap tissues (7) (in Microsoft Excel worksheet file (.xls)).

Supplementary 2. Figures 1-2: structural analysis of the *MhyYABBY* genes and encoded proteins. (1) Alignment of *M. hypopitys YABBY* genes (a) and encoded putative proteins (b); (2) novel amino acid motifs predicted in the sequence of the MhyYABBY proteins. (a)–(g) Logos created from aligned sequences and copied directly from MEME graphically represent amino acid conservation: FIL-H (a); FIL-I (b); YAB5-Cm (c); CRC-E (d); INO-B (e); INO-C (f); and INO-D (g). The height of the letters in each stack indicates the relative frequency of individual residues at the position (in portable document format (.pdf)).

References

[1] A. A. Anderberg, C. Rydin, and M. Kallersjo, "Phylogenetic relationships in the order Ericales s.l.: analyses of molecular data from five genes from the plastid and mitochondrial genomes," *American Journal of Botany*, vol. 89, no. 4, pp. 677–687, 2002.

[2] J. R. Leake, "The biology of myco-heterotrophic ("saprophytic") plants," *New Phytologist*, vol. 127, no. 2, pp. 171–216, 1994.

[3] E. A. Devers, J. Teply, A. Reinert, N. Gaude, and F. Krajinski, "An endogenous artificial microRNA system for unraveling the function of root endosymbioses related genes in *Medicago truncatula*," *BMC Plant Biology*, vol. 13, no. 1, p. 82, 2013.

[4] G. D. Wallace, "Studies of the Monotropoidiae (Ericaceae): taxonomy and distribution," *The Wassman Journal of Biology*, vol. 33, pp. 1–88, 1975.

[5] V. S. F. T. Merckx, J. V. Freudenstein, J. Kissling et al., "Taxonomy and classification," in *Mycoheterotrophy*, V. S. F. T. Merckx, Ed., pp. 19–101, The Biology of Plants Living on Fungi. Springer Science+Buisness Media, New York, 2013.

[6] W. N. Stewart and G. W. Rothwell, *Paleobotany and the Evolution of Plants*, Cambridge University Press, New York, 1993.

[7] Q. C. B. Cronk, "Plant evolution and development in a post-genomic context," *Nature Reviews. Genetics*, vol. 2, no. 8, pp. 607–619, 2001.

[8] J. L. Bowman, Y. Eshed, and S. F. Baum, "Establishment of polarity in angiosperm lateral organs," *Trends in Genetics*, vol. 18, no. 3, pp. 134–141, 2002.

[9] D. Beerling and A. Fleming, "Zimmermann's telome theory of megaphyll leaf evolution: a molecular and cellular critique," *Current Opinion in Plant Biology*, vol. 10, no. 1, pp. 4–12, 2007.

[10] M. Snow and R. Snow, "The dorsiventrality of leaf primordia," *New Phytologist*, vol. 58, no. 2, pp. 188–207, 1959.

[11] J. R. McConnell and M. K. Barton, "Leaf polarity and meristem formation in *Arabidopsis*," *Development*, vol. 125, no. 15, pp. 2935–2942, 1998.

[12] K. Lynn, A. Fernandez, M. Aida et al., "The *PINHEAD/ZWILLE* gene acts pleiotropically in *Arabidopsis* development and has overlapping functions with the *ARGONAUTE1* gene," *Development*, vol. 126, no. 3, pp. 469–481, 1999.

[13] A. Y. Husbands, D. H. Chitwood, Y. Plavskin, and M. C. P. Timmermans, "Signals and prepatterns: new insights into organ polarity in plants," *Genes & Development*, vol. 23, no. 17, pp. 1986–1997, 2009.

[14] T. Yamaguchi, A. Nukazuka, and H. Tsukaya, "Leaf adaxial-abaxial polarity specification and lamina outgrowth: evolution and development," *Plant & Cell Physiology*, vol. 53, no. 7, pp. 1180–1194, 2012.

[15] X. Sun, Z. Feng, and L. Meng, "Ectopic expression of the *Arabidopsis ASYMMETRIC LEAVES2-LIKE5* (*ASL5*) gene in cockscomb (*Celosia cristata*) generates vascular-pattern modifications in lateral organs," *Plant Cell, Tissue and Organ Culture (PCTOC)*, vol. 110, no. 1, pp. 163–169, 2012.

[16] T. Huang, Y. Harrar, C. Lin et al., "*Arabidopsis* KANADI1 acts as a transcriptional repressor by interacting with a specific *cis*-element and regulates auxin biosynthesis, transport, and signaling in opposition to HD-ZIPIII factors," *The Plant Cell*, vol. 26, no. 1, pp. 246–262, 2014.

[17] P. Merelo, E. B. Paredes, M. G. Heisler, and S. Wenkel, "The shady side of leaf development: the role of the REVOLUTA/KANADI1 module in leaf patterning and auxin-mediated growth promotion," *Current Opinion in Plant Biology*, vol. 35, pp. 111–116, 2017.

[18] E. M. Meyerowitz, "Genetic control of cell division patterns in developing plants," *Cell*, vol. 88, no. 3, pp. 299–308, 1997.

[19] S. K. Floyd and J. L. Bowman, "Gene expression patterns in seed plant shoot meristems and leaves: homoplasy or homology?," *Journal of Plant Research*, vol. 123, no. 1, pp. 43–55, 2010.

[20] T. Yamada, S.'y. Yokota, Y. Hirayama, R. Imaichi, M. Kato, and C. S. Gasser, "Ancestral expression patterns and evolutionary diversification of *YABBY* genes in angiosperms," *The Plant Journal*, vol. 67, no. 1, pp. 26–36, 2011.

[21] J. L. Bowman, "The *YABBY* gene family and abaxial cell fate," *Current Opinion in Plant Biology*, vol. 3, no. 1, pp. 17–22, 2000.

[22] E. Kanaya, N. Nakajima, and K. Okada, "Non-sequence-specific DNA binding by the FILAMENTOUS FLOWER protein from *Arabidopsis thaliana* is reduced by EDTA," *The Journal of Biological Chemistry*, vol. 277, no. 14, pp. 11957–11964, 2002.

[23] C. Finet, S. K. Floyd, S. J. Conway, B. Zhong, C. P. Scutt, and J. L. Bowman, "Evolution of the *YABBY* gene family in seed plants," *Evolution & Development*, vol. 18, no. 2, pp. 116–126, 2016.

[24] H. L. Liu, Y. Y. Xu, Z. H. Xu, and K. Chong, "A rice *YABBY* gene, *OsYABBY4*, preferentially expresses in developing vascular tissue," *Development Genes and Evolution*, vol. 217, no. 9, pp. 629–637, 2007.

[25] R. Sarojam, P. G. Sappl, A. Goldshmidt et al., "Differentiating *Arabidopsis* shoots from leaves by combined *YABBY* activities," *Plant Cell*, vol. 22, no. 7, pp. 2113–2130, 2010.

[26] A. M. R. de Almeida, R. Yockteng, J. Schnable, E. R. Alvarez-Buylla, M. Freeling, and C. D. Specht, "Co-option of the polarity gene network shapes filament morphology in angiosperms," *Scientific Reports*, vol. 4, no. 1, 2014.

[27] K. Morioka, R. Yockteng, A. M. R. Almeida, and C. D. Specht, "Loss of *YABBY2*-like gene expression may underlie the evolution of the laminar style in Canna and contribute to floral morphological diversity in the Zingiberales," *Frontiers in Plant Science*, vol. 6, 2015.

[28] D. R. Kelley, D. J. Skinner, and C. S. Gasser, "Roles of polarity determinants in ovule development," *The Plant Journal*, vol. 57, no. 6, pp. 1054–1064, 2009.

[29] C. Bartholmes, O. Hidalgo, and S. Gleissberg, "Evolution of the *YABBY* gene family with emphasis on the basal eudicot

Eschscholzia californica (Papaveraceae)," *Plant Biology*, vol. 14, no. 1, pp. 11–23, 2012.

[30] A. Z. Worden, J.-H. Lee, T. Mock et al., "Green evolution and dynamic adaptations revealed by genomes of the marine picoeukaryotes Micromonas," *Science*, vol. 324, no. 5924, pp. 268–272, 2009.

[31] S. Sawa, K. Watanabe, K. Goto et al., "*FILAMENTOUS FLOWER*, a meristem and organ identity gene of *Arabidopsis*, encodes a protein with a zinc finger and HMG-related domains," *Genes & Development*, vol. 13, no. 9, pp. 1079–1088, 1999.

[32] J. F. Golz, M. Roccaro, R. Kuzoff, and A. Hudson, "GRAMINIFOLIA promotes growth and polarity of *Antirrhinum* leaves," *Development*, vol. 131, no. 15, pp. 3661–3670, 2004.

[33] J. L. Bowman and D. R. Smyth, "*CRABS CLAW*, a gene that regulates carpel and nectary development in *Arabidopsis*, encodes a novel protein with zinc finger and helix-loop-helix domains," *Development*, vol. 126, no. 11, pp. 2387–2396, 1999.

[34] Y. Eshed, S. F. Baum, and J. L. Bowman, "Distinct mechanisms promote polarity establishment in carpels of *Arabidopsis*," *Cell*, vol. 99, no. 2, pp. 199–209, 1999.

[35] J. Y. Lee, S. F. Baum, S. H. Oh, C. Z. Jiang, J. C. Chen, and J. L. Bowman, "Recruitment of CRABS CLAW to promote nectary development within the eudicot clade," *Development*, vol. 132, no. 22, pp. 5021–5032, 2005.

[36] S. Orashakova, M. Lange, S. Lange, S. Wege, and A. Becker, "The CRABS CLAW ortholog from California poppy *(Eschscholzia californica*, Papaveraceae*)*, EcCRC, is involved in floral meristem termination, gynoecium differentiation and ovule initiation," *The Plant Journal*, vol. 58, no. 4, pp. 682–693, 2009.

[37] H. Li, W. Liang, Y. Hu et al., "Rice MADS6 interacts with the floral homeotic genes *SUPERWOMAN1*, *MADS3*, *MADS58*, *MADS13*, and *DROOPING LEAF* in specifying floral organ identities and meristem fate," *Plant Cell*, vol. 23, no. 7, pp. 2536–2552, 2011.

[38] W. Sun, W. Huang, Z. Li, H. Lv, H. Huang, and Y. Wang, "Characterization of a *Crabs Claw* gene in basal eudicot species *Epimedium sagittatum* (Berberidaceae)," *International Journal of Molecular Sciences*, vol. 14, no. 1, pp. 1119–1131, 2013.

[39] S. C. Baker, K. Robinson-Beers, J. M. Villanueva, J. C. Gaiser, and C. S. Gasser, "Interactions among genes regulating ovule development in *Arabidopsis thaliana*," *Genetics*, vol. 145, no. 4, pp. 1109–1124, 1997.

[40] J. M. Villanueva, J. Broadhvest, B. A. Hauser, R. J. Meister, K. Schneitz, and C. S. Gasser, "INNER NO OUTER regulates abaxial-adaxial patterning in *Arabidopsis* ovules," *Genes & Development*, vol. 13, no. 23, pp. 3160–3169, 1999.

[41] M. T. Juarez, R. W. Twigg, and M. C. P. Timmermans, "Specification of adaxial cell fate during maize leaf development," *Development*, vol. 131, no. 18, pp. 4533–4544, 2004

[42] W. Zhao, H. Y. Su, J. Song, X. Y. Zhao, and X. S. Zhang, "Ectopic expression of *TaYAB1*, a member of *YABBY* gene family in wheat, causes the partial abaxialization of the adaxial epidermises of leaves and arrests the development of shoot apical meristem in *Arabidopsis*," *Plant Science*, vol. 170, no. 2, pp. 364–371, 2006.

[43] M. Dai, Y. Hu, Y. Zhao, and D. X. Zhou, "Regulatory networks involving *YABBY* genes in rice shoot development," *Plant Signaling & Behavior*, vol. 2, no. 5, pp. 399-400, 2007.

[44] Q. Chen, A. Atkinson, D. Otsuga, T. Christensen, L. Reynolds, and G. N. Drews, "The *Arabidopsis FILAMENTOUS FLOWER* gene is required for flower formation," *Development*, vol. 126, no. 12, pp. 2715–2726, 1999.

[45] B. Cong, L. S. Barrero, and S. D. Tanksley, "Regulatory change in YABBY-like transcription factor led to evolution of extreme fruit size during tomato domestication," *Nature Genetics*, vol. 40, no. 6, pp. 800–804, 2008.

[46] H. Nakayama, T. Yamaguchi, and H. Tsukaya, "Expression patterns of AaDL, a CRABS CLAW ortholog in *Asparagus asparagoides (Asparagaceae)*, demonstrate a stepwise evolution of *CRC/DL* subfamily of *YABBY* genes," *American Journal of Botany*, vol. 97, no. 4, pp. 591–600, 2010.

[47] S. Sawa, T. Ito, Y. Shimura, and K. Okada, "*FILAMENTOUS FLOWER* controls the formation and development of *Arabidopsis* inflorescences and floral meristems," *Plant Cell*, vol. 11, no. 1, pp. 69–86, 1999.

[48] M. I. Bidartondo and T. D. Bruns, "Extreme specificity in epiparasitic Monotropoideae (Ericaceae): widespread phylogenetic and geographical structure," *Molecular Ecology*, vol. 10, no. 9, pp. 2285–2295, 2001.

[49] M. A. Filyushin, N. M. Reshetnikova, E. Z. Kochieva, and K. G. Skryabin, "Polymorphism of sequences and the secondary structure of b/c intron of mitochondrial gene *nad1* in *Monotropa hypopitys* and related Ericaceae species," *Biology Bulletin*, vol. 43, no. 3, pp. 271–275, 2016.

[50] M. A. Filyushin, E. Z. Kochieva, and K. G. Skryabin, "Polymorphism of the chloroplast gene *rps2* in parasitic plant *Monotropa hypopitys* L. from the European Russian populations," *Russian Journal of Genetics*, vol. 53, no. 3, pp. 400–405, 2017.

[51] N. V. Ravin, E. V. Gruzdev, A. V. Beletsky et al., "The loss of photosynthetic pathways in the plastid and nuclear genomes of the non-photosynthetic mycoheterotrophic eudicot *Monotropa hypopitys*," *BMC Plant Biology*, vol. 16, Suppl 3, p. 238, 2016.

[52] G. E. Beatty and J. Provan, "High clonal diversity in threatened peripheral populations of the yellow bird's nest (*Hypopitys monotropa*; syn. *Monotropa hypopitys*)," *Annals of Botany*, vol. 107, no. 4, pp. 663–670, 2011.

[53] A. V. Beletsky, M. A. Filyushin, E. V. Gruzdev et al., "De novo transcriptome assembly of the mycoheterotrophic plant *Monotropa hypopitys*," *Genomics Data*, vol. 11, pp. 60-61, 2017.

[54] B. Langmead and S. L. Salzberg, "Fast gapped-read alignment with Bowtie 2," *Nature Methods*, vol. 9, no. 4, pp. 357–359, 2012.

[55] J. C. Oliveros and Venny, "An interactive tool for comparing lists with Venn's diagrams (2007–2015)," http://bioinfogp.cnb.csic.es/tools/venny/index.html.

[56] S. Babicki, D. Arndt, A. Marcu et al., "Heatmapper: web-enabled heat mapping for all," *Nucleic Acids Research*, vol. 44, no. W1, pp. W147–W153, 2016.

[57] T. L. Bailey and C. Elkan, "Fitting a mixture model by expectation maximization to discover motifs in biopolymers," *Proceedings, International Conference on Intelligent Systems for Molecular Biology*, vol. 2, pp. 28–36, 1994.

[58] M. A. Larkin, G. Blackshields, N. P. Brown et al., "Clustal W and Clustal X version 2.0," *Bioinformatics*, vol. 23, no. 21, pp. 2947-2948, 2007.

[59] F. Tajima and M. Nei, "Estimation of evolutionary distance between nucleotide sequences," *Molecular Biology and Evolution*, vol. 1, no. 3, pp. 269–285, 1984.

[60] K. Tamura and S. Kumar, "Evolutionary distance estimation under heterogeneous substitution pattern among lineages," *Molecular Biology and Evolution*, vol. 19, no. 10, pp. 1727–1736, 2002.

[61] K. Tamura, M. Nei, and S. Kumar, "Prospects for inferring very large phylogenies by using the neighbor-joining method," *Proceedings of the National Academy of Sciences*, vol. 101, no. 30, pp. 11030–11035, 2004.

[62] S. Kumar, G. Stecher, and K. Tamura, "MEGA7: molecular evolutionary genetics analysis version 7.0 for bigger datasets," *Molecular Biology and Evolution*, vol. 33, no. 7, pp. 1870–1874, 2016.

[63] B. Li and C. N. Dewey, "RSEM: accurate transcript quantification from RNA-Seq data with or without a reference genome," *BMC Bioinformatics*, vol. 12, no. 1, p. 323, 2011.

[64] R. G. Franks, Z. Liu, and R. L. Fischer, "SEUSS and LEUNIG regulate cell proliferation, vascular development and organ polarity in *Arabidopsis* petals," *Planta*, vol. 224, no. 4, pp. 801–811, 2006.

[65] S. Nole-Wilson and B. A. Krizek, "AINTEGUMENTA contributes to organ polarity and regulates growth of lateral organs in combination with YABBY genes," *Plant Physiology*, vol. 141, no. 3, pp. 977–987, 2006.

[66] S. M. Doyle, A. Haeger, T. Vain et al., "An early secretory pathway mediated by GNOM-LIKE 1 and GNOM is essential for basal polarity establishment in *Arabidopsis thaliana*," *Proceedings of the National Academy of Sciences of the United States of America*, vol. 112, no. 7, pp. E806–E815, 2015.

[67] H. R. Pires, M. M. Monfared, E. A. Shemyakina, and J. C. Fletcher, "ULTRAPETALA trxG genes interact with KANADI transcription factor genes to regulate *Arabidopsis* gynoecium patterning," *Plant Cell*, vol. 26, no. 11, pp. 4345–4361, 2014.

[68] J. Bou-Torrent, M. Salla-Martret, R. Brandt et al., "ATHB4 and HAT3, two class II HD-ZIP transcription factors, control leaf development in *Arabidopsis*," *Plant Signaling & Behavior*, vol. 7, no. 11, pp. 1382–1387, 2012.

[69] N. Raikhel, "Nuclear targeting in plants," *Plant Physiology*, vol. 100, no. 4, pp. 1627–1632, 1992.

[70] R. G. H. Immink, I. A. N. Tonaco, S. de Folter et al., "SEPALLATA3: the "glue" for MADS box transcription factor complex formation," *Genome Biology*, vol. 10, no. 2, p. R24, 2009.

[71] S. Wicke, K. F. Müller, C. W. dePamphilis, D. Quandt, S. Bellot, and G. M. Schneeweiss, "Mechanistic model of evolutionary rate variation en route to a nonphotosynthetic lifestyle in plants," *Proceedings of the National Academy of Sciences of the United States of America*, vol. 113, no. 32, pp. 9045–9050, 2016.

[72] S. W. Graham, V. K. Y. Lam, and V. S. F. T. Merckx, "Plastomes on the edge: the evolutionary breakdown of mycoheterotroph plastid genomes," *New Phytologist*, vol. 214, no. 1, pp. 48–55, 2017.

[73] K. A. Kron, W. S. Judd, P. F. Stevens et al., "Phylogenetic classification of Ericaceae: molecular and morphological evidence," *Botanical Review*, vol. 68, no. 3, pp. 335–423, 2002.

[74] T. W. A. Braukmann, M. B. Broe, S. Stefanović, and J. V. Freudenstein, "On the brink: the highly reduced plastomes of nonphotosynthetic Ericaceae," *New Phytologist*, vol. 216, no. 1, pp. 254–266, 2017.

[75] E. V. Gruzdev, A. V. Mardanov, A. V. Beletsky, E. Z. Kochieva, N. V. Ravin, and K. G. Skryabin, "The complete chloroplast genome of parasitic flowering plant *Monotropa hypopitys*: extensive gene losses and size reduction," *Mitochondrial DNA part B Resources*, vol. 1, no. 1, pp. 212-213, 2016.

[76] M. D. Logacheva, M. I. Schelkunov, V. Y. Shtratnikova, M. V. Matveeva, and A. A. Penin, "Comparative analysis of plastid genomes of non-photosynthetic Ericaceae and their photosynthetic relatives," *Scientific Reports*, vol. 6, no. 1, article 30042, 2016.

[77] M. I. Bidartondo, "The evolutionary ecology of myco-heterotrophy," *New Phytologist*, vol. 167, no. 2, pp. 335–352, 2005.

[78] V. Buchanan-Wollaston, T. Page, E. Harrison et al., "Comparative transcriptome analysis reveals significant differences in gene expression and signalling pathways between developmental and dark/starvation-induced senescence in *Arabidopsis*," *The Plant Journal*, vol. 42, no. 4, pp. 567–585, 2005.

[79] H. Zhang and C. Zhou, "Signal transduction in leaf senescence," *Plant Molecular Biology*, vol. 82, no. 6, pp. 539–545, 2013.

[80] H. Tsukaya, "Leaf shape: genetic controls and environmental factors," *The International Journal of Developmental Biology*, vol. 49, no. 5-6, pp. 547–555, 2005.

[81] H. Tsukaya, "Mechanism of leaf-shape determination," *Annual Review of Plant Biology*, vol. 57, no. 1, pp. 477–496, 2006.

[82] N. A. Eckardt, "*YABBY* genes and the development and origin of seed plant leaves," *Plant Cell*, vol. 22, no. 7, p. 2103, 2010.

[83] S. Mathews and E. M. Kramer, "The evolution of reproductive structures in seed plants: a re-examination based on insights from developmental genetics," *The New Phytologist*, vol. 194, no. 4, pp. 910–923, 2012.

[84] K. R. Siegfried, Y. Eshed, S. F. Baum, D. Otsuga, G. N. Drews, and J. L. Bowman, "Members of the *YABBY* gene family specify abaxial cell fate in *Arabidopsis*," *Development*, vol. 126, no. 18, pp. 4117–4128, 1999.

[85] Y. Wang, H. Huang, Y. Ma, J. Fu, L. Wang, and S. Dai, "Construction and de novo characterization of a transcriptome of *Chrysanthemum lavandulifolium*: analysis of gene expression patterns in floral bud emergence," *Plant Cell, Tissue and Organ Culture (PCTOC)*, vol. 116, no. 3, pp. 297–309, 2014.

[86] H. Q. Han, Y. Liu, M. M. Jiang, H. Y. Ge, and H. Y. Chen, "Identification and expression analysis of *YABBY* family genes associated with fruit shape in tomato (*Solanum lycopersicum* L.)," *Genetics and Molecular Research*, vol. 14, no. 2, pp. 7079–7091, 2015.

[87] J. Kim, J. Yang, R. Yang, R. C. Sicher, C. Chang, and M. L. Tucker, "Transcriptome analysis of soybean leaf abscission identifies transcriptional regulators of organ polarity and cell fate," *Frontiers in Plant Science*, vol. 7, 2016.

[88] C. Machida, A. Nakagawa, S. Kojima, H. Takahashi, and Y. Machida, "The complex of ASYMMETRIC LEAVES (AS) proteins plays a central role in antagonistic interactions of genes for leaf polarity specification in *Arabidopsis*," *Wiley Interdisciplinary Reviews: Developmental Biology*, vol. 4, no. 6, pp. 655–671, 2015.

[89] O. A. Shulga, A. V. Shchennikova, A. V. Beletsky et al., "Transcriptome-wide characterization of the MADS-box family in pinesap *Monotropa hypopitys* reveals flowering conservation in non-photosynthetic myco-heterotrophs," *Journal of Plant Growth Regulation*, pp. 1–16, 2017.

[90] J. F. Emery, S. K. Floyd, J. Alvarez et al., "Radial patterning of *Arabidopsis* shoots by class III HD-ZIP and KANADI genes," *Current Biology*, vol. 13, no. 20, pp. 1768–1774, 2003.

[91] M. K. Simon, D. J. Skinner, T. L. Gallagher, and C. S. Gasser, "Integument development in *Arabidopsis* depends on interaction of YABBY protein INNER NO OUTER with co-activators and co-repressors," *Genetics*, vol. 207, no. 4, pp. 1489–1500, 2017.

[92] Y. Eshed, A. Izhaki, S. F. Baum, S. K. Floyd, and J. L. Bowman, "Asymmetric leaf development and blade expansion in *Arabidopsis* are mediated by KANADI and YABBY activities," *Development*, vol. 131, no. 12, pp. 2997–3006, 2004.

[93] M. I. Stahle, J. Kuehlich, L. Staron, A. G. von Arnim, and J. F. Golz, "YABBYs and the transcriptional corepressors LEUNIG and LEUNIG_HOMOLOG maintain leaf polarity and meristem activity in *Arabidopsis*," *Plant Cell*, vol. 21, no. 10, pp. 3105–3118, 2009.

[94] O. Bonaccorso, J. E. Lee, L. Puah, C. P. Scutt, and J. F. Golz, "FILAMENTOUS FLOWER controls lateral organ development by acting as both an activator and a repressor," *BMC Plant Biology*, vol. 12, no. 1, p. 176, 2012.

[95] C. La Rota, J. Chopard, P. Das et al., "A data-driven integrative model of sepal primordium polarity in *Arabidopsis*," *Plant Cell*, vol. 23, no. 12, pp. 4318–4333, 2011.

[96] I. Pekker, J. P. Alvarez, and Y. Eshed, "Auxin response factors mediate *Arabidopsis* organ asymmetry via modulation of KANADI activity," *Plant Cell*, vol. 17, no. 11, pp. 2899–2910, 2005.

[97] B. J. Reinhart, T. Liu, N. R. Newell et al., "Establishing a framework for the ad/abaxial regulatory network of *Arabidopsis*: ascertaining targets of class III homeodomain leucine zipper and KANADI regulation," *Plant Cell*, vol. 25, no. 9, pp. 3228–3249, 2013.

[98] M. Nakata, N. Matsumoto, R. Tsugeki, E. Rikirsch, T. Laux, and K. Okada, "Roles of the middle domain-specific *WUSCHEL-RELATED HOMEOBOX* genes in early development of leaves in *Arabidopsis*," *Plant Cell*, vol. 24, no. 2, pp. 519–535, 2012.

[99] H. Iwakawa, M. Iwasaki, S. Kojima et al., "Expression of the *ASYMMETRIC LEAVES2* gene in the adaxial domain of *Arabidopsis* leaves represses cell proliferation in this domain and is critical for the development of properly expanded leaves," *The Plant Journal*, vol. 51, no. 2, pp. 173–184, 2007.

[100] B. Xu, Z. Li, Y. Zhu et al., "*Arabidopsis* genes AS1, AS2, and JAG negatively regulate boundary specifying genes to promote sepal and petal development," *Plant Physiology*, vol. 146, no. 2, pp. 566–575, 2008.

[101] M. Iwasaki, H. Takahashi, H. Iwakawa et al., "Dual regulation of *ETTIN* (*ARF3*) gene expression by AS1-AS2, which maintains the DNA methylation level, is involved in stabilization of leaf adaxial-abaxial partitioning in *Arabidopsis*," *Development*, vol. 140, no. 9, pp. 1958–1969, 2013.

[102] N. Prunet, P. Morel, A. M. Thierry et al., "REBELOTE, SQUINT, and ULTRAPETALA1 function redundantly in the temporal regulation of floral meristem termination in *Arabidopsis thaliana*," *Plant Cell*, vol. 20, no. 4, pp. 901–919, 2008.

[103] A. V. Shchennikova, O. A. Shulga, E. Z. Kochieva et al., "Homeobox genes encoding WOX transcription factors in the flowering parasitic plant *Monotropa hypopitys*," *Russian Journal of Genetics: Applied Research*, vol. 7, no. 7, pp. 781–788, 2017.

[104] Y. Matsumura, H. Iwakawa, Y. Machida, and C. Machida, "Characterization of genes in the ASYMMETRIC LEAVES2/LATERAL ORGAN BOUNDARIES (AS2/LOB) family in *Arabidopsis thaliana*, and functional and molecular comparisons between AS2 and other family members," *The Plant Journal*, vol. 58, no. 3, pp. 525–537, 2009.

[105] E. Semiarti, Y. Ueno, H. Tsukaya, H. Iwakawa, C. Machida, and Y. Machida, "The *ASYMMETRIC LEAVES2* gene of *Arabidopsis thaliana* regulates formation of a symmetric lamina, establishment of venation and repression of meristem-related homeobox genes in leaves," *Development*, vol. 128, no. 10, pp. 1771–1783, 2001.

[106] M. J. Prigge and S. E. Clark, "Evolution of the class III HD-Zip gene family in land plants," *Evolution & Development*, vol. 8, no. 4, pp. 350–361, 2006.

[107] A. S. Chanderbali, F. He, P. S. Soltis, and D. E. Soltis, "Out of the water: origin and diversification of the LBD gene family," *Molecular Biology and Evolution*, vol. 32, no. 8, pp. 1996–2000, 2015.

[108] T. Toriba, K. Harada, A. Takamura et al., "Molecular characterization the *YABBY* gene family in *Oryza sativa* and expression analysis of *OsYABBY1*," *Molecular Genetics and Genomics*, vol. 277, no. 5, pp. 457–468, 2007.

[109] F. Vandelook and J. A. Van Assche, "Temperature conditions control embryo growth and seed germination of *Corydalis solida* (L.) Clairv., a temperate forest spring geophyte," *Plant Biology (Stuttgart, Germany)*, vol. 11, no. 6, pp. 899–906, 2009.

[110] J. R. Leake, S. L. McKendrick, M. Bidartondo, and D. J. Read, "Symbiotic germination and development of the myco-heterotroph *Monotropa hypopitys* in nature and its requirement for locally distributed *Tricholoma* spp," *New Phytologist*, vol. 163, no. 2, pp. 405–423, 2004.

[111] C. Fourquin, A. Primo, I. Martínez-Fernández, E. Huet-Trujillo, and C. Ferrándiz, "The CRC orthologue from *Pisum sativum* shows conserved functions in carpel morphogenesis and vascular development," *Annals of Botany*, vol. 114, no. 7, pp. 1535–1544, 2014.

[112] M. K. Simon, L. A. Williams, K. Brady-Passerini, R. H. Brown, and C. S. Gasser, "Positive- and negative-acting regulatory elements contribute to the tissue-specific expression of INNER NO OUTER, a YABBY-type transcription factor gene in *Arabidopsis*," *BMC Plant Biology*, vol. 12, no. 1, p. 214, 2012.

[113] M. Boter, J. F. Golz, S. Giménez-Ibañez, G. Fernandez-Barbero, J. M. Franco-Zorrilla, and R. Solano, "FILAMENTOUS FLOWER is a direct target of JAZ3 and modulates responses to Jasmonate," *Plant Cell*, vol. 27, no. 11, pp. 3160–3174, 2015.

6

Identification of Novel *ARSA* Mutations in Chinese Patients with Metachromatic Leukodystrophy

Li Chen,[1,2] Huifang Yan,[1] Binbin Cao,[1] Ye Wu,[1] Qiang Gu,[1] Jiangxi Xiao,[3] Yanling Yang,[1] Huixia Yang,[4] Zhen Shi,[1] Zhixian Yang,[1] Hong Pan,[5] Xingzhi Chang,[1] Junya Chen,[4] Yu Sun,[4] Yuehua Zhang,[1] Xiru Wu,[1] Yuwu Jiang,[1] and Jingmin Wang[1]

[1]*Department of Pediatrics, Peking University First Hospital, Beijing 100034, China*
[2]*Department of Pediatric Cardiology, Beijing Anzhen Hospital, Capital Medical University, Beijing 100029, China*
[3]*Department of Radiology, Peking University First Hospital, Beijing 100034, China*
[4]*Department of Obstetrics and Gynecology, Peking University First Hospital, Beijing 100034, China*
[5]*Department of Central Laboratory, Peking University First Hospital, Beijing 100034, China*

Correspondence should be addressed to Jingmin Wang; wang66jm@163.com

Academic Editor: Margarita Hadzopoulou-Cladaras

Objective. Metachromatic leukodystrophy (MLD) is an inherited disease caused by a deficiency of the enzyme arylsulfatase A (ARSA) that leads to severe physiologic and developmental problems. Our study is aimed at elucidating the clinical and genetic characteristics of Chinese MLD patients. *Methods.* Clinical data of 21 MLD patients was collected. All coding exons of *ARSA* and their flanking intronic sequences were amplified by polymerase chain reaction and subjected to direct sequencing. *Results.* All 21 patients were diagnosed with MLD clinically and genetically, out of which 17 patients were late infantile and 4 were juvenile types. A total of 34 *ARSA* mutations, including 28 novel mutations (22 missense, 1 splicing, 1 nonsense, 3 small insertions, and 1 small deletion mutation) and 6 known mutations (5 missense and 1 small insertion mutation), were identified. Prenatal diagnosis was performed for four pedigrees. One fetus was a patient, two fetuses were carriers, and two were wild type. *Conclusions.* The present study discovered 28 novel *ARSA* mutations and widely expanded the mutation spectrum of *ARSA*. Four successful prenatal diagnoses provided critical information for MLD families.

1. Introduction

Metachromatic leukodystrophy (MLD) is a genetic disorder caused by the deficiency of the enzyme arylsulfatase A (ARSA) in lipid metabolism [1–3]. It is estimated that the overall incidence of autosomal recessive MLD is 1 : 40,000–1 : 100,000 [1]. The classical symptoms are presented mainly as progressive physical and mental deterioration, clumsiness, frequent falls, toe walking, slurred speech, weakness, hypotonia, and seizures. MLD patients could be classified into three subtypes, including late infantile, juvenile, and adult forms based on the age of the first symptom onset [1]. The late infantile form usually manifests in the second year of life. The juvenile form, with an onset between 4 and 15 years, is further subdivided into early juvenile and late juvenile depending on whether the onset is before or after 6 years of age. The adult form has an onset older than 15 years of age [1, 4–6]. Magnetic resonance imaging and biochemical assays are commonly used for diagnosis [7–9]. Measuring ARSA enzymatic activity in leukocytes from whole blood is the standard biochemical procedure [10]. However, diagnosis of MLD would be based on *ARSA* diagnosis. *ARSA* is located on chromosome 22q13 containing eight exons and is transcribed into three mRNA species with a total length of 3.2 kb. Currently, the Human Gene Mutation Database (http://www.hgmd.cf.ac.uk/ac/gene.php?gene=ARSA) has reported a total of 217 *ARSA* mutations. Among them are c.459+1G>A and c.1277C>T (pPro426Leu), which occur more frequently in the European population with over 200 reported mutations in MLD patients [10]. The former causes a splice donor site mutation, which occurs frequently in late infantile patients, whereas the latter appears in adult or juvenile patients [10]. In addition to pathogenic mutations, an ARSA pseudodeficiency allele, such as c.1049A>G, leads to lower ARSA activity, which results in a partial mistargeting of the enzyme [11]. The aim of this study was to elucidate the clinical and genetic characteristics of Chinese MLD patients. Additionally, for MLD pedigrees diagnosed genetically, we performed prenatal diagnosis.

2. Materials and Methods

2.1. Patients. Twenty-one MLD patients (Pt1-Pt21) from twenty-one unrelated families were enrolled in this study at the Department of Pediatrics at Peking University First Hospital. All patients were clinically diagnosed with MLD based on the manifestation, classical MRI feature, and ARSA enzyme deficiency in leukocytes. Out of 21 patients, thirteen patients were male (61.90%) and eight patients were female (38.10%). The median age was 32 months (minimum 20 months, maximum 11 years old). Clinical characteristics of patients were collected, such as onset age, symptoms, neurological findings, and brain MRI findings. Our study was conducted with the patients' understanding and consent. The study was approved by the Ethics Committee of Peking University First Hospital, and informed consent was obtained from the patients' legal guardians.

2.2. Biochemical Studies. Patient studies were performed using low-temperature assays of arylsulfatase A activity determination in leukocytes [12], and the patients demonstrated low ARSA activity (<10% normal range).

2.3. Mutation Analysis. Genomic DNA was extracted from peripheral blood leukocytes of patients and their parents according to the standard protocols of the QIAamp DNA blood mini kit (Qiagene Inc.). Using reported PCR primers and annealing temperatures [13], *ARSA* was amplified and the PCR products were purified by a DNA purification kit. PCR conditions were as follows: 100 ng of patient DNA was amplified in 10 μL of buffer containing 0.5 μL of 5 mmol/L dNTP, 50 ng of each primer, 5 μL of 2x GC buffer (Takara, Dalian, China), and 1 μL of Taq DNA polymerase (Tiangen, Beijing, China). Then, the PCR products were subject to an automated sequencer. The samples collected before 2013 were sequenced by the Sanger sequencing method, whereas samples collected after 2013 were sequenced by PANO-seq. The putative mutations were confirmed using duplicate PCR products or digested PCR products. The effect of the amino acid changes in ARSA was predicted by the web server PolyPhen-2 with the HumDiv model (http://genetics.bwh.harvard.edu/pph2/).

For prenatal diagnosis, chorionic villus biopsy (CVS) or amniocentesis was performed for fetuses in four MLD families (Pt5, Pt7, Pt15, and Pt18) to obtain samples. A Promega Wizard® Genomic DNA Purification Kit (A1120) was used to extract the genomic DNA.

2.4. RNA Isolation from Peripheral Blood Cells and Detection of ARSA mRNA. Total RNA was prepared from blood cells using the TRIzol reagent (Invitrogen, Waltham, MA, USA) according to the manufacturer's instructions. Five hundred nanograms of RNA was reverse transcribed using oligo d(T)$_{20}$, and the complementary DNA was added to the quantitative real-time PCR assays.

3. Results

3.1. Clinical Finding. Clinical data were collected and are shown in Table 1. Ages at onset ranged from 6 months to 5 years old, with a median age of 25 months. Out of 21 patients, the main complaints from the 10 patients were motor regression, 4 with both motor and intelligence regression, 6 showing development retardation, and 1 with abnormal walking posture. Symmetrical abnormalities in deep white matter (WM) (Figure 1) were observed in MRI images in all 21 patients. ARSA activity was measured in 14 patients, all of which were lower than 10% of the normal range. A Babinski sign was positive in all patients except Pt1, Pt5, and Pt8.

All patients were clinically diagnosed with MLD according to the clinical manifestations described above (Table 1). Out of 21 patients, 17 late infantile-type and 4 juvenile-type patients were diagnosed based on their onset age of younger than 30 mon and older than 30 mon, respectively.

3.2. Genetic Findings. Thirty-four mutations, including 28 novel and 6 known mutations, were found in *ARSA* (Table 2). The 28 novel mutations included 22 missense mutations, 3 small insertions, 1 splicing mutation, 1 small deletion, and 1 nonsense mutation. Among these novel mutations, the mutation c.1130_1132delTCT caused a three base pair deletion in *ARSA*, while the mutation c.954G>A, pTrp318Term produced a premature termination code, the mutations c.1344_1345insCC, c.302_303insG, and c.1428_1429insC caused one or two base pair insertion in *ARSA*, and the mutations c.1108-20A>G, c.465G>A (p.Lys125ProfsX17), and IVS3+2T>C led to splicing and amino acid changes in the protein. Mutations c.1172T>G (p.Val391Gly), c.827C>T (p.Thr276Met), c.925G>A (pGlu309Lys), and c.1130_1132delTCT were all detected in 2 patients, respectively. After predicting the effect of the amino acid change in ARSA by the web server PolyPhen-2 with the HumDiv model, we found that 28 mutations were probably damaging to the protein activity with a high score and specificity. Six known mutations have been reported, including c.917C>T, c.827C>T, c.179_180insCA, c.257G>A, c.925G>A, and c.302G>T.

Out of 21 patients, 2 (Pt2 and Pt3) harbored homozygous mutations in *ARSA* and 19 had compound heterozygous mutations in *ARSA*. Two patients had homozygous mutations of *ARSA* that were inherited from their parents. All 19 patients with compound heterozygous mutations of *ARSA* had mutations that were inherited from their parents, except for Pt14 and Pt19. The *ARSA* c.465G>A (p.Lys125ProfsX17) mutation in Pt14 was inherited from the father with a heterozygous variation on one allele and the mother with a wild-

Table 1: Summary of major clinical features in 21 Chinese MLD patients.

Pt. ID	Sex	Age onset	Family history	Symptoms at onset	Neurological findings	ARSA enzyme activity	Brain MRI findings
1	Female	5 yrs	–	Psychomotor deterioration, motor regression, mental deterioration	Spastic paraparesis	+	Symmetrical deep WM abnormalities
2	Male	3 yrs	–	Psychomotor deterioration, motor regression	–		Symmetrical deep WM abnormalities
3	Male	6 mon	–	Motor retardation, nystagmus	Nystagmus, peripheral neuropathy		Symmetrical deep WM abnormalities and basal ganglia changes
4	Male	17 mon	–	Psychomotor deterioration, motor regression	Peripheral neuropathy		Symmetrical deep WM abnormalities
5	Male	18 mon	–	Psychomotor deterioration, motor regression	Peripheral neuropathy	+	Symmetrical deep WM abnormalities
6	Female	4 yrs	–	Psychomotor deterioration, motor regression	Dysarthria	+	Symmetrical deep WM abnormalities
7	Male	14 mon	–	Psychomotor deterioration, motor regression, strabismus, nystagmus	Peripheral neuropathy	+	Symmetrical deep WM abnormalities
8	Male	15 mon	–	Psychomotor deterioration, motor regression	Peripheral neuropathy		Symmetrical deep WM abnormalities
9	Female	6 mon	–	Motor retardation	–		Symmetrical deep WM abnormalities
10	Male	16 mon	–	Psychomotor deterioration, motor regression, mental deterioration	Peripheral neuropathy		Symmetrical deep WM abnormalities
11	Female	16 mon	–	Psychomotor deterioration, motor regression	Peripheral neuropathy	+	Symmetrical deep WM abnormalities
12	Female	41 mon	–	Psychomotor deterioration, motor regression	Peripheral neuropathy		Symmetrical deep WM abnormalities
13	Female	7 mon	–	Motor retardation	Peripheral neuropathy	+	Symmetrical deep WM abnormalities
14	Male	18 mon	–	Psychomotor deterioration, motor regression, mental deterioration, strabismus	Peripheral neuropathy	+	Symmetrical deep WM abnormalities
15	Male	27 mon	–	Psychomotor deterioration, motor regression	Peripheral neuropathy	+	Symmetrical deep WM abnormalities
16	Female	28 mon	–	Psychomotor deterioration, motor regression	Peripheral neuropathy	+	Symmetrical deep WM abnormalities
17	Female	12 mon	–	Psychomotor deterioration, motor regression	Peripheral neuropathy	+	Symmetrical deep WM abnormalities
18	Male	12 mon	–	Psychomotor deterioration, motor regression	Peripheral neuropathy	+	Symmetrical deep WM abnormalities
19	Male	15 mon	–	Psychomotor deterioration, motor regression, nystagmus	Nystagmus, peripheral neuropathy	+	Symmetrical deep WM abnormalities
20	Male	12 mon	–	Psychomotor deterioration, motor regression	Dysarthria, peripheral neuropathy	+	Symmetrical deep WM abnormalities
21	Male	15 mon	–	Psychomotor deterioration, motor regression	Peripheral neuropathy	+	Symmetrical deep WM abnormalities

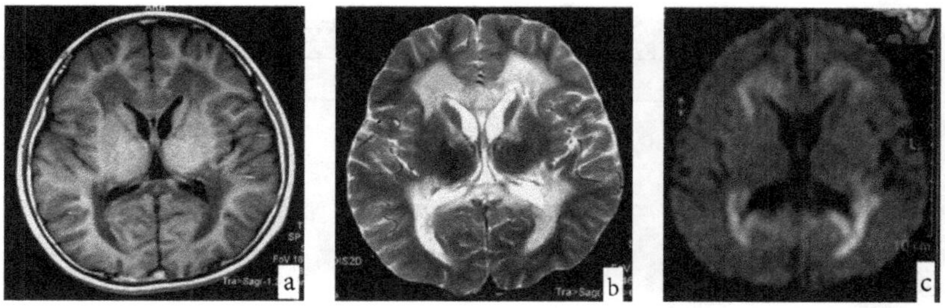

(a) (b) (c)

FIGURE 1: Magnetic resonance imaging (MRI) shows symmetrical deep lesions located in periventricular white matter, which was low signal in T1WI (a), high signal in T2WI (b), and high signal in DWI (c).

type mutation on the other allele. Pt19 has spontaneous splicing mutation IVS3+2T>C.

In Pt5, Pt7, Pt15, and Pt18 MLD family genetic diagnoses, prenatal diagnosis was performed for fetuses after the second or third pregnancy (Table 3). All samples were tested by a short tandem repeat (STR) linkage analysis to exclude contamination from the mother's tissue. The results showed that the fetuses had wild-type *ARSA* in the Pt5 (c.917C>T and c.827C>T) and Pt15 (c.44G>T and c.610C>G) cases. While one fetus in the Pt7 MLD family had a compound heterozygous mutation (c.1130_1132delTCT and c.1238A>G), another fetus in Pt7 carried one heterozygous variation (c.1238A>G (p.Asp413Gly)). The compound heterozygous mutations c.302_303insG and c.1428_1429insC were detected in fetuses of Pt18's mother.

4. Discussion

MLD is a kind of lysosomal storage disorder due to the deficiency of the ARSA enzyme, which is involved in the metabolism of membrane sulfatides into galactosylceramide. Progressive demyelination and dysfunction of the peripheral and central nervous systems is the symptom of this disease as the undegraded sulfatides require time to accumulate in oligodendrocytes and Schwann cells. In this study, 21 patients presented typical clinical symptoms, including motor regression, motor and intelligence regression, development retardation, and abnormal walking posture. They all had typical brain MRI findings. Fourteen patients had lower than 10% normal activity of the ARSA outcome. According to the above clinical manifestations described, all patients were clinically diagnosed as MLD. All the patients' ages at onset ranged from 6 months to 5 years old. Out of 21 patients, 17 patients of younger than 24 mon were diagnosed as the late infantile type and the remaining 4 patients as juvenile type.

ARSA, the disease-causing gene of MLD, is located on chromosome 22q13, has a total length of 3.2 kb, contains eight exons, and is transcribed into three mRNA species. To date, the Human Gene Mutation Database (http://www.hgmd.cf.ac.uk/ac/gene.php?gene=ARSA) has reported a total of 217 *ARSA* mutations, including 161 missense mutations, 14 splicing-site mutations, 20 small deletions, 12 small insertions, 4 small indels, 2 gross deletions, 2 complex rearrangements, 1 gross insertion, and 1 regulatory mutation. Here, 34 *ARSA* mutations were identified, which included 28 novel and 6 known mutations [4–9, 14]. Because most mutations in our study are novel mutations, the diagnosis of MLD should be confirmed not only by low ARSA activity but also by increased sulfatiduria. In our study, 28 novel mutations, including 22 missense mutations, 3 small insertions, 1 splicing mutation, 1 small deletion, and 1 nonsense mutation, and 6 reported mutations (c.917C>T, c.827C>T, c.179_180insCA, c.257G>A, c.925G>A, and c.302G>T) were found [4–9, 14].

For the 28 novel mutations, we predicted the effect of amino acid changes of the 22 missense mutations in ARSA by the web server PolyPhen-2 with the HumDiv model (Table 4). The scores of mutations, including 251C>A (pPro84Gln), 427T>C (p.Phe143Leu), 640G>A (p.Ala214Thr), and 754T>C (p.Ser252Pro), were 0.991–0.999. Therefore, we predicted that they could cause damage to the function of the ARSA protein. The mutational sensitivity of 427T>C (p.Phe143Leu), 640G>A (p.Ala214Thr), and 754T>C (p.Ser252Pro) was 0.14, 0.69, and 0.55, respectively. All the scores of the other missense mutations were 1, with a sensitivity of 0 and specificity of 1. They included the mutation of the terminal C domain. Therefore, we predicted that they could destroy the function of the ARSA protein. The mutation c.1130_1132delTCT caused a three base pair deletion in *ARSA*, and the mutation c.954G>A, pTrp318Term produced a premature termination codon, while the mutations c.1344_1345insCC, c.302_303insG, and c.1428_1429insC inserted one or two base pairs in *ARSA*. We predicted that the mutations c.1108-20A>G, c.465G>A (p.Lys125ProfsX17), and IVS3+2T>C resulted in splicing, which could cause damage to the function of ARSA protein. Based on this analysis, the 28 mutations are probably damaging to protein activity. All 6 known mutations, including c.917C>T, c.827C>T, c.179_180insCA, c.257G>A, c.925G>A, and c.302G>T, could destroy the function of the ARSA protein and were considered disease-causing mutations.

Through further analysis, we found that mutations c.251C>A, c.1049T>A, c.244C>T, c.296G>T c.911A>T, c.1238A>G, and c.925G>A lead to changes of the chemical properties in residues. The crystal structure of ARSA and the substrate also indicated that p.Lys304 is one of the key residues that interacted with the ester oxygen atom (O_4) of the substrate in the reaction pocket [15] (Figure 2). Other mutations, which do not change the chemical properties of

TABLE 2: The *ARSA* genotypes in 21 Chinese MLD patients.

Pt. ID	Mutation	Genetic	Reported	Heritage
1	c.251C>A (pPro84Gln)	Hetero	Novel	Father
1	c.1172T>G (p.Val391Gly)*	Hetero	Novel	Mother
2	c.1172T>G (p.Val391Gly)*	Homo	Novel	Father and mother
3	c.960G>A (p.Trp320Term)	Homo	Novel	Father and mother
4	c.911A>T (p.Lys304Ile)	Hetero	Novel	Father
4	c.1049T>A (p.Leu350Gln)	Hetero	Novel	Mother
5	c.917C>T (P.Thr306Met)	Hetero	Reported	Father
5	c.827C>T (P.Thr276Met)*	Hetero	Reported	Mother
6	c.925G>A (p.Glu309Lys)*	Hetero	Reported	Father
6	c.427T>C (p.Phe143Leu)	Hetero	Novel	Mother
7	c.1130_1132delTCT*	Hetero	Novel	Father
7	c.1238A>G (p.Asp413Gly)	Hetero	Novel	Mother
8	c.244C>T (p.Arg82Trp)	Hetero	Novel	Father
8	c.179_180insCA	Hetero	Reported	Mother
9	c.1130_1132delTCT*	Hetero	Novel	Father
9	c.853C>G (p.Asp283Glu)	Hetero	Novel	Mother
10	c.218C>T (p.Pro73Leu)	Hetero	Novel	Father
10	c.827C>T (P.Thr276Met)*	Hetero	Reported	Mother
11	c.32T>C (p.Leu11Pro)	Hetero	Novel	Mother
11	c.1108-20A>G	Hetero	Novel	Father
12	c.257G>A (p.Arg86Gln)	Hetero	Reported	Mother
12	c.482T>C (p.Leu161Pro)	Hetero	Novel	Father
13	c.925G>A (pGlu309Lys)*	Hetero	Reported	Mother
13	c.302G>T (pGly101Val)	Hetero	Reported	Father
14	c.465G>A (p.Lys125ProfsX17)	Hetero	Novel	Father
14	ARSA del?			Mother
15	c.610C>G (p.Arg204Gly)	Hetero	Novel	Mother
15	c.44G>T (p.Gly15Val)	Hetero	Novel	Father
16	c.640G>A (p. Ala214Thr)	Hetero	Novel	Mother
16	c.893G>T (p.Gly298Val)	Hetero	Novel	Father
17	c.754T>C (p.Ser252Pro)	Hetero	Novel	Father
17	c.1344_1345insCC	Hetero	Novel	Mother
18	c.302_303insG	Hetero	Novel	Father
18	c.1428_1429insC	Hetero	Novel	Mother
19	c.1160G>T (p.387Gly>Val)	Hetero	Novel	Mother
19	IVS3+2T>C	Hetero	Novel	Spontaneous
20	c.830T>C (p.Leu277Pro)	Hetero	Novel	Father
20	c.383T>C (p.Leu128Pro)	Hetero	Novel	Mother
21	c.466G>C (p.Gly156Arg)	Hetero	Novel	Father
21	c.629T>C (pLeu210Pro)	Hetero	Novel	Mother

* indicates that mutations were detected more than once in this study.

the residues, could also be important if they were localized near the reaction pocket. In human ARSA, lysine 123, lysine 302, serine 150, and histidine 229 were identified as the key residues that form the reaction pocket and interact with the substrate [15]. Here, the mutation p.Thr306Met near p.Lys302 in Pt4 and Pt5 resulted in the loss of enzyme activity in vitro. The mutation p.Asp283Glu in Pt9, which also does not change the chemical properties of the residue, also leads to MLD in previous studies. It suggests that p.Asp283Glu results in a loss of the interaction between ARSA and the substrate.

Unlike the reported studies in the European population, no patients were detected with mutations c.459+1G>A and c.1277C>T (pPro426Leu), which occurred with high

Table 3: The prenatal diagnosis of four families.

Family	Family number	*ARSA* mutation	*ARSA* mutation
Pt5	Proband	c.917C>T (P.Thr306Met) (heterozygous)	c.827C>T (P.Thr276Met) (heterozygous)
	Father	c.917C (wild type)	c.827C>T (P.Thr276Met) (heterozygous)
	Mother	c.917C>T (P.Thr306Met) (heterozygous)	c.827C (wild type)
	Fetus	c.917C (wild type)	c.827C (wild type)
Pt7	Proband	c.1130_1132delTCT (p.Phe377del) (heterozygous)	c.1238A>G (p.Asp413Gly) (heterozygous)
	Father	c.1130_1132TCT (wild type)	c.1238A>G (p.Asp413Gly) (heterozygous)
	Mother	c.1130_1132delTCT (p.Phe377del) (heterozygous)	c.1238A (wild type)
	Fetus 1	c.1130_1132delTCT (p.Phe377del) (heterozygous)	c.1238A>G (p.Asp413Gly) (heterozygous)
	Fetus 2	c.1130_1132TCT (wild type)	c.1238A>G (p.Asp413Gly) (heterozygous)
Pt15	Proband	c.610C>G (p.Arg204Gly) (heterozygous)	c.44G>T (p.Gly15Val) (heterozygous)
	Father	c.610C (wild type)	c.44G>T (p.Gly15Val) (heterozygous)
	Mother	c.610C>G (p.Arg204Gly) (heterozygous)	c.44G (wild type)
	Fetus	c.610C (wild type)	c.44G (wild type)
Pt18	Proband	c.302_303insG (p.L102Pfs) (heterozygous)	c.1428_1429insC (p.S477Qfs) (heterozygous)
	Father	c.302_303insG (p.L102Pfs) (heterozygous)	c.1428_1429CA (wild type)
	Mother	c.302_303GC (wild type)	c.1428_1429insC (p.S477Qfs) (heterozygous)
	Fetus	c.302_303insG (p.L102Pfs) (heterozygous)	c.1428_1429insC (p.S477Qfs) (heterozygous)

Table 4: Functional prediction of mutation in *ARSA* with amino acid changing by PolyPhen-2.

Mutation	Function	Score	Sensitivity	Specificity
251C>A (pPro84Gln)	Probably damaging	0.911	0	0.94
1172T>G (p.Val391Gly)	Probably damaging	1	0	1
911A>T (p.Lys304Ile)	Probably damaging	1	0	1
1049T>A (p.Leu350Gln)	Probably damaging	1	0	1
917C>T (P.Thr306Met)	Probably damaging	1	0	1
827C>T (P.Thr276Met)	Probably damaging	1	0	1
925G>A (p.Glu309Lys)	Probably damaging	1	0	1
427T>C (p.Phe143Leu)	Probably damaging	0.999	0.14	0.99
1238A>G (p.Asp413Gly)	Probably damaging	1	0	1
244C>T (p.Arg82Trp)	Probably damaging	1	0	1
853C>G (p.Asp283Glu)	Probably damaging	1	0	1
218C>T (p.Pro73Leu)	Probably damaging	1	0	1
827C>T (P.Thr276Met)	Probably damaging	1	0	1
257G>A (p.Arg86Gln)	Probably damaging	1	0	1
925G>A (pGlu309Lys)	Probably damaging	1	0	1
302G>T (pGly101Val)	Probably damaging	1	0	1
640G>A (p. Ala214Thr)	Probably damaging	0.994	0.69	0.97
893G>T (p.Gly298Val)	Probably damaging	1	0	1
754T>C (p.Ser252Pro)	Probably damaging	0.996	0.55	0.98
1160G>T (p.387 Gly>ValV)	Probably damaging	1	0	1
830T>C (p.Leu277Pro)	Probably damaging	1	0	1
383T>C (p.Leu128Pro)	Probably damaging	1	0	1
466G>C (p.Gly156Arg)	Probably damaging	1	0	1
629T>C (pLeu210Pro)	Probably damaging	1	0	1

frequency, according to the publication. The known mutation c.917C>T (p.Thr306Met) was first reported in Europe [16]. Mutation c.827C>T (PThr276Met) was found in a high frequency of MLD patients of Lebanese or Arabic descent with a high degree of consanguinity and a common ethnic origin [17]. In contrast to these studies, no

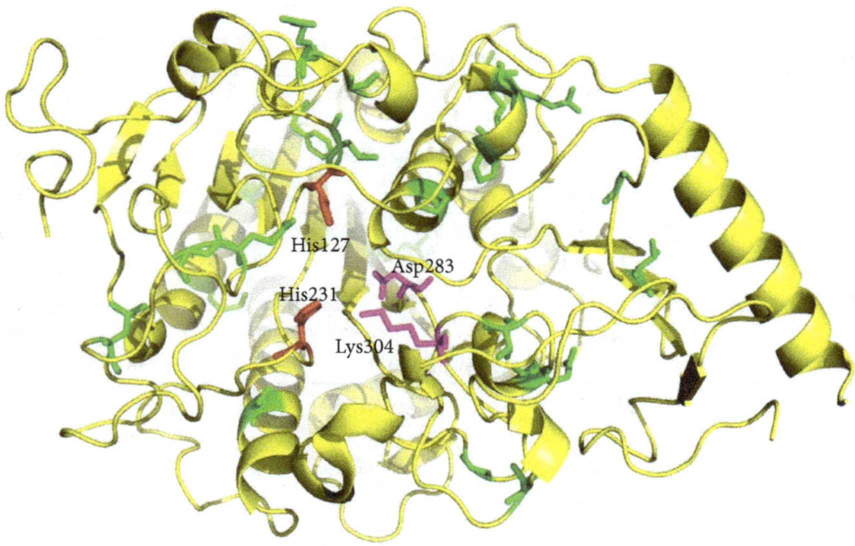

Figure 2: The illustration of *ARSA* 3D structure. Red and purple colors indicate the key residues to form reaction pocket. Green colors indicate the missense mutations in this study.

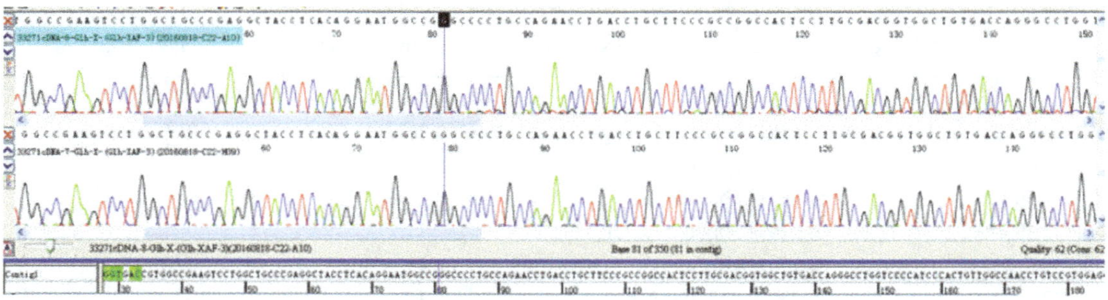

Figure 3: *ARSA* RNA-cDNA of Pt19.

highly frequent mutation was found in the current study of the Chinese population.

In 21 MLD patients, 2 (Pt2 and Pt3) had homozygous and 19 had compound heterozygous mutations of *ARSA*. They were all consistent with autosomal recessive inheritance. Two patients had homozygous mutations of *ARSA* that were inherited from their parents. All 19 patients with compound heterozygous mutations of *ARSA* had mutations that were inherited from their parents, except Pt14 and Pt19. The mutation of *ARSA* (c.465G>A (p.Lys125ProfsX17)) in Pt14 was inherited from the father with a heterozygous variation on one allele and the mother with a wild-type allele. Pt19 has the spontaneous splicing mutation IVS3+2T>C.

In Pt19, the mutation c.1160G>T was inherited from his mother and the mutation IVS3+2T>C was a spontaneous splicing mutation. After sequencing the cDNA of *ARSA* from Pt19, it was suggested that this mutation causes exon 3 skipping (Figure 3). Interestingly, mutation IVS3+2T>C could not be detected in Pt19's parents. It indicated that this mutation may be a de novo mutation or that his father was a chimera. Pt14 (c.465G>A) had a compound heterozygous mutation, which has not been reported. His father (c465G>A) had a heterozygous mutation, but his mother had a wild-type allele (Figure 4). The 465 locus was the last base in the second exon of *ARSA*. This synonymous mutation was located between the second exon and second intron. Therefore, we considered that it might affect splicing. Based on the analysis of the child's RNA, we found that after the change of the 465 locus, RNA indeed produced a splicing mutation, which presented as a splicing jump of the 371–465 loci in the second exon (involved 95 bases). In this family, children with a homozygous mutation in the 465 locus might come from the maternal chromosome deletion in this region. Therefore, we performed a fluorescence quantitative PCR for the 465 site and found that the expression of the reference gene in the generation was consistent with the control group. However, the expression of the *ARSA* region was half that of the control group, which confirmed that the proband with the 465 locus had a heterozygous deletion (Figure 5). To determine the deletion region, we designed several primers to amplify the fragment of *ARSA*, including exons 1–3, exons 1–5, and exons 1–8. Interestingly, the proband, father, and mother showed only one band. Thus, we

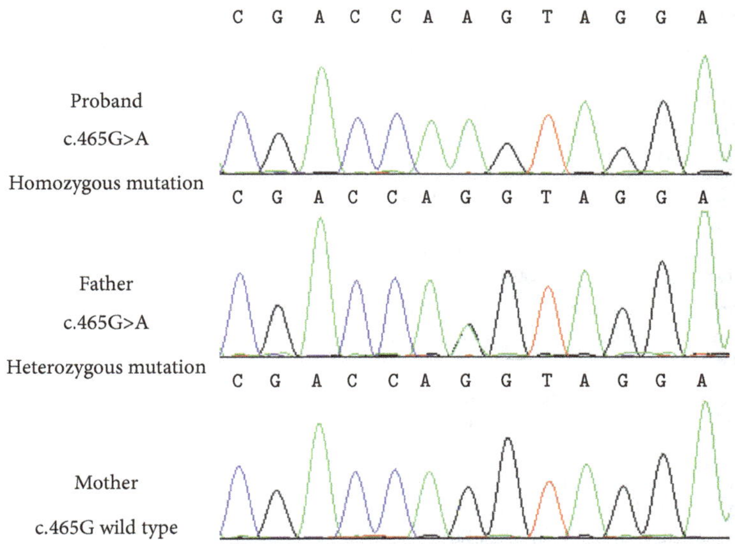

FIGURE 4: The *ARSA* mutation of Pt14, his father, and mother.

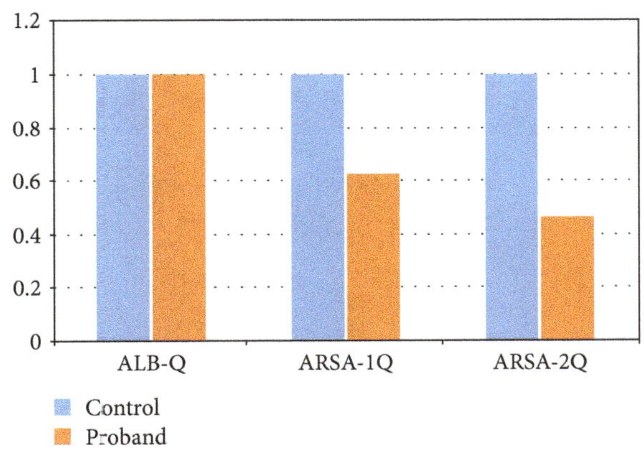

FIGURE 5: The fluorescent quantitation PCR result of the *ARSA* 465 site in proband (blue stands for reference gene, and orange stands for *ARSA*).

supposed that the proband of the maternal chromosomes contained the entire *ARSA* heterozygous deletion. Therefore, these two mutations (IVS3+2T>C and c.465G>A) are important not only for patient counseling but also for the evaluation of prenatal diagnosis.

Prenatal diagnosis was important for MLD families with *ARSA* mutations. We performed *ARSA* mutation analysis for four MLD families in this study. There was a 25% risk of recurrence and 25% risk of female carriers in all the MLD families with an autosomal recessive genotype. In the Pt5 and Pt15 MLD families, *ARSA* without a mutation provided critical information for parents carrying heterozygous *ARSA* mutations without MLD. This is valuable for prenatal diagnosis in the second pregnancy.

5. Conclusions

In summary, 21 Chinese patients were clinically and genetically diagnosed with MLD and analyzed clinically and genetically; four MLD families had additional prenatal diagnoses. MLD is usually caused by the lack of the important enzyme ARSA and results in damage to the nervous system, kidneys, gallbladder, and other organs. In this study, we first found that the patients showed classic clinical symptoms, typical brain MRI findings, and low ARSA enzyme activity of MLD, consistent with previous reports. Furthermore, DNA sequencing detected that all patients carried *ARSA* mutations, including 28 novel and 6 reported mutations. We first reported one coding region mutation that influenced splicing (c.465G>A (p.Lys125ProfsX17)). Prenatal diagnosis was successfully carried out on five fetuses. This study provides more information on critical mutations of *ARSA* in the Chinese population for MLD diagnosis and treatment in the future.

Conflicts of Interest

The authors declare that there is no conflict of interest regarding the publication of this article.

Acknowledgments

The authors would like to thank the patients and families described. This work was supported in part by the National Key Research and Development Program of China (no. 2016YFC1306201 and no. 2016YFC0901505), Beijing Key Laboratory of Molecular Diagnosis and Study on Pediatric Genetic Diseases (no: Z0317), and Key Laboratory of Neuroscience, Ministry of Education/National Health and Family Planning Commission, Peking University, Beijing 100083, China.

References

[1] A. Biffi, G. Lucchini, A. Rovelli, and M. Sessa, "Metachromatic

leukodystrophy: an overview of current and prospective treatments," *Bone Marrow Transplantation*, vol. 42, pp. S2–S6, 2008.

[2] J. Kreysing, K. Figura, and V. Gieselmann, "Structure of the arylsulfatase A gene," *European Journal of Biochemistry*, vol. 191, no. 3, pp. 627–631, 1990.

[3] F. Matthes, S. Stroobants, D. Gerlach et al., "Efficacy of enzyme replacement therapy in an aggravated mouse model of metachromatic leukodystrophy declines with age," *Human Molecular Genetics*, vol. 21, no. 11, pp. 2599–2609, 2012.

[4] V. Gieselmann, "Metachromatic leukodystrophy: genetics, pathogenesis and therapeutic options," *Acta Paediatrica*, vol. 97, no. s457, pp. 15–21, 2008.

[5] V. Gieselmann and I. Krageloh-Mann, "Metachromatic leukodystrophy – an update," *Neuropediatrics*, vol. 41, no. 1, pp. 1–6, 2010.

[6] C. Kehrer, S. Groeschel, B. Kustermann-Kuhn et al., "Language and cognition in children with metachromatic leukodystrophy: onset and natural course in a nationwide cohort," *Orphanet Journal of Rare Diseases*, vol. 9, no. 1, p. 18, 2014.

[7] F. Eichler, W. Grodd, E. Grant et al., "Metachromatic leukodystrophy: a scoring system for brain MR imaging observations," *American Journal of Neuroradiology*, vol. 30, no. 10, pp. 1893–1897, 2009.

[8] A. Martin, C. Sevin, C. Lazarus, C. Bellesme, P. Aubourg, and C. Adamsbaum, "Toward a better understanding of brain lesions during metachromatic leukodystrophy evolution," *American Journal of Neuroradiology*, vol. 33, no. 9, pp. 1731–1739, 2012.

[9] J. P. van der Voorn, P. J. Pouwels, W. Kamphorst et al., "Histopathologic correlates of radial stripes on MR images in lysosomal storage disorders," *American Journal of Neuroradiology*, vol. 26, no. 3, pp. 442–446, 2005.

[10] P. Luzi, M. A. Rafi, H. Z. Rao, and D. A. Wenger, "Sixteen novel mutations in the arylsulfatase A gene causing metachromatic leukodystrophy," *Gene*, vol. 530, no. 2, pp. 323–328, 2013.

[11] S. A. Patil and G. H. Maegawa, "Developing therapeutic approaches for metachromatic leukodystrophy," *Drug Design, Development and Therapy*, vol. 7, pp. 729–745, 2013.

[12] J. W. Rip and B. A. Gordon, "A simple spectrophotometric enzyme assay with absolute specificity for arylsulfatase A," *Clinical Biochemistry*, vol. 31, no. 1, pp. 29–31, 1998.

[13] J. Wang, W. Zhang, H. Pan et al., "ARSA gene mutations in five Chinese metachromatic leukodystrophy patients," *Pediatric Neurology*, vol. 36, no. 6, pp. 397–401, 2007.

[14] P. Poeppel, M. Habetha, A. Marcao, H. Bussow, L. Berna, and V. Gieselmann, "Missense mutations as a cause of metachromatic leukodystrophy. Degradation of arylsulfatase A in the endoplasmic reticulum," *The FEBS Journal*, vol. 272, no. 5, pp. 1179–1188, 2005.

[15] R. von Bulow, B. Schmidt, T. Dierks, K. von Figura, and I. Uson, "Crystal structure of an enzyme-substrate complex provides insight into the interaction between human arylsulfatase A and its substrates during catalysis," *Journal of Molecular Biology*, vol. 305, no. 2, pp. 269–277, 2001.

[16] S. Grossi, S. Regis, C. Rosano et al., "Molecular analysis of *ARSA* and *PSAP* genes in twenty-one Italian patients with metachromatic leukodystrophy: identification and functional characterization of 11 novel *ARSA* alleles," *Human Mutation*, vol. 29, no. 11, pp. E220–E230, 2008.

[17] J. S. Harvey, P. V. Nelson, W. F. Carey, E. F. Robertson, and C. P. Morris, "An arylsulfatase A (ARSA) missense mutation ($T_{274}M$) causing late-infantile metachromatic leukodystrophy," *Human Mutation*, vol. 2, no. 4, pp. 261–267, 1993.

7

Discovery of Novel Caeridins from the Skin Secretion of the Australian White's Tree Frog, *Litoria caerulea*

Lei Li,[1] Qing Wu,[2] Xi Wang,[1] Huimin Lu,[1] Xinping Xi,[1] Mei Zhou,[1] Chris J. Watson,[3] Tianbao Chen,[1] and Lei Wang[1]

[1]*Natural Drug Discovery Group, School of Pharmacy, Queen's University, Belfast BT9 7BL, UK*
[2]*School of Chinese Materia Medica, Beijing University of Chinese Medicine, Beijing 102488, China*
[3]*Centre for Experimental Medicine, School of Medicine, Dentistry and Biomedical Sciences, Queen's University, Belfast BT9 7BL, UK*

Correspondence should be addressed to Lei Li; lei.li@qub.ac.uk and Mei Zhou; m.zhou@qub.ac.uk

Academic Editor: Marco Gerdol

Abundant biologically active peptides have been discovered from frog skin secretions, a rich natural source of bioactive compounds with great potential in drug discovery. In this study, three Caeridin peptides, namely, Caeridin-1, S5-Caeridin-1, and Caeridin-a1, were discovered from the skin secretion of the Australian White's tree frog, *Litoria caerulea*, for the first time, by means of combining transcriptomic and peptidomic analyses. It also represents the first report on bioactive Caeridins since this family of peptides was initially studied 20 years ago. Chemically synthetic versions of each natural Caeridin demonstrated promising bioactivities either on rat smooth muscles or against microbial growth. Specifically, Caeridin-1 produced contraction of rat bladder smooth muscle, while S5-Caeridin-1 induced relaxation of rat ileum smooth muscle, both at nanomolar concentrations. Moreover, Caeridin-a1 was shown to potently inhibit the growth of the planktonic Gram-positive bacteria *Staphylococcus aureus* (*S. aureus*), methicillin-resistant *S. aureus* (MRSA), and *Enterococcus faecalis* (*E. faecalis*), the Gram-negative bacterium, *Escherichia coli* (*E. coli*), and the yeast, *Candida albicans* (*C. albicans*). The discovery of these Caeridins may induce further intensive and systematic studies of frog skin peptides to promote the discovery of natural templates as lead compounds for drug discovery and therapeutic application.

1. Introduction

In recent years, peptide and peptide-based drugs have been developing fast and have pioneered an essential area in the pharmaceutical industry [1]. Peptide-based medicinal therapy could offer benefits over other biologics, while demanding less costs in production [1]. Recent trends suggest that the application of therapeutic peptides is becoming more popular, and it has extended to the treatment of many diseases, such as infection, cancer, enzyme deficiency disorders, protein-dysfunction disorders, and even genetic and degenerative diseases [2].

Frog skin with poisonous glands can produce secretions in a holocrine manner and release complex chemical contents [3–5]. On the one hand, peptides and proteins contained in these secretions have been considered to play different but important roles in regulating frogs' physiological functions and defense against predators and microorganisms. On the other hand, they have been proven to hold hidden treasures to supplement drug development [6]. However, with the observed global decline in the frog population during the 1980s, including species extinction, it became apparent that this potentially valuable natural resource needed environmental protection and animal conservation to ensure peptide discovery and acquisition from these natural products could continue and play a role in the future development of novel therapeutic peptides [7, 8].

The Australian White's tree frog, *Litoria caerulea* (*L. caerulea*), belongs to the tree frog family Hylidae. Most members of this tree frog family can produce multiple defensive and regulatory chemical compounds in their skin secretions to provide protective survival advantages in their natural environment. Compared with the peptide discovery research on other tree frogs, previous studies on the skin secretions of *L. caerulea* were not as productive in terms of peptide quantities acquired and in the limited identification of peptides with a range of bioactive properties. The original study investigating peptides derived from *L. caerulea* was conducted in 1993 by Waugh et al. [9] who reported the discovery of 25 peptides from the secretions of the parotoid glands, including Caerulein 1, eighteen Caerins, and six Caeridins (Caeridins 1–6), the latter also being evidenced later in four other Australian frog species, including, *L. gilleni*, *L. splendida*, *L. xanthomera*, and *L. chloris*. In 1997, Steinborner et al. [10] reported the discovery of Caeridin 7.1 from the dorsal glandular skin extract of the brown tree frog, *Litoria ewingii*. Interestingly, with the Caeruleins showing hypotensive bioactivity and the Caerins antimicrobial activity, the Caeridins, however, were not reported to display any structural and functional resemblance to the Caerulein and Caerin peptides. These small Caeridin peptides (about 1000–1500 Da), comprising of 12–15 amino acid residues, share the same sequence -Gly-Leu-Leu/Phe- at the N-terminal ends and -Leu/Ile (NH_2) at the C-terminal ends. Each of this Caeridins may be well fitted into an α-helical wheel with clearly delineated hydrophilic and hydrophobic zones [9]. The possession of dominant hydrophobic zones may suggest appreciable hydrophobic moments of these Caeridins and consequentially may bind more easily to the surface of biological membranes [11]. However, despite their distinct structural features, none of the Caeridins was reported to have any bioactive functions within these previous studies, which might explain why there has been no research reported on Caeridins over the last 20 years. Here, we report the isolation and structural characterisation of Caeridin-1 and two novel Caeridins from the skin secretion of *L. caerulea*, which were obtained through an ecofriendly method described previously [12], using parallel transcriptomic and peptidomic analyses, as well as through examining their biological activities using synthetic versions in a variety of bioassays.

2. Results

2.1. "Shotgun" Cloning of cDNAs Encoding Caeridins. Preprocaeridin cDNAs were consistently cloned from the skin secretion cDNA library, and each encoded a single copy of Caeridin, which were named Caeridin-1, S5-Caeridin-1, and Caeridin-a1 shown in Figure 1. The putative signal peptide sequences were localized to the acidic spacer peptide and were highly identical in structure. The latter encoding regions were flanked at the N-terminus by classical propeptide convertase-processing sites involving a propeptide convertase with -Lys-Arg- (-KR-) cleavage specificity, which suggested that there were mutations in the structure. Alignments of the full-length nucleic acid sequences and the translated open-reading frame amino acid sequences are shown in Figure 2 and highlight that preprocaeridins exhibit high degrees of both nucleotide and peptide primary structural similarities and conservation. Transcripts of Caeridin-1, S5-Caeridin-1, and Caeridin-a1 were registered with and deposited at http://www.ebi.ac.uk/ena/data/view/ permanently under the accession numbers LT852758, LT852759, and LT853865, respectively. Caeridin-1, S5-Caeridin-1, and Caeridin-a1 were compared with other Caeridins isolated from *L. caeulea*, and their amino acid sequences showed certain degrees of similarities suggested by the amino acid alignment in Figure 3.

2.2. Fractionation and Primary Structural Identification of Caeridins. The computed molecular masses of the deduced primary structures of Caeridin-1, S5-Caeridin-1, and Caeridin-a1 located within the open-reading frames of respective cloned cDNAs were used to detect each Caeridin within reversed-phase HPLC fractions as shown in Figure 4, which were subjected to mass spectrometric analysis. The primary structures of each Caeridin were confirmed by tandem mass spectrometric (MS/MS) fragmentation as shown in Figure 5 and Table 1. The existence of C-terminal amidation of each Caeridin was confirmed with the appropriated positioned G residue playing as the amide donor.

2.3. Secondary Structural Identification and Physicochemical Property Analysis of Caeridins. The purified synthetic replicates of Caeridins were subjected to secondary structural identification. From the CD spectra, these positive bands at 190 nm and double-negative bands at 208 nm and 222 nm in the 10 mM NH_4Ac with 50% TFE solution indicated that Caeridin-1, S5-Caeridin-1, and Caeridin-a1 adopted a typical α-helical secondary structure in a membrane mimic environment (Figure 6). Multiple CD analysis methods were used and processed by the K2D3 online portal to measure the α-helicity of these Caeridins (Table 2) [13].

The newly discovered Caeridins together with the previously reported Caeridins were subjected to the analysis of physicochemical properties by the Heliquest Analysis online bioinformatic tool [14]. The hydrophobicity, hydrophobic moment (amphipathicity), net charge, and charged residues of these Caeridins were determined and are shown in Table 3.

2.4. Smooth Muscle Pharmacological Activities of Caeridins. The purified synthetic replicates of Caeridins were subjected to smooth muscle pharmacological activity tests. Caeridin-1 was found to induce contractions on rat bladder smooth muscle, and S5-Caeridin-1 was found to have relaxation effects on rat ileum smooth muscle, at various concentrations, as shown in Figure 7. The half effective concentration (EC_{50}) of Caeridin-1 inducing rat bladder smooth muscle contractions was 5.6 nM (5.6×10^{-9} M), and EC_{50} of S5-Caeridin-1 causing rat ileum smooth muscle relaxations was 9.5 nM (9.5×10^{-9} M). The maximum contraction effect of Caeridin-1 on bladder smooth muscle preparations was 0.3 g in tension change, and the maximum relaxant effect of S5-Caeridin-1 on ileum smooth muscle preparations was 0.2 g in tension change. Caeridin-a1 did not show any smooth muscle activities.

2.5. Antimicrobial Activities of Caeridins. The purified synthetic replicates of Caeridins were subjected to antimicrobial activity tests, with melittin and bradykinin as positive and negative controls, respectively. Melittin originally discovered from the venom of the honeybee (*Apis mellifera*) has potent antimicrobial activities. The inclusion of melittin as the positive peptide in the antimicrobial activity tests set a reference

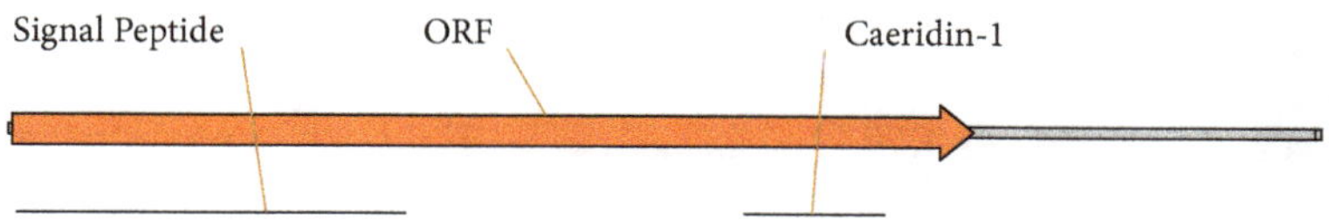

Caeridin-1

```
           M  A  S  L  K  K  S  L  F  L  V  L  F  L  G  L  V
  1        ATGGCTTCCT TGAAGAAATC TCTTTTCCTT GTACTATTTC TTGGATTAGT
           TACCGAAGGA ACTTCTTTAG AGAAAAGGAA CATGATAAAG AACCTAATCA
            S  L  S  I  C  E  K  E  K  R  Q  N  G  E  D  E
 51        TTCTCTTTCC ATCTGTGAAA AAGAGAAAAG ACAGAATGGA GAGGATGAGG
           AAGAGAAAGG TAGACACTTT TTCTCTTTTC TGTCTTACCT CTCCTACTCC
           D  E  A  E  K  E  E  E  S  E  E  E  S  E  E  K  R
101        ATGAAGCTGA GAAAGAGGAA GAAAGTGAAG AGGAAAGTGA AGAGAAGAGA
           TACTTCGACT CTTTCTCCTT CTTTCACTTC TCCTTTCACT TCTCTTCTCT
            G  L  L  D  G  L  L  G  T  L  G  L  G  K  K  R  *
151        GGTCTGCTAG ATGGTCTTCT TGGAACTCTT GGTCTTGGAA AAAAACGTTA
           CCAGACGATC TACCAGAAGA ACCTTGAGAA CCAGAACCTT TTTTTGCAAT
            *
201        AGAAAATGTA ATGCAACTAA AAGGAGCACA CATTATTAAA GAAAATTTAA
           TCTTTTACAT TACGTTGATT TTCCTCGTGT GTAATAATTT CTTTTAAATT
251        CTAAAAAAAA AAAAAAAAAA AAAAAAAAAA
           GATTTTTTTT TTTTTTTTTT TTTTTTTTTT
```

(a)

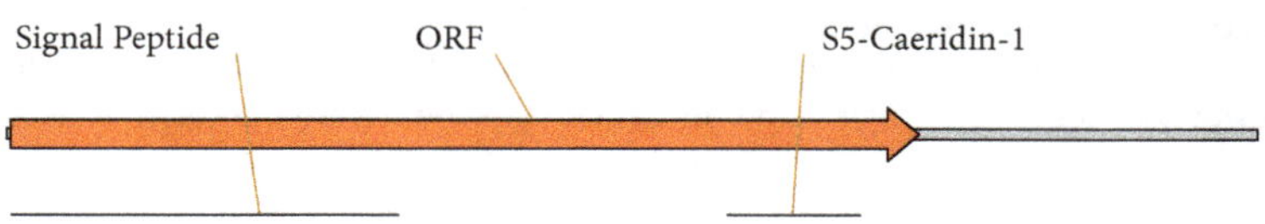

S5-Caeridin-1

```
           M  A  S  L  K  K  S  L  F  L  V  L  F  L  G  L  V
  1        ATGGCTTCCT TGAAGAAATC TCTTTTCCTT GTACTATTTC TTGGATTAGT
           TACCGAAGGA ACTTCTTTAG AGAAAAGGAA CATGATAAAG AACCTAATCA
            S  L  S  I  C  E  K  E  K  R  Q  N  E  E  D  E
 51        TTCTCTTTCC ATCTGTGAAA AAGAGAAAAG ACAGAATGAA GAGGATGAGG
           AAGAGAAAGG TAGACACTTT TTCTCTTTTC TGTCTTACTT CTCCTACTCC
           D  E  A  E  K  E  E  E  S  E  E  E  S  E  E  K  R
101        ATGAAGCTGA GAAAGAGGAA GAAAGTGAAG AGGAAAGTGA AGAGAAGAGA
           TACTTCGACT CTTTCTCCTT CTTTCACTTC TCCTTTCACT TCTCTTCTCT
            G  L  L  D  S  L  L  G  G  L  G  L  G  K  K  R  *
151        GGTCTGCTAG ATAGTCTTCT TGGAGGTCTT GGTCTTGGAA AAAAACGTTA
           CCAGACGATC TATCAGAAGA ACCTCCAGAA CCAGAACCTT TTTTTGCAAT
            *
201        AGAAAATGTA ATGCAACTAA AAGGAGCACA CATTATTAAA GAAAATTTAA
           TCTTTTACAT TACGTTGATT TTCCTCGTGT GTAATAATTT CTTTTAAATT
251        CTGAAAAAAA AAAAAAAAAA AAAAAAAAAA
           GACTTTTTTT TTTTTTTTTT TTTTTTTTTT
```

(b)

FIGURE 1: Continued.

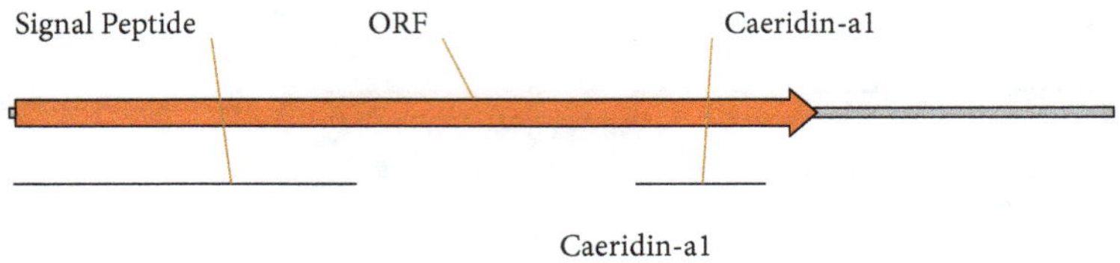


```
         M  A  F  L  K  K  S  L  F  L  V  L  F  L  G  L  V
1        ATGGCTTTCT TGAAGAAATC TCTTTTCCTT GTACTATTCC TGGGATTAGT
         TACCGAAAGA ACTTCTTTAG AGAAAAGGAA CATGATAAGG ACCCTAATCA
          S  L  S  I  C  E  K  E  K  R  Q  N  E  E  D  E
51       TTCCCTTTCC ATCTGTGAAA AAGAGAAAAG ACAGAATGAA GAGGACGAGG
         AAGGGAAAGG TAGACACTTT TTCTCTTTTC TGTCTTACTT CTCCTGCTCC
         D  E  N  E  A  E  E  E  S  E  E  G  S  E  E  K  R
101      ATGAAAATGA GGCAGAGGAA GAAAGTGAAG AGGGAAGTGA AGAGAAGAGA
         TACTTCGACT CTTTCTCCTT CTTTCACTTC TCCTTTCACT TCTCTTCTCT
          G  L  F  D  I  V  G  K  L  L  G  G  L  G  L  G  R
151      GGTCTGTTTG ATATTGTTGG AAAGCTTCTT GGAGGTCTTG GATTAGGACG
         CCAGACAAAC TATAACAACC TTTCGAAGAA CCTCCAGAAC CTAATCCTGC
          *  *
201      TTAAGAAAAT GTAACATTTT ACAACTAAAA GGAGCACAAT TATCAATAAT
         AATTCTTTTA CATTGTAAAA TGTTGATTTT CCTCGTGTTA ATAGTTATTA
251      TGTGCCAAAA TTATATTAAA GCATATTTAC CTGACAAAAA AAAAAAAAAA
         ACACGGTTTT AATATAATTT CGTATAAATG GACTGTTTTT TTTTTTTTTT
301      AAAAAAAAAA A
         TTTTTTTTTT T
```

(c)

FIGURE 1: Nucleotide sequences of full-length cDNA clones encoding open-reading frames of (a) preprocaeridin-1, (b) preproS5-caeridin-1, and (c) preprocaerdin-a1 cloned from a *Litoria caerulea* (*L. caerulea*) skin secretion cDNA library. Putative signal peptide sequences are double-underlined, mature Caeridin sequences are single-underlined, and stop codons are indicated by asterisks.

for the evaluation of the antimicrobial activity of the novel peptides [15–17]. Bradykinin derived from the frog skin extract of the European grass frog (*Rana temporaria*) is the counterpart to mammalian bradykinins with no obvious antimicrobial activities; thus, it is incorporated as the negative peptide in the antimicrobial activity tests [18–20]. Caeridin-a1 exhibited different levels of antimicrobial activities against the growth of the tested microorganisms (Table 4) and demonstrated the most potent antimicrobial activity against *Staphylococcus aureus* (*S. aureus*). It also inhibited the growth of antibiotic-resistant bacteria, methicillin-resistant *S. aureus* (MRSA), and *Enterococcus faecalis* (*E. faecalis*), but slightly less potent than that against *S. aureus*. It also had antimicrobial activities against the Gram-negative bacterium, *Escherichia coli* (*E. coli*) and the yeast, *Candida albicans* (*C. albicans*). Caeridin-1 and S5-Caeridin-1 did not demonstrate any antimicrobial activities on these tested microorganisms.

2.6. Membrane Permeability Studies. The novel antimicrobial peptide Caeridin-a1, melittin, and bradykinin were tested in the membrane permeability studies. Melittin is a canonical membrane interactive peptide, so it was included as the positive peptide [15–17]. Bradykinin was referred to as the negative peptide for it is known to be inactive on membrane interaction and permeabilisation [18–20].

2.6.1. Membrane Permeability Studies on S. aureus. The peptide-induced permeabilisation of the cytoplasmic membranes of *S. aureus* was detected using SYTOX Green, a counterpart stain of nucleic acid and an indicator of dead cells. The cell membrane of *S. aureus* remained intact after being treated with the purified synthetic Caeridin-a1 (MIC) for 2 h, but became massively compromised after being treated with the purified synthetic Caeridin-a1 (MBC) for the same period of time, as shown in Figure 8. This phenomena suggested that Caeridin-a1 (MIC) failed to induce permeabilisation to the cytoplasmic membrane of *S. aureus* within the 2 h treatment. After the 2 h treatment, the positive peptide melittin produced approximately 50% membrane permeability at MIC (1 μM) and 100% membrane permeability at MBC (2 μM). The negative peptide bradykinin, Caeridin-1, and S5-Caeridin-1 barely showed membrane-permeabilising activity up to 512 μM.

2.6.2. Membrane Permeability Tests on E. coli. The peptide-induced membrane permeabilisation of *E. coli* was determined by the release of β-galactosidase activity into culture medium MHB (2% lactose) using ONPG as the substrate. As shown in Figure 9, the purified synthetic Caeridin-a1 (MIC and MBC) induced inner membrane permeabilisation within 30 min given that the ONPG entered the cytoplasm and was degraded by β-galactosidase, generating

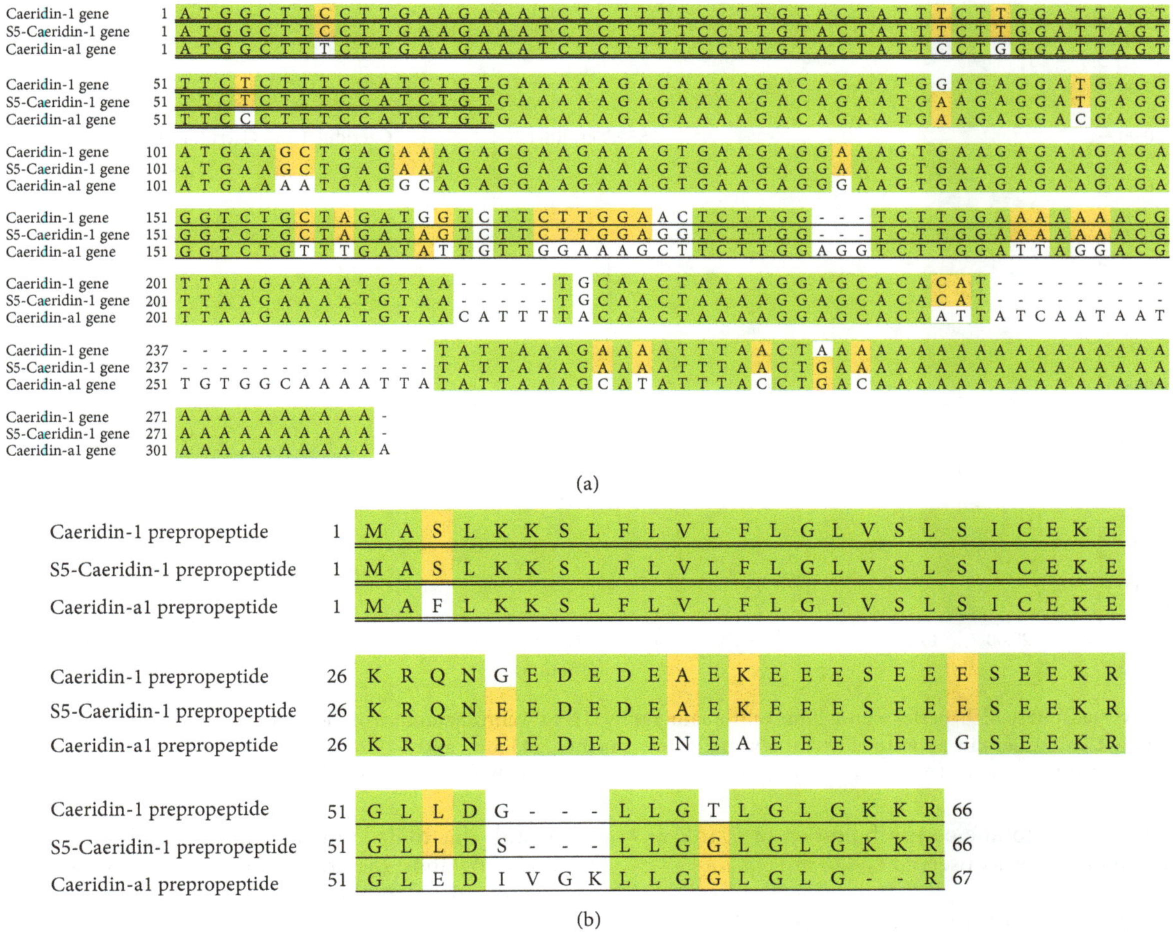

FIGURE 2: Alignments of (a) nucleotide sequences of cloned preprocaeridin cDNAs and (b) translated preprocaeridin open-reading frames established MEGA6. The identical and conservative regions are highlighted in green and yellow, respectively. Putative signal peptide sequences encoding regions and putative signal peptide sequences are double-underlined, and mature Caeridin sequences encoding regions and mature Caeridins are single-underlined.

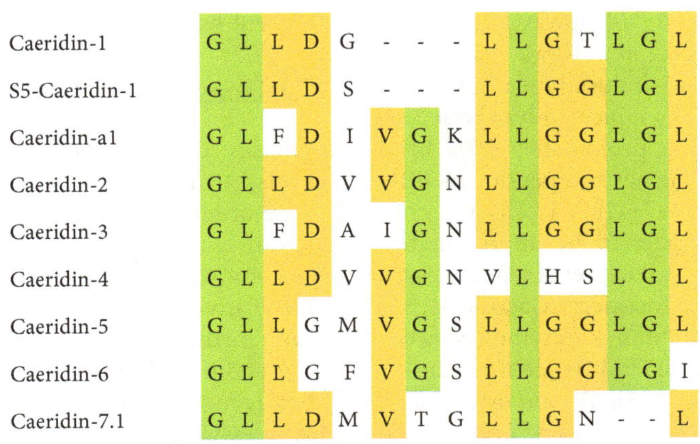

FIGURE 3: Alignments of amino acid sequences of Caeridins identified from *L. caerulea*. The identical and conservative amino acids were highlighted in green and orange, respectively.

o-nitrophenol that produced an absorbance at 420 nm. The Caeridin-a1-induced membrane permeabilisation of *E. coli* was not as strong as the positive peptide melittin, evidenced by the higher absorbance at 420 nm, indicating an increased production of o-nitrophenol by ONPG hydrolysis. Caeridin-1, S5-Caeridin-1, and the negative peptide bradykinin

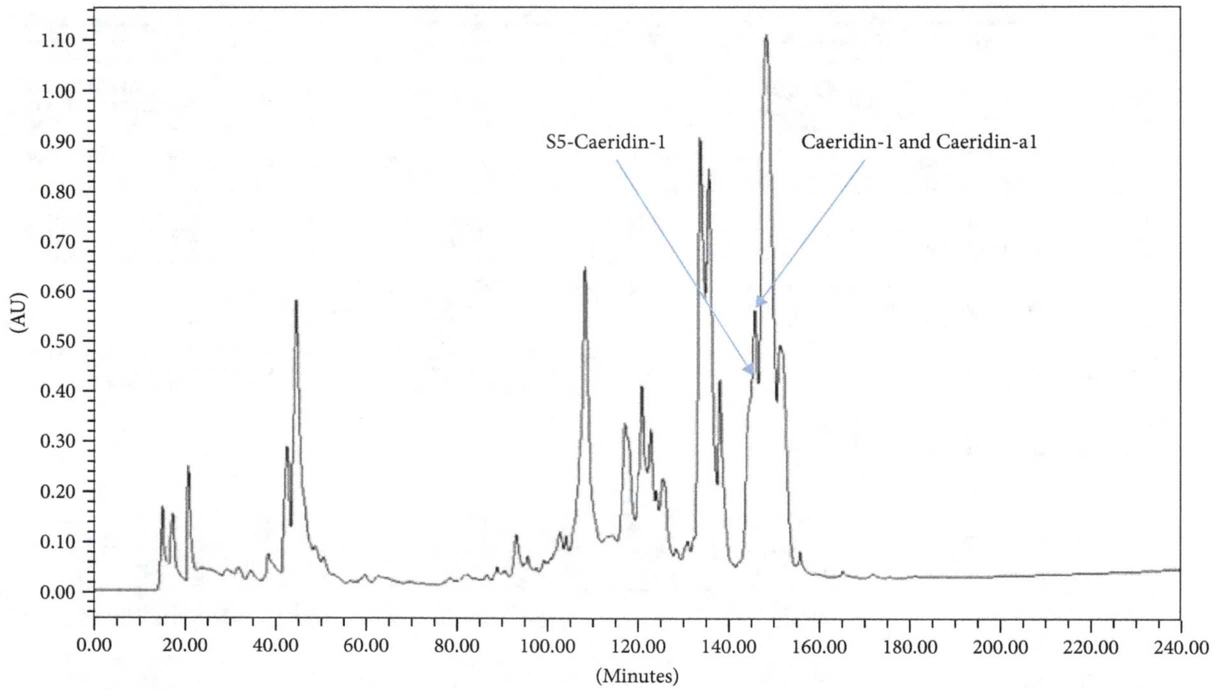

FIGURE 4: Reversed-phase high-performance liquid chromatography (HPLC) chromatogram of the skin secretion from *L. caerulea*. Caeridin-1, S5-Caeridin-1, and Caeridin-a1 were eluted at approximately 144 min, 143 min, and 144 min, respectively. The *y*-axis shows the relative absorbance in absorbance units (AU) at a wavelength of 214 nm, and the *x*-axis shows the retention time in minutes.

failed to induce membrane permeabilisation of *E. coli*, as no absorbance increase was observed at 420 nm.

2.6.3. Membrane Permeability Study on Human Microvascular Endothelial Cell Line, HMEC-1. The peptide-induced membrane permeabilisation of normal human cells was evaluated by the release of lactate dehydrogenase (LDH), a cytosolic enzyme, from the damaged membrane of the human microvascular endothelial cell line HMEC-1. After treatment with Caeridin-1, S5-Caeridin-1, and negative peptide bradykinin at concentrations of 10^{-7} M to 10^{-4} M, lower than 5% of membrane lysis was resulted (Figures 10(a), 10(b), and 10(e)). In comparision, Caeridin-a1-induced membrane permeabilisation of HMEC-1 cells was indicated by about 50% of LDH release at 100 μM and approximately 10% LDH release at 10 μM (Figure 10(c)). The positive peptide melittin showed membrane permeability in HMEC-1 cells at a concentration as low as 0.1 μM, suggested by approximately 10% LDH release (Figure 10(d)). As the level of LDH released into the medium is indicative of cytotoxicity, Caeridin-1, S5-Caeridin-1, and negative peptide bradykinin showed weak cytotoxicity to HMEC-1 cells at 100 μM, while Caeridin-a1 and positive peptide melittin showed significant cytotoxicity at 100 μM. However, at the lower tested concentrations, no cytotoxicity was detected with Caeridins and negative peptide bradykinin.

2.7. Haemolysis Induced by Caeridins. Haemolytic activities of Caeridins, positive peptide melittin, and negative peptide bradykinin were studied using horse erythrocytes as shown in Figure 11. Melittin was incorporated as the positive peptide due to its known haemolytic activities. The amphibian bradykinin, the counterpart to mammalian Bradykinin, has so little haemolytic activities that it was included in the haemolysis test as the negative peptide [15–20]. The incorporation of melittin and bradykinin as positive and negative peptide provided evidence on the evaluation of the haemolytic activities of the novel peptides. Bradykinin, Caeridin-1, and S5-Caeridin-1 exhibited weak haemolytic activities at the tested concentrations, while Caeridin-a1 exhibited 18% haemolysis at 32 μM, and positive peptide Melittin showed eminent haemolysis from 1 μM.

3. Discussion

Peptides could offer multiple benefits over other drug treatments, such as higher potency, higher selectivity, a broader range of targets, potentially lower toxicity, lower accumulation in tissues, more abundance in chemical and biological diversity, discoverability at amino acid and nucleotide levels, and relatively low cost in production [2]. Thus, the future of many medicinal treatments could depend on the use of peptides and peptide-based drugs [21]. Frog skin secretions have been recognised as a valuable source of abundant bioactive peptides which are generally categorised and grouped into cytolytic peptides and pharmacologically active peptides [6]. Given their bioactive potential, there are now global efforts to minimise the observed global decline in frog populations through frog protection and environmental conservation strategies, along with the development of improved animal-friendly methods to conduct peptide discovery experiments from frog skin secretions [22–24].

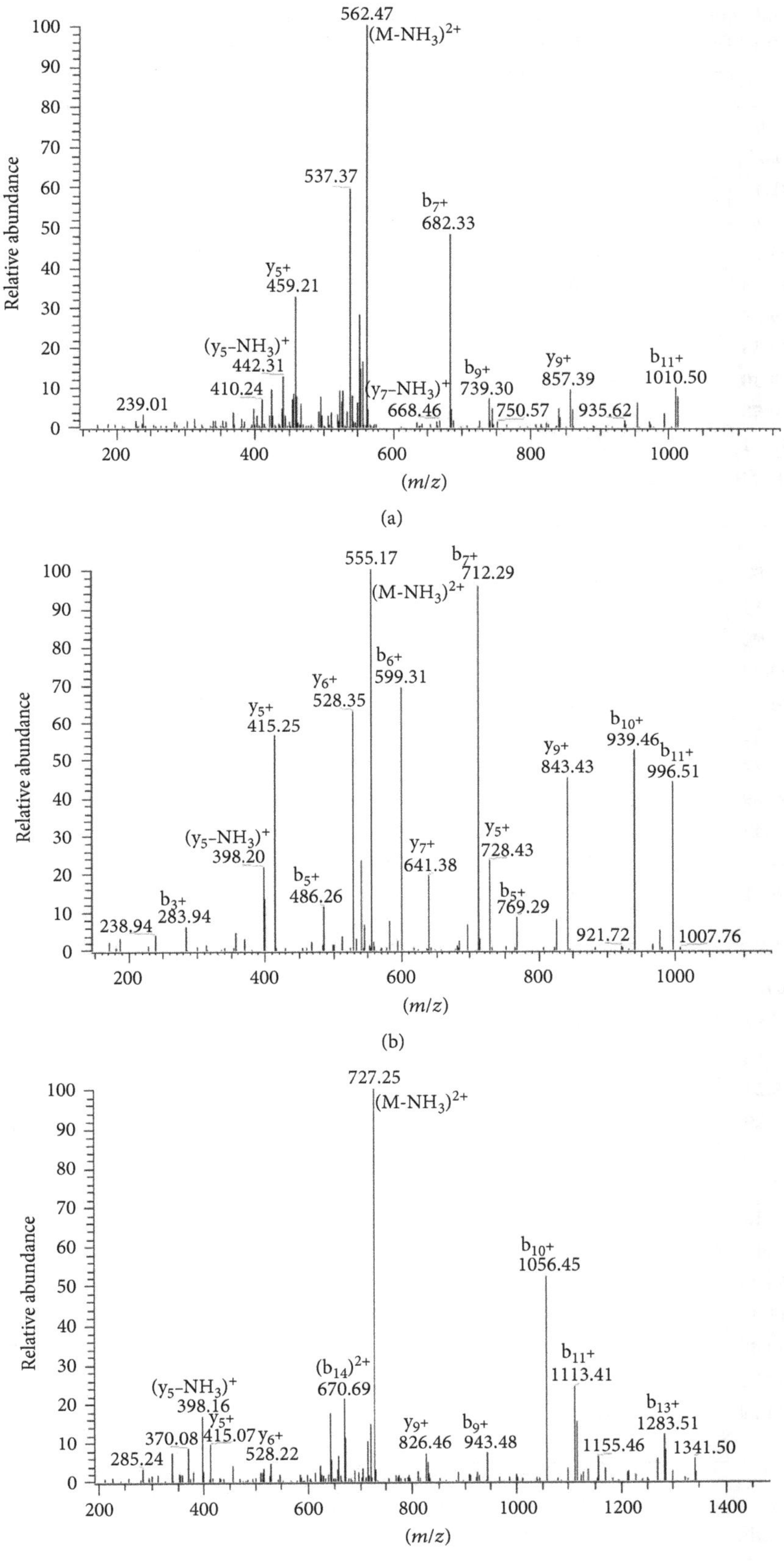

FIGURE 5: Annotated tandem mass (MS/MS) fragmentation spectrum of (a) Caeridin-1, (b) S5-Caeridin-1, and (c) Caeridin-a1. The observed *b* ions and *y* ions are labelled in blue and red, respectively.

Table 1: Predicted singly and doubly charged *b* ions and *y* ions arising from MS/MS fragmentation of (a) Caeridin-1, (b) S5-Caeridin-1, and (c) Caeridin-a1. Actual fragment ions observed following MS/MS fragmentation are indicated in bold and italic.

Number 1	b(1+)	b(2+)	Seq.	y(1+)	y(2+)	Number 2
			(a)			
1	58.03	29.52	G			12
2	**171.11**	86.06	L	1083.68	*542.34*	11
3	**284.20**	142.60	L	*970.59*	485.80	10
4	**399.27**	**200.12**	D	*857.39*	*429.26*	9
5	**456.25**	**228.63**	G	*742.48*	*371.74*	8
6	**569.33**	**285.17**	L	*685.46*	*343.23*	7
7	**682.33**	**341.71**	L	*572.38*	*286.69*	6
8	**739.30**	**370.22**	G	*459.21*	230.15	5
9	**840.58**	420.74	T	*402.27*	201.64	4
10	**953.41**	**477.29**	L	*301.22*	151.12	3
11	**1010.50**	**505.80**	G	*188.05*	94.57	2
12			L-Amidated	131.12	66.06	1
			(b)			
1	58.03	29.52	G			12
2	**171.11**	86.06	L	1069.66	*535.33*	11
3	**283.94**	142.60	L	*956.58*	478.79	10
4	**399.22**	200.12	D	*843.43*	*422.25*	9
5	**486.26**	243.63	S	*728.43*	364.74	8
6	**599.31**	300.17	L	*641.38*	321.22	7
7	**712.42**	356.72	L	*528.35*	264.68	6
8	**769.29**	385.23	G	*415.25*	208.14	5
9	**826.47**	413.74	G	*358.25*	179.63	4
10	**939.46**	**470.28**	L	*301.22*	151.12	3
11	**996.51**	**498.79**	G	*188.14*	94.57	2
12			L-Amidated	131.12	66.06	1
			(c)			
1	58.03	29.52	G			15
2	**171.11**	86.06	L	1413.88	*707.44*	14
3	**318.18**	160.09	F	*1300.79*	*650.90*	13
4	**433.21**	217.61	D	*1153.73*	577.67	12
5	**546.29**	274.15	I	*1038.70*	519.86	11
6	**645.36**	323.68	V	*925.62*	463.31	10
7	**702.38**	352.19	G	*826.46*	413.78	9
8	**830.48**	415.7	K	*769.53*	385.27	8
9	**943.48**	472.3	L	*641.38*	321.22	7
10	**1056.45**	528.8	L	*528.22*	264.68	6
11	**1113.41**	557.3	G	*415.07*	208.14	5
12	**1169.48**	585.9	G	*358.25*	179.63	4
13	**1283.51**	**642.2**	L	*301.22*	151.12	3
14	**1340.79**	**670.7**	G	*188.05*	94.57	2
15			L-Amidated	131.12	66.06	1

Within this study, we have successfully constructed a cDNA library of the lyophilized skin secretion of the Australian White's tree frog, *L. caerulea*. *L. caerulea*, a member of the tree frog family Hylidae, was selected for analysis as it has been understudied to date and has multiple skin-defensive and regulatory compounds with significant pharmaceutical potentials that warrant further investigation [25, 26]. So far, three major peptide families have been identified from *L. caerulea*, namely, Caerins, Caeruleins, and Caeridins [9]. Caerins were found similar in primary structures with conserved -Gly-Leu- at the N-terminus, and they possess sufficient physiochemical characteristics to carry out

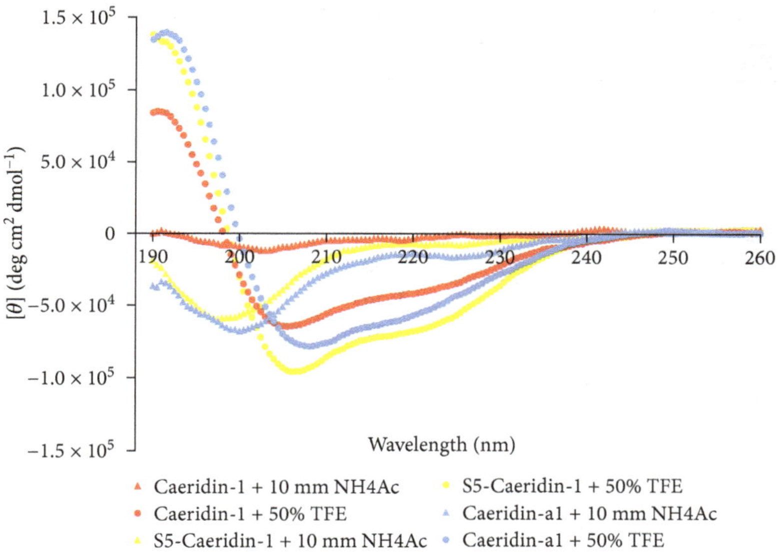

FIGURE 6: CD spectra under different conditions measured from 260 nm to 190 nm and plotted as mean residue ellipticity [θ] versus wavelength.

TABLE 2: Secondary structures of Caeridin-1, S5-Caeridin-1, and Caeridin-a1 deduced from CD spectra by K2D3 programme.

Peptide	Method	α	β
Caeridin-1	K2D3	86.68%	0.02%
S5-Caeridin-1	K2D3	94.98%	0.01%
Caeridin-a1	K2D3	95.13%	0.02%

α = helices; β = strands.

antimicrobial activities [27]. Caeruleins, similar to the mammalian cholecystokinin-8 (CCK-8) and hexagastrin, were proven as powerful pharmacological peptides inducing gastrointestinal smooth muscle contractions and analgesic and hormonal activities [28]. Caeridins, unfortunately, were reported to have neither antimicrobial activities nor pharmacologically active activities [9]. From the constructed cDNA, we have consistently cloned three different preprocaeridin cDNAs by means of a 3′-RACE strategy, using a degenerate sense primer pool designed to a conserved nucleotide sequence in the 5′ untranslated region of homologous Caerin cDNA. These clones contained full-length open-reading frames of preprocaeridins. The signal peptide primary structures and nucleic acid sequences are highly conserved and some are identical, and the conserved trends extend to the acidic prosegments, but not very obviously conserved when it comes to mature peptide encoding regions, as shown in Figure 2. All these mature peptides deduced from cloned skin secretion cDNAs were found to have a C-terminal amidation in the corresponding reversed-phase HPLC fractions by the subsequent mass spectrometric analysis as shown in Figures 5 and 6. These mature peptides were subsequently compared with the protein database of the National Center of Biotechnology Information (NCBI), and it was found that one of these mature peptides was identical to Caeridin-1 with unknown transcriptomic information [9], here also called Caeridin-1. The peptide with two amino acids different from Caeridin-1 was called S5-Caeridin-1, and the final peptide with 80% identity to Caeridin-2 and Caeridin-3 was called Caeridin-a1.

The chemical synthetic replicates of Caeridin-1, S5-Caeridin-1, and Caeridin-a1 were tested through various bioassays. Antimicrobial activities of these Caeridins were screened on standard bacteria and drug-resistant bacteria given that peptides have previously shown potential to be developed as conventional antibiotic alternatives to unravel the global antimicrobial resistance. Smooth muscle is of great significance in physiological modulation; for example, the smooth muscle in the uterus helps in parturition, the smooth muscle in the bladder helps in uresis, the smooth muscle in arteries determines blood flow, and the smooth muscle in the digestive tract modulate nutrients and wastes mobility. Therefore, pharmacological activities of these Caeridins were tested on different rat smooth muscles.

Caeridin-1 and S5-Caeridin-1, deficient of physiochemical properties, were found to have no effects on membrane interaction and disruption and thus failed to show antimicrobial activities [29]. However, Caeridin-1 was found to mediate a contraction of rat urinary bladder smooth muscle with EC_{50} = 5.6 nM and a maximum contraction of 0.3 g in tension change with no haemolytic effects. S5-Caeridin-1 was found to induce a relaxation of rat ileum smooth muscle with EC_{50} = 9.5 nM and a maximum relaxation of 0.2 g in tension change with no haemolysis. The discovery of Caeridin-1 and S5-Caeridin-1 provided templates for developing urinary system and digestive system regulatory drugs, but the mechanisms of action need further investigations.

Caeridin-a1 was recognised to be neutral in charge, with a basic amino acid, -Lys-, and an acidic amino acid, -Asp-, and adopted α-helical conformation in the membrane-mimetic environment. Caeridin-a1 was found to be inhibitory against the growth of normal microorganisms, the

Table 3: Physicochemical properties of Caeridins determined by Heliquest.

Peptides	Hydrophobicity	Hydrophobic moment	Net charge	Charged residues
Caeridin-1	0.807	0.417	−1	ASP 1
S5-Caeridin-1	0.783	0.413	−1	ASP 1
Caeridin-a1	0.77	0.477	0	ASP 1, LYS 1
Caeridin-2	0.751	0.45	−1	ASP 1
Caeridin-3	0.735	0.489	−1	ASP 1
Caeridin-4	0.725	0.421	−1	ASP 1
Caeridin-5	0.841	0.369	0	none
Caeridin-6	0.878	0.377	0	none
Caeridin-7.1	0.757	0.615	−1	ASP 1

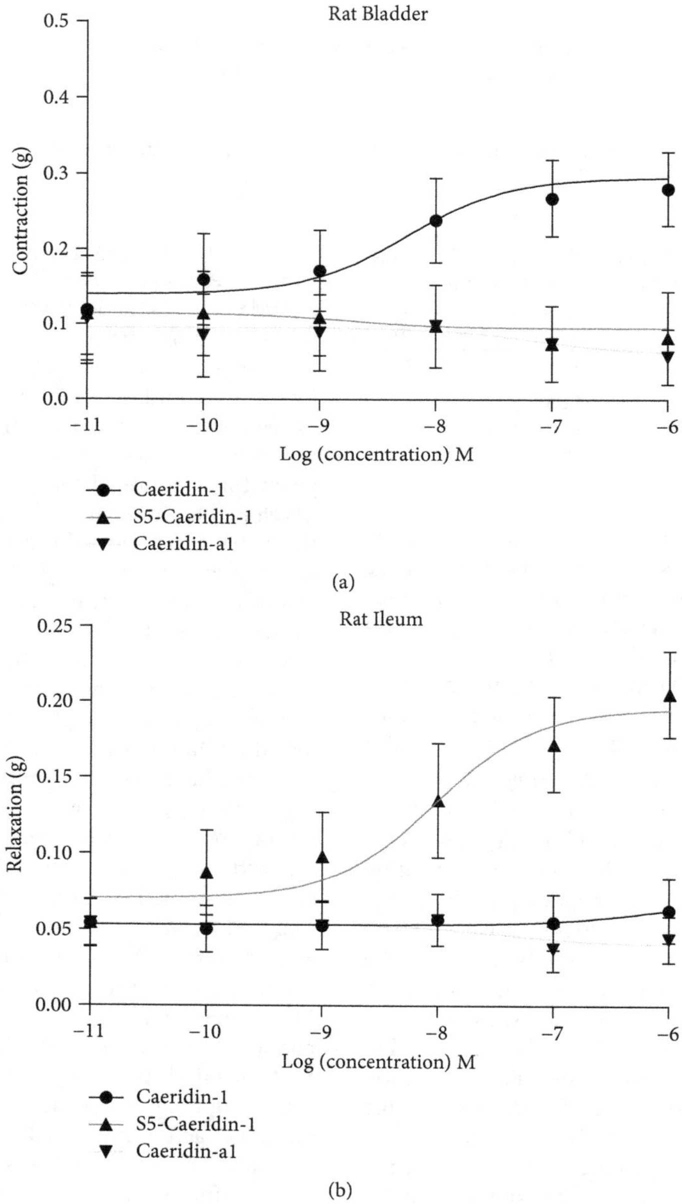

Figure 7: Dose-response curves of (a) synthetic Caeridins using rat urinary bladder smooth muscle preparations and (b) synthetic Caeridins on rat ileum smooth muscle preparations. Each point represents the mean and standard error of mean (SEM) of 5 determinations.

Table 4: Minimum inhibitory concentration (MIC) and minimum bactericidal concentration (MBC) of Caeridins, melittin, and bradykinin against different tested microorganisms.

Peptides	MIC/MBC (μM)				MIC/MFC (μM)
	S. aureus NCTC10788	MRSA NCTC12493	*E. faecalis* NCTC12697	*E. coli* NCTC10418	*C. albicans* NCYC1467
Caeridin-a1	8/16	16/32	32/64	32/64	32/64
S5-Caeridin-1	>512/>512	>512/>512	>512/>512	>512/>512	>512/>512
Caeridin-1	>512/>512	>512/>512	>512/>512	>512/>512	>512/>512
Melittin	1/2	2/4	2/4	4/8	1/8
Bradykinin	>512/>512	>512/>512	>512/>512	>512/>512	>512/>512

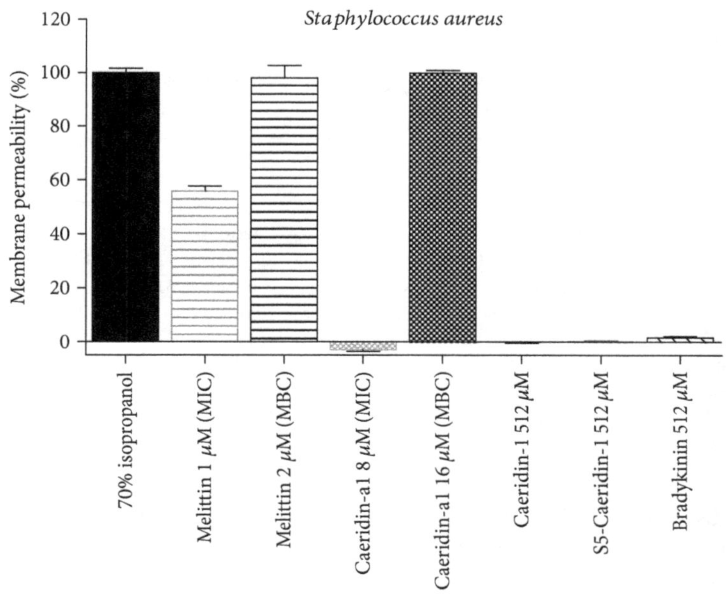

Figure 8: Membrane permeabilisation of *S. aureus* treated with melittin (MIC = 1 μM and MBC = 2 μM), Caeridin-a1 (MIC = 8 μM and MBC = 16 μM), Caeridin-1 (512 μM), S5-Caeridin-1 (512 μM), bradykinin (512 μM), and 70% isopropanol for 2 h. The degree of membrane permeability was calculated in comparison with the maximum membrane permeabilisation control (70% isopropanol-treated). The error bars represent SEM of three independent experiments and each experiment with five replicates.

Gram-positive bacterium, *S. aureus*, the Gram-negative bacterium, *E. coli*, and the yeast, *C. albicans*, as well as against MRSA and β-lactam-resistant bacterium *E. faecalis* which are hard to be controlled by conventional antibiotics and tend to cause more virulent infections. Antimicrobial susceptibility tests suggest that Caeridin-a1 showed better antimicrobial potency against the Gram-positive bacteria, *S. aureus*, MRSA, and *E. faecalis*, with MIC at 8 μM, 16 μM, and 32 μM, respectively, but less potent against the Gram-negative bacterium, *E. coli*, with MIC at 32 μM. As the membrane disruption is recognised as one of the mechanisms of antimicrobial actions of antimicrobial peptides, membrane permeability tests were performed on the standard Gram-positive bacterium (*S. aureus*) and Gram-negative bacterium (*E. coli*). The Caeridin-a1-induced membrane permeabilisation of *S. aureus* was monitored only at its MBC (16 μM), indicating that Caeridin-a1 at its MIC = 8 μM was not able to permeabilise the cytoplasmic membrane within 2 h. However, Caeridin-a1 induced membrane permeabilisation of *E. coli* within 30 min at both its MIC (32 μM) and MBC (64 μM) against *E. coli*. The significant membrane-disruptive effect of Caeridin-a1 was also observed in HMEC-1 cells at 100 μM and was weakened with the decrease of the peptide concentration. These results suggest that the swift membrane permeability depends on the administrated concentration of Caeridin-a1. Caeridin-a1 at its higher concentrations could rapidly and unspecifically interact and disrupt the cytoplasmic membrane leading to cell lysis, but Caeridin-a1 at its lower concentrations may perform antimicrobial activities by other mechanisms or may need longer time to induce membrane permeabilisation. Caeridin-a1 was also found to induce haemolysis at higher concentrations, whereby 18% haemolysis was detected at 32 μM, which consequently decreases its potential therapeutic value. However, the size of Caeridin-a1 is relatively small with only 15 amino acid residues, and therefore, Caeridin-a1 could be used as a template for developing smaller antimicrobial peptides after rational structural modification, which may be of interest to the pharmaceutical industry due to reduced cost of development and easier manufacturing processes.

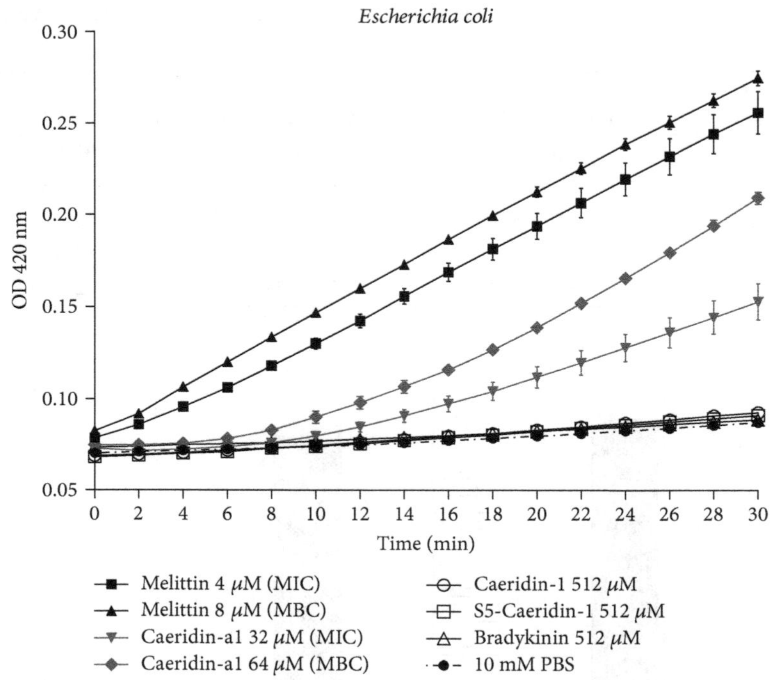

FIGURE 9: Membrane permeabilisation of *E. coli* treated with Caeridin-a1, Caeridin-1, S5-Caeridin-1, melittin, and bradykinin. The hydrolysis of ONPG generated by the release of cytoplasmic β-galactosidase of *E. coli* after treatment at MIC and MBC of Caeridin-a1, MIC and MBC of melittin, and maximum tested concentration of Caeridin-1, S5-Caeridin-1, and bradykinin was measured spectroscopically at 420 nm UV for 30 min. Each point represents mean absorbance and SEM of three independent experiments and each experiment with five replicates.

Although the Caeridins discovered in this study share a similar amino acid composition and structural elements with each other and with the Caeridins reported previously, the bioactivities they exhibited are different. Only Caeridin-a1 was found to demonstrate potent antimicrobial and haemolytic effects, but Caeridin-1, S5-Caeridin-1, and other reported Caeridins were not. This could be explained by the variation in the membrane-disruptive action of peptides. The Gram-positive bacterial cell walls contain anionic teichoic acids and lipoteichoic acids. The anionic lipopolysaccharides are widely distributed in the outer leaflet of the outer membrane of Gram-negative bacteria. The negatively charged glycosaminoglycans and gangliosides are composed in mammalian cells exterior to the cytoplasmic membrane. Firstly, amphipathic peptides with positively charged residues are attracted to the vicinity of the cell surface by electrostatic actions. Secondly, peptides attach to the cell membrane via electrostatic and hydrophobic interactions. Then they undergo conformational transition to bind with the membrane and form transmembrane pores/channels or translocate phospholipid bilayers, thus destroying the membrane integrity. Because of the abundance of the acidic phospholipids, such as cardiolipin and phosphatidylglycerol, and the absence of sterol, in the cytoplasmic membrane of bacteria, amphipathic peptides with cationic residues are able to attach to the bacterial membrane by both electrostatic interaction and hydrophobic interaction. However, the mammalian cell membranes only have acidic phospholipids at the inner facet of the phospholipid bilayers, suggesting that peptides are more able to interact with the mammalian cell membrane by hydrophobic interaction. In addition to phospholipids, the numerous sterols embedded in the bilayers impede the peptide from better binding with the membrane and triggering further membrane disruption, which is the same on red blood cells. The cationic residue, -Lys-, in the amino acid composition of Caeridin-a1 may facilitate the peptide being better attracted to the cell surface and interacting with the cell membrane by electrostatic effects, though the net charge of Caeridin-a1 was 0 due to the presence of the anionic residue, -Asp-. In addition to electrostatic interaction, the high amphipathicity and hydrophobicity of Caeridin-a1 may also promote the peptide to bind to the cell membrane by hydrophobic interaction. Then, Caeridin-a1 may conform to an α-helical structure to bind with the membrane and form transmembrane pores, thus destroying the membrane. Therefore, Caeridin-a1 was able to disrupt the non-sterol-protected bacterial cell membrane at lower concentrations, and sterol/cholesterol-protected mammalian cell membrane and red blood cell membrane at higher concentrations, showing antimicrobial effects on the tested microorganisms, cytotoxic effects on HMEC-1 cells, and haemolytic effects on the horse erythrocytes. Regarding the other Caeridins, as shown in Table 3, the net charge of Caeridin-1, S5-Caeridin-1, Caeridin-2-4, and Caeridin-7.1 is −1 due to the presence of an acidic residue, -Asp-, and the net charge of Caeridin-5 and Caeridin-6 is zero because of the lack of charged residues. Therefore, these Caeridins are expected to fail to undergo electrostatic attraction and interaction with the membrane. They showed no membrane disruption in the tested microorganisms, mammalian cells, and horse red blood cells.

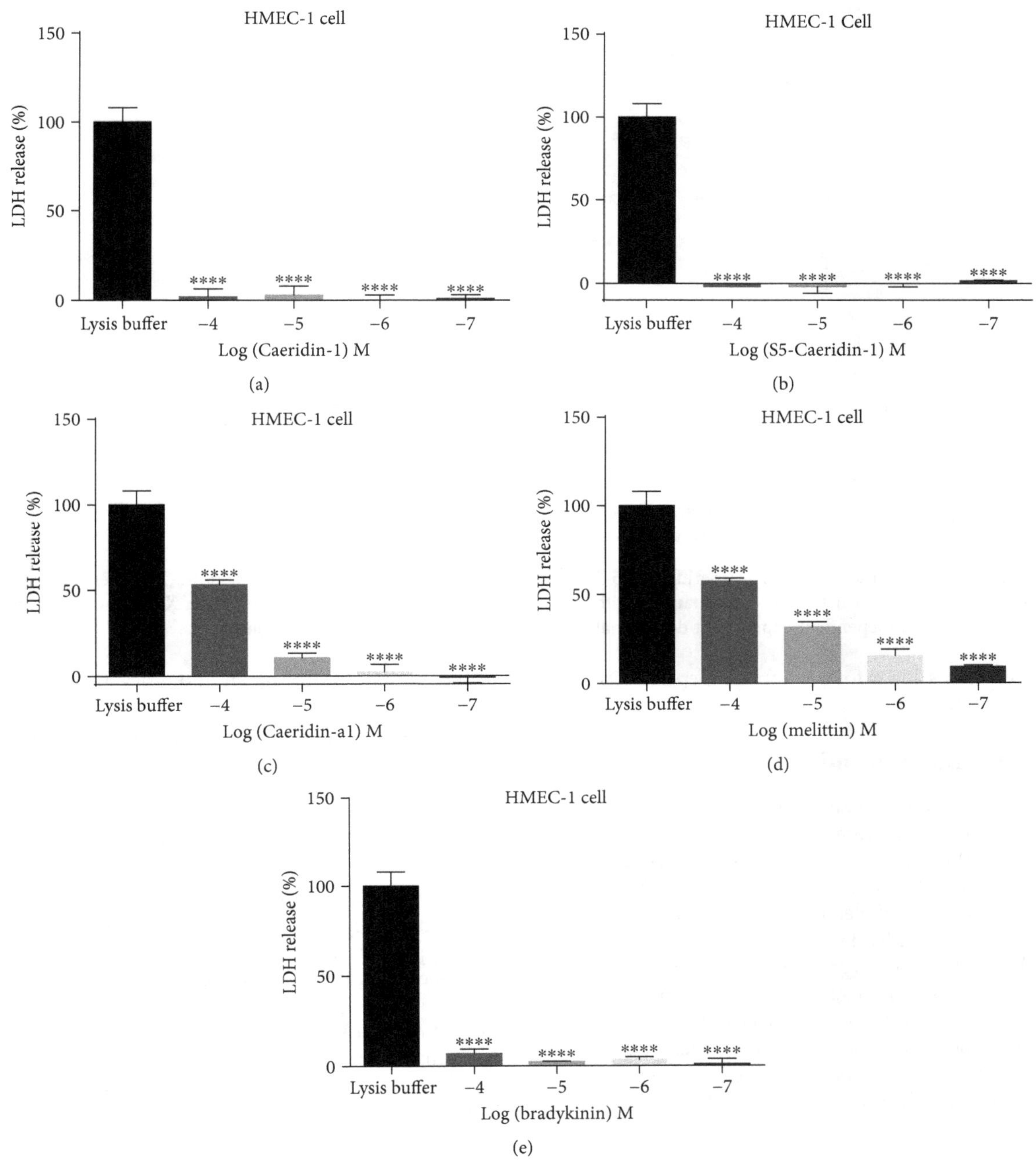

FIGURE 10: Membrane permeabilisation of human microvascular endothelial cell line, HMEC-1, reflected by the level of LDH released into the medium, after being treated by (a) Caeridin-1, (b) S5-Caeridin-1, (c) Caeridin-a1, (d) positive peptide melittin, and (e) negative peptide Bradykinin. The level of LDH release was calculated in comparison with the maximum LDH release control group (lysis buffer-treated)). $^{****}p < 0.0001$ represents a significant change in the level of LDH release, compared with the maximum LDH release control group (lysis buffer-treated), using one-way analysis of variance followed by Dunnett's multiple comparisons tests. The error bar represents SEM of three independent experiment determinations and each experiment with triplicates.

Although no antimicrobial effects were detected, Caeridin-1 and S5-Caeridin-1 were presumed to be mediated by the G protein-coupled receptors to induce contraction and relaxation of smooth muscle preparations with no haemolysis and cytotoxicity. However, Caeridin-a1 with strong membrane disruptive activity would also cause damage to smooth muscle cell membrane; thus, no smooth muscle activities were found with Caeridin-a1.

The discovery of Caeridin-1, S5-Caeridin-1, and Caeridin-a1 reinforced the multiple bioactivities of the Caeridin peptide family. Although they are very similar with other reported Caeridins, certain amino acid mutations were identified. It is these small mutations in amino acid sequences and compositions that affect physicochemical characteristics and functioning structures and lead to big differences in bioactivities. Therefore, optimal amino acid sequence selections from

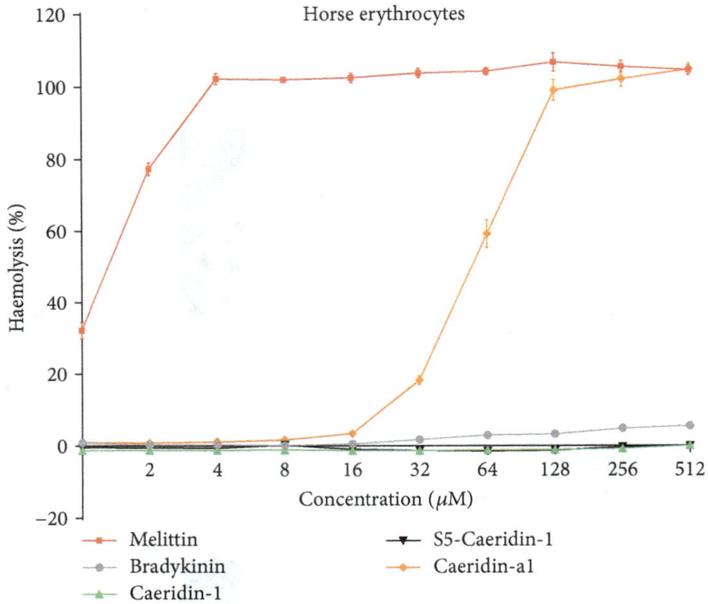

FIGURE 11: Haemolytic activities of synthetic Caeridin-1, S5-Caeridin-1, Caeridin-a1, melittin, and bradykinin. The level of haemolysis tested on horse red blood cells was calculated in comparison with the maximum haemolysis control (1% Triton X-100-treated). The error bar represents SEM of three independent experiment determinations and each experiment with five replicates.

frog skin secretions lay a foundation for peptide discovery and development.

4. Materials and Methods

4.1. Collection of Frog Skin Secretions. The adult Australian frogs ($n = 4$), *L. caerulea*, were captive bred and kept for three months before the collection of their skin secretions. Their defensive skin secretions were collected from their dorsal skin by mild transdermal electrical stimulation according to the technique reported by Tyler et al. [30], then lyophilised and stored at −20°C prior to analysis. Sampling of skin secretion was performed by Mei Zhou under UK Animal (Scientific Procedures) Act 1986, project license PPL 2800, issued by the Department of Health, Social Services and Public Safety, Northern Ireland. Procedures had been vetted by the IACUC of Queen's University Belfast and approved on 19 February 2016.

4.2. "Shotgun" Cloning of Novel Caeridin Precursor-Encoding cDNAs from the Skin Secretion-Derived cDNA Library of L. caerulea. The cDNA library of *L. caerulea* was established through the reverse transcription of the mRNA. The mRNA was extracted from the lyophilised skin secretion by using Dynabeads® mRNA DIRECT™ Kit (Life Technologies, Oslo, Norway). The extracted mRNA was reverse transcribed to synthesise the first-strand cDNA by following the instructions described by the BD SMART™ RACE cDNA Amplification Kit (Clontech, Palo Alto, CA, USA) to establish the cDNA library. The cDNA library was subjected to 3′-RACE (rapid amplification of cDNA ends) procedures to obtain full-length preprocaeridin nucleic acid sequence data with a SMART-RACE Kit (Clontech, Palo Alto, CA, USA) according to the manufacturer, which involved a NUP primer supplied with the kit and a degenerate sense primer pool (S1: 5′-GVCCTTGTAAAGACCAAVCATGGCTT-3′) designed to a segment of the 5′-untranslated region of Caerin cDNAs cloned from the skin secretion of *L. caerulea* (EMBL accession Number AY218778-82). This approach was applied as the segment of the 5′-untranslated region employed was found to be highly conserved among the homologous peptide cDNAs [31]. An additional advantage of this strategy was that the entire open-reading frame of these transcripts could be deduced through a single 3′-RACE procedure. This PCR cycling procedure was carried out as follows: initial denaturation step: 60 s at 94°C; 40 cycles of amplification: denaturing for 30 s at 94°C; primer annealing for 30 s at 61°C; and strand extension for 180 s at 72°C. RACE-PCR products were analysed by DNA agarose gel electrophoresis, purified, and cloned in a pGEM®-T Easy vector system (Promega Corporation, Southampton, UK), and the selected samples were sequenced by an ABI 3730 DNA automated analyser (Applied Biosystems, Foster City, CA, USA). The nucleic acid sequences were translated into amino acid sequences by the ExPASy Translate Tool online portal. The deduced preprocaeridins were analysed using the Blast Alignment Search Tool (BLAST) of the National Centre for Biotechnology Information (NCBI) by comparing to peptides with known amino acid sequences in the protein database. Alignments of similar regions were established by MEGA 6. The nucleotide sequences of the cDNA encoding the novel Caeridins were registered with the European Nucleotide Archive (ENA) browser at http://www.ebi.ac.uk/embl/genomes/submission/.

4.3. Chromatographic Isolation and Primary Structural Characterisation of Novel cDNA-Deduced Caeridins from the Skin Secretion of L. caerulea. The lyophilised skin secretion of *L. caerulea* dissolved in solution A (99.95% H_2O, 0.05% trifluoroacetic acid (TFA) was fractionated by reversed-phase HPLC using a Waters Binary pump HPLC system (Waters, Milford, MA, USA) fitted with an analytical column (Jupiter, C5, 300 Å, 5 μm, 250 mm × 4.6 mm, Phenomenex, Macclesfield, Cheshire, UK) and eluted with a linear gradient formed from 100% solution A (99.95% H_2O, 0.05% TFA) to 100% solution B (80.00% acetonitrile, 19.95% H_2O, 0.05% TFA) in 240 min at a flow rate of 1000 μl/min. Fractions were collected at each minute, and the effluent was continuously detected at $\lambda = 214$ nm. The dead volume between column and fraction collector was minimal (20 μl).

The molecular masses of peptides in each fraction were analysed by matrix-assisted laser desorption/ionisation time-of-flight mass spectrometry (MALDI-TOF MS) on a linear time-of-flight Voyager DE mass spectrometer (Voyager DE, Perspective Biosystems, Foster City, CA, USA) in a positive detection mode using α-cyano-4-hydroxycinnamic acid matrix to construct a mass spectral library of the *L. caerulea* skin secretion peptide. The instrument was calibrated in the range of 1–4 kDa, and the accuracy of mass determinations was ±0.1%. The computed molecular masses of predicted mature Caeridins deduced from encoded cDNA were used to interrogate the mass spectral library to identify the putative Caeridins. The skin secretion fractions containing the putative Caeridins diluted in solution A (99.90% H_2O, 0.10% formic acid) were pumped directly onto an analytical HPLC column (Jupiter, C18, 300 Å, 5 μm, 150 mm × 4.6 mm, Phenomenex, Macclesfield, Cheshire, UK) connected to an LCQ Fleet ESI ion-trap mass spectrometer (Thermo Fisher, San Jose, CA, USA) and eluted from 100% solution A (99.90% H_2O, 0.10% formic acid) to 100% solution B (19.90% H_2O, 80.00% acetonitrile, and 0.10% formic acid) in 135 min at a flow rate of 20 μl/min. Mass analysis was performed in a positive ion mode with acquired spectra in the range of 500–2000 *m*/*z* with >50% relative intensity during HPLC-MS. Parameters for electrospray ionisation ion-trap mass spectrometry (ESI/MS) were spray voltage +4.5 kV, drying gas temperature 320°C, drying gas flow 200 μl/min, and maximum accumulation time (ion trap) 350 ms. The first mass analysis was performed in full-scan mode, then peptide ions with >50% relative intensity were selected for fragmentation by collision-induced dissociation (CID), to generate *b* and *y* ions detected in a second mass analysis. The instrument was controlled by Xcalibur software (Thermo Fisher, San Jose, CA, USA), and data analysis was performed using Proteome Discoverer 1.0 (Thermo Fisher, San Jose, CA, USA). Sequest™ algorithm was employed to compare the acquired fragment ion profiles with the theoretical fragment ions generated from a FASTA database specific for this species built by "shotgun" cloning (as described above) to confirm the amino acid sequences of individual Caeridins.

4.4. Solid-Phase Peptide Synthesis. The novel Caeridins, positive peptide Melittin, and negative peptide Bradykinin were synthesised by solid-phase Fmoc chemistry using a Tribute automated solid phase peptide synthesiser 4 (Protein Technologies, Tucson, AZ, USA). After cleavage from resin and deprotection, each crude peptide was obtained. The amino acid sequences of these peptides are summarised in Table 5. The authenticity of the synthetic peptides was established by HPLC purification and confirmed by MALDI-TOF (Voyager DE, Perspective Biosystems, Foster City, CA, USA) and an LCQ Fleet electrospray ion-trap mass spectrometer (Thermo Fisher Scientific, San Francisco, CA, USA).

4.5. Secondary Structural Determination by Circular Dichroism (CD). CD spectra were obtained at 20°C using a 1 mm high-precision quartz cell (Hellma Analytics, Essex, UK) with a JASCO J-815 CD spectrometer (Jasco, Essex, UK). The measurement range was from 260 nm to 190 nm at a scanning speed of 200 nm/min. The bandwidth was 1 nm, and the data pitch was 0.5 nm. Caeridins were, respectively, dissolved in 10 mM NH_4Ac or 10 mM NH_4A c with 50% TFE to reach a final concentration of 100 μM. The CD spectra were analysed by a K2D3 online portal programme [13].

4.6. Rat Smooth Muscle Tests. Female Wistar rats (250–300 g) were euthanized by CO_2 asphyxiation followed by cervical dislocation under appropriate Home Office (UK) animal licenses. Full urinary bladder and ileum were carefully removed from the abdomen, then placed in cold Kreb's solution (118 mM NaCl, 4.7 mM KCl, 25 mM $NaHCO_3$, 1.15 mM NaH_2PO_4, 2.5 mM $CaCl_2$, 1.1 mM $MgCl_2$, and 5.6 mM glucose), equilibrated with a carbogen mixture (95% O_2, 5% CO_2) [32]. The prepared smooth muscles linked with transducers were placed in 2 ml capacity organ baths with Kreb's solution flowing through at 2 ml/min and maintained at 37°C with constant bubbling of carbogen gas mixture (95% O_2, 5% O_2) and equilibrated for 1 h prior to analysis. Then the viability of these smooth muscle preparations was tested by 60 mM KCl. Caeridins were tested at sequential molar concentrations in the range from 10^{-6} M to 10^{-11} M. Dose-response curves were depicted for each Caeridin on individual smooth muscle preparations ($n = 5$). Changes in tension of the smooth muscle preparations were detected by pressure transducers connected to a PowerLab System (ADInstruments Pty Ltd.). Data were recorded and calculated to obtain the means of responses and standard error of means (SEM). The dose-response curve of each Caeridin was established using best-fit algorithms through the data analysis package provided by GraphPad Prism (Version 6.0; GraphPad Software Inc., San Diego, CA, USA), and the comparative EC_{50} value for each peptide was depicted.

4.7. Antimicrobial Tests. The antimicrobial activities were tested by the broth microdilution method and evaluated by the determination of MIC values against different tested microorganisms, the Gram-positive bacteria, *S. aureus* (NCTC 10788), *E. faecalis* (NCTC12697), and MRSA (NCTC 12493); the Gram-negative bacterium, *E. coli* (NCTC 10418); and the yeast, *C. albicans* (NCYC 1467). The microorganisms were cultured to their log phase in MHB and diluted to 5×10^5 CFU/ml then subsequently treated with peptide solutions of gradient concentrations. After incubation at 37°C overnight, the growth of microorganisms was

TABLE 5: The amino acid compositions of novel Caeridins, positive peptide melittin, and negative peptide bradykinin.

Peptide	Amino acid sequence
Caeridin-1	H_2N-GLLDGLLGTLGL-$CONH_2$
S5-Caeridin-1	N_2H-GLLDSLLGGLGL-$CONH_2$
Caeridin-a1	H_2N-GLFDIVGKLLGGLGL-$CONH_2$
Melittin	H_2N-GIGAVLKVLTTGLPALISWIKRKRQQ-$CONH_2$ [15–17]
Bradykinin	H_2N-RPPGFSPFR-COOH [18–20]

detected at a wavelength of 550 nm using a Synergy HT plate reader (Biolise BioTek ELx808, Winooski, VT, USA). The MICs were determined as the lowest concentration of peptide where growth was detectable.

4.8. Cell Membrane-Permeable Tests. The cell membrane permeability of *S. aureus* was tested with SYTOX Green Nucleic Acid Stain (Life Technologies, Carlsbad, CA, USA). The log-phase *S. aureus* was pelleted and suspended in 5% TSB in 0.85% NaCl solution to achieve 1×10^8 CFU/ml cells which was then treated with peptide solutions at 37°C for 2 h. The maximum membrane-compromised *S. aureus* was produced by 70% isopropanol treatment. Subsequently, cells were mixed with 5 μM SYTOX Green stain and incubated in the dark for 10 min. The fluorescence was measured in a 37°C Synergy HT plate reader (Biolise BioTek ELx808, Winooski, VT, USA) at excitation and emission wavelengths of 480 nm and 528 nm, respectively. The membrane permeability was calculated as

$$\text{Membrane permeability}(\%) = \frac{(F - F_0)}{(F_1 - F_0)} \times 100\%, \quad (1)$$

where F represents the processed fluorescence intensity with peptide treatments, F_0 represents the processed fluorescence intensity of untreated *S. aureus*, and F_1 represents the processed fluorescence intensity in the maximum membrane-compromised *S. aureus* (70% isopropanol-treated). The fluorescence intensities of background were subtracted prior to calculation.

The membrane permeabilsation of *E.coli* was determined by measuring the release of β-galactosidase activity from *E. coli* into the culture medium with ONPG as the substrate. The log-phase *E. coli* was grown in MHB medium containing 2% lactose at 37°C, and the cells were pelleted. The cell pellets were suspended in 10 mM PBS (pH 7.4) containing 1.5 mM ONPG and diluted to 1×10^8 CFU/ml. Aliquots of cells were incubated with different concentrations of peptides at 37°C. Permeabilisation was measured at 420 nm every 2 min from 0 to 30 min, which reflects ONPG influx into the cells working as an indicator of the permeability of the tested peptides.

The membrane permeabilisation of the human microvascular endothelial cell line, HMEC-1, was evaluated using the Pierce LDH Cytotoxicity Assay Kit (Thermo Fisher Scientific, San Francisco, CA, USA). HMEC-1 cells were cultured in MCDB-131 medium (Life Technology, Paisley, UK) with 10% FBS, 10 ng/ml epidermal growth factor (EGF) (Life Technology, Paisley, UK), 10 mM L-glutamine (Life Technology, Paisley, UK), and 1% penicillin-streptomycin (Sigma-Aldrich, St. Louise, MO, USA), at 37°C under 5% CO_2. Cells were seeded in a 96-well plate at 5000 cells per well and cultured overnight to 80% confluency. Then, the cells were treated with different peptide solutions of gradient concentrations from 10^{-7} M to 10^{-4} M for 24 h at 37°C, 5% CO_2, and H_2O-treated cells were made as the spontaneous LDH release control. Lysis buffer was added to cells and incubated for 45 min at 37°C, 5% CO_2, to produce the lysis buffer-treated cells which were referred to as maximum LDH release control. Then 50 μl of each sample supernatant was transferred to a 96-well flat-bottom plate in triplicate wells, to which 50 μl of reaction mixture was added, incubated, and protected from light at room temperature for 30 min. After addition of 50 μl Stop Solution to each sample well and removal of bubbles present in sample wells, the absorbance at 490 nm and 680 nm was measured. The absorbance values at 680 nm were subtracted from the absorbance values at 490 nm before calculation of the level of LDH release (cytotoxicity). The level of LDH release was calculated as

$$\text{LDH release}(\%) = \frac{(L - L_0)}{(L_1 - L_0)} \times 100\%, \quad (2)$$

where L refers to sample-treated LDH activity, L_0 to spontaneous (H_2O-treated) LDH activity, and L_1 to maximum (lysis buffer-treated) LDH activity.

Data analysis was performed using one-way analysis of variance followed by Dunnett's multiple comparison tests with maximum LDH release control in GraphPad Prism (Version 6.0; GraphPad Software Inc., San Diego, CA, USA).

4.9. Haemolytic Tests. The haemolysis of Caeridins was tested on prepared defrinated horse red blood cells (TCS Biosciences Ltd., Buckingham, UK) by incubating equal volumes of peptide solutions with erythrocyte suspensions at 37°C for 2 H. Maximum haemolysis control was included by incubating equal volumes of 1% Triton X-100 (Sigma-Aldrich, St. Louis, MO, USA) with erythrocyte suspensions. Lysis of erythrocytes was detected by the measurement of supernatant absorbance at a wavelength of 550 nm using a Synergy HT plate reader (Biolise BioTek ELx808, Winooski, VT, USA). The level of haemolysis was as calculated as

$$\text{Haemolysis}(\%) = \frac{(A - A_0)}{(A_1 - A_0)} \times 100\%, \quad (3)$$

where A represents the absorbance with peptide treatments, A_0 represents the average absorbance in untreated erythrocytes, and A_1 represents the average absorbance in the maximum haemolysis control (1% Triton X-100-treated).

5. Conclusions

The discovery of Caeridin-1, S5-Caeridin-1, and Caeridin-a1 not only contributed templates for drug design but also provided evidence that peptides from frog skin secretions possess bioactive properties and uphold expectations that peptides derived from natural resources may contribute to the clinical development of novel therapeutics.

Abbreviations

HPLC: High-performance liquid chromatography
MIC: Minimum inhibitory concentration
MBC: Minimum bactericidal concentration
CID: Collision-induced dissociation
MALDI-TOF MS: Matrix-assisted laser desorption/ionisation time-of-flight mass spectrometry
ESI/MS: Electrospray ionisation ion-trap mass spectrometry.

Disclosure

An early progress of this work was presented in the Drug Discovery & Therapy World Congress: https://www.ddtwc.com/abstracts/Lei-Li.php.

Conflicts of Interest

The authors declare no conflict of interest.

Authors' Contributions

Tianbao Chen, Mei Zhou, Qing Wu, and Lei Wang conceived and designed the experiments; Lei Li, Xi Wang, and Huimin Lu performed the experiments; Lei Li and Lei Wang analysed the data; Lei Wang contributed reagents/materials/analysis tools; Lei Li wrote the paper; Chris J. Watson, Xinping Xi, and Mei Zhou revised the paper".

References

[1] D. J. Craik, D. P. Fairlie, S. Liras, and D. Price, "The future of peptide-based drugs," *Chemical Biology & Drug Design*, vol. 81, no. 1, pp. 136–147, 2013.

[2] S. Sachdeva, "Peptides as 'drugs': the journey so far," *International Journal of Peptide Research and Therapeutics*, vol. 23, no. 1, pp. 49–60, 2017.

[3] C. Lacombe, C. Cifuentes-Diaz, I. Dunia, M. Auber-Thomay, P. Nicolas, and M. Amiche, "Peptide secretion in the cutaneous glands of South American tree frog *Phyllomedusa bicolor*: an ultrastructural study," *European Journal of Cell Biology*, vol. 79, no. 9, pp. 631–641, 2000.

[4] B. T. Clarke, "The natural history of amphibian skin secretions, their normal functioning and potential medical applications," *Biological Reviews*, vol. 72, no. 3, pp. 365–379, 1997.

[5] R. C. Toledo and C. Jared, "Cutaneous granular glands and amphibian venoms," *Comparative Biochemistry and Physiology Part A: Physiology*, vol. 111, no. 1, pp. 1–29, 1995.

[6] E. König, O. R. P. Bininda-Emonds, and C. Shaw, "The diversity and evolution of anuran skin peptides," *Peptides*, vol. 63, pp. 96–117, 2015.

[7] P. C. Eterovick, A. C. O. de Queiroz Carnaval, D. M. Borges-Nojosa, D. L. Silvano, M. V. Segalla, and I. Sazima, "Amphibian declines in Brazil: an overview," *Biotropica*, vol. 37, no. 2, pp. 166–179, 2005.

[8] J. W. Gibbon, D. E. Scott, T. J. Ryan et al., "The global decline of reptiles, déjà vu amphibians: reptile species are declining on a global scale. Six significant threats to reptile populations are habitat loss and degradation, introduced invasive species, environmental pollution, disease, unsustainable use, and global climate change," *Bioscience*, vol. 50, no. 8, pp. 653–666, 2000.

[9] R. J. Waugh, D. J. M. Stone, J. H. Bowie, J. C. Wallace, and M. J. Tyler, "Peptides from Australian frogs. Structures of the caeridins from *Litoria caerulea*," *Journal of the Chemical Society, Perkin Transactions*, vol. 1, no. 5, pp. 573–576, 1993.

[10] S. T. Steinborner, J. H. Bowie, M. J. Tyler, and J. C. Wallace, "An unusual combination of peptides from the skin glands of Ewing's tree frog, Litoria ewingi. Sequence determination and antimicrobial activity," *Australian Journal of Chemistry*, vol. 50, no. 9, pp. 889–894, 1997.

[11] H. De Loof, M. Rosseneu, R. Brasseur, and J.-M. Ruysschaert, "Functional differentiation of amphiphilic helices of the apolipoproteins by hydrophobic moment analysis," *Biochimica et Biophysica Acta (BBA) - Protein Structure and Molecular Enzymology*, vol. 911, no. 1, pp. 45–52, 1987.

[12] M. Zhou, T. Chen, B. Walker, and C. Shaw, "Novel frenatins from the skin of the Australasian giant white-lipped tree frog, *Litoria infrafrenata*: cloning of precursor cDNAs and identification in defensive skin secretion," *Peptides*, vol. 26, no. 12, pp. 2445–2451, 2005.

[13] C. Louis-Jeune, M. A. Andrade-Navarro, and C. Perez-Iratxeta, "Prediction of protein secondary structure from circular dichroism using theoretically derived spectra," *Proteins: Structure, Function, and Bioinformatics*, vol. 80, no. 2, pp. 374–381, 2012.

[14] R. GAUTIER, D. DOUGUET, B. ANTONNY, and G. Drin, "HELIQUEST: a web server to screen sequences with specific α-helical properties," *Bioinformatics*, vol. 24, no. 18, pp. 2101-2102, 2008.

[15] E. Habermann, "Bee and wasp venoms," *Science*, vol. 177, no. 4046, pp. 314–322, 1972.

[16] E. Habermann, "Melittin-structure and activity," *Natural Toxins*, pp. 173–181, 1980.

[17] J. Chen and W. R. Lariviere, "The nociceptive and antinociceptive effects of bee venom injection and therapy: a double-edged sword," *Progress in Neurobiology*, vol. 92, no. 2, pp. 151–183, 2010.

[18] Anastasi, V. Erspamer, and G. Bertaccini, "Occurrence of bradykinin in the skin of *Rana temperaria*," *Comparative Biochemistry and Physiology*, vol. 14, no. 1, pp. 43–52, 1965.

[19] J. M. Conlon and U. Aronsson, "Multiple bradykinin-related peptides from the skin of the frog, *Rana temporaria*," *Peptides*, vol. 18, no. 3, pp. 361–365, 1997.

[20] J. M. Conlon, "Bradykinin and its receptors in non-mammalian vertebrates," *Regulatory Peptides*, vol. 79, no. 2-3, pp. 71–81, 1999.

[21] P. Vlieghe, V. Lisowski, J. Martinez, and M. Khrestchatisky, "Synthetic therapeutic peptides: science and market," *Drug Discovery Today*, vol. 15, no. 1-2, pp. 40–56, 2010.

[22] L. F. Skerratt, L. Berger, R. Speare et al., "Spread of chytridiomycosis has caused the rapid global decline and extinction of frogs," *EcoHealth*, vol. 4, no. 2, pp. 125–134, 2007.

[23] T. B. Hayes, P. Falso, S. Gallipeau, and M. Stice, "The cause of global amphibian declines: a developmental endocrinologist's perspective," *Journal of Experimental Biology*, vol. 213, no. 6, pp. 921–933, 2010.

[24] J. C. Z. Woinarski, A. A. Burbidge, and P. L. Harrison, "Ongoing unraveling of a continental fauna: decline and extinction of Australian mammals since European settlement," *Proceedings of the National Academy of Sciences of the United States of America*, vol. 112, no. 15, pp. 4531–4540, 2015.

[25] M. A. Apponyi, T. L. Pukala, C. S. Brinkworth et al., "Host-defence peptides of Australian anurans: structure, mechanism of action and evolutionary significance," *Peptides*, vol. 25, no. 6, pp. 1035–1054, 2004.

[26] D. Vanhoye, F. Bruston, P. Nicolas, and M. Amiche, "Antimicrobial peptides from hylid and ranin frogs originated from a 150-million-year-old ancestral precursor with a conserved signal peptide but a hypermutable antimicrobial domain," *European Journal of Biochemistry*, vol. 270, no. 9, pp. 2068–2081, 2003.

[27] S. T. Steinborner, G. J. Currie, J. H. Bowie, J. C. Wallace, and M. J. Tyler, "New antibiotic caerin 1 peptides from the skin secretion of the Australian tree frog *Litoria chloris*. Comparison of the activities of the caerin 1 peptides from the genus *Litoria*," *The Journal of Peptide Research*, vol. 51, no. 2, pp. 121–126, 1998.

[28] A. Anastasi, V. Erspamer, and R. Exdean, "Isolation and amino acid sequence of caerulein, the active decapeptide of the skin of *Hyla caerulea*," *Archives of Biochemistry and Biophysics*, vol. 125, no. 1, pp. 57–68, 1968.

[29] H. Jenssen, P. Hamill, and R. E. W. Hancock, "Peptide antimicrobial agents," *Clinical Microbiology Reviews*, vol. 19, no. 3, pp. 491–511, 2006.

[30] M. J. Tyler, D. J. M. Stone, and J. H. Bowie, "A novel method for the release and collection of dermal, glandular secretions from the skin of frogs," *Journal of Pharmacological and Toxicological Methods*, vol. 28, no. 4, pp. 199-200, 1992.

[31] T. Chen, C. Scott, L. Tang, M. Zhou, and C. Shaw, "The structural organization of aurein precursor cDNAs from the skin secretion of the Australian green and golden bell frog, *Litoria aurea*," *Regulatory Peptides*, vol. 128, no. 1, pp. 75–83, 2005.

[32] Y. Wu, L. Wang, C. Lin et al., "Vasorelaxin: a novel arterial smooth muscle-relaxing eicosapeptide from the skin secretion of the Chinese piebald odorous frog (*Odorrana schmackeri*)," *PLoS One*, vol. 8, no. 2, article e55739, 2013.

8

Revisiting Key Entry Routes of Human Epidemic Arboviruses into the Mainland Americas through Large-Scale Phylogenomics

Túlio De Lima Campos,[1] Ricardo Durães-Carvalho,[2] Antonio Mauro Rezende,[3] Otávio Valério de Carvalho,[2] Alain Kohl,[4] Gabriel Luz Wallau,[5] and Lindomar José Pena[2]

[1]*Bioinformatics Core Facility, Aggeu Magalhães Institute (IAM), Oswaldo Cruz Foundation (FIOCRUZ), Recife PE, Brazil*
[2]*Department of Virology, Aggeu Magalhães Institute (IAM), Oswaldo Cruz Foundation (FIOCRUZ), Recife PE, Brazil*
[3]*Department of Microbiology, Aggeu Magalhães Institute (IAM), Oswaldo Cruz Foundation (FIOCRUZ), Recife PE, Brazil*
[4]*MRC-University of Glasgow Centre for Virus Research, Glasgow G61 1QH, UK*
[5]*Department of Entomology, Aggeu Magalhães Institute (IAM), Oswaldo Cruz Foundation (FIOCRUZ), Recife PE, Brazil*

Correspondence should be addressed to Gabriel Luz Wallau; gabriel.wallau@cpqam.fiocruz.br and Lindomar José Pena; lindomarvet10@gmail.com

Academic Editor: Corey Nislow

The rapid worldwide spread of chikungunya (CHIKV), dengue (DENV), and Zika (ZIKV) viruses have raised great international concern. Knowledge about the entry routes and geographic expansion of these arboviruses to the mainland Americas remain incomplete and controversial. Epidemics caused by arboviruses continue to cause socioeconomic burden globally, particularly in countries where vector control is difficult due to climatic or infrastructure factors. Understanding how the virus circulates and moves from one country to another is of paramount importance to assist government and health officials in anticipating future epidemics, as well as to take steps to help control or mitigate the spread of the virus. Through the analyses of the sequences of arbovirus genomes collected at different locations over time, we identified patterns of accumulated mutations, being able to trace routes of dispersion of these viruses. Here, we applied robust phylogenomic methods to trace the evolutionary dynamics of these arboviruses with special focus on Brazil, the epicenter of these triple epidemics. Our results show that CHIKV, DENV-1–4, and ZIKV followed a similar path prior to their first introductions into the mainland Americas, underscoring the need for systematic arboviral surveillance at major entry points of human population movement between countries such as airports and seaports.

1. Introduction

Mosquito-borne viruses such as chikungunya (CHIKV), dengue (DENV), and Zika (ZIKV) viruses have rapidly spread across the globe in recent years, causing large-scale outbreaks in countries of the Southern Hemisphere including Brazil [1, 2]. DENV and ZIKV are members of the genus Flavivirus within the Flaviviridae family, which also includes other species such as yellow fever virus (YFV), Japanese encephalitis virus (JEV), Spondweni virus (SPOV), and West Nile virus (WNV) [3]. CHIKV is a member of the Alphavirus genus in the family Togaviridae [4]. DENV, ZIKV, and CHIKV genomes consist of a +ssRNA of approximately 11 kb in length. Their main vectors are the anthropophilic mosquitoes of the *Aedes* genus, which are widespread across tropical and subtropical regions of the world [5].

Clinically, it is often difficult to differentiate DENV, CHIKV, and ZIKV infections as they have similar manifestations, which include fever, exanthema, and arthralgia [6, 7]. During the acute phase of infection, the diseases can be diagnosed and differentiated using molecular methods, such as real-time reverse transcription polymerase chain reaction [8]. CHIKV causes a febrile illness characterized by sudden onset, backache, headache, photophobia, and rash; unlike

DENV and ZIKV fever, CHIKV infection is associated with high rates of symptomatic infections and with recurrent polyarthralgias, which can be disabling [6]. CHIKV has only one serotype, but four viral genotypes have been identified to date (East-Central-South African (ECSA), West African, Asian, and the Indian Ocean lineage (IOL)) [9]. Dengue is the most prevalent arboviral disease of humans worldwide. It is estimated that 390 million infections occur every year and several billion people are at risk of infection [10]. The disease is caused by four closely related but antigenically distinct serotypes of DENV (DENV-1–4). For DENV-1 and DENV-2, five and six genotypes have been described, respectively, whereas DENV-3 and DENV-4 have been subdivided into four genotypes [11].

Initially discovered in 1947, ZIKV was a relatively unknown pathogen until 2007, when a large outbreak occurred in Yap Island, a part of the Federated States of Micronesia. ZIKV reemerged in 2013 in French Polynesia and rapidly disseminated throughout the Pacific [12–14]. In March 2015, Brazil reported autochthonous transmission of ZIKV for the first time in mainland South America and the virus has since spread throughout the Americas. ZIKV infection has already been reported in approximately 60 countries or territories in different continents [15]. Many regions around the world are now considered at risk of infection [16]. The Brazilian ZIKV epidemic changed the world's perspective on this neglected virus because of the dramatic increase of microcephaly in newborns [17, 18]. There is by now compelling evidence that ZIKV is involved in the etiology of this condition, as well as other severe congenital defects such as intracranial calcifications, ventricular system dilation, and neuronal migration disorders [19]. In adults, severe neurological complications such as myelitis, meningoencephalitis, and Guillain-Barré syndrome (GBS) have been associated with ZIKV infection [20, 21]. To date, there is a single ZIKV serotype, but two major viral lineages (African and Asian) have been identified through phylogenetic analyses of genome sequences [22].

Previous evolutionary analyses have demonstrated that the ZIKV strains circulating in Brazil belong to the Asian lineage [23]. However, the entry routes of this pathogen into the Americas remain unclear and very disputed. One of the hypotheses is that ZIKV entered in Brazil during the 2014 World Cup (June 12th to July 13th), brought in by African travelers, though this hypothesis is not compatible with the circulating virus in Brazil [24, 25]. Another hypothesis states that ZIKV introduction followed an international canoe event in August of 2014 held in Rio de Janeiro, which hosted competitors from various ZIKV-endemic Pacific countries [12]. Molecular clock and phylogenetic analysis of seven ZIKV sequences from patients in Brazil suggest a single introduction into Brazil between May and December 2013, coinciding with increased air travel from ZIKV endemic areas and outbreaks in the Pacific Ocean [23].

With respect to the geographic expansion of other arboviruses, previous phylogenetic studies have suggested a common entry route into South America for the CHIKV Asian genotype and for some lineages of DENV-1, 2, and 4 outbreaks: from Asia to Oceania, then emerging in Central America and the Caribbean Islands [9], [26–31]. DENV-3 lineages (BR-I, BR-II, and BR-III) were probably imported from the Lesser Antilles (Caribbean) [32]. Here, we applied in-depth phylogenomic analyses to trace and retrace the potential dissemination routes of the most important emerging and reemerging arboviruses until they reached South America. We also studied the spatial arrangements of these viruses. Our data suggests that ZIKV followed a similar route as CHIKV and DENV until its first emergence on mainland South America, placing Central America and the Caribbean as the main entry pathway of human pathogenic arboviruses into this continent.

2. Materials and Methods

2.1. Sequence Database. Initially, whole-genome sequences of all ZIKV, CHIKV, DENV-1–4, genotypes, and variants isolated from several countries/regions of North/South/Central America, Africa, Asia, and Oceania collected from 1944 to 2016 (covering a period of 72 years) were retrieved from GenBank in October 2016. Then, sequences were filtered by known location, isolation source, and sampling date (see below). Only whole genome sequences were used instead of specific gene fragments or partial genomes since short sequences have a limited number of genomic locations available for analysis. Specifically, in the context of an epidemic, the amplification of limited genomic regions may not detect mutations that will be informative for the phylogenetic analysis of viral spread. Sequences containing degenerate bases, information such as virus isolated from "unknown hosts" and "laboratory strains," duplicate sequences, and sequences from clones and recombinants were discarded from our analysis. The six datasets (ZIKV, CHIKV, DENV-1, DENV-2, DENV-3, and DENV-4) were aligned independently using MUSCLE v3.8.31 [33]. Each individual alignment was used in subsequent analyses.

2.2. Phylogenetic Analysis. To analyze the global phylogenies for each DENV-1–4 (DENV-1 (1560), DENV-2 (1132), DENV-3 (857), DENV-4 (161)), CHIKV (262), and ZIKV (63), phylogenetic reconstructions were performed with all complete genomes available until October 2016 using the maximum likelihood (ML) method implemented in FastTree v.2.1.7 software [34] with the standard implementation of GTR + CAT with 20 gamma distribution parameters and a mix of nearest-neighbor interchanges (NNI) and subtree-prune-regraft (SPR).

The Python environment available in the ETE3 programming interface was applied for the analysis and visualization of phylogenomic data [35]. The presence of a phylogenetic signal was investigated through the likelihood mapping analysis using TREE-PUZZLE v.5.2 (available at: http://www.tree-puzzle.de) [36, 37]. The Pairwise Homoplasy Index (PHI) test for recombination was conducted with the SplitsTree v.4.10 following the default settings [38]. The reliability of the nodes was analyzed by Shimodaira-Hasegawa- (SH-) like test support values with 1000 replications [39].

To complement our ZIKV analysis due to the increasing amount of new ZIKV genomes available monthly in the databases analyzed, we constructed an updated phylogenetic tree. We downloaded all full ZIKV genomes from the ViPR database [40] deposited until May 2017. Next, we performed and obtained an alignment using MUSCLE v3.8.31 [33]. A phylogenetic tree was generated (285 full genomes) using MrBayes [41] (GTR + I, 1 million generations).

2.3. Phylogenomics, Phylogeography, and Spatiotemporal Analyses. Comprehensive phylogenomic, phylogeographic, and spatiotemporal analyses to study the geographic spread pattern of different arboviruses (and genotypes) were performed using a Bayesian evolutionary framework analysis through the Metropolis-Hasting Markov Chain Monte Carlo (MCMC) algorithm implemented in the Bayesian Evolutionary Analysis Sampling Trees (BEAST) software package, v2.4.340. The datasets used for Bayesian analyses were based on large ML trees using the following criteria: only sequences that clustered with flaviviruses isolated in South America and sequences of viruses isolated from humans (GenBank information: "/host = *Homo sapiens*") were selected. For depicting the changes related to the effective population size (Ne) over time, the coalescent parameter Bayesian Skyline Plot (BSP) was applied imposing a strict or relaxed molecular clock (with log-normal distribution rates). The Markov models of nucleotide substitution (ZIKV = GTR + I; CHIKV = GTR + I; and DENV-1 = GTR + Γ, DENV-2 = GTR + I + Γ, DENV-3 = GTR + Γ, and DENV-4 = GTR + I + Γ) utilized were indicated by the software jModelTest v.2.1.641 [42].

The MCMC algorithm was run up to 1 billion generations, with sampling every 100,000 generations, for each molecular clock model. The good mixing of the MCMC was determined by effective sample size (ESS) values ≥ 200, and the convergence of parameters was assessed through of Tracer v1.6 software (http://beast.bio.ed.ac.uk) with 10% burn-in. The marginal likelihood for each clock model was obtained using the path sampling and stepping-stone algorithm [43, 44]. Different clock models were compared using Bayes factors for all viruses' genomic alignments. A strict molecular clock for CHIKV, DENV-1, and DENV-4 and a relaxed molecular clock for ZIKV, DENV-2, and DENV-3 were selected as the most likely models to represent the viral demographic history and were subsequently used in the following analysis. The set of trees with well-supported clades were observed through DensiTree v.2.2.544 [45]. The posterior distribution of the maximum clade credibility (MCC) tree was summarized by TreeAnnotator (implemented in BEAST v.2.4.2 package) and visualized in the FigTree v.1.4.2 software (http://tree.bio.ed.ac.uk/software/figtree). Phylogenetic uncertainty was estimated by the 95% highest probability density (HPD) intervals. Analyses of phylogeographic patterns and viral transmission networks during outbreaks were also generated by the minimum spanning tree (MST) generated by the PHYLOViZ v.2.0 software [46]. The MST uses the maximum parsimony principle that evolution should be explained with as few events as possible [47].

3. Results and Discussion

The ongoing rapid spread and severe disease outcomes associated with the CHIKV, DENV, and ZIKV epidemics in several countries of South America and the Pacific region have caused significant global public health concerns. Knowledge of the geographic expansion of these viruses and their pathway into mainland South America remains incomplete and controversial. In this study, we traced the entry and dispersal routes for these arboviruses to the New World, especially to Brazil, which is at the epicenter of these triple arbovirus epidemics.

Our phylogenetic analyses show the presence of East Central/South African (ECS) and Asian/Caribbean (AC) genotypes circulating in Brazil (Figure 1). Considering only the Asian genotype, CHIKV caused an outbreak in the Caribbean (Martinique) in 2013 [32] and in September 2014 autochthonous and nonautochthonous CHIKV Asian-genotype infections were reported in Oiapoque (northern Brazil) and Recife (northeastern Brazil) [9]. The analyses suggest that CHIKV from Martinique reached the north and northeast of Brazil in 2014 (Figure 1).

In Latin America, DENV resurged during the 1960s in the Caribbean and Venezuela and in the 1970s in Colombia. Before 1975, only DENV-2 and DENV-3 were circulating in the Americas, causing epidemic waves in the Caribbean (1969) and in Jamaica and Puerto Rico (1963). In 1977, DENV-1 caused epidemics in Jamaica and Cuba, and in 1978, in Venezuela and Puerto Rico. In subsequent years, it spread to the Caribbean, Central America, and parts of South America. This path was also followed by DENV-4 in 1981 [48]. DENV outbreaks were commonly reported in port cities from the Caribbean and North, Central, and South American regions, and these were mostly related to trade activities [49, 50]. As shown in Figure 2, we analyzed the flows and dispersion of the main DENV genotypes between Central and South American countries as detailed below.

The time-scaled Bayesian MCC tree shows the origin and spread of DENV-1 genotype V. We observed two well-supported and independent introductions of DENV from the Asian region (Indian sequences are the most likely ancestral ones) to Brazil, Venezuela, Colombia, and Argentina (Clade I, highlighted in red), one from India to Brazil (Clade II, highlighted in green), and the last one from India to Brazil and Argentina passing through the British Virgin Islands (Tortola, Caribbean region) belonging to Clade III (highlighted in blue) (Figure 2(a)). Our data on DENV-1 genotype V (the most prevalent genotype circulating in the Americas), agree partly with Bruycker-Nogueira et al. [51], in that all DENV-1 genotype V sequences from South America investigated in this study have an ancient Asian origin. However, we suggest three independent introductions into South America with at least one in Tortola, one of the largest and most populated Islands in the Caribbean (Figure 2(a)).

Introduction of DENV-2 (Asian/American (AS/AM) lineage) into South America occurred through the Caribbean, originating from the Dominican Republic between 1997 and 2002 (highlighted in red) and Puerto Rico (highlighted

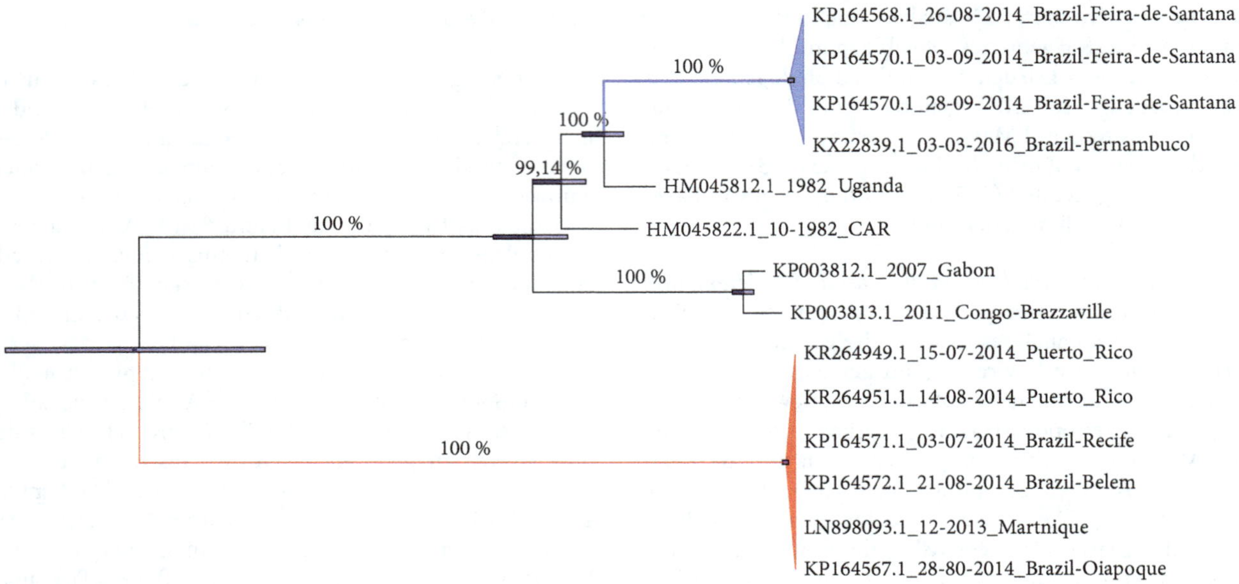

FIGURE 1: Phylogenetic reconstructions of CHIKV whole genome sequences. Bayesian MCC tree enforcing a strict molecular clock. The values along the tree branches represent the posterior distribution. The horizontal bars show phylogenetic uncertainty. The branches in blue and red represent Central/East African and Asian CHIKV genotypes, respectively. Abbreviation: CAR = Central African Republic.

in green and blue). The first introduction from Puerto Rico occurred probably in the 1980s, with further ones in 1987 and 1995 (Figure 2(b)). In the case of DENV-2, the origin and evolutionary dynamics of lineage II (AS/AM) into South America originated with three key introductions, all of them involving countries of the Caribbean: Puerto Rico (responsible for two introductory events) and the Dominican Republic. Considering that lineages I and II were predominant in epidemics in the Caribbean during the 1980s and 1990s [52], we assume that the DENV-2 lineages in South America have a Caribbean origin (Figure 2(b)).

Next, we traced the dispersion of DENV-3 lineages (BR-I, BR-II, BR-III, and BR-IV (originating from the Caribbean/South American regions)) to South America. Importation of DENV-3 lineage BR-I into Brazil and Paraguay was from Puerto Rico (highlighted in red) and the BR-II lineage from Trinidad and Tobago (highlighted in green) (Figure 2(c)). DENV-3 BR-III (highlighted in blue) and IV (BRA/VEN in bold) lineages were probably imported from the Philippines (KU050695.1 isolated in 1956) and the United States of America (JQ922554.1 isolated in 1963), which explains the Asian and American relationships of the subsequent lineages (Figure 2(c)). In addition, our analyses suggest that the introduction of the BR-I genotype into South America followed a similar route of DENV-2 with an entry point in Puerto Rico from where it spread to Brazilian states such as São Paulo (southeast), Pará and Rondônia (north), Pernambuco/Maranhão (northeast), and then Paraguay (Asunción and Juan Caballero). Genotypes II and III were introduced from Trinidad and Tobago in the early 2000s, whereas genotype IV was probably introduced into Brazil from Venezuela (Figure 2(c)). Considering the clustering among the South American countries, Brazil and Paraguay, shown in red and green clades (Figure 2(c)), we would also suggest the classification of South American I and II (SA-I and II) lineages rather than BR-I and II.

Genotypes I and II of this virus were probably imported into South America from Puerto Rico. The red (Brazil), green (Brazil, Colombia, and Venezuela), and blue (Brazil and Venezuela) clades represent DENV-4 genotype II (Figure 2(d)). DENV-4 genotype I represented by the BRA (isolate from the Bahia State, northeastern Brazil) was likely introduced into the Americas via Cambodia (KHM), thus representing a DENV of Asian origin (Figure 2(d)). We also believe that this classification may also be extended to the III and IV genotypes. Analysis of DENV-4, and more specifically genotype II, indicated two main independent introductions into South America between 1990 and 1996. The first was via Cambodia to the Bahia State (northeastern Brazil) and the second in 1996 from Puerto Rico followed by several entries from Venezuela to different Brazilian states (Figure 2(d)). Previous work [26] already reported the possible introduction of DENV-4 genotype I into Brazil from Asia and our data indicate that genotype II followed the same route. The complexity of DENV population dynamics in South America may be associated with the multiple introduction and dissemination of new DENV variants in this region, vector density [5], and the increased flow of humans between countries of the Caribbean and Central and South America [53].

Although our data does not match perfectly with the findings of Allicock et al. in 2012 [31] about the original source country of Central America/Caribbean islands, mainly due to the different samples used as well as an inherent sampling bias introduced when analyzing only complete genomes, the general picture holds true showing that those regions are most of the time the first point of entry from which pathogenic arboviruses subsequently spread to the

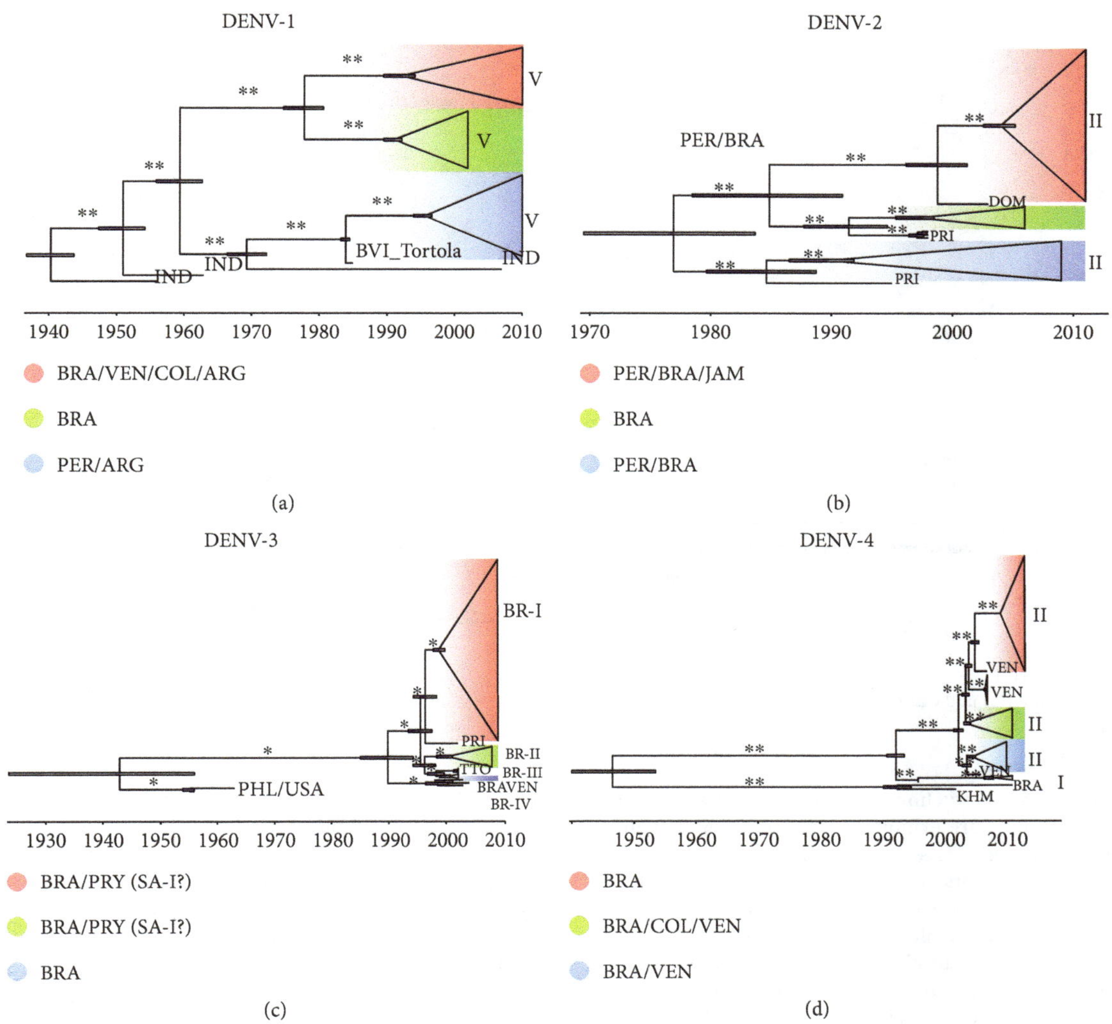

FIGURE 2: Bayesian maximum clade credibility (MCC) phylogenetic tree of DENVs' whole genomes. MCC trees were used to represent population dynamics over time of different genotypes of DENV-1 (a), DENV-2 (b), DENV-3 (c), and DENV-4 (d). The branch length is scaled in time enforcing strict (DENV-1 and DENV-4) and relaxed (DENV-2 and DENV-3) molecular clocks. The colors highlighted in the MCC trees represent different groups of countries and the horizontal blue bars represent phylogenetic uncertainty. DENV genotypes and lineages are indicated in each panel by roman numerals. One asterisk represents ≥86% and two ≥99% of posterior probability values. In (c) (DENV-3), the naming of Brazilian genotypes BR-I and BR-II as South America I (SA-I) and South America II (SA-II) is suggested to better represent the geographic locations of the isolates. Country abbreviations: ARG: Argentina; BRA: Brazil; BVI_Tortola = British Virgin Island, Tortola; COL: Colombia; DOM: Dominican Republic; IND: India; JAM: Jamaica; KHM: Cambodia; PRI: Puerto Rico; PRY: Paraguay; TTO: PER: Peru; PHL: Philippines; Trinidad and Tobago; USA: United States of America; VEN: Venezuela.

mainland South America. In addition, we also showed that multiple independent entrances occurred from Central American countries and/or Caribbean islands to South America right after reported outbreaks. Perhaps the link between new entrance events and corresponding outbreaks are due to the highest DENV diversity present in the Caribbean region which increases the probability of new variants' emergence into South America. In summary, these introductions altered viral population dynamics in South America as it was observed in the 90s and 2010 (DENV-1 genotype V); in 2009 and 2010 onwards (DENV-2 lineages II—(AS/AM)); in 1999-2000; in 2010 (DENV-3 BR I, II, and III lineages, Caribbean/South American); and in 2013-2014 (DENV-4 genotypes I and II) in Figure 2.

On the other hand, ZIKV was only detected in South America in 2015, and few studies have been carried out to map its transmission dynamics. Here, possible pathways taken by ZIKV to reach different countries in mainland South America were traced. To begin, we performed minimum spanning tree (MST) analysis in order to clarify the short-range transmission route. Our MST analysis centers on the 2014 Haitian ZIKV strain [54] as the most closely related to the French Polynesian ZIKV isolate and thus as the most likely source of the outbreak in Central America and the Caribbean before spreading to Brazil and other South American countries (Figure 3(a)).

Phylogenetic data from MCC and Densitree, which show the frequency of clades and nodes clustering, demonstrates

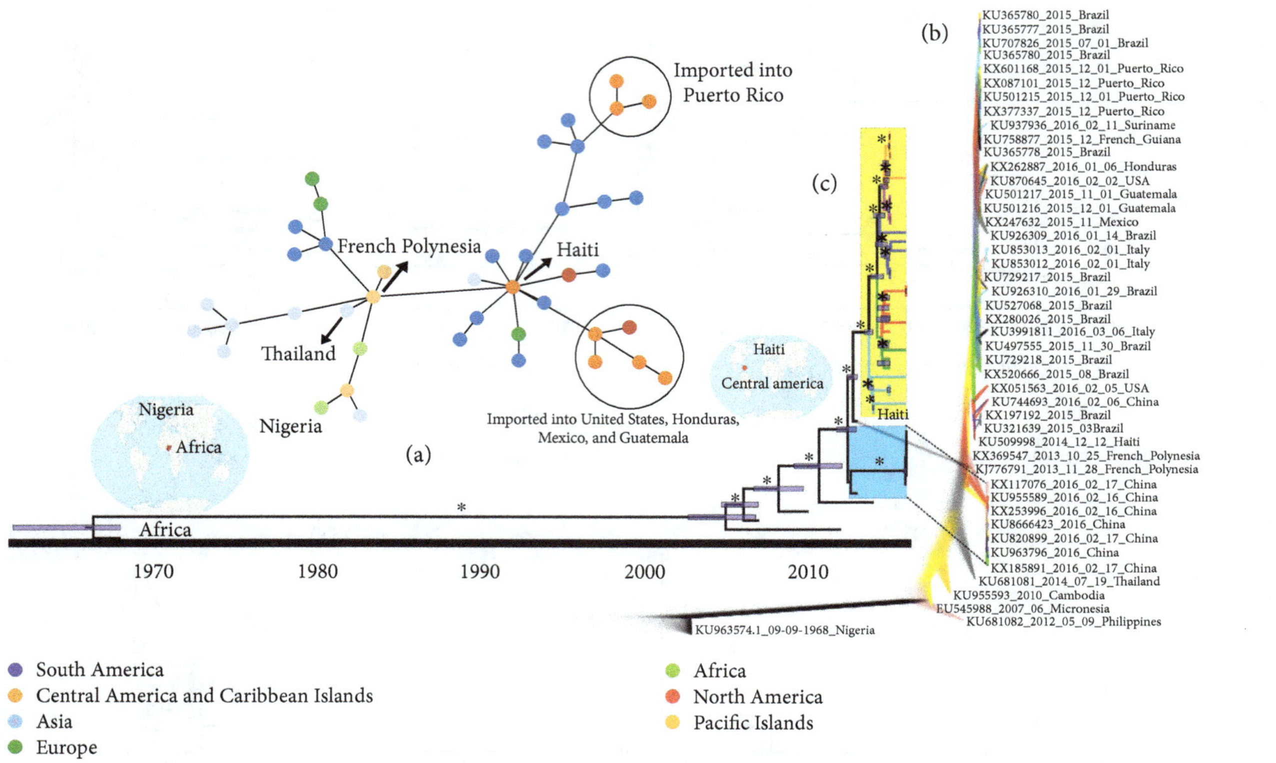

FIGURE 3: Bayesian reconstruction of the spatiotemporal spread pattern of ZIKV. The transmission networks during ZIKV outbreaks as well as its origin and spread are exhibited by the minimum spanning tree (MST) analysis. Countries and nodes were colored following the continent legend (a). DensiTree analysis shows the frequency of clades and nodes from the Bayesian approach. Well-supported branches are indicated by solid colors. The sequences in bold indicate ZIKV-associated microcephaly and those in italics indicate that ZIKV was imported from Venezuela to China (b). Time-scaled Bayesian MCC phylogenetic tree of ZIKV full-length genome sequences enforcing a relaxed molecular clock. The blue and yellow rectangles show two well-supported monophyletic clades that help to elucidate the dynamics and dissemination of ZIKV to different countries. Roman numerals in front of the yellow rectangles represent our hypothesis for different ZIKV variants that are currently circulating in South America. The horizontal blue bars in the tree indicate the phylogenetic uncertainty and the asterisks the posterior probability values greater than or equal to 98% (c).

that all Brazilian ZIKV isolates presented high phylogenetic relationships with several isolates from Central America and the Caribbean region (Figure 3(b)), reinforcing this important link for ZIKV spread across the Americas. Our results suggest multiple and simultaneous ZIKV introductions into South America, as shown in the time-scaled MCC phylogenetic tree (Figure 3(c)). Intriguingly, however, is a 2014 Haitian ZIKV isolate (GenBank accession number KU509998) clustering with French Polynesian isolates (GenBank accession number KX369547-2013) at the bottom of the tree and from which other South American ZIKV isolates appear to have descended from.

The subsequent dynamics in South America possibly triggered successive events responsible for the generation of different ZIKV variants, as exhibited by the multiple clades in the MCC tree (Figure 3(c)). Thus, the presence of multiple modern ZIKV variants circulating in South America could be considered as a potential division into ZIKV-I to ZIKV-VI variants. Few base substitutions per site were found between and within ZIKV variants, but they were sufficient to allow the division of ZIKV isolates into different clades (Supplementary Tables S1 and S2). Taken together, these results suggest that ZIKV followed the same route of the Asian genotype of CHIKV and DENV-1–4: from Asia to Oceania, then emerging in Central America and the Caribbean Islands, and finally ending up in South America (Figure 4). Similar results were also found by analyzing only NS5 data where a broad set of samples are available [55]. Twelve Brazilian cities including three from the northeastern region hosted FIFA World Cup matches in July 2014. Before this event, imported cases of CHIKV were confirmed with most being travelers arriving from Haiti and the Dominican Republic. Among the imported CHIKV cases, there were Brazilian military officers who were on a peacekeeping mission in Haiti [9]. After the event, an explosion of officially confirmed cases of CHIKV and ZIKV took place in Brazil.

The migration of thousands of Haitians to Brazil since the 2010 earthquake has been another important human population flow that might be associated with the introduction of new pathogens [53]. Our results identified the 2014 Haitian ZIKV strain [54] as the most closely related to the French Polynesian isolate placing Haiti as a possible entry point of ZIKV into the Americas, following the same path as DENV and CHIKV as previously stated [56]. It is important to note that this single Haitian genomic sequence could suggest that ZIKV arrived in the Caribbean islands but did

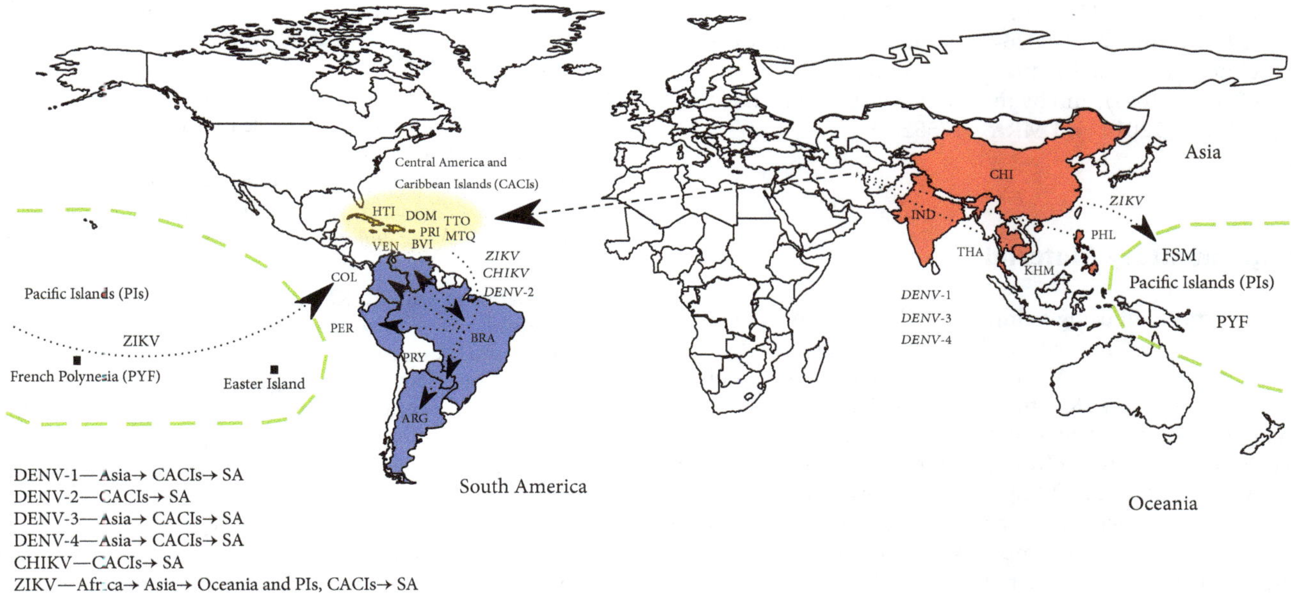

FIGURE 4: Route map of arboviruses assessed in the study. The colors and arrows highlight the entry routes of DENV, CHIKV, and ZIKV into mainland South America. On the bottom left, there is a summary of routes traced in the study. Abbreviations: ARG: Argentina; BRA: Brazil; BVI = British Virgin Island; CHI: CACIs: Central America and Caribbean islands; China; COL: Colombia; DOM: Dominican Republic; FSM: Federated States of Micronesia; HTI: Haiti; IND: India; KHM: Cambodia; MTQ: Martinique; PIs: Pacific Islands; PER: Peru; PHL: Philippines; PRI: Puerto Rico; PRY: Paraguay; PYF: French Polynesia; SA: South America; THA: Thailand; TTO: Trinidad and Tobago; VEN: Venezuela.

not spread from there, which is in agreement with the absence of ZIKV epidemics data from the Caribbean islands back in 2014/2015. However, a recently published new genome from a patient from the USA who travelled to Haiti (KX051563_2016_02_05_USA) confirms that ZIKV is still circulating in the Caribbean islands [55]. The presence of this strain could be explained by the further spread of ZIKV from Brazil to the Caribbean islands after the Brazilian outbreak; however, a new Bayesian analysis with all available genomes until May 2017 confirmed that this ZIKV strain is closely related with the 2014 Haitian ZIKV sample, that is, this genome confirms that ZIKV first offset in the Americas is still circulating in the Caribbean islands (Supplementary tree).

This introduction boosted the generation of modern South American ZIKV variants, exhibited by the multiple and statistically well-defined clades present in the time-scaled Bayesian MCC tree. This has led us to propose a subclassification of the ZIKV Asian lineage into at least six variants (ZIKV-I to VI) that are currently circulating in Latin America. Further studies investigating the possible phenotypic differences among these variants are warranted.

4. Conclusions

In conclusion, Central America and the Caribbean are important entry routes of human pathogenic arboviruses into South America. Our results underscore the need of the observance of strict biosecurity procedures at the main countries' entry sites and warrant systematic arbovirus surveillance to monitor its circulation and evolution as the viruses spread to other areas of the world. Monitoring arbovirus presence from mosquito samples collected regularly from key areas is a suggestion in addition to human clinical diagnostics.

Conflicts of Interest

The authors declare that they have no competing interests.

Authors' Contributions

Túlio De Lima Campos, Ricardo Durães-Carvalho, Gabriel Luz Wallau, and Lindomar José Pena designed the study. Túlio De Lima Campos, Ricardo Durães-Carvalho, and Gabriel Luz Wallau performed the experiments. Túlio De Lima Campos, Ricardo Durães-Carvalho, Antonio Mauro Rezende, Otávio Valério de Carvalho, Gabriel Luz Wallau, Alain Kohl, and Lindomar José Pena analyzed the data. Túlio De Lima Campos, Ricardo Durães-Carvalho, Antonio Mauro Rezende, Otávio Valério de Carvalho, Gabriel Luz Wallau, Alain Kohl, and Lindomar José Pena wrote the paper. Túlio De Lima Campos and Ricardo Durães-Carvalho contributed equally to this work.

Acknowledgments

This work was supported by the following Brazilian agencies: the Pernambuco State Research Foundation (FACEPE,

Grant nos. APQ-0154-2.12/16; APQ-1776-2.12/15, APQ-0078-2.02/16, and DCR-0026-2.12/16), the National Counsel of Technological and Scientific Development (CNPq, Process no. 300700/2016-4), and by the UK Medical Research Council (MC_UU_12014 and MR/N017552/1).

Supplementary Materials

Bayesian tree built using as input all ZIKV genome sequences described in the Materials and Methods section along with the ones deposited at VIPR until May 2017. This analysis confirms that the ZIKV first offset in the Americas keeps circulating in the Caribbean Islands. The analysis was performed with 285 full ZIKV genomes showing the African and Asian genotypes. Orange branches are genomes previous to ZIKV epidemics, green branches are genomes from Brazilian samples, and samples with blue names are ZIKV genomes from Haiti. The red collapsed clade denotes the polytomic clade responsible for the ZIKV epidemics in the Americas and exportation to other parts of the world. *(Supplementary Materials)*

References

[1] J. Patterson, M. Sammon, and M. Garg, "Dengue, Zika and chikungunya: emerging arboviruses in the New World," *Western Journal of Emergency Medicine*, vol. 17, no. 6, pp. 671–679, 2016.

[2] A. Wilder-Smith, D. J. Gubler, S. C. Weaver, T. P. Monath, D. L. Heymann, and T. W. Scott, "Epidemic arboviral diseases: priorities for research and public health," *The Lancet Infectious Diseases*, vol. 17, no. 3, pp. e101–e106, 2017.

[3] S. M. Best, "Flaviviruses," *Current Biology*, vol. 26, no. 24, pp. R1258–R1260, 2016.

[4] J. C. Rupp, K. J. Sokoloski, N. N. Gebhart, and R. W. Hardy, "Alphavirus RNA synthesis and non-structural protein functions," *Journal of General Virology*, vol. 96, no. 9, pp. 2483–2500, 2015.

[5] M. U. G. Kraemer, M. E. Sinka, K. A. Duda et al., "The global distribution of the arbovirus vectors *Aedes aegypti* and *Ae. albopictus*," *eLife*, vol. 4, 2015.

[6] S. L. Beltrán-Silva, S. S. Chacón-Hernández, E. Moreno-Palacios, and J. Á. Pereyra-Molina, "Clinical and differential diagnosis: dengue, chikungunya and Zika," *Revista Médica del Hospital General de México*, vol. 81, no. 3, pp. 146–153, 2018.

[7] D. E. Carey, "Chikungunya and dengue: a case of mistaken identity?," *Journal of the History of Medicine and Allied Sciences*, vol. 26, no. 3, pp. 243–262, 1971.

[8] J. J. Waggoner and B. A. Pinsky, "Zika virus: diagnostics for an emerging pandemic threat," *Journal of Clinical Microbiology*, vol. 54, no. 4, pp. 860–867, 2016.

[9] M. R. T. Nunes, N. R. Faria, J. M. de Vasconcelos et al., "Emergence and potential for spread of chikungunya virus in Brazil," *BMC Medicine*, vol. 13, no. 1, p. 102, 2015.

[10] S. Bhatt, P. W. Gething, O. J. Brady et al., "The global distribution and burden of dengue," *Nature*, vol. 496, no. 7446, pp. 504–507, 2013.

[11] T. Ernst, S. McCarthy, G. Chidlow et al., "Emergence of a new lineage of dengue virus type 2 identified in travelers entering Western Australia from Indonesia, 2010–2012," *PLOS Neglected Tropical Diseases*, vol. 9, no. 1, article e0003442, 2015.

[12] D. Musso, "Zika virus transmission from French Polynesia to Brazil," *Emerging Infectious Diseases*, vol. 21, no. 10, article 1887, 2015.

[13] D. Gatherer and A. Kohl, "Zika virus: a previously slow pandemic spreads rapidly through the Americas," *Journal of General Virology*, vol. 97, no. 2, pp. 269–273, 2016.

[14] J. Tognarelli, S. Ulloa, E. Villagra et al., "A report on the outbreak of Zika virus on Easter Island, South Pacific, 2014," *Archives of Virology*, vol. 161, no. 3, pp. 665–668, 2016.

[15] Centers for Disease Control and Prevention, *All Countries & Territories with Active Zika Virus Transmission*, Centers for Disease Control and Prevention, 2016.

[16] J. P. Messina, M. U. G. Kraemer, O. J. Brady et al., "Mapping global environmental suitability for Zika virus," *eLife*, vol. 5, 2016.

[17] J. Mlakar, M. Korva, N. Tul et al., "Zika virus associated with microcephaly," *New England Journal of Medicine*, vol. 374, no. 10, pp. 951–958, 2016.

[18] T. V. B. de Araújo, R. A. . A. Ximenes, D. . B. Miranda-Filho et al., "Association between microcephaly, Zika virus infection, and other risk factors in Brazil: final report of a case-control study," *The Lancet Infectious Diseases*, vol. 18, no. 3, pp. 328–336, 2018.

[19] F. Krauer, M. Riesen, L. Reveiz et al., "Zika virus infection as a cause of congenital brain abnormalities and Guillain-Barré syndrome: systematic review," *PLOS Medicine*, vol. 14, no. 1, artilce e1002203, 2017.

[20] V. M. Cao-Lormeau, A. Blake, S. Mons et al., "Guillain-Barré syndrome outbreak associated with Zika virus infection in French Polynesia: a case-control study," *Lancet*, vol. 387, no. 10027, pp. 1531–1539, 2016.

[21] T. Yoshihara, M. Morita, and T. Ishii, "Ultrastructure and three-dimensional imaging of epimyoepithelial islands in benign lymphoepithelial lesions," *Oto-Rhino-Laryngology*, vol. 252, no. 2, pp. 106–111, 1995.

[22] Z. Gong, Y. Gao, and G. Z. Han, "Zika virus: two or three lineages?," *Trends in Microbiology*, vol. 24, no. 7, pp. 521-522, 2016.

[23] N. R. Faria, R. d. S. d. S. Azevedo, M. U. G. Kraemer et al., "Zika virus in the Americas: early epidemiological and genetic findings," *Science*, vol. 352, no. 6283, pp. 345–349, 2016.

[24] F. S. Salvador and D. M. Fujita, "Entry routes for Zika virus in Brazil after 2014 World Cup: new possibilities," *Travel Medicine and Infectious Disease*, vol. 14, no. 1, pp. 49–51, 2016.

[25] C. Zanluca, V. C. A. Melo, A. L. P. Mosimann, G. I. V. Santos, C. N. D. Santos, and K. Luz, "First report of autochthonous transmission of Zika virus in Brazil," *Memórias do Instituto Oswaldo Cruz*, vol. 110, no. 4, pp. 569–572, 2015.

[26] F. L. de Melo, C. M. Romano, and P. M. d. A. Zanotto, "Introduction of dengue virus 4 (DENV-4) genotype I into Brazil from Asia?," *PLoS Neglected Tropical Diseases*, vol. 3, no. 4, article e390, 2009.

[27] R. Rico-Hesse, "Molecular evolution and distribution of dengue viruses type 1 and 2 in nature," *Virology*, vol. 174, no. 2, pp. 479–493, 1990.

[28] J. M. G. Araújo, R. M. R. Nogueira, H. G. Schatzmayr, P. M. d. A. Zanotto, and G. Bello, "Phylogeography and evolutionary history of dengue virus type 3," *Infection, Genetics and Evolution*, vol. 9, no. 4, pp. 716–725, 2009.

[29] C. V. F. Carrington, J. E. Foster, O. G. Pybus, S. N. Bennett, and E. C. Holmes, "Invasion and maintenance of dengue virus type 2 and type 4 in the Americas," *Journal of Virology*, vol. 79, no. 23, pp. 14680–14687, 2005.

[30] J. E. Foster, S. N. Bennett, C. V. F. Carrington, H. Vaughan, and W. O. McMillan, "Phylogeography and molecular evolution of dengue 2 in the Caribbean basin, 1981–2000," *Virology*, vol. 324, no. 1, pp. 48–59, 2004.

[31] O. M. Allicock, P. Lemey, A. J. Tatem et al., "Phylogeography and population dynamics of dengue viruses in the Americas," *Molecular Biology and Evolution*, vol. 29, no. 6, pp. 1533–1543, 2012.

[32] J. M. Galvão de Araújo, G. Bello, H. Romero, and R. M. Ribeiro Nogueira, "Origin and evolution of dengue virus type 3 in Brazil," *PLoS Neglected Tropical Diseases*, vol. 6, no. 9, p. e1784, 2012.

[33] R. C. Edgar, "MUSCLE: multiple sequence alignment with high accuracy and high throughput," *Nucleic Acids Research*, vol. 32, no. 5, pp. 1792–1797, 2004.

[34] M. N. Price, P. S. Dehal, and A. P. Arkin, "FastTree 2—approximately maximum-likelihood trees for large alignments," *PloS One*, vol. 5, no. 3, article e9490, 2010.

[35] J. Huerta-Cepas, F. Serra, and P. Bork, "ETE 3: reconstruction, analysis, and visualization of phylogenomic data," *Molecular Biology and Evolution*, vol. 33, no. 6, pp. 1635–1638, 2016.

[36] K. Strimmer and A. Haeselervon, "Likelihood-mapping: a simple method to visualize phylogenetic content of a sequence alignment," *Proceedings of the National Academy of Sciences of the United States of America*, vol. 94, no. 13, pp. 6815–6819, 1997.

[37] H. A. Schmidt, K. Strimmer, M. Vingron, and A. von Haeseler, "TREE-PUZZLE: maximum likelihood phylogenetic analysis using quartets and parallel computing," *Bioinformatics*, vol. 18, no. 3, pp. 502–504, 2002.

[38] T. C. Bruen, H. Philippe, and D. Bryant, "A simple and robust statistical test for detecting the presence of recombination," *Genetics*, vol. 172, no. 4, pp. 2665–2681, 2005.

[39] M. Anisimova and O. Gascuel, "Approximate likelihood-ratio test for branches: a fast, accurate, and powerful alternative," *Systematic Biology*, vol. 55, no. 4, pp. 539–552, 2006.

[40] B. E. Pickett, E. L. Sadat, Y. Zhang et al., "ViPR: an open bioinformatics database and analysis resource for virology research," *Nucleic Acids Research*, vol. 40, no. D1, pp. D593–D598, 2012.

[41] F. Ronquist and J. P. Huelsenbeck, "MrBayes 3: Bayesian phylogenetic inference under mixed models," *Bioinformatics*, vol. 19, no. 12, pp. 1572–1574, 2003.

[42] D. Darriba, G. L. Taboada, R. Doallo, and D. Posada, "jModelTest 2: more models, new heuristics and parallel computing," *Nature Methods*, vol. 9, no. 8, p. 772, 2012.

[43] G. Baele, P. Lemey, and S. Vansteelandt, "Make the most of your samples: Bayes factor estimators for high-dimensional models of sequence evolution," *BMC Bioinformatics*, vol. 14, no. 1, p. 85, 2013.

[44] G. Baele, W. L. S. Li, A. J. Drummond, M. A. Suchard, and P. Lemey, "Accurate model selection of relaxed molecular clocks in Bayesian phylogenetics," *Molecular Biology and Evolution*, vol. 30, no. 2, pp. 239–243, 2012.

[45] R. Bouckaert and J. Heled, "DensiTree 2: seeing trees through the forest," 2014.

[46] M. Nascimento, A. Sousa, M. Ramirez, A. P. Francisco, J. A. Carriço, and C. Vaz, "PHYLOViZ 2.0: providing scalable data integration and visualization for multiple phylogenetic inference methods," *Bioinformatics*, vol. 33, no. 1, pp. 128-129, 2016.

[47] A. P. Francisco, M. Bugalho, M. Ramirez, and J. A. Carriço, "Global optimal eBURST analysis of multilocus typing data using a graphic matroid approach," *BMC Bioinformatics*, vol. 10, no. 1, p. 152, 2009.

[48] M. G. Guzman and G. Kouri, "Dengue and dengue hemorrhagic fever in the Americas: lessons and challenges," *Journal of Clinical Virology*, vol. 27, no. 1, pp. 1–13, 2003.

[49] O. B. Dick, J. L. San Martin, R. H. Montoya, J. del Diego, B. Zambrano, and G. H. Dayan, "The history of dengue outbreaks in the Americas," *The American Journal of Tropical Medicine and Hygiene*, vol. 87, no. 4, pp. 584–593, 2012.

[50] S. B. Halstead, "Dengue in the Americas and Southeast Asia: do they differ?," *Revista Panamericana de Salud Pública*, vol. 20, no. 6, pp. 407–415, 2006.

[51] F. de Bruycker-Nogueira, D. Mir, F. B. dos Santos, and G. Bello, "Evolutionary history and spatiotemporal dynamics of DENV-1 genotype V in the Americas," *Infection, Genetics and Evolution*, vol. 45, pp. 454–460, 2016.

[52] D. Mir, H. Romero, L. M. Fagundes de Carvalho, and G. Bello, "Spatiotemporal dynamics of DENV-2 Asian-American genotype lineages in the Americas," *PLoS ONE*, vol. 9, no. 6, p. e98519, 2014.

[53] T. Rawlinson, A. M. Siqueira, G. Fontes et al., "From Haiti to the Amazon: public health issues related to the recent immigration of Haitians to Brazil," *PLoS Neglected Tropical Diseases*, vol. 8, no. 5, article e2685, 2014.

[54] J. Lednicky, V. M. Beau de Rochars, M. el Badry et al., "Zika virus outbreak in Haiti in 2014: molecular and clinical data," *PLOS Neglected Tropical Diseases*, vol. 10, no. 4, article e0004687, 2016.

[55] N. M. Iovine, J. Lednicky, K. Cherabuddi et al., "Coinfection with Zika and Dengue-2 viruses in a traveler returning from Haiti, 2016: clinical presentation and genetic analysis," *Clinical Infectious Diseases*, vol. 64, no. 1, pp. 72–75, 2017.

[56] D. Musso, V. M. Cao-Lormeau, and D. J. Gubler, "Zika virus: following the path of dengue and chikungunya?," *Lancet*, vol. 386, no. 9990, pp. 243-244, 2015.

9

Whole-Exome Sequencing Identifies One De Novo Variant in the *FGD6* Gene in a Thai Family with Autism Spectrum Disorder

Chuphong Thongnak,[1,2,3] Areerat Hnoonual,[4] Duangkamol Tangviriyapaiboon,[5] Suchaya Silvilairat,[6] Apichaya Puangpetch,[1,2] Ekawat Pasomsub,[7,8] Wasun Chantratita,[7,8] Pornprot Limprasert,[9,10] and Chonlaphat Sukasem[1,2]

[1]*Division of Pharmacogenomics and Personalized Medicine, Department of Pathology, Faculty of Medicine Ramathibodi Hospital, Mahidol University, Bangkok, Thailand*
[2]*Laboratory for Pharmacogenomics, Somdech Phra Debaratana Medical Center (SDMC), Ramathibodi Hospital, Bangkok, Thailand*
[3]*Department of Clinical Pathology, Faculty of Medicine Vajira Hospital, Navamindradhiraj University, Bangkok, Thailand*
[4]*Graduate Program in Biomedical Sciences, Faculty of Medicine, Prince of Songkla University, Songkhla, Thailand*
[5]*Rajanagarindra Institute of Child Development, Chiang Mai, Thailand*
[6]*Division of Pediatric Cardiology, Department of Pediatrics, Faculty of Medicine, Chiang Mai University, Chiang Mai, Thailand*
[7]*Excellence Center for Genomic Medicine, Faculty of Medicine Ramathibodi Hospital, Mahidol University, Bangkok, Thailand*
[8]*Virology and Molecular Microbiology Unit, Department of Pathology, Faculty of Medicine Ramathibodi Hospital, Mahidol University, Bangkok, Thailand*
[9]*Human Genetics Unit, Department of Pathology, Faculty of Medicine, Prince of Songkla University, Hat Yai, Songkhla, Thailand*
[10]*Faculty of Medicine, King Mongkut's Institute of Technology Ladkrabang, Bangkok, Thailand*

Correspondence should be addressed to Pornprot Limprasert; lpornpro@yahoo.com and Chonlaphat Sukasem; chonlaphat.suk@mahidol.ac.th

Academic Editor: Mohamed Salem

Autism spectrum disorder (ASD) has a strong genetic basis, although the genetics of autism is complex and it is unclear. Genetic testing such as microarray or sequencing was widely used to identify autism markers, but they are unsuccessful in several cases. The objective of this study is to identify causative variants of autism in two Thai families by using whole-exome sequencing technique. Whole-exome sequencing was performed with autism-affected children from two unrelated families. Each sample was sequenced on SOLiD 5500xl Genetic Analyzer system followed by combined bioinformatics pipeline including annotation and filtering process to identify candidate variants. Candidate variants were validated, and the segregation study with other family members was performed using Sanger sequencing. This study identified a possible causative variant for ASD, c.2951G>A, in the *FGD6* gene. We demonstrated the potential for ASD genetic variants associated with ASD using whole-exome sequencing and a bioinformatics filtering procedure. These techniques could be useful in identifying possible causative ASD variants, especially in cases in which variants cannot be identified by other techniques.

1. Introduction

Autism spectrum disorder (ASD) is a neurodevelopmental disorder described by abnormalities in two domains: (1) social interaction and communication/language and (2) restricted and repetitive behavior. A study of the prevalence of autism in Thailand showed that it affects 9.9 children per 10,000 populations for children 1–5 years old [1]. According to the *Diagnostic and Statistical Manual of Mental Disorders, Fifth Edition (DSM-5)*, autism subtypes, autistic disorder, Asperger's disorder, pervasive developmental disorder not otherwise specified (PDD-NOS), and childhood disintegrative disorder (CDD), are merged to a single term "autism spectrum disorder." Rett disorder that was previously assigned as an autism subtype under DSM-IV criteria was classified as a separate disorder [2, 3].

Autism disturbs information processing in the brain by changing connection and organization of nerve cells and their synapses, the cause of which is not well understood [4]. ASD has a strong genetic root, although the genetics of autism is unclear and complex whether autism can have explained more by major effects of rare mutations or by common genetic variants with rare multigene interactions [5, 6]. Interactions of several genes, epigenetic factors, and the environment increase the complexity of etiology and physiology in ASD [7]. A number of researchers have attempted to explore the genetic basis of autism to apply in diagnosis. Studies to find candidate genes that are associated with ASD have used cytogenetics, targeted sequencing, and genome-wide association; identifying a large amount of genes may be responsible [5, 8] such as *SHANK3* on chromosome 22p16.3 and *NRXN1* and *NLGN3/4* on chromosomes Xq13 and Xq22.33 [9]. Several candidate genes have been located but cannot explain genetic mechanisms of ASD.

High-throughput sequencing has less bias for common variants than other techniques; therefore, it is an effective tool to discover genetic variants at genome-wide level and capable to apply in personalized medicine [10]. Exome sequencing is one high-throughput sequencing application that has been widely used in medical research, especially to find disease candidate genes in the last decade [11, 12]. It can expedite diagnosis through the sequencing of all exon or coding regions of the genome simultaneously [13]. In the field of autism, exome sequencing was also used to find genetic causes or related variation [14–16]. These showed the potential and possibility of the exome sequencing applications in genetic study of autism.

This study aims at applying whole-exome sequencing in identifying possible autism causative variants in two Thai families. Three de novo variants were detected in one family by using analysis pipeline, which was executed by integration of familial data, exome sequencing, bioinformatics, and segregation analysis.

2. Materials and Methods

2.1. Subjects. The subjects in this study were autism-affected children and normal individuals from two unrelated families. The family number 1 consisted of unaffected parents (1.I-1 and 1.I-2) and two children with autism (1.II-1 and 1.II-2). The proband (1.II-1) was featured in a previous cohort study [17]. The family number 2 had three generations, of whom four out of seven members were available for DNA study, an unaffected grandmother (2.I-2), an unaffected mother (2.II-2), and two children with autism (2.III-2 and 2.III-3). The proband of the family number 2 (2.III-2) had also been included in another study [18]. The clinical information of both families is shown in Supplementary 1. The pedigree for both families is shown in Figure 1. All samples had negative results when screened for common autism-related disorders in a previous microarray study. Genomic DNA was obtained from the Department of Pathology, Faculty of Medicine, Prince of Songkla University. All samples from patients were prepared for whole-exome sequencing while available samples from other family members were prepared for segregation analysis. Written informed consent was obtained from adult individuals themselves and for all children. All procedures were approved by the Institutional Review Board (MURA2012/02/SN1) of the Faculty of Medicine Ramathibodi Hospital, Mahidol University, and the Faculty of Medicine, Prince of Songkla University (EC48/364-006-3).

2.2. Exome Sequencing and Data Analysis. All samples passed quality verification by measurement of DNA concentration by Nanodrop ND1000 and measurement of DNA fragmentation by agarose gel electrophoresis. DNA samples of the samples 1.II-1, 1.II-2, 2.III-2, and 2.III-3 were prepared for whole-exome sequencing following standard SOLiD 5500xl (Applied Biosystems, California, USA) protocols. Each sample was diluted to three micrograms followed by fragmentation process using Covaris s220 (Covaris, USA) sonication machine. Agencourt beads were used to select for defined fragment size, which was measured using Agilent Bioanalyzer. TargetSeq Exome and Custom Enrichment System (Invitrogen, California, USA) was used to capture all fragments containing exons.

Whole-exome sequencing was performed on the SOLiD 5500xl system with 150 bp paired-end read. Primary analysis was performed on SOLiD instrument control software (SOLiD ICS) followed by the secondary analysis, mapping and variants calling, on LifeScope Genomic Analysis server. The human reference genome assembly hg19 (GRCh37) was used as a reference sequence in this analysis process.

Tertiary analysis, which includes quality filtering, annotation, data aggregation, and multisample processing, was performed with a variant call format (VCF) file from the secondary analysis. By using Golden Helix SVS software, all variants which have genotype quality or coverage depths lower than 20 and 10, respectively, were filtered out. Noncoding and synonymous variants were removed after annotation with the UCSC KnownGenes database. High-frequency variants commonly found in the general population were filtered out by annotation with 1000 Genomes Project Phase 3 [19] and an in-house exome database which consists of 172 Thai individuals. Variants that have minor allele frequency (MAF) greater than 2% were excluded in this step. Functional prediction databases were used to assess an effect of variants on deleterious encoded protein function. The database for this process was dbNSFP NS Functional Predictions v2.3 that compiled prediction scores from several prediction algorithms including SIFT, Polyphen2, MutationTaster, LRT, MutationAssessor, GERP ++, phastCons, PhyloP, FATHMM, PROVEAN, MetaSVM, MetaLR, CADD, VEST3, and SiPhy and other related information [20]. This step removed variants that were predicted not to affect protein function by any algorithms. Filtering by gene list was performed to emphasize variants on genes that

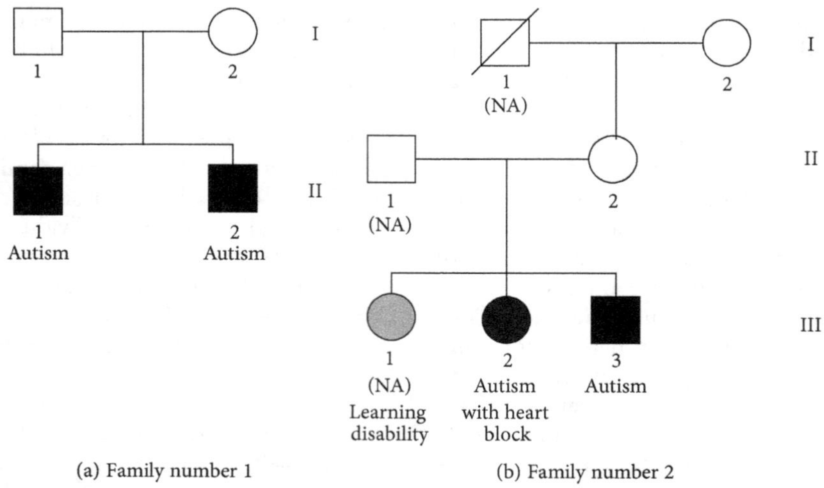

FIGURE 1: Pedigree of the family with autism indicated by the dark symbol. Gray symbol indicates learning disability phenotype. Females are indicated by circles and males by squares.

are most possibly related to ASD. The gene list consisted of reported ASD-associated genes from public databases, AutDB and AutKB [21, 22]. Furthermore, brain function and signaling-associated gene panels from Enlis Genome Research software (Enlis, Berkeley, CA) were included. Finally, genotype filtering according to inheritance mode was executed. Genotypes shared between 2 siblings for each family (both homozygous for the recessive allele, both heterozygous for de novo dominant allele) were identified as candidate variants. Two or more heterozygous variants in the same gene indicate a compound heterozygous inheritance model. A summary of the steps to filter variants is shown in Figure 2. Additionally, the Human Gene Mutation Database (HGMD) was used to identify known pathogenic variants [23].

2.3. Variant Validations and Segregation Analysis. Candidate variants from whole-exome sequencing were validated by Sanger sequencing. Along with this validation process, segregation analyses were performed on the family members. Primer3 v0.4.0, a web-based tool, was used to designed primers for amplification and sequencing [24]. Sequencing reactions were performed using Applied Biosystems 3130 DNA Analyzer (Life Technologies, Carlsbad, CA, USA).

3. Results

All samples from the autistic patients were successfully sequenced and variants detected with an average depth over 60× coverage. For family number 1, 66,840 and 52,066 variants were detected in samples 1.II-1 and 1.II-2 while 79,609 and 52,519 variants were detected in samples 2.III-2 and 2.III-3 for family number 2, respectively. VCF files for samples in the same family were imported together into Golden Helix SVS software for posterior annotation and filtering. A total of 13,190 variants for family number and 13,827 variants for family number 2 remained after removal of low-quality, noncoding, and synonymous variants. Afterwards, a filtering pipeline for minor allele frequencies, inheritance models, and a candidate gene filter was executed to reduce the number of variants to 21 and 25 variants for family numbers 1 and 2, respectively. Next, variant annotation and prioritization with the deleterious function prediction database resulted in 10 variants for family number 1 and 11 variants for family number 2. These variants were indicated as candidate variants. Finally, all candidate variants underwent validation and segregation analysis by Sanger sequencing. A total of 124 reactions of amplification and sequencing were achieved. In this process, variants that had discordant results between whole-exome sequencing and Sanger sequencing or had invalid segregation result with other family members were excluded. The candidate variants which passed in the validation process are shown in Table 1. All candidate variants of family number 1 failed in this step whereas three variants of family number 2 passed and were identified as final candidate variants for ASD in this family. The list of final candidate variants is shown in Table 2.

All three final candidate variants were nonsynonymous missense variants with genotype heterozygous in both autism-affected siblings (2.III-2 and 2.III-3) while showing homozygous wild-type in other unaffected family members. These variants consisted of c.2014G>A (NM_004836) in the *EIF2AK3* gene, c.2951G>A (NM_018351) in the *FGD6* gene, and c.6119A>G (NM_001170629) in the *CHD8* gene. Chromatograms of all variants in all available subjects are shown in Figure 3.

In addition, associations between the candidate variants and ASD were studied by using each ASD-affected individual from the family number 1 and the family number 2 plus 14 ASD-affected individuals from another project as the case group and 224 ASD-unaffected Thai individuals as the control group. Fisher's exact test was used for statistical testing in this study. The association study results revealed that only one of the three final candidate variants, c.2951G>A in the *FGD6* gene, had a statistically significant association with ASD ($P = 0.0041$). The details and results of the association study are shown in Supplementary 2.

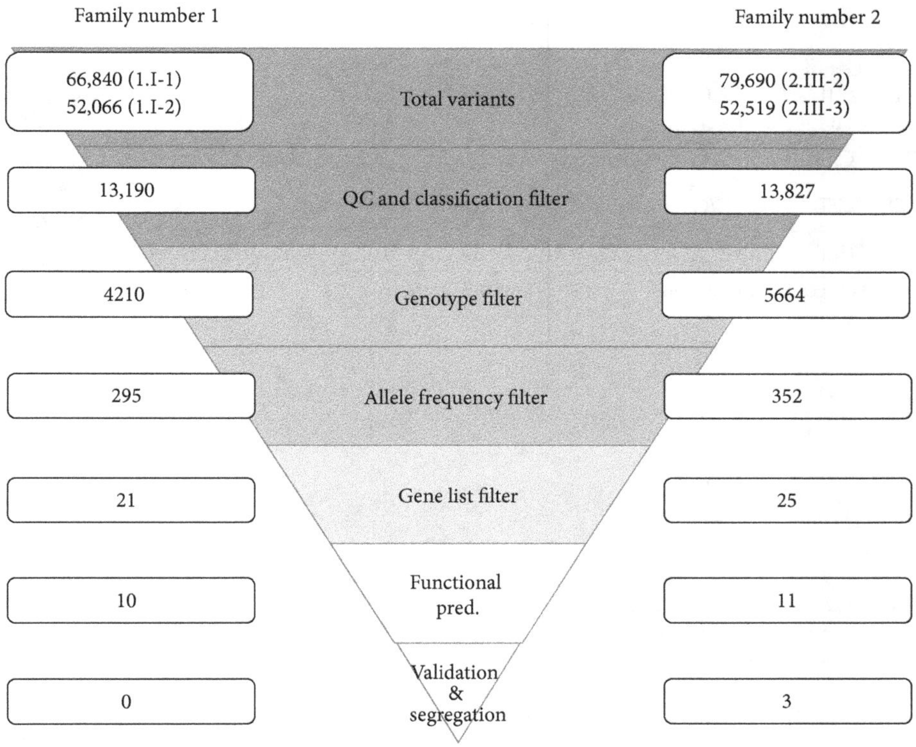

FIGURE 2: Filtering procedure of variants obtained by whole-exome sequencing 2nd data analysis. The number indicates an amount of variants passed for each step.

TABLE 1: Candidate variants from whole-exome sequencing which passed in the validation process. The variants which passed in the segregation process are presented with bold letters.

Position	Gene	Classification	Transcript	HGVS Coding	rsID 138	Family
Chr2: 73675228	*ALMS1*	Ins	NM_015120	c.1570_1571insCTC	rs34628045	1, 2
Chr2: 88876094	***EIF2AK3***	**Nonsyn SNV**	**NM_004836**	**c.2014G>A**	**rs35226268**	**2**
Chr4: 84240515	*HPSE*	Nonsyn SNV	NM_001199830	c.481A>C	rs75865093	1
Chr5: 140183237	*PCDHA3*	Nonsyn SNV	NM_031497	c.2455A>G	rs17844265	2
Chr5: 156721864	*CYFIP2*	Frameshift Ins	NM_001037333	c.279_280insC	rs5872508	1, 2
Chr6: 132270417	*CTGF*	Frameshift Del	NM_001901	c.1037delG	rs373467469	2
Chr7: 82581489	*PCLO*	Ins	NM_033026	c.8780_8781insTGA	rs10630259	1
Chr9: 117853022	*TNC*	Frameshift Del	NM_002160	c.276delG	rs944510	1
Chr12: 95531341	***FGD6***	**Nonsyn SNV**	**NM_018351**	**c.2951G>A**		**2**
Chr14: 21861835	***CHD8***	**Nonsyn SNV**	**NM_001170629**	**c.6119A>G**	**rs148494847**	**2**
Chr17: 26101336	*NOS2*	Nonsyn SNV	NM_000625	c.1423G>A		2
Chr21: 40883672	*SH3BGR*	Ins	NM_001001713	c.356_357insAGA	rs3831201	1

4. Discussion

This study was designed to perform exome sequencing to identify possible ASD causal variants in two Thai families and successfully identified three variants for one family, family number 2. For family number 1, all 10 variants did not pass validation and segregation. This indicates that possible causal variants may be lost in the prior process. Exome capturing using beads is one factor that highly impacts loss of target fragments [25]. Moreover, the possible causal variant may be passed over if it is located within a noncoding region or in genes that are not in the gene list used in the filtering process. Replication or whole-genome sequencing will be needed for this family.

Among 3 ASD candidate genes for the family number 2, *CHD8* encodes chromodomain helicase DNA binding protein 8 which functions as a transcription repressor by transforming chromatin structure. The binding of *CHD8* product and beta-catenin motivates negative regulation of the Wnt signaling pathway, which functions essentially in

Table 2: List of final candidate variants.

Genomic coordinates	Genotype*				AA change	Gene	MAF		SIFT	Functional prediction	
	2.III-2	2.III-3	2.II-2	2.I-2			1 kG (ASN)	Thai		Polyphen2 HumVar	MutationTaster
Chr2: 88876094	C/T	C/T	C/C	C/C	p.Glu672Lys	*EIF2AK3*	0.02	0.0067	Tolerated	Benign	Disease causing
Chr12: 95531341	C/T	C/T	C/C	C/C	p.Cys984Tyr	*FGD6*	0	0	Tolerated	Probably damaging	Disease causing
Chr14: 21861835	C/T	C/T	T/T	T/T	p.Asp2040Gly	*CHD8*	0.02	0.0067	Tolerated	Benign	Disease causing

*Genotypes in this table are FWD genotype while genotypes in HGVS are REV genotype.

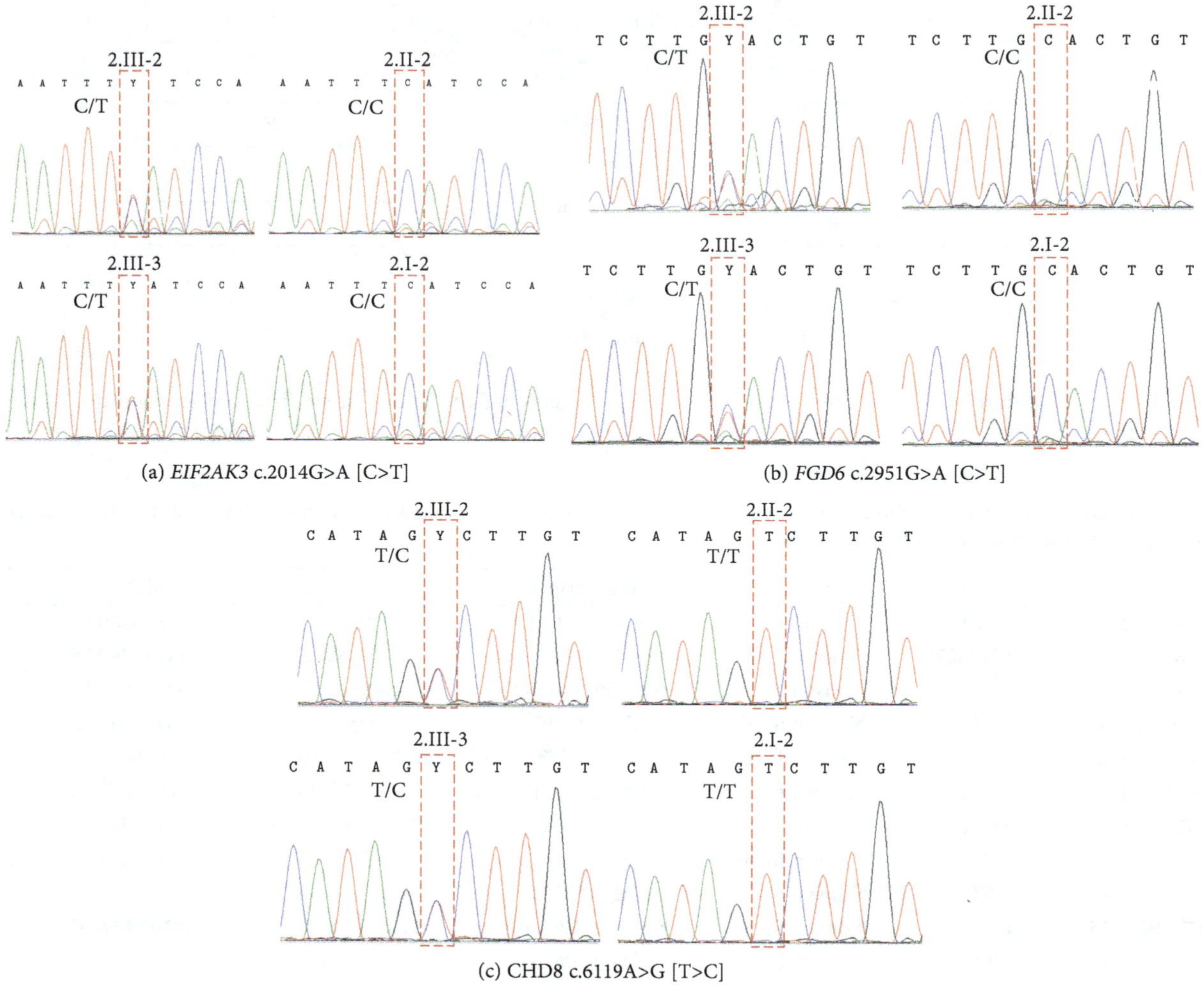

Figure 3: Chromatograms of 2 heterozygous missense variants in *EIF2AK3* (a), *FGD6* (b), and *CHD8* (c) gene. Letters in [] indicate complementary (FWD) alleles.

vertebrate early development and morphogenesis. This gene is one of the genes that have frequently been reported to be associated with ASD. This study found one possible de novo variant: c.6119A>G results in replacement of asparagine by glycine (p.Asp2040Gly) on this gene. Rare de novo mutations in this gene have been reported in individuals with ASD [26]. A study of balanced chromosomal abnormality in autism revealed disruption of genes including *CHD8* [27]. The association between the mutation in *CHD8* and ASD has been reported previously [28, 29]. Merner et al. reported 14 variants, including a de novo frameshift mutation, p.Asn2092Lysfs $*$ 2, that results in premature stop codon leading to the loss of 212 amino acids of the protein in *CHD8* by all exon genotyping technique of 142 autistic individuals with macrocephaly, psychopathology, infantile hypotonia, lack of major ID, and speech delay [30]. Wilkinson et al. have reported de novo loss-of-function mutations in 12 individuals with ASD by whole-exome sequencing [31]. Moreover, *CHD8* knock-down by siRNA-mediated techniques revealed that insufficiency of this gene leads to changed expression of 1715 genes involved in the neuronal development pathway and previously reported to be candidate genes for ASD [31].

EIF2AK3 encodes eukaryotic translation initiation factor 2-alpha kinase 3, also known as protein kinase R- (PKR-) like endoplasmic reticulum kinase or PRKR-like endoplasmic reticulum kinase [32]. This study found a variant, c.2014G>A, which results in replacement of glutamine by lysine (p.Glu672Lys) on this gene. Patients who have homozygous mutations in this gene develop Wolcott-Rallison syndrome, a disorder with infant-onset diabetes mellitus, hepatic and renal dysfunction, developmental delay or mental retardation, osteopenia, and multiple epiphyseal dysplasia [33]. *EIF2AK3* mutant mice revealed multiple phenotypes consistent with cognitive and information processing deficits, which are one feature of ASD [34].

The last one, *FGD6*, encodes FYVE, RhoGEF, and PH domain-containing 6, a protein-coding factor. This study found a variant, c.2951G>A, which results in replacement of cytosine by tyrosine (p.Cys984Tyr) on this gene. *FGD6* is an important paralog of the *FGD1* gene which is associated with syndromic autism, where some patients are affected with a particular syndrome that presents as autism. Specifically, patients with Aarskog-Scott syndrome (AAS) carry rare mutations in the *FGD1* gene [35, 36].

From an additional association study of the 3 final candidate variants (Supplementary 2), we found that only the variant in the *FGD6* gene, c.2951G>A, had a statistically significant association with ASD ($P = 0.0041$) and was found only in the cases but not in the controls. This result suggests that the *FGD6* variant might be an autism-associated variant. However, this association study was performed with only a small sample, and further studies with a larger number of cases and controls are needed to verify this finding.

Additionally, since one autism-affected child, 2.III-2, of family number 2 had heart block along with ASD, a cause of this heart defect was explored in our previous publication. The previous result revealed that autism and heart block in this family are coincidental, not likely to be Timothy syndrome [18].

Although whole-exome sequencing is an effective tool for the study of genetic variation, its limitation is that it sequences only coding regions of genome. Noncoding variants and structural variants were discounted in this study. Furthermore, functional prediction of the novel variants in this study was found by bioinformatics processes. Further study is required to determine biological function.

5. Conclusion

In conclusion, this study revealed the utility of whole-exome sequencing and a bioinformatics analysis process for identifying possible causative variants of ASD. An additional association study of the 3 final candidate variants (Supplementary 2) found that only the variant in the *FGD6* gene, c.2951G>A, had a statistically significant association with ASD. This suggests that the *FGD6* variant might be an ASD-associated variant. However, we performed the association study in only a small number of cases and controls; therefore, further studies in larger samples are needed to verify this finding.

Conflicts of Interest

The authors have no relevant affiliations or financial involvements with any organization or entity with a financial interest in or financial conflict with the subject matter or materials discussed in the manuscript.

Acknowledgments

This study was supported by grants from the (1) Pharmacogenomics for Autistic Child Project, Khoon Poom Foundation, the Project in Her Royal Highness Princess Ubonratana Rajakanya Siriwatana Bhanawadee; (2) National Research Council of Thailand; (3) Faculty of Medicine Ramathibodi Hospital; and (4) Excellence Center for Genomic Medicine, Faculty of Medicine Ramathibodi Hospital, Mahidol University, Bangkok, Thailand. This study was also partially supported by the National Center for Genetic Engineering and Biotechnology (BIOTEC) Grant no. BT-B-01-MG-18-4814 and the Faculty of Medicine, Prince of Songkla University (48/364-006-3). The authors thank all the staffs in the Laboratory for Pharmacogenomics, Somdech Phra Debaratana Medical Center (SDMC), Ramathibodi Hospital, and Human Genetics Unit, Department of Pathology, Faculty of Medicine, Prince of Songkla University.

References

[1] S. Poolsuppasit, B. Panyayong, D. Liknapichitkul, P. Serisathien, and W. Chutha, "Holistic care for Thai autism," *Journal of Mental Health of Thailand*, vol. 13, no. 1, pp. 10–16, 2005.

[2] American Psychiatric Association, *Diagnostic and Statistical Manual of Mental Disorders: DSM-5*, American Psychiatric Association, Arlington, VA, USA, Fifth edition, 2013.

[3] M. C. Lai, M. V. Lombardo, B. Chakrabarti, and S. Baron-Cohen, "Subgrouping the autism "spectrum": reflections on DSM-5," *PLoS Biology*, vol. 11, no. 4, article e1001544, 2013.

[4] S. E. Levy, D. S. Mandell, and R. T. Schultz, "Autism," *The Lancet*, vol. 374, no. 9701, pp. 1627–1638, 2009.

[5] B. S. Abrahams and D. H. Geschwind, "Advances in autism genetics: on the threshold of a new neurobiology," *Nature Reviews Genetics*, vol. 9, no. 5, pp. 341–355, 2008.

[6] J. D. Buxbaum, "Multiple rare variants in the etiology of autism spectrum disorders," *Dialogues in Clinical Neuroscience*, vol. 11, no. 1, pp. 35–43, 2009.

[7] H. V. Ratajczak, "Theoretical aspects of autism: causes—a review," *Journal of Immunotoxicology*, vol. 8, no. 1, pp. 68–79, 2011.

[8] C. Betancur, "Etiological heterogeneity in autism spectrum disorders: more than 100 genetic and genomic disorders and still counting," *Brain Research*, vol. 1380, pp. 42–77, 2011.

[9] C. Lintas and A. M. Persico, "Autistic phenotypes and genetic testing: state-of-the-art for the clinical geneticist," *Journal of Medical Genetics*, vol. 46, no. 1, pp. 1–8, 2009.

[10] T. Tucker, M. Marra, and J. M. Friedman, "Massively parallel sequencing: the next big thing in genetic medicine," *American Journal of Human Genetics*, vol. 85, no. 2, pp. 142–154, 2009.

[11] M. Choi, U. I. Scholl, W. Ji et al., "Genetic diagnosis by whole exome capture and massively parallel DNA sequencing," *Proceedings of the National Academy of Sciences of the United States of America*, vol. 106, no. 45, pp. 19096–19101, 2009.

[12] M. J. Bamshad, S. B. Ng, A. W. Bigham et al., "Exome sequencing as a tool for Mendelian disease gene discovery," *Nature Reviews Genetics*, vol. 12, no. 11, pp. 745–755, 2011.

[13] K. A. Schrader, A. Heravi-Moussavi, P. J. Waters et al., "Using next-generation sequencing for the diagnosis of rare disorders: a family with retinitis pigmentosa and skeletal abnormalities," *The Journal of Pathology*, vol. 225, no. 1, pp. 12–18, 2011.

[14] B. J. O'Roak, P. Deriziotis, C. Lee et al., "Exome sequencing in sporadic autism spectrum disorders identifies severe de novo mutations," *Nature Genetics*, vol. 43, no. 6, pp. 585–589, 2011.

[15] S. J. Sanders, M. T. Murtha, A. R. Gupta et al., "De novo mutations revealed by whole-exome sequencing are strongly associated with autism," *Nature*, vol. 485, no. 7397, pp. 237–241, 2012.

[16] B. M. Neale, Y. Kou, L. Liu et al., "Patterns and rates of exonic de novo mutations in autism spectrum disorders," *Nature*, vol. 485, no. 7397, pp. 242–245, 2012.

[17] C. Charalsawadi, W. Maisrikhaw, V. Praphanphoj et al., "A case with a ring chromosome 13 in a cohort of 203 children with non-syndromic autism and review of the cytogenetic literature," *Cytogenetic and Genome Research*, vol. 144, no. 1, pp. 1–8, 2014.

[18] C. Thongnak, P. Limprasert, D. Tangviriyapaiboon et al., "Exome sequencing identifies compound heterozygous mutations in *SCN5A* associated with congenital complete heart block in the Thai population," *Disease Markers*, vol. 2016, Article ID 3684965, 10 pages, 2016.

[19] The 1000 Genomes Project Consortium, G. R. Abecasis, A. Auton et al., "An integrated map of genetic variation from 1,092 human genomes," *Nature*, vol. 491, no. 7422, pp. 56–65, 2012.

[20] X. Liu, X. Jian, and E. Boerwinkle, "dbNSFP v2.0: a database of human non-synonymous SNVs and their functional predictions and annotations," *Human Mutation*, vol. 34, no. 9, pp. E2393–E2402, 2013.

[21] S. N. Basu, R. Kollu, and S. Banerjee-Basu, "AutDB: a gene reference resource for autism research," *Nucleic Acids Research*, vol. 37, Supplement 1, pp. D832–D836, 2009.

[22] L. M. Xu, J. R. Li, Y. Huang, M. Zhao, X. Tang, and L. Wei, "AutismKB: an evidence-based knowledgebase of autism genetics," *Nucleic Acids Research*, vol. 40, no. D1, pp. D1016–D1022, 2012.

[23] P. D. Stenson, M. Mort, E. V. Ball, K. Shaw, A. D. Phillips, and D. N. Cooper, "The human gene mutation database: building a comprehensive mutation repository for clinical and molecular genetics, diagnostic testing and personalized genomic medicine," *Human Genetics*, vol. 133, no. 1, pp. 1–9, 2014.

[24] A. Untergasser, I. Cutcutache, T. Koressaar et al., "Primer3—new capabilities and interfaces," *Nucleic Acids Research*, vol. 40, no. 15, article e115, 2012.

[25] J. S. Parla, I. Iossifov, I. Grabill, M. S. Spector, M. Kramer, and W. R. McCombie, "A comparative analysis of exome capture," *Genome Biology*, vol. 12, no. 9, article R97, 2011.

[26] B. J. O'Roak, L. Vives, S. Girirajan et al., "Sporadic autism exomes reveal a highly interconnected protein network of de novo mutations," *Nature*, vol. 485, no. 7397, pp. 246–250, 2012.

[27] M. E. Talkowski, J. A. Rosenfeld, I. Blumenthal et al., "Sequencing chromosomal abnormalities reveals neurodevelopmental loci that confer risk across diagnostic boundaries," *Cell*, vol. 149, no. 3, pp. 525–537, 2012.

[28] R. Bernier, C. Golzio, B. Xiong et al., "Disruptive *CHD8* mutations define a subtype of autism early in development," *Cell*, vol. 158, no. 2, pp. 263–276, 2014.

[29] A. Sugathan, M. Biagioli, C. Golzio et al., "*CHD8* regulates neurodevelopmental pathways associated with autism spectrum disorder in neural progenitors," *Proceedings of the National Academy of Sciences of the United States of America*, vol. 111, no. 42, pp. E4468–E4477, 2014.

[30] N. Merner, B. Forgeot d'Arc, S. C. Bell et al., "A de novo frameshift mutation in chromodomain helicase DNA-binding domain 8 (CHD8): a case report and literature review," *American Journal of Medical Genetics Part A*, vol. 170, no. 5, pp. 1225–1235, 2016.

[31] B. Wilkinson, N. Grepo, B. L. Thompson et al., "The autism-associated gene chromodomain helicase DNA-binding protein 8 (*CHD8*) regulates noncoding RNAs and autism-related genes," *Translational Psychiatry*, vol. 5, no. 5, article e568, 2015.

[32] Y. Shi, J. An, J. Liang et al., "Characterization of a mutant pancreatic eIF-2α kinase, PEK, and co-localization with somatostatin in islet delta cells," *The Journal of Biological Chemistry*, vol. 274, no. 9, pp. 5723–5730, 1999.

[33] F. Durocher, R. Faure, Y. Labrie, L. Pelletier, I. Bouchard, and R. Laframboise, "A novel mutation in the *EIF2AK3* gene with variable expressivity in two patients with Wolcott–Rallison syndrome," *Clinical Genetics*, vol. 70, no. 1, pp. 34–38, 2006.

[34] M. A. Trinh, H. Kaphzan, R. C. Wek, P. Pierre, D. R. Cavener, and E. Klann, "Brain-specific disruption of the eIF2α kinase PERK decreases ATF4 expression and impairs behavioral flexibility," *Cell Reports*, vol. 1, no. 6, pp. 676–688, 2012.

[35] T. Kaname, K. Yanagi, N. Okamoto, and K. Naritomi, "Neurobehavioral disorders in patients with Aarskog–Scott syndrome affected by novel *FGD1* mutations," *American Journal of Medical Genetics Part A*, vol. 140A, no. 12, pp. 1331-1332, 2006.

[36] D. Pinto, A. T. Pagnamenta, L. Klei et al., "Functional impact of global rare copy number variation in autism spectrum disorders," *Nature*, vol. 466, no. 7304, pp. 368–372, 2010.

10

Cerebellar lncRNA Expression Profile Analysis of SCA3/MJD Mice

Zhe Long,[1,2] Tianjiao Li,[1] Zhao Chen,[1] Yun Peng,[1] Chunrong Wang,[1] Xiaocan Hou,[1] Hongyu Yuan,[1] Puzhi Wang,[1] Yue Xie,[1] Lang He,[1] Xin Zhou,[1] Huirong Peng,[1] Rong Qiu,[3] Kun Xia,[4] Beisha Tang,[1,4,5,6,7,8,9] and Hong Jiang[1,4,6,10]

[1]*Department of Neurology, Xiangya Hospital, Central South University, Changsha, Hunan 410008, China*
[2]*Sydney Medical School and the Brain & Mind Institute, The University of Sydney, 94 Mallett St, Camperdown, NSW 2050, Australia*
[3]*School of Information Science and Engineering, Central South University, Changsha, Hunan 410083, China*
[4]*Laboratory of Medical Genetics, Central South University, Changsha, Hunan 410078, China*
[5]*National Clinical Research Center for Geriatric Diseases, Xiangya Hospital, Central South University, Changsha, Hunan 410008, China*
[6]*Key Laboratory of Hunan Province in Neurodegenerative Disorders, Central South University, Changsha, Hunan 410008, China*
[7]*Parkinson's Disease Center of Beijing Institute for Brain Disorders, Beijing 100069, China*
[8]*Collaborative Innovation Center for Brain Science, Shanghai 200032, China*
[9]*Collaborative Innovation Center for Genetics and Development, Shanghai 200433, China*
[10]*Xinjiang Medical University, Xinjiang 830011, China*

Correspondence should be addressed to Hong Jiang; jianghong73868@126.com

Academic Editor: Elena Pasyukova

Spinocerebellar ataxia type 3 (SCA3) or Machado-Joseph disease (MJD) is the most common autosomal dominant spinocerebellar ataxia in China with highly clinical heterogeneity, such as progressive cerebellar ataxia, dysarthria, pyramidal signs, external ophthalmoplegia, dysphagia, and distal muscle atrophy. It is caused by the abnormal expansion of CAG repeats in a coding region of *ATXN3*. However, by focusing on the *ATXN3* itself cannot fully explain the heterogeneous clinical features of SCA3/MJD. With the discovery of the increasing number of long noncoding RNAs (lncRNAs) that are believed to be involved in spinocerebellar ataxia type 8 (SCA8) and Huntington disease (HD), we wonder whether the lncRNAs are differentially expressed in the SCA3/MJD patients compared to the nonpatients. As the first step, we used lncRNA-Seq to investigate differential expression of the lncRNAs in the SCA3/MJD mice. Two known lncRNAs, n297609 and n297477, and a novel lncRNA TCONS_00072962 have been identified in SCA3/MJD mice with abnormal expression. The first discovery of the novel lncRNA TCONS_00072962 enriched the lncRNA expression profile in the SCA3/MJD mouse model.

1. Introduction

PolyQ diseases is a group of disorders caused by CAG repeat expansions within the, respectively, responsible genes, including Huntington disease (HD), dentatorubral-pallidoluysian atrophy (DRPLA), spinocerebellar ataxias (SCA1, SCA2, SCA3/Machado-Joseph disease, SCA6, SCA7, and SCA17) [1–3], and the recently discovered Huntington disease-like 2 (HDL2) [4, 5]. Among these, the SCA3/MJD is an autosomal dominantly inherited disorder with high clinical heterogeneity, such as progressive cerebellar ataxia, dysarthria, pyramidal signs, external ophthalmoplegia, dysphagia, and distal muscle atrophy, with wide range of age of onset (AO) from 4 to 75 years old [6, 7].

The SCA3/MJD, the most common case, accounts for 62.64% of autosomal dominant spinocerebellar ataxia in China [8]. The abnormal expansion of CAG in the causative gene ATXN3 coding region causes SCA3/MJD. Healthy individuals usually have 12–40 CAG repeats, while SCA3/MJD patients over 51 repeats [9, 10]. The abnormally translated polyQ tract leads to a conformational change in ATXN3, resulting in alternations of protein properties, including stability, subcellular location, and easier aggregation [11]. These alternations further lead to loss or gain of function and cause pathogenic effects. To explain the toxic effects, several hypotheses of pathogenic mechanisms, not mutually exclusive, have been presented, including aggregate formation [1, 11–13], disturbance of cellular protein and Ca2+ homeostasis [13–15], dysregulation of transcription [15, 16], axonal transport deficits [17, 18], impairment of mitochondrial function [15, 19, 20], and abnormal neuronal signalling [11].

Long noncoding RNA (lncRNA) is defined the nontranslatable RNA with the length of 200 nucleotides or above. The lncRNAs used to be regarded as the transcriptional "noise," the products of RNA polymerase II transcription, and did not have the biological function. However, the emerging evidence has proved their significant roles in the regulation of gene transcription, posttranscriptional regulation, and epigenetic regulation [21, 22]. Previous studies suggested that lncRNAs regulate the gene expression and transcriptional processes by several different functional mechanisms. Some show function as transcriptional regulation in cis or trains, some as an organization of nuclear domains, and others as regulation of proteins or RNA molecules. All the evidence indicated that lncRNAs have great potential to impact physiological and pathological processes. Furthermore, it has been found that some transcripts of lncRNA encode small proteins [23], making the noncoding inappropriate any longer to name this class of RNA.

In recent years, accumulating studies have found that lncRNAs are associated with neurodegenerative diseases. Spinocerebellar ataxia type 8 (SCA8), a kind of slowly progressive ataxia, is caused by the abnormal expansion of (CTG)n within the responsible gene ATXN8. A study proposes that the pathogenesis of SCA8 involves both protein and RNA gain-of-function mechanisms. (CTG)n-expanded ATXN8 encodes a pathogenic protein, and the antisense strand encodes CUG-enriched lncRNA ATXN8OS which is deposited in the nucleus and activates alternative splicing, resulting in an alternation of the expression of GABA-A transport factor 4 (GAT4/Gabt4) and finally loss of the GABAergic inhibition [24].

In a separate study, the expression of lncRNA was compared between the normal brain tissue and the brain tissue of patients with Huntington disease. A total of 35 upregulated and 146 downregulated lncRNA molecules were identified, and NEAT1 was selected by Bioinformatics. Based on the cell-level experiments, it was found that overexpression of NEAT1 was significantly resistant to H_2O_2-induced cellular damage, providing a new potential strategy for clinical treatment of the Huntington's disease [25].

To further explore the pathogenesis of SC3/MJD at RNA level, the lncRNAs specifically expressed in SCA3/MJD mice were investigated in this study.

2. Materials and Methods

2.1. SCA3/MJD Mice. SCA3/MJD mouse model (B6; CBA-Tg (ATXN3*) 84.2Cce/IbezJ; ID: 012075) from Jackson Laboratory was used, and the second generation was used in this study. The CAG repeats in the first generation mice are 84, and the *ATXN3* gene is widely expressed in various organs of the body, including the cerebellum, cerebral cortex, heart, lung, spleen, liver, and skeletal muscle [26, 27]. The SCA3/MJD adult mice (32 weeks old) of the second generation, in which carrying *ATXN3* positive rate is about 50%, and comparable age, number, and weight wild-type mice were used for experimental analysis. The study was approved by the Ethics Committee in Xiangya Hospital of Central South University.

2.2. Validation of Genotype of SCA3/MJD Mice. Validation of genotype was conducted in the second generation. Polymerase chain reaction (PCR), agarose gel electrophoresis, and capillary electrophoresis sequencing were used for genotype validation. Genomic DNA was extracted from mice tails. CAG repeats were amplified using a pair of primers 5′-CCAGTGACTACTTTGATTCG-3′ (forward) 5′-TGGCCTTTCACATGGATGTGAA-3′ (reverse). The amplification reactions contained 1 μL genomic DNA (50 ng/μL), 0.2 μL rTaq DNA polymerase (Takara, Japan), 0.2 μL dNTPs, 0.2 μL of each primer (100 ng/μL), 7.2 μL sterile water, and 1.0 μL 10x buffer (TaKaRa, Japan), for a total of 10 μL. The amplification was performed in Mastercyclers (Eppendorf AG, 22331 Hamburg, Germany) under the following conditions: initial denaturation at 95.0°C for 5 minutes, followed by 38 cycles of 95.0°C for 30 seconds, 59.0°C for 30 seconds, and 72.0°C for 30 seconds. PCR products were detected by 1% agarose gel electrophoresis (120 v, 30 min), and the results of PCR amplification were observed on imaging system after 15 minutes of ethidium bromide (EB) staining. Capillary electrophoresis sequencing was used for testing the repeats number of (CAG)n and performed on ABI 3730XL DNA Analyzer (Applied Biosystems, Foster City, CA, USA).

2.3. Validation of Phenotype of SCA3/MJD Mice. The phenotype was validated using the footprint and rotating tests. For footprint pattern analysis, the hind paws of mice were painted with black ink and the forepaws were painted with red ink. The mice walked along a narrow corridor paved with white paper. Pretraining was conducted for one week before the formal test. Mice were tested three times with 5-minute intervals. Stride length, hind paw width, front paw width, and front/hind footprint overlap were measured. For rotation, mice were placed on a rotating rod and must maintain its balance. The interval from the start of the rod rotating to the mice falling from the rotating rod was recorded. Mice were tested on separate trials at fixed speeds including 10 r/min and 20 r/min.

2.4. lncRNA-Seq. lncRNA-Seq, a high-throughput sequencing, was performed in BGI. After extracting the total RNA from mice cerebellum (three SCA3/MJD mice versus three wild-type mice), mRNA and noncoding RNAs are enriched by removing rRNA from the total RNA. By using the fragmentation buffer, the mRNAs and noncoding RNAs are fragmented into short fragments (about 200~500 nt), then

Table 1: The results of rotation test.

Rotation speed	Time (seconds)	Wild-type mice ($n = 6$)	SCA3/MJD mice ($n = 6$)	p value
10 r/min	Mean ± SD	250.067 ± 9.487	104.400 ± 8.902	0.023
20 r/min	Mean ± SD	196.467 ± 8.126	34.600 ± 6.710	0.002

The wild-type mice performed much better than the SCA3/MJD mice in the rotation test.

the first-strand cDNA is synthesized by random hexamer-primer using the fragments as templates, and dTTP is substituted by dUTP during the synthesis of the second strand. Short fragments are purified and resolved with EB buffer for end reparation and single nucleotide A (adenine) addition. After that, the short fragments are connected with adapters, and then the second strand is degraded using UNG (uracil-N-glycosylase) finally [28]. After agarose gel electrophoresis, the suitable fragments are selected for the PCR amplification as templates. During the QC steps, Agilent 2100 Bioanalyzer and ABI StepOnePlus real-time PCR system are used in quantification and qualification of the sample library. At last, the library could be sequenced using Illumina HiSeqTM 2000 or other sequencers when necessary. Differential expression analysis for both predicted novel lncRNA and lncRNA from the database has proceeded. It was compared between SCA3/MJD and wild-type mice (one by one, grouped randomly) through the Cuffdiff software to calculate the FPKM value of the gene or transcript in both samples and to detect the presence of differential expression. Also, group differential expression analysis was also performed by NOIseq method.

2.5. Quantitative Real-Time PCR. The total RNA from mice cerebellum (six SCA3/MJD mice versus six wild-type mice) was reversely transcribed to cDNA with a kit (Thermo Scientific, RevertAid First Strand cDNA #K1622). The 44 most differentially expressed lncRNAs were screened for further validation by qRT-PCR assays (Maxima SYBR Green qPCR Master Mix, CFX96, Bio-Rad, USA). β-Actin was used as an internal reference in the qRT-PCR analyses. The primers (see Table S1 in the Supplementary Material) were designed using Primer3 (http://bioinfo.ut.ee/primer3-0.4.0/). qRT-PCR assay was performed in triplicate in a volume of 20 μL containing 1 μL of cDNA. The relative expression level of lncRNA was calculated using the $2^{-\Delta\Delta Ct}$ method and Ct > 35 were excluded. A Wilcoxon rank sum test was used for statistical analyses, and $p < 0.05$ was considered statistical significance.

2.6. Bioinformatics Analysis. We conducted some biological analysis of the differentially expressed lncRNAs by search and comparison of databases such as NONCODE v5 (http://www.noncode.org/index.php), FANTOM5 (http://fantom.gsc.riken.jp/5/), STRING v.10.0 (http://version10.string-db.org/), and Gene Ontology Consortium (http://www.geneontology.org/) to predict the location, distribution, and function of these lncRNAs.

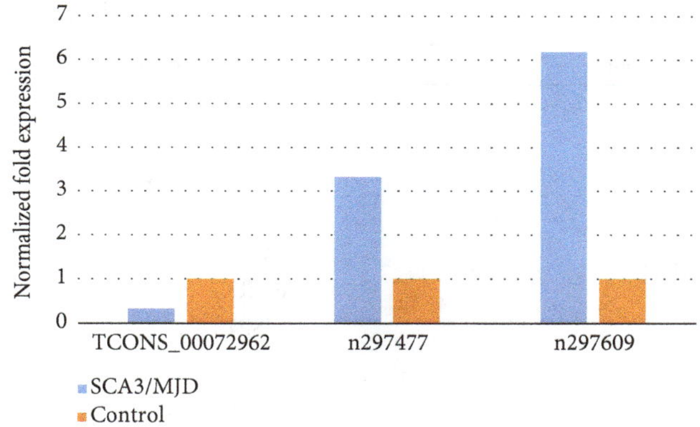

Figure 1: The expression level of the three lncRNAs.The expression level of three lncRNAs was different between wild-type and SCA3/MJD mice with statistical significance. The two known lncRNAs, n297477 and n297609, were upregulated in the cerebellum of SCA3/MJD mice. The expression level was increased by 3.329 times ($p = 0.016$) and 6.182 times ($p = 0.041$), respectively. The novel lncRNA TCONS_00072962 was downregulated ($p = 0.036$) in the cerebellum of SCA3/MJD mice, which was nearly one-third of that in wild-type mice.

3. Results

3.1. Genotyping and Phenotyping of the SCA3/MJD Mice. The SCA3/MJD mice that carry the ATXN3 were genotyped and phenotyped. As a result, the ATXN3 carrying rate was about 50% in the second generation of SCA3/MJD mice, by Jackson Laboratory's report. The number of CAG repeats was 84 or 83 in the second generation SCA3/MJD mice. In rotation test, the interval of mice keeping balance on the rotating rod was significantly different between the SCA3/MJD and wild-type mice no matter the rotation speed was 10 r/min or 20 r/min (see Table 1), indicating that the balance and motor abilities of the SCA3/MJD mice were much worse relative to the control mice. Morphologically, wider hind paw width was evident in SCA3/MJD mice.

Abnormal gait was observed by analyzing the footprint pattern. In contrast to the control mice's straight line with regular alternating gait, the SCA3/MJD mice showed unstable movements in a way that weaved from side to side when walked along the narrow corridor. Altogether suggest that the SCA3/MJD mice have similar clinical manifestations with the SCA3/MJD patients.

3.2. lncRNA-Seq. By using RNA-seq, 10,443 of novel and 13,395 of known lncRNAs were detected in 3 of the SCA3/MJD and 3 of the WT mice. One by one differential expression analysis showed total 2964 upregulated and 4376 downregulated lncRNAs, respectively, in the three experimental groups. More specifically, there were 745, 776, and 1250 of the upregulated lncRNAs in three groups, in contrast to the significantly increased numbers of 1285, 1065, and 1600, the downregulated lncRNAs, respectively, in three groups. Further group differential expression analysis found 193 of the upregulated and 467 of the downregulated lncRNAs in the SCA3/MJD mice. The lncRNAs with differential

Table 2: The summary of genetic association.

Gene name	Associated gene name	Proteins of associated gene
ATXN3	VCP	Valosin containing protein
	UBC	Ubiquitin C
	KCTD10	Potassium channel tetramerisation domain containing 10
	RAD23A	RAD23 homolog A
	RAD23B	RAD23 homolog B
	PARK2	Parkinson protein 2
	USP13	Ubiquitin specific peptidase 13
	UBE4B	Ubiquitination factor E4B
	STUB1	STIP1 homology and U-box containing protein 1
	SERPINC1	Serpin peptidase inhibitor, clade C (antithrombin), member 1
UBC	PSMD4	Proteasome (prosome, macropain) 26S subunit, non-ATPase 4
	HSP90AA1	Heat shock protein 90 kDa alpha (cytosolic), class A member 1
	HGS	Hepatocyte growth factor-regulated tyrosine kinase substrate
	PSMC2	Proteasome (prosome, macropain) 26S subunit, ATPase 2
	TSG101	Tumor susceptibility gene 101
	UBE2D2	Ubiquitin-conjugating enzyme E2D 2
	PSMD14	Proteasome (prosome, macropain) 26S subunit, non-ATPase 14
	CUL1	Cullin 1
	RPS27A	Ribosomal protein S27a (156 aa)

expression identified by the one by one differential expression analysis in two or three groups were chosen for further validation. The lncRNAs were sorted based on their FPKM and p values in the group differential expression analysis. For both differential expression analyses, a total of 44 lncRNAs were chosen for further experiments because they were statistically significant.

3.3. qRT-PCR. lncRNA number TCONS_00031478 was excluded due to Ct > 35. In the rest 43 of the chosen lncRNAs, 3 of the lncRNAs were validated to be differential expression between SCA3/MJD and WT mice cerebellum. It turned out that two of the three are the known lncRNAs number n297477 ($p = 0.016$) and number n297609 ($p = 0.041$) (NONCODE v5.0: http://www.noncode.org/), and the remaining one belonged to a novel lncRNA (number TCONS_00072962) ($p = 0.036$). $2^{-\triangle\triangle Ct}$ method was used for calculating the relative expression level of each lncRNA. When compared with wild-type mice, the two known lncRNAs, n297477 and n297609, were upregulated by 3.329-fold and 6.182-fold, respectively, in the cerebellum of SCA3/MJD mice (see Figure 1), while the expression level of novel lncRNA TCONS_00072962 was downregulated in the cerebellum of SCA3/MJD mice, which was nearly one-third of that in control mice.

4. Discussion

In this study, 2964 upregulated lncRNAs and 4376 downregulated lncRNAs were identified specifically in the SCA3/MJD mice using lncRNA-Seq analysis. Additionally, differentially expressed three lncRNAs, including one novel lncRNA and two known lncRNAs, were further characterized.

The lncRNA n297477 is transcribed from the antisense strand of the chr11: 6270375–6271530, which is highly expressed in mouse heart, hippocampus, liver, lung, spleen, and thymus. According to the records in database FANTOM5, n297477 is considered one of the transcripts of the *TMED4* gene whose promoter is located in the sense strand at position 142–392 (chr11: 6270517–6270767) and TATA box starts at 377 (chr11: 6270712) (http://www-bimas.cit.nih.gov/molbio/proscan/). According to the database STRING v.10.0, Tmed4 and ubiquitin C may be functional partners to each other, whereas the ubiquitin C might be a functional partner of ataxin3 encoded by the SCA3/MJD pathogenic gene *ATXN3* (Table 2; http://version10.string-db.org/cgi/network.pl?all_channels_on=1&block_structure_pics_in_bubbles=0&direct_neighbor=1&hide_disconnected_nodes=0&hide_node_labels=0&network_display_mode=svg&network_flavor=evidence&targetmode=proteins&identifier=9606.ENSP00000376965; http://version10.string-db.org/cgi/network.pl?all_channels_on=1&block_structure_pics_in_bubbles=0&direct_neighbor=1&hide_disconnected_nodes=0&hide_node_labels=0&network_display_mode=svg&network_flavor=evidence&targetmode=proteins&identifier=9606.ENSP00000404042).

Given that the ubiquitin-proteasome system (UPS) is involved in the pathogenesis of SCA3/MJD [29], it is plausible to speculate that lncRNA may participate in gene expression regulation when it is located close to the transcription start site of the promoter region. Since the lncRNA n297477 and TMED4 gene starting site are separated by 16 bp in the middle, and the gene promoter region is located in the lncRNA coding region, it is highly possible that the n297477 participates in TMED4 expression regulation. We assume that the altered expression of the TMED4 affects its interaction with UBC, which affects the efficiency of ubiquitin C in the ubiquitin-proteasome system. Thus, the lncRNA n297477 may be involved in the regulation of ataxin-3 protein degradation by regulating UPS. The elevated expression of n297477 in the SCA3/MJD mice may serve as a response to the abnormal degradation of toxic ataxin-3.

The lncRNA n297609 is transcribed from the antisense strand of the chr9: 44047488–43856016, which is highly expressed in mouse heart, hippocampus, liver, lung, spleen, and thymus. According to the records in database FANTOM5, n297609 is considered one of the transcripts of the *THY1* gene which is highly homologous to humans. The lncRNA n297609 mainly participates in the protein phosphorylation (GO: 0006468), cell adhesion (GO: 0007155), cellular response to heat (GO: 0034605), positive regulation of transcription (GO: 0045893), and negative regulation of cell migration (GO: 0030336). Previous studies pointed out that protein casein kinase 2- (CK2-) dependent phosphorylation can control the stability, nuclear localization, and aggregation of ataxin-3 [30]. Thus, it is logical to assume that the lncRNA n297609 may be involved in biological processes, such as protein phosphorylation, and protein casein kinase 2- (CK2-) dependent phosphorylation plays a crucial role in SCA3/MJD pathophysiology [30]. However, this assumption requires further validation.

Moreover, we also found a novel lncRNA TCONS_00072962 with the genomic location at chr7:119737479–119737966, contributing to the expansion of lncRNA expression profiles of the mouse.

5. Conclusions

In summary, we used lncRNA-Seq to profile cerebellar expression in SCA3/MJD mice and identified three potential lncRNAs significantly associated with the disease. These identified lncRNAs will be beneficial for the further understanding of cerebellum gene coexpression network correlating with disease progression. Furthermore, investigation on the lncRNA-associated neuroprotective factors remains to be necessary for the elucidation of the therapeutic target implication.

Conflicts of Interest

The authors declare no conflict of interest.

Authors' Contributions

Zhe Long and Tianjiao Li contributed equally to this work.

Acknowledgments

This study was supported by the National Natural Science Foundation of China (nos. 81471156 and 81771231 to Hong Jiang; nos. 81430023 and 81130021 to Beisha Tang; no. 31401135 to Rong Qiu), National Key Research and Development Program of China (nos. 2016YFC0901504 and 2016YFC0905100 to Hong Jiang; no. 2016YFC1306000 to Beisha Tang), Clinical Research Funds of Xiangya Hospital (no. 2014L03 to Hong Jiang), the Clinical and Rehabilitation Fund of Peking University Weiming Biotech Group (no. xywm2015I10 to Hong Jiang), Youth Foundation of Xiangya Hospital (no. 2017Q03 to Zhao Chen), and Independent Exploration and Innovation Project of Graduate Student of Central South University (no. 1053320170177 to Tianjiao Li). The authors want to give the sincerest thanks to all those who helped in the study.

References

[1] A. Ashkenazi, C. F. Bento, T. Ricketts et al., "Polyglutamine tracts regulate beclin 1-dependent autophagy," *Nature*, vol. 545, no. 7652, pp. 108–111, 2017.

[2] S. Esteves, S. Duarte-Silva, and P. Maciel, "Discovery of therapeutic approaches for polyglutamine diseases: a summary of recent efforts," *Medicinal Research Reviews*, vol. 37, no. 4, pp. 860–906, 2016.

[3] H. L. Paulson, V. G. Shakkottai, H. B. Clark, and H. T. Orr, "Polyglutamine spinocerebellar ataxias - from genes to potential treatments," *Nature Reviews. Neuroscience*, vol. 18, no. 10, pp. 613–626, 2017.

[4] I. Paradisi, V. Ikonomu, and S. Arias, "Huntington disease-like 2 (HDL2) in Venezuela: frequency and ethnic origin," *Journal of Human Genetics*, vol. 58, no. 1, pp. 3–6, 2013.

[5] C. Wang, Z. Chen, F. Yang et al., "Analysis of the GGGGCC repeat expansions of the C9orf72 gene in SCA3/MJD patients from China," *PLoS One*, vol. 10, no. 6, article e0130336, 2015.

[6] X. Li, H. Liu, P. L. Fischhaber, and T. S. Tang, "Toward therapeutic targets for SCA3: insight into the role of Machado-Joseph disease protein ataxin-3 in misfolded proteins clearance," *Progress in Neurobiology*, vol. 132, pp. 34–58, 2015.

[7] H. Peng, C. Wang, Z. Chen et al., "The *APOE* ε2 allele may decrease the age at onset in patients with spinocerebellar ataxia type 3 or Machado-Joseph disease from the Chinese Han Population," *Neurobiology of Aging*, vol. 35, no. 9, pp. 2179.e15–2179.e18, 2014.

[8] Z. Chen, P. Wang, C. Wang et al., "Updated frequency analysis of spinocerebellar ataxia in China," *Brain*, vol. 141, no. 4, article e22, 2018.

[9] C. Wang, H. Peng, J. Li et al., "Alteration of methylation status in the *ATXN3* gene promoter region is linked to the SCA3/MJD," *Neurobiology of Aging*, vol. 53, pp. 192.e5–192.e10, 2017.

[10] Z. Chen, C. Zheng, Z. Long et al., "(CAG)n loci as genetic modifiers of age-at-onset in patients with Machado-Joseph disease from mainland China," *Brain*, vol. 139, no. 8, article e41, 2016.

[11] C. Costa Mdo and H. L. Paulson, "Toward understanding Machado-Joseph disease," *Progress in Neurobiology*, vol. 97, no. 2, pp. 239–257, 2012.

[12] K. Seidel, S. Siswanto, M. Fredrich et al., "Polyglutamine aggregation in Huntington's disease and spinocerebellar ataxia type

3: similar mechanisms in aggregate formation," *Neuropathology and Applied Neurobiology*, vol. 42, no. 2, pp. 153–166, 2016.

[13] K. Rajamani, J. W. Liu, C. H. Wu et al., "N-Butylidenephthalide exhibits protection against neurotoxicity through regulation of tryptophan 2, 3 dioxygenase in spinocerebellar ataxia type 3," *Neuropharmacology*, vol. 117, pp. 434–446, 2017.

[14] M. Tada, M. Nishizawa, and O. Onodera, "Roles of inositol 1,4,5-trisphosphate receptors in spinocerebellar ataxias," *Neurochemistry International*, vol. 94, pp. 1–8, 2016.

[15] X. Chen, J. Wu, Y. Luo et al., "Expanded polyglutamine-binding peptoid as a novel therapeutic agent for treatment of Huntington's disease," *Chemistry & Biology*, vol. 18, no. 9, pp. 1113–1125, 2011.

[16] A. H. Chou, Y. L. Chen, C. C. Chiu et al., "T1-11 and JMF1907 ameliorate polyglutamine-expanded ataxin-3-induced neurodegeneration, transcriptional dysregulation and ataxic symptom in the SCA3 transgenic mouse," *Neuropharmacology*, vol. 99, pp. 308–317, 2015.

[17] S. Gunawardena, L.-S. Her, R. G. Brusch et al., "Disruption of axonal transport by loss of huntingtin or expression of pathogenic polyQ proteins in Drosophila," *Neuron*, vol. 40, no. 1, pp. 25–40, 2003.

[18] B. Malik, N. Nirmalananthan, L. G. Bilsland et al., "Absence of disturbed axonal transport in spinal and bulbar muscular atrophy," *Human Molecular Genetics*, vol. 20, no. 9, pp. 1776–1786, 2011.

[19] J. C. Chang, S. L. Wu, F. Hoel et al., "Far-infrared radiation protects viability in a cell model of spinocerebellar ataxia by preventing polyQ protein accumulation and improving mitochondrial function," *Scientific Reports*, vol. 6, no. 1, article 30436, 2016.

[20] J. Y. Hsu, Y. L. Jhang, P. H. Cheng et al., "The truncated C-terminal fragment of mutant ATXN3 disrupts mitochondria dynamics in spinocerebellar ataxia type 3 models," *Frontiers in Molecular Neuroscience*, vol. 10, p. 196, 2017.

[21] F. P. Marchese, I. Raimondi, and M. Huarte, "The multidimensional mechanisms of long noncoding RNA function," *Genome Biology*, vol. 18, no. 1, p. 206, 2017.

[22] J. Beermann, M. T. Piccoli, J. Viereck, and T. Thum, "Non-coding RNAs in development and disease: background, mechanisms, and therapeutic approaches," *Physiological Reviews*, vol. 96, no. 4, pp. 1297–1325, 2016.

[23] F. Kopp and J. T. Mendell, "Functional classification and experimental dissection of long noncoding RNAs," *Cell*, vol. 172, no. 3, pp. 393–407, 2018.

[24] M. L. Moseley, T. Zu, Y. Ikeda et al., "Bidirectional expression of CUG and CAG expansion transcripts and intranuclear polyglutamine inclusions in spinocerebellar ataxia type 8," *Nature Genetics*, vol. 38, no. 7, pp. 758–769, 2006.

[25] J. S. Sunwoo, S. T. Lee, W. Im et al., "Altered expression of the long noncoding RNA NEAT1 in Huntington's disease," *Molecular Neurobiology*, vol. 54, no. 2, pp. 1577–1586, 2017.

[26] C. K. Cemal, C. Huxley, and S. Chamberlain, "Insertion of expanded CAG trinucleotide repeat motifs into a yeast artificial chromosome containing the human Machado-Joseph disease gene," *Gene*, vol. 236, no. 1, pp. 53–61, 1999.

[27] C. K. Cemal, C. J. Carroll, L. Lawrence et al., "YAC transgenic mice carrying pathological alleles of the MJD1 locus exhibit a mild and slowly progressive cerebellar deficit," *Human Molecular Genetics*, vol. 11, no. 9, pp. 1075–1094, 2002.

[28] D. Parkhomchuk, T. Borodina, V. Amstislavskiy et al., "Transcriptome analysis by strand-specific sequencing of complementary DNA," *Nucleic Acids Research*, vol. 37, no. 18, article e123, 2009.

[29] F. Mori, M. Nishie, Y. S. Piao et al., "Accumulation of NEDD8 in neuronal and glial inclusions of neurodegenerative disorders," *Neuropathology and Applied Neurobiology*, vol. 31, no. 1, pp. 53–61, 2005.

[30] T. Mueller, P. Breuer, I. Schmitt, J. Walter, B. O. Evert, and U. Wüllner, "CK2-dependent phosphorylation determines cellular localization and stability of ataxin-3," *Human Molecular Genetics*, vol. 18, no. 17, pp. 3334–3343, 2009.

11

Metagenomic Approaches for Understanding New Concepts in Microbial Science

Luana de Fátima Alves,[1] Cauã Antunes Westmann,[2] Gabriel Lencioni Lovate,[1] Guilherme Marcelino Viana de Siqueira,[1] Tiago Cabral Borelli,[3] and María-Eugenia Guazzaroni[3]

[1] *Department of Biochemistry, Faculdade de Medicina de Ribeirão Preto, University of São Paulo, Ribeirão Preto, SP, Brazil*
[2] *Department of Cell Biology, Faculdade de Medicina de Ribeirão Preto, University of São Paulo, Ribeirão Preto, SP, Brazil*
[3] *Department of Biology, Faculdade de Filosofia, Ciências e Letras de Ribeirão Preto, University of São Paulo, Ribeirão Preto, SP, Brazil*

Correspondence should be addressed to María-Eugenia Guazzaroni; meguazzaroni@gmail.com

Academic Editor: Henry Heng

Over the past thirty years, since the dawn of metagenomic studies, a completely new (micro) universe was revealed, with the potential to have profound impacts on many aspects of the society. Remarkably, the study of human microbiome provided a new perspective on a myriad of human traits previously regarded as solely (epi-) genetically encoded, such as disease susceptibility, immunological response, and social and nutritional behaviors. In this context, metagenomics has established a powerful framework for understanding the intricate connections between human societies and microbial communities, ultimately allowing for the optimization of both human health and productivity. Thus, we have shifted from the old concept of microbes as harmful organisms to a broader panorama, in which the signal of the relationship between humans and microbes is flexible and directly dependent on our own decisions and practices. In parallel, metagenomics has also been playing a major role in the prospection of "hidden" genetic features and the development of biotechnological applications, through the discovery of novel genes, enzymes, pathways, and bioactive molecules with completely new or improved biochemical functions. Therefore, this review highlights the major milestones over the last three decades of metagenomics, providing insights into both its potentialities and current challenges.

1. Introduction

About thirty years ago, in 1986, Pace and collaborators [1] proposed, for the first time, the revolutionary idea of cloning DNA directly from environmental samples to analyze the complexity of natural microbial populations (Figure 1, indicated as M2). The adopted strategy was based on shotgun cloning of 16S rRNA genes using purified DNA from natural samples. At that time, authors stressed that although the DNA was originated from a mixed population of microorganisms, the methodology allowed the recovery and subsequent sequencing of individual rRNA genes. Thus, by evaluating complete or partial rRNA sequences, the composition of the original microbial populations could be retrieved.

Around ten years later, in 1998, the term "metagenome" appeared, when Handelsman and collaborators [2] described the importance of soil microorganisms as sources for new natural compounds (Figure 1, indicated as M6). According to them, a new frontier in science was emerging—the mining for novel chemical compounds from uncultured microorganisms, which comprises more than 99% of the microbial diversity [3]. This new concept in microbial science opened the mind of the scientific community with respect to the astonishingly large catalogue of biochemical functions available in nature remaining to be discovered.

Currently, metagenomics is subdivided into two major approaches, which target different aspects of the local microbial community associated with a determined environment.

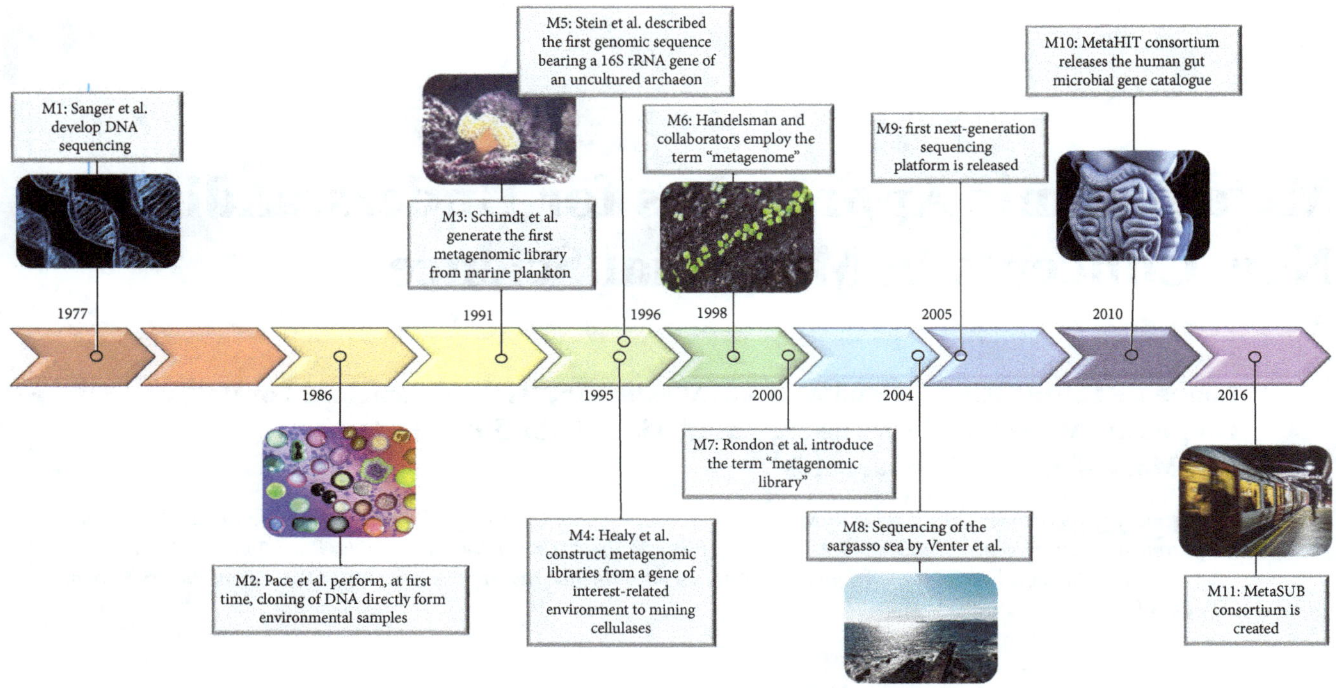

Figure 1: Timeline of the major advancements in metagenomics. Timeline highlighting important developments in the field over the last 40 years, since Sanger sequencing (M1), and over the last 30 years, since the first published metagenomic experiment (M2). The main metagenomic milestones are shown as M1–M11 (all of them are highlighted in the text where they were mentioned).

In the first one, the so-called structural metagenomic approach, the main focus is to study the structure of the uncultivated microbial population, which can be expanded to other properties, such as the reconstruction of the complex metabolic network established between community members (Figure 2) [4, 5]. In this sense, the microbial community structure can be defined as the population composition and its dynamics in a specific ecosystem, in response to selective pressures and spatiotemporal parameters. The study of the community structure allows a deeper understanding about the relationships between the individual components that build a community and is essential for deciphering ecological or biological functions among its members [5, 6]. In a different manner, the functional metagenomic approach aims to identify genes that code for a function of interest, which involves the generation of expression libraries with thousands of metagenomic clones followed by activity-based screenings (Figure 2) [7, 8].

It is important to highlight that 16S rRNA gene surveys are often referred to as metagenomic studies, although they are not. In the 16S rRNA gene analysis, the study is focused on a single gene used as a taxonomic marker (Figure 2). On the other hand, structural metagenomics aims to investigate the genomes of the microbial community members. In this sense, the later approach allows the overall reconstruction of the community structure, potentially revealing metabolic pathways of the whole microbiome and assigning minor or major geoecological roles to community members [4, 6, 9, 10]. We highlight that 16S rRNA surveys and metagenomics are not mutually exclusive; on the contrary, approaches that establish a link between 16S rRNA analyses to genes or metabolic pathways have been shown to be useful in determining the functional potential of a microbiome [11–13]. Thus, the combination of these complementary strategies allows for a deeper exploration of relevant biological questions in microbial ecology such as "who are the members of the community?" and "what are their functional roles?"

At the beginning of the metagenomic studies, the use of Sanger sequencing technology [14] provided important progresses in the field [15–17] (Figure 1, indicated as M1). However, the advent of next-generation sequencing (NGS) technologies capable of sequencing millions of DNA fragments simultaneously, at a low cost, greatly bolstered the field [18–21] (Figure 1, indicated as M9). Comparatively, NGS platforms can recover up to 5000 Mb of DNA sequence per day with costs at about 0.50$/Mb, while Sanger sequencing methodology allows about 6 Mb of DNA sequence to be created per day with costs at about 1000 higher [22].

This review focuses on how metagenomics has contributed to gain scientific comprehension in many different areas of knowledge. In this manner, milestones in metagenomics have ranged from findings with significant biotechnological impact to unexpected outcomes with high biomedical relevance, shining a light on hidden molecular components and on the connections between microbial communities and complex diseases [23–26]. We also discuss the current boundaries of the field that should be overcome for the achievement of conceptual advances in microbial science.

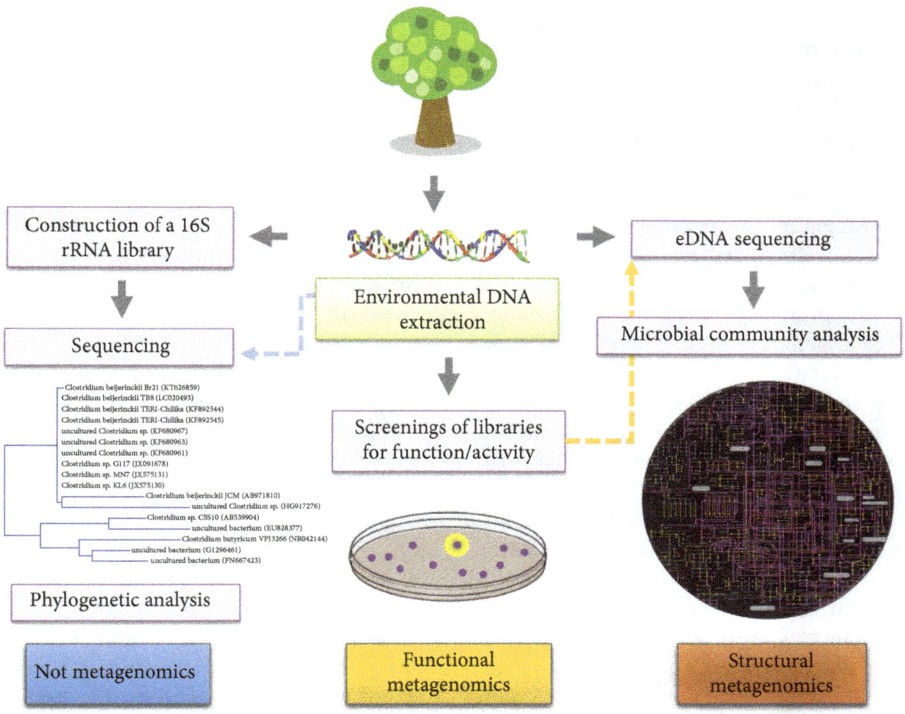

Figure 2: The metagenomics framework and its two main approaches. Both structural and functional metagenomic approaches are the main strategies for exploring key ecological and biotechnological features in environmental samples, respectively. Additionally, 16S rRNA gene surveys can work in synergy with metagenomics for further understanding of microbial ecology.

2. Milestones in Metagenomics

In 1991, Schmidt and collaborators generated the first metagenomic library using DNA from marine picoplankton [27], and, some years later, Healy et al. constructed metagenomic libraries from an enriched consortia sampled from cellulose digesters to mine genes encoding cellulases [28] (Figure 1, indicated as M3 and M4, resp.). In this context, the idea of screening metagenomic libraries from specific environments was introduced, allowing that the number and properties of the retrieved genes (e.g., enzymes) could be correlated to the conditions of the source environments. However, only in 2000, Rondon and collaborators [29] coined the term "metagenomic libraries," by generating libraries in BACs (bacterial artificial chromosome) using DNA from soil samples (Figure 1, indicated as M7). Furthermore, the authors also performed phylogenetic analysis of 16S rRNA sequences and identified a number of clones expressing heterologous genes in functional screenings using *Escherichia coli* as host.

Since then, a large amount of data has been generated using metagenomic approaches and impacting different areas of high applicability in our society (Figure 2). Here, we highlighted the main milestones in metagenomic studies that defined the field in distinct contexts. Although many of the advances in the field can be credited to novel sequencing approaches and the development of new computational methodologies to analyze the generated data, in this review, we have focused on highlighting biological discoveries through metagenomics rather than describing these more technical strategies, which have been addressed in recent reviews [30–32].

2.1. Genomic and Taxonomic Novelties in Environments. Ever since the proposal of using molecular markers such as 16S (or 18S) rRNA for comparing species similarity, in the late 1970s, followed by the outlining of a third "urkingdom" (the Archaebacteria), our knowledge of organism relatedness has taken a great leap forward. Taxonomic classification started to rely on a "comparative approach that can measure the degree of difference in comparable structures," which not only allowed a more resolved phylogeny and a less biased organization based on the *Prokaryotae* versus *Eukaryotae* dichotomy but also made it possible to better understand how life on earth has come to be [33]. In this context, metagenomic approaches have been used to generate data of novel genomes from otherwise uncultivated organisms, deepening the framework of genomic tools available for comparison and study.

The shotgun sequencing of the Sargasso Sea waters, by Venter and colleagues in 2004 [34], is one of the most illustrative examples of how metagenomics are a feasible way to accumulate genomic knowledge (Figure 1, indicated as M8). In this study, Venter and collaborators have recovered almost 1.5 Gbp of microbial DNA sequences from microbial populations of three marine sites using en masse whole genome shotgun sequencing from filtered sea water. This leads to the finding of almost 70 thousand novel genes among the roughly 1.2 million genes by ORF (open reading frame) identification and alignment of the putative protein products. Among the main findings, the researchers described a novel ammonia oxidation pathway uninhibited by UV light, putative genes for trace metal resistance such as arsenate and copper, and, additionally, up to 782 new proteorhodopsin-like receptors genes. The latter added an insight to the exotic marine photoheterotrophic lifestyle first described by Béjà and colleagues only a few years before [35]. Further, the metagenomic study of the Sargasso Sea allowed the identification of 148 previously unknown rRNA bacterial phylotypes. However, this probably was an underestimation of the environment's total genomic pool, given that the difference in the number of rRNA coding genes within the rRNA operon between species may result in biases in PCR studies with under representation of some of the community constituents, as also discussed by Klappenbach and colleagues [36].

In this context, a recent study based on metagenomic data has shown that about 10% of environmental bacterial or archaeal sequences might not be recovered when using a targeted PCR survey with the most common primers for SSU rRNA [37]. For instance, a study led by Brown and colleagues in 2015 [38] first described the Candidate Phyla Radiation (CPR) bacterial lineage comprising at least 25 new bacterial phyla, which make up to at least 15% to the Bacteria domain. These novel organisms lack many biosynthetic pathways and possess unusual features in their small genomes, such as self-splicing introns within the rRNA genes and novel ribosome structure, providing insight to the organelle's evolution trajectory. It is worth noting that all complete CPR genomes curated in the work have only one copy of the 16S rRNA gene and many of the organisms would evade detection by 16S rRNA gene amplicon surveys, even though corresponding to such a high percentage of bacterial diversity in environments. In their work, Venter and collaborators have attempted to address this issue and to better elucidate the phylogenetic relationships in the recovered genetic material by employing six other phylogenetic markers, such as heat shock protein 70 (HSP70) and elongation factor Tu (EF-Tu), as well as several methods to estimate species diversity, which have resulted in the estimation of over 1000 species [34].

Metagenomic studies also shed a light on challenges in the field of evolutionary biology, such as in the understanding of sexual reproduction as a constraint on genomic variation [39]. Previous to the metagenomic era, it was believed that many microbial species should be genomic clones—asexual reproduction was assumed to produce identical clones. Yet, metagenomic analyses unexpectedly revealed that most microbial species were not clonal [34]. Thus, asexual reproduction present in bacteria should increase genetic variation providing evolutionary diversity for future environmental challenges [40]. This finding was essential to support the conclusion that sexual reproduction acts as a constraint on genomic and epigenetic variation, thereby limiting adaptive evolution [39].

In recent years, new sampling and sequencing techniques allowed researchers to further explore life diversity improving phylogenetic, genomic, and ecological notions previously established. Rinke and colleagues, for instance, revealed the "microbial dark matter" in a single-cell genomics study that comprised over 20 major uncultivated archaeal and bacterial lineages [41]. They first described archaeal sigma factors and the first reported lateral gene transfer from a eukaryote to an archaeon. Another insight of how the archaeal/eukaryotic relationship came to be and how the modern eukaryotic cell was formed was found by Spang and collaborators [42], who have identified a candidate archaeal phylum which forms a monophyletic group with eukaryotes. It also possesses genes encoding proteins similar to those related to cell shape formation processes in eukaryotic cells, which might suggest sophisticated membrane remodeling and vesicular trafficking processes in eukaryotic cells even before the acquisition of mitochondrion.

2.2. Innovative Functions in Uncultivable Microbes. Sequenced-based analysis in metagenomics can be accomplished, overall, by following one of the following paths: (i) sequencing all clones with a phylogenetic marker indicating the potential taxonomic source of the DNA fragment or (ii) sequencing random fragments until a gene of interest is found followed by sequencing of the adjacent regions to find taxonomic markers. The former method was developed by Stein's group and described the first genomic sequence bearing a 16S rRNA gene of an uncultured archaeon (Figure 1, indicated as M5) [43]. This provided the first highlights of the metagenomics capabilities for unravelling novel genes, functions, and taxonomic groups. In this study, with a marine picoplankton assemblage collected at eastern North Pacific, a 38.5 kbp fragment containing an archaeal 16S rRNA gene was isolated for the first time. Among other features, the discovery of an RNA helicase and a glutamate semialdehyde aminotransferase, which were still unknown in archaeal organisms, was reported. The metagenomic library in this study had an average fragment size of 40 kb and was based on fosmid backbones. The screening was very labor intensive, relying on Southern blotting with phylogenetic probes for 16S rRNA gene, followed by sequencing of selected clones through automated Sanger or shotgun methods.

It did not take long for the same group to push the boundaries and further advance the incipient field of metagenomics [35, 44]. From the surface waters of the Californian coast, a marine planktonic bacterial assemblage was collected for metagenomics analysis. Hitherto, bacteriorhodopsins were considered unique features of halophilic archaea; however, the small ribosome subunit gene revealed its source as a gammaproteobacterium (uncultivable bacterial SAR86 lineage). It was also shown that the protein was functional when cloned into *E. coli*, presenting similar kinetics to the archaeal-related cognates [35]. In this study, both metagenomic library and fragment average sizes were much larger than in the previous study and the screening process was accelerated (6240 screened clones with an average size of 80 kb) [44]. These improvements were only possible due to the establishment of BACs and PCR-based screening as new tools for metagenomic studies. Thus, it did not take long for novel proteorhodopsins to be detected in other populations of planktonic marine bacteria [45]. This was considered the first great "metagenomic success," allowing the adoption of these new techniques among laboratories around the world.

2.3. Deciphering and Rebuilding Microbial Communities. By starting with an extremely simple microbial community—an acid mine drainage (AMD) microbial community—evidenced by an initial group-specific fluoresce in situ hybridization (FISH) assay, Tyson and collaborators [46] were able to perform the first assembly of genomes directly from environmental samples. In this study, they have obtained two nearly complete genomes of *Leptospirillum* group II and *Ferroplasma* type II and partially recovered another three genomes. The initial characterization of this microenvironment, a biofilm with rather extreme conditions like very acidic pH (approximately 0.83), revealed the presence of archaea (*Thermoplasmatales*) and bacteria (*Leptospirillum*, *Sulfobacillus*, and *Acidimicrobium*), with the domination of *Leptospirillum* group II. The low diversity of the system was reported as possibly related to the extreme environmental conditions. Afterwards, the DNA sequencing of a biofilm sample suggested adaptive molecular traits of the community to survive in this environment—such as homologous recombination forming mosaic genome types—and metabolic adaptations, such as abundance of genes related to ferrous iron oxidation.

In other metagenomic studies, Tringe et al. [5] performed comparisons of the composition and functionality of microbial communities from two nutrient-poor and two enriched-nutrient environments. This approach was mainly concerned with gene function rather than genome composition, overcoming limitations to genome assembly from complex environments. Authors showed that gene function and structure differ in nutrient-limited (Sargasso Sea and AMD) versus nutrient-abundant (Minnesota farm soil and deep-sea "whale fall" carcasses) environments. Some gene functions were exclusive to specific environments, for instance, (i) cellobiose phosphorylases were only found at the agricultural soil and (ii) light-driven proton pumps are only found at the Sargasso Sea samples, whereas no photoreceptors were found at the deep-sea samples.

The development of functional approaches in large scales also provided novel insights into communities' key metabolic process. A comparative functional profiling of 9 biomes was performed by Dinsdale and collaborators [47], describing how different biological traits play essential roles in each environment. For instance, authors showed an abundance of virulence genes in fish- and terrestrial-animal-associated metagenomes in comparison to the other biomes. In contrast, virome analysis of the total number of biomes showed a more uniform gene functional composition due to phages playing similar roles in different environments [47].

2.4. City-Scale Molecular Profile of DNA. The last decades of metagenomics have shed a light on the human microbiome and its profound influence on a wide range of aspects which were previously regarded as solely (epi-) genetically encoded, such as diseases susceptibility, immunological response, and social and nutritional behaviors [48–52]. Although there is still much to learn in this subject, recent studies are targeting not only the inside but also the external part of the human world, also known as microbiomes from human-built environments [53–55]. In this context, a myriad of anthropocentric utensils and physical spaces has been assessed, such as kitchen sponges [56], dollar bills [57], ATMs (Automated Teller Machines) [58], homes [59], hospitals [60], subways [61, 62], food production sites [63], and even the International Space Station [64]. The main goal of this new research branch is to provide a framework for understanding the relationship between human societies and microbial communities, ultimately optimizing both human health and productivity.

A seminal research in the context of urban areas was conducted by Afshinnekoo et al. [61], which sampled different features of New York subways and reported that nearly 1700 microbial taxa were dominated mostly by human skin bacteria and to a lesser extent by microbes from the human gastrointestinal and urogenital tracts. Almost half the DNA present on the subway surfaces matched no known organism. Although results showed that the bacteria found in the subways were mostly harmless, several pathogenic agents, including fragments of the plague and anthrax genomes were detected. This was the beginning of an international consortium called The Metagenomics and Metadesign of Subways and Urban Biomes (MetaSUB) that has been sampling urban microbiomes throughout the world [62] (Figure 1, indicated as M11). An important scientific agenda was also launched in 2017, the microbiomes of the built environment [65]. Its main objective is to assess the current state of knowledge on indoor microbiomes and also to map out research agendas and advise government agencies on how living spaces can be designed "to support occupant health and well-being." Other studies have assessed complementary aspects of this matter, such as the influence of landscape connectivity in microbial diversity [66], the influence of green areas in urban spaces [67, 68], and the consequences of excessive time expenditure indoors in the context of both human health and environmental microbiomes [54, 69].

Altogether, those studies allowed the depiction of a few important conclusions regarding human-built spaces and microbiomes, further explained by Stephens et al. [70]: (i) culture-independent methods are essential for those surveys, (ii) indoor spaces often harbor unique microbial communities, tightly related to the indoor sources—mostly humans and pets, (iii) building design and operation can directly modulate indoor microbial communities, and (iv) it is possible to optimize human health by exposure to certain microbial groups. Consequently, society moved from the old concept of microbes as harmful organisms to a new view in which the interaction between humans and microorganisms can be flexible and directly dependent on our own decisions and practices. Further studies shall reveal novel rules on "good living standards" for both humans and microbes in built environments.

2.5. The Human Microbiome. The concept of the human microbiome, embracing the idea that human beings are highly susceptible to the microbial communities that live in and on our bodies, was an indubitable milestone in metagenomics with large repercussion in many areas. In this sense, scientific contributions involving metagenomic approaches rapidly highlighted the evidences showing that these microbiomes play key roles in human health and disease.

The human gut microbiota—the collection of microorganisms that compose the human gut—is composed by up to 10^{14} microorganisms [71, 72] including bacteria, viruses, fungi, and protozoa. Deciphering the function and composition of our microbiome—the collective genomes of the microbial community resident—is a challenge that has been explored by researchers in a series of initiatives like the Human Microbiome Project (HMP), the Integrative Human Microbiome Project (iHMP), and the MetaHIT (METAgenomics of the Human Intestinal Tract) [73–75] (Figure 1, indicated as M10). The findings of these projects have provided valuable data about the function of the human microbiome in different tracts (e.g., nasal, oral, skin, gastrointestinal, and urogenital). Particularly, advances in molecular biology procedures, next-generation DNA sequencing, and *omics* techniques have allowed to access not only to the microbial genetic diversity but also to the understanding of the physiology and the lifestyle of our microbiome. In this sense, it was demonstrated that gut microorganisms perform several elemental functions like synthesis of essential amino acids and vitamins and processing of cellulosic material [76], playing an important role in a number of human health aspects [77].

In a series of studies coming out of the Gordan lab at Washington University School of Medicine in St. Louis, Ley and collaborators [78] showed that obesity has a microbial component. To explore the relation between gut microbial ecology and body fat in humans, authors studied 12 obese people, who were randomly assigned to two types of low-calorie diet. The composition of their gut microbiota was monitored over the course of one year by sequencing 16S ribosomal RNA genes from stool samples [78]. Obtained data showed that two groups of beneficial bacteria were dominant in the human gut, the Bacteroidetes and the Firmicutes. In addition, they showed that the percentage of Bacteroidetes correlated with the percentage of loss weight. In other studies [79], they found that the transplantation of the microbiota from obese mice to lean mice could lead to an increase of body fat in transplanted mice when compared with transplantation from lean mice microbiota.

The other important outcome from the human gut microbiome studies was regarding antibiotic resistance. In order to determine an "antibiotic resistance potential," Forslund et al. [80] performed a quantitative gut metagenomic analysis of known resistance genes from people of three countries. In this study, authors showed that antibiotic resistance gene abundance in the general human population is correlated with the length of antibiotic usage in livestock [80–82]. In another study, Raymond and coworkers [83] showed that the initial gut microbiome affects its recomposition after antibiotics treatment. They administrated two second-generation antibiotics (cephalosporin and cefprozil) in healthy individuals and showed that antibiotics altered the microbiome of healthy volunteers in an interindividual manner, allowing the emergence of potentially pathogenic *Enterobacteriaceae*—in a subgroup of patients—probably related to a decreased initial gut microbiome diversity in those individuals.

Besides obesity and antibiotic resistance, the human gut microbiome has been associated with several diseases as type 2 diabetes, cardiovascular and inflammatory bowel diseases, and even cancer [84]. Some studies have also associated gut microbiome with intestinal immunity. It has been shown that a healthy microbiota improves local expression of a Toll-like receptor (TLR) [85] which recognizes the PAMPs (pathogen-associated molecular patterns) expressed by a broad range of infective agents and improves the percentage of antigen-presenting cells, differentiated T cells, and lymphoid follicles [86, 87]. Besides the local immunity, the gut microbiota affects the systemic immunity by increasing splenic $CD4^+$ T cells and systematic antibody expression [88]. Consequently, the global role of the gut microbiota in intestinal immunity has increased the interest of the scientific community in developing techniques that improve human health by manipulating the gut microbiota.

Nowadays, researchers are exploiting these important results for medical applications; for instance, fecal microbiota transplantation (FMT) has been used to eliminate *Clostridium difficile* recurrent infection by transplantation of healthy microbiota in human patients [89]. Besides that, FMT has been successfully used in treatment of inflammatory bowel disease, functional gastrointestinal disorders, hepatic encephalopathy, obesity, and metabolic syndrome [24, 90].

2.6. Biomedical Significance. Findings from studies of the gut microbiome shed a light over a number of diseases directly impacted by it, becoming a promising scope for advances in understanding and treating of complex diseases. Among them, Crohn's disease [91], rheumatoid arthritis [92], obesity [93, 94], type 1 and type 2 diabetes [95–97], breast cancer [98, 99], and atherosclerosis [100] can be cited as associated to the gut microbiome. Thus, researchers are interested in finding biomarkers and microbiome-based signatures for use in diagnostics, prognostics, and treatment of patients with diseases related to the human microbiome, describing important targets with biomedical significance that could be useful for public health.

In this context, Yu and collaborators [101] performed a metagenome-wide association study (MGWAS) using stool samples from 74 Chinese individuals with colorectal cancer and 54 controls, aiming to identify noninvasive biomarkers for colorectal cancer. Authors found that, besides known colorectal cancer-associated species such as *Fusobacterium nucleatum* and *Peptostreptococcus stomatis*, the other 20 microbial gene markers could differentiate colorectal cancer from control patients. In order to define a "worldwide" signature for colorectal cancer identification, they validated four of these gene markers in Danish-, French-, and Austrian-published cohorts. This result indicates that the four biomarkers validated in individuals from different countries might be used to early diagnosis of colorectal cancer even in different populations with different gut microbiome structures and is a promise for early noninvasive diagnosis of the disease. Later, Yu's group [102] developed a new diagnostic tool for colorectal cancer using the four biomarkers validated in 2015. For this, they applied a probe-based duplex qPCR assay for quantifying these bacterial genes and showed that one of these genes can discriminate colorectal cancer from control individuals with 77.7% of sensitivity and 79.5% of specificity. Moreover, combining these four bacterial genes with a fecal immunochemical test improved the diagnostic, displaying a sensitivity of 92.8% and a specificity of 81.5%.

In the same way, Pascal et al. [103] defined a microbial signature for Crohn's disease. They performed a MGWAS using samples from 2045 individuals from four countries (Spain, Belgium, UK, and Germany) and found eight groups of microorganisms that could be used to discriminate between Crohn' disease and ulcerative colitis (the two main inflammatory bowel diseases that share many immunologic, epidemiologic, and clinical features). Then, they developed an algorithm that showed specificity of approximately 90% of Crohn's disease detection when compared with healthy control, anorexia, ulcerative colitis, and inflammatory bowel syndrome patients. Similarly, Loomba et al. 2017 [104] defined a gut microbiome signature for the diagnosis of advanced fibrosis in individuals with nonalcoholic fatty liver disease using a metagenomic analysis. Analogously, a similar approach was used to distinguish between type 2 diabetes individuals and nondiabetic controls [97]. By means of a sequence-based profiling metagenomic approach, authors showed that type 2 diabetes individuals were characterized by an increase of opportunist pathogens, an enrichment of sulphate reduction genes and oxidative stress resistance genes, and a decrease in butyrate-producing species. Taken together, these gut metagenomic markers might become a powerful tool for the diagnosis of patients with the disease.

2.7. Mining of Microbial Genes In Vivo. New outcomes from the studies of the human microbiome inspired novel biological questions related to the microbial fitness at diverse human tracts. Thus, elucidating the set of genes involved in the colonization and maintenance of the intestinal microbiota would provide valuable tools to further engineering of probiotics or to enhance the survival of certain microorganisms directly related to healthy conditions.

In this way, the *in vivo* temporal functional metagenomics approach, developed by Yaung and collaborators [105], allowed them to mine microbial genes associated to microbial fitness in the mammalian gastrointestinal tract. The temporal sequencing platform employed in this study allowed the detection of genes that confer microbial fitness using a mouse as a model host. Thus, authors inoculated germ-free mice with *E. coli* transformed with a library composed of 2–5 kb fragments of the *Bacteroides thetaiotaomicron* (Bt) complete genome. Through a 28-day experiment of kinetic monitoring of the enriched clones inside the mice's gastrointestinal tract, they were able to identify genes related to colonization and maintenance of the bacteria inside the tract. The study revealed that different sets of genes were enriched in the pool of bacteria at contrasting times. For instance, during the early phase of colonization, genes responsible for the synthesis of polysaccharides and lipopolysaccharides (LPS) were significantly present in the pool, expanding *E. coli's* LPS biosynthesis repertoire with the acquired Bt biosynthetic genes. Those changes in the LPS synthesis could provide the *E. coli* with different antigenicity and enhance resistance to barriers for colonization in the gut. Nevertheless, a distinct set of genes was present in the long-term experiment, mostly related to sugar metabolism or transport. Furthermore, except for a mutation in the *galK* chromosomal gene, the recipient strain maintained its genetic stability [105].

The work is the first one to use temporal-functional metagenomics to describe how temporal data can contribute to the discovery of genes with functions of interest, once most of the genes involved in the GI tract community fitness would not be found if the data was from only an endpoint [105]. Temporal approaches like these could also bring interesting insights about interaction dynamics and fitness of microbes in other environments (such as bacteria associated to plant growth or parasitic interactions), unraveling the genes involved in a microbial community structure and metabolism. Moreover, identified genes could be used as novel drug target genes in pathogenic bacteria.

2.8. Biotechnological Impact. Biotechnology is one of the most favored fields by the metagenomic era. As microorganisms are the major source of biocatalysts for industrial purposes, increasing the repertoire of biochemical transformations available for biotechnological solutions is of high relevance [106]. Since Healy and collaborators [28] introduced the idea of constructing metagenomic libraries from a gene of interest-related environment, functional and sequence-based metagenomics have been shown to be effective in the identification of novel genes that confer resistance to extreme conditions, enzymes, antibiotics, and other bioactive molecules derived from a variety of environments (Table 1) [107–111]. Furthermore, functional metagenomic approaches made it possible to identify novel biological parts with specific activities without the need for isolation and cultivation of microorganisms.

In this context, a report from Ferrer and collaborators [112] should be highlighted. Authors constructed a metagenomic library of DNA from a cow rumen in a phage lambda vector and performed functional screenings for different carbohydrate-active enzymes. Considering that cow rumen microorganisms are specialized in degradation of cellulosic plant material, the sample used for library construction should be enriched in biomass-degrading genes. The success of the approach was exposed by the high rate of recovering of clones with different hydrolytic activities (22 clones), being among them are 9 esterases, 12 endo-β-1,4-glucanase, and 1 cyclodextrinase. Moreover, after DNA sequencing analysis of the enzymes, 8 could not be found in any sequenced genomic data, revealing that 36% of the recovered enzymes were completely new.

Other interesting works related to exploring functional genes in leaf-cutter ant fungus gardens were carried out to determine enzymes and pathways involved in symbiosis between leaf-cutter ants and their cultivar. By using metagenomic approaches, authors have determined the microbial

Table 1: Genes discovered through metagenomic approaches with high biotechnological potential.

Function/gene target	DNA source	Library size	Screening method*	Number of hits found	Biotechnological relevance	Reference
Enzymes						
Esterases, endo-β-1,4-glucanases, and cyclodextrinase	Cow rumen	1.1 Gb	Function based	22	Eight enzymes (36%) were entirely new	[112]
Laccase	Water from South China Sea	1.4 Gb	Sequencing based	1	High chloride resistance and ability to decolorize industrial dyes	[136]
Naphthalene dioxygenase	Oil-contaminated soil	294 Mb	Function based	2	Applicable in oil-contaminated soil/water	[137]
Oxygenases	Artificially polluted soil	5.2 Gb	Function based	29	Applicable in oil-contaminated soil/water	[126]
Cutinases	Leaf-branch compost	735 Mb	Function based	19	Potential application in polyethylene terephthalate (PET) degradation	[138]
Phenol hydroxylases and catechol 2,3-dioxygenases	Wastewater treatment plant	495 Mb	Function based	413	Potential use in aromatic compound degradation	[139]
Carboxylesterase	Marine water	~1.3 Gb	Function based	95	Cold-active and salt-resistant enzyme	[140]
Cellulase/esterase	Water lakes	1.86 Gb	Function based	3	New cellulase	[141]
Cellulase	Soil	Not found	Function based	1	Halo- and thermotolerant enzyme	[142]
β-Glucosidase	Hydrothermal spring water	Not found	Function based	1	Thermotolerant and heath-active enzyme	[143]
Lipase/protease/ hemolysins/biosurfactants	Slaughterhouse drain	~884 Mb	Function based	22	Antimicrobial activity	[144]
Genes that confer resistance to extreme conditions						
Acid resistance genes	Plankton and rhizosphere from Tinto River	~2.3 Gb	Function based	15	Genes involved in acid resistance	[118]
Nickel resistance genes	Rhizosphere of E. andevalensis from Tinto River	2.15 Gb	Function based	13	Genes related to nickel resistance	[117]
Salt resistance genes	Brine and rhizosphere from Es Trenc saltern	2.15 Gb	function-based	11	Genes conferring salt resistance	[116]
Arsenic resistance genes	Headwater from Tinto River	151 Mb	Function based	18	Genes involved in arsenic resistance	[145]
Regulatory sequences						
Constitutive promoters	Soil from secondary Atlantic Forest	~500 Mb	Function based	33	Use as "biobricks"	[135]
Pathways/systems/operons						
Naphthalene-degrading system	Naphthalene-contaminated groundwater	~283 Mb	Sequencing based	3	Pollutant-degrading enzyme systems	[146]
Dioxygenase-degrading cluster	Forest soil	260–815 bp	Sequencing based	11	Degrading phenoxyalkanoic acid (PAA) herbicides avoiding groundwater contamination	[147]
NRPS biosynthetic pathway	Tunicate consortium in Florida Keys	~280 Mb	Sequencing based	1	ET-743 biosynthetic pathway; anticancer molecule	[148]
Bioactive molecules						
Pigmentation producing and antibacterial activity	Soil	Not found	Function based	45	Potential new molecules to be used as antibiotics	[125]

Table 1: Continued.

Function/gene target	DNA source	Library size	Screening method*	Number of hits found	Biotechnological relevance	Reference
Turbomycin A and B	Soil	~1 Gb	Function based	3	Antibiotic activity	[16]
Antimicrobial small molecules	Soil	~720 Mb	Function based	4	Antibiotic activity	[149]

*All genes discovered through sequencing-based methodologies were experimentally tested for their related functions.

composition of the fungus gardens [113, 114] and how the plant biomass degrading process in this microbial community occurs, showing a number of novel cellulases involved in it [113–115].

Alternative relevant metagenomic studies take advantage of the genetic potential of microbe inhabitants of extreme environments, such as high or low temperatures, salinity, acidity, pressure, radiation, or high concentrations of heavy metals (Table 1) [116–118]. Deciphering microbial diversity and metabolic activities of microorganisms in extreme conditions reveals the biochemical strategies used by them to survive under extreme conditions. This, in turn, can be used to expand the capability of survival of bacteria used in industrial processes. In addition, those organisms are interesting sources for enzymatic activities with specific and unusual features. Table 1 summarises some of the most noticeable genes that have been discovered until now using the strategies described above.

3. Conclusions and Perspectives

Over the last three decades, since the first studies using the concept of metagenomics, extraordinary advances in the field have been achieved. Collective intelligence from a plethora of experts in diverse fields (such as biologists, biochemists, geneticists, physicists, and computer scientists) was imperative for answering central biological questions and for bringing biotechnological solutions in a myriad of different fields.

Understanding properties such as structure, diversity, richness, and dynamics of microbial communities is essential for unravelling the underlying processes that govern the organization of those systems. However, for a more comprehensive analysis, it is crucial to integrate information from both the microbial community and the environment it is embedded in. The macrodynamics of physical spaces and the interactions between their components directly modulate the microbiological universe (and vice versa). Thus, understanding the physical, chemical, and relational aspects of an environment can provide insightful predictions about its microinhabitants, whereas the reverse process, depicting an environment from a collection of "microbial footprints," although more challenging, is also attainable [119].

In this context, one of the most proximal models of the study is our own and the built spaces we create and live in. From our bodies to our cities and far away, the Earth is heavily packed with microbes [120–122]. A "reference man" (one who is 70 kilograms, 20–30 years old, and 1.7 meters tall) contains on average about 30 trillion human cells and 39 trillion bacteria [123] and emits bacteria at rates of over a million biological particles per hour [124]. Then, it is daunting to ask can we understand the "maketh men" through its microbiome? Nowadays, we know that our personal ecosystem of microbes is shed on everything we touch and everyplace we go as "molecular echoes." Thus, can we trace back an individual lifetime through metagenomics? What about the lifetime of a whole society? In reverse, can we use metagenomics to guide the rational design of novel-built environments (indoors and outdoors) for artificially selecting microbial communities, which will ultimately contribute to human health? The answer to all those exciting questions is yes, we can—at least, partially—as it was described in diverse examples along this revision, which are enabling us to move towards the right answers.

In contrast, despite the notable milestones reached in the field, there are still crucial challenges that need to be faced in order to delineate new conceptual advances in microbial science. Development of novel bioinformatic tools specific for metagenomic analysis is still necessary, once the next-generation sequencing platforms are generating an increasing amount of data that is not directly proportional to its biological significance. That is, there is an enormous quantity of information in sequenced data that need to be transformed into biological understating. In the near future, huge data processing and analysis in an integrated way with data already known will be the main challenge of the field [30].

On the other hand, the success of the function-based screenings depends on factors like the size of the gene in metagenomic DNA, its abundance in the samples, the efficiency of the screening method, the host vector system used, and the heterologous expression of the gene. Nowadays, after overcoming of some critical limitations related to host-biased screenings, researchers have used alternative hosts besides *E. coli* to perform the screenings [125–130]. For this, the use of broad host range vectors, able to replicate in different hosts, is required. Although a large collection of broad host range vectors is available [131], we still need to create *a la carte* vectors specific for some microorganisms that are essential to certain industrial uses. In this sense, strategies involving synthetic biological approaches [132] have been crucial to develop new smart screening methods. Engineering biological circuits—the so-called biosensors—have accelerated the identification of positive activities in metagenomic screenings including millions of clones. Interesting examples are the substrate-induced gene expression (SIGEX) and the product-induced gene expression (PIGEX) approaches [17, 133] or a riboswitch-based selection system initially constructed for mining thiamine uptake functions [134], but generalizable to other compounds. Furthermore, developing new methods for expanding the search space of functional metagenomics from enzymes to novel genetic elements such as regulators, promoters, and cis-regulatory sequences is very important for both mining biological parts and understanding their natural diversity [17, 130, 133, 135].

In summary, current challenges in metagenomics that need to be addressed can be divided into two main groups: (i) the development of novel bioinformatic tools and (ii) the generation of novel molecular tools. The first group comprises the necessity of dealing with the colossal amount of information delivered from novel sequencing approaches, as previously described. Thus, it is imperative to transform the metagenomic information overload into biological understanding. The second challenge is related to the generation of novel molecular approaches such as merging metagenomic and synthetic biology to delineate novel strategies for activity-driven screening. Existing functional screening methods usually have low rates of gene target identification. Therefore, the construction of novel synthetic circuits able to detect enzymatic activities—or other target gene output—present in the cloned metagenomic fragments is essential to improve the screening efficiency of metagenomic libraries. In this manner, by combining the collective efforts of specialists for overcoming the previously described challenges, it will be possible to integrate emerging concepts and dive deeper into the universe of metagenomics expanding the current knowledge in a myriad of areas. Shedding a light on the "hidden" world of uncultured microorganisms—and its inherent biochemical treasures—shall tell us stories not only about the multitude of Microverses that surround us but also about ourselves.

Conflicts of Interest

The authors have no conflict of interest to declare.

Acknowledgments

The authors are grateful to the anonymous reviewer for her/his careful reading of our manuscript and her/his many insightful comments and suggestions. This work was supported by the National Counsel of Technological and Scientific Development (Conselho Nacional de Desenvolvimento Científico e Tecnológico 472893/2013-0) and by Young Research Wards by the Sao Paulo State Foundation (Fundação de Amparo à Pesquisa do Estado de São Paulo, Award no. 2015/04309-1). Luana de Fátima Alves, Cauã Antunes Westmann, and Tiago Cabral Borelli are beneficiaries of Fundação de Amparo à Pesquisa do Estado de São Paulo fellowships (Award nos. 2016/06323-4, 2016/05472-6, and 2017/20818-9, resp.).

References

[1] N. R. Pace, D. A. Stahl, D. J. Lane, and G. J. Olsen, "The analysis of natural microbial populations by ribosomal RNA sequences," in *Advances in Microbial Ecology*, M. K. Cou, Ed., pp. 1–55, Springer, Boston, MA, USA, 1986.

[2] J. Handelsman, M. R. Rondon, S. F. Brady, J. Clardy, and R. M. Goodman, "Molecular biological access to the chemistry of unknown soil microbes: a new frontier for natural products," *Chemistry & Biology*, vol. 5, no. 10, pp. R245–R249, 1998.

[3] R. D. Sleator, C. Shortall, and C. Hill, "Metagenomics," *Letters in Applied Microbiology*, vol. 47, no. 5, pp. 361–366, 2008.

[4] J. Handelsman, "Metagenomics: application of genomics to uncultured microorganisms," *Microbiology and Molecular Biology Reviews*, vol. 69, no. 1, pp. 195–195, 2005.

[5] S. G. Tringe, C. von Mering, A. Kobayashi et al., "Comparative metagenomics of microbial communities," *Science*, vol. 308, no. 5721, pp. 554–557, 2005.

[6] J. M. Vieites, M. E. Guazzaroni, A. Beloqui, P. N. Golyshin, and M. Ferrer, "Metagenomics approaches in systems microbiology," *FEMS Microbiology Reviews*, vol. 33, no. 1, pp. 236–255, 2009.

[7] C. Schmeisser, H. Steele, and W. R. Streit, "Metagenomics, biotechnology with non-culturable microbes," *Applied Microbiology and Biotechnology*, vol. 75, no. 5, pp. 955–962, 2007.

[8] M. E. Guazzaroni, R. Silva-Rocha, and R. J. Ward, "Synthetic biology approaches to improve biocatalyst identification in metagenomic library screening," *Microbial Biotechnology*, vol. 8, no. 1, pp. 52–64, 2015.

[9] M. E. Guazzaroni, A. Beloqui, P. N. Golyshin, and M. Ferrer, "Metagenomics as a new technological tool to gain scientific knowledge," *World Journal of Microbiology and Biotechnology*, vol. 25, no. 6, pp. 945–954, 2009.

[10] S. Louca, M. F. Polz, F. Mazel et al., "Function and functional redundancy in microbial systems," *Nature Ecology & Evolution*, vol. 2, no. 6, pp. 936–943, 2018.

[11] M. Li, B. Wang, M. Zhang et al., "Symbiotic gut microbes modulate human metabolic phenotypes," *Proceedings of the National Academy of Sciences*, vol. 105, no. 6, pp. 2117–2122, 2008.

[12] M. G. I. Langille, J. Zaneveld, J. G. Caporaso et al., "Predictive functional profiling of microbial communities using 16S rRNA marker gene sequences," *Nature Biotechnology*, vol. 31, no. 9, pp. 814–821, 2013.

[13] D. E. Hunt, Y. Lin, M. J. Church et al., "Relationship between abundance and specific activity of bacterioplankton in open ocean surface waters," *Applied and Environmental Microbiology*, vol. 79, no. 1, pp. 177–184, 2013.

[14] F. Sanger, G. M. Air, B. G. Barrell et al., "Nucleotide sequence of bacteriophage phi X174 DNA," *Nature*, vol. 265, no. 5596, pp. 687–695, 1977.

[15] M. Breitbart, I. Hewson, B. Felts et al., "Metagenomic analyses of an uncultured viral community from human feces," *Journal of Bacteriology*, vol. 185, no. 20, pp. 6220–6223, 2003.

[16] D. E. Gillespie, S. F. Brady, A. D. Bettermann et al., "Isolation of antibiotics turbomycin a and B from a metagenomic library of soil microbial DNA," *Applied and Environmental Microbiology*, vol. 68, no. 9, pp. 4301–4306, 2002.

[17] T. Uchiyama, T. Abe, T. Ikemura, and K. Watanabe, "Substrate-induced gene-expression screening of environmental metagenome libraries for isolation of catabolic genes," *Nature Biotechnology*, vol. 23, no. 1, pp. 88–93, 2005.

[18] P. J. Turnbaugh, V. K. Ridaura, J. J. Faith, F. E. Rey, R. Knight, and J. I. Gordon, "The effect of diet on the human gut microbiome: a metagenomic analysis in humanized gnotobiotic

mice," *Science Translational Medicine*, vol. 1, no. 6, p. 6ra14, 2009.

[19] S. Sunagawa, L. P. Coelho, S. Chaffron et al., "Structure and function of the global ocean microbiome," *Science*, vol. 348, no. 6237, article 1261359, 2015.

[20] A. Klindworth, E. Pruesse, T. Schweer et al., "Evaluation of general 16S ribosomal RNA gene PCR primers for classical and next-generation sequencing-based diversity studies," *Nucleic Acids Research*, vol. 41, no. 1, pp. e1–11, 2013.

[21] A. Oulas, C. Pavloudi, P. Polymenakou et al., "Metagenomics: tools and insights for analyzing next-generation sequencing data derived from biodiversity studies," *Bioinformatics and Biology Insights*, vol. 9, pp. BBI.S12462–BBI.S12488, 2015.

[22] M. Kircher and J. Kelso, "High-throughput DNA sequencing - concepts and limitations," *BioEssays*, vol. 32, no. 6, pp. 524–536, 2010.

[23] E. Le Chatelier, MetaHIT consortium, T. Nielsen et al., "Richness of human gut microbiome correlates with metabolic markers," *Nature*, vol. 500, no. 7464, pp. 541–546, 2013.

[24] D. Kao, B. Roach, H. Park et al., "Fecal microbiota transplantation in the management of hepatic encephalopathy," *Hepatology*, vol. 63, no. 1, pp. 339-340, 2016.

[25] N. Qin, F. Yang, A. Li et al., "Alterations of the human gut microbiome in liver cirrhosis," *Nature*, vol. 513, no. 7516, pp. 59–64, 2014.

[26] H. K. Pedersen, V. Gudmundsdottir, H. B. Nielsen et al., "Human gut microbes impact host serum metabolome and insulin sensitivity," *Nature*, vol. 535, no. 7612, pp. 376–381, 2016.

[27] T. M. Schmidt, E. F. DeLong, and N. R. Pace, "Analysis of a marine picoplankton community by 16S rRNA gene cloning and sequencing," *Journal of Bacteriology*, vol. 173, no. 14, pp. 4371–4378, 1991.

[28] F. G. Healy, R. M. Ray, H. C. Aldrich, A. C. Wilkie, L. O. Ingram, and K. T. Shanmugam, "Direct isolation of functional genes encoding cellulases from the microbial consortia in a thermophilic, anaerobic digester maintained on lignocellulose," *Applied Microbiology and Biotechnology*, vol. 43, no. 4, pp. 667–674, 1995.

[29] M. R. Rondon, P. R. August, A. D. Bettermann et al., "Cloning the soil metagenome: a strategy for accessing the genetic and functional diversity of uncultured microorganisms," *Applied and Environmental Microbiology*, vol. 66, no. 6, pp. 2541–2547, 2000.

[30] A. Escobar-Zepeda, A. V.-P. de León, and A. Sanchez-Flores, "The road to metagenomics: from microbiology to DNA sequencing technologies and bioinformatics," *Frontiers in Genetics*, vol. 6, 2015.

[31] D. D. Roumpeka, R. J. Wallace, F. Escalettes, I. Fotheringham, and M. Watson, "A review of bioinformatics tools for bioprospecting from metagenomic sequence data," *Frontiers in Genetics*, vol. 8, 2017.

[32] F. P. Breitwieser, J. Lu, and S. L. Salzberg, "A review of methods and databases for metagenomic classification and assembly," *Briefings in Bioinformatics*, pp. 1–15, 2017.

[33] C. R. Woese and G. E. Fox, "Phylogenetic structure of the prokaryotic domain: the primary kingdoms," *Proceedings of the National Academy of Sciences*, vol. 74, no. 11, pp. 5088–5090, 1977.

[34] J. C. Venter, K. Remington, J. F. Heidelberg et al., "Environmental genome shotgun sequencing of the Sargasso Sea," *Science*, vol. 304, no. 5667, pp. 66–74, 2004.

[35] O. Béjà, L. Aravind, E. V. Koonin et al., "Bacterial rhodopsin: evidence for a new type of phototrophy in the sea," *Science*, vol. 289, no. 5486, pp. 1902–1906, 2000.

[36] J. A. Klappenbach, J. M. Dunbar, and T. M. Schmidt, "rRNA operon copy number reflects ecological strategies of bacteria," *Applied and Environmental Microbiology*, vol. 66, no. 4, pp. 1328–1333, 2000.

[37] E. A. Eloe-Fadrosh, N. N. Ivanova, T. Woyke, and N. C. Kyrpides, "Metagenomics uncovers gaps in amplicon-based detection of microbial diversity," *Nature Microbiology*, vol. 1, no. 4, 2016.

[38] C. T. Brown, L. A. Hug, B. C. Thomas et al., "Unusual biology across a group comprising more than 15% of domain Bacteria," *Nature*, vol. 523, no. 7559, pp. 208–211, 2015.

[39] R. Gorelick and H. H. Q. Heng, "Sex reduces genetic variation: a multidisciplinary review," *Evolution*, vol. 65, no. 4, pp. 1088–1098, 2011.

[40] H. H. Q. Heng, "Elimination of altered karyotypes by sexual reproduction preserves species identity," *Genome*, vol. 50, no. 5, pp. 517–524, 2007.

[41] C. Rinke, P. Schwientek, A. Sczyrba et al., "Insights into the phylogeny and coding potential of microbial dark matter," *Nature*, vol. 499, no. 7459, pp. 431–437, 2013.

[42] A. Spang, J. H. Saw, S. L. Jørgensen et al., "Complex archaea that bridge the gap between prokaryotes and eukaryotes," *Nature*, vol. 521, no. 7551, pp. 173–179, 2015.

[43] J. L. Stein, T. L. Marsh, K. Y. Wu, H. Shizuya, and E. F. DeLong, "Characterization of uncultivated prokaryotes: isolation and analysis of a 40-kilobase-pair genome fragment from a planktonic marine archaeon," *Journal of Bacteriology*, vol. 178, no. 3, pp. 591–599, 1996.

[44] O. Béjà, M. T. Suzuki, E. V. Koonin et al., "Construction and analysis of bacterial artificial chromosome libraries from a marine microbial assemblage," *Environmental Microbiology*, vol. 2, no. 5, pp. 516–529, 2000.

[45] O. Béjà, E. N. Spudich, J. L. Spudich, M. Leclerc, and E. F. DeLong, "Proteorhodopsin phototrophy in the ocean," *Nature*, vol. 411, no. 6839, pp. 786–789, 2001.

[46] G. W. Tyson, J. Chapman, P. Hugenholtz et al., "Community structure and metabolism through reconstruction of microbial genomes from the environment," *Nature*, vol. 428, no. 6978, pp. 37–43, 2004.

[47] E. A. Dinsdale, R. A. Edwards, D. Hall et al., "Functional metagenomic profiling of nine biomes," *Nature*, vol. 452, no. 7187, pp. 629–632, 2008.

[48] J. Lloyd-Price, A. Mahurkar, G. Rahnavard et al., "Strains, functions and dynamics in the expanded human microbiome project," *Nature*, vol. 550, pp. 61–66, 2017.

[49] J. A. Gilbert, R. A. Quinn, J. Debelius et al., "Microbiome-wide association studies link dynamic microbial consortia to disease," *Nature*, vol. 535, no. 7610, pp. 94–103, 2016.

[50] A. L. Kau, P. P. Ahern, N. W. Griffin, A. L. Goodman, and J. I. Gordon, "Human nutrition, the gut microbiome and the immune system," *Nature*, vol. 474, no. 7351, pp. 327–336, 2011.

[51] E. A. Archie and J. Tung, "Social behavior and the microbiome," *Current Opinion in Behavioral Sciences*, vol. 6, pp. 28–34, 2015.

[52] T. Kuntz and J. Gilbert, "Does the brain listen to the gut?," *eLife*, vol. 5, 2016.

[53] A. J. Prussin and L. C. Marr, "Sources of airborne microorganisms in the built environment," *Microbiome*, vol. 3, no. 1, p. 78, 2015.

[54] S. W. Kembel, E. Jones, J. Kline et al., "Architectural design influences the diversity and structure of the built environment microbiome," *The ISME Journal*, vol. 6, no. 8, pp. 1469–1479, 2012.

[55] M. H. Y. Leung and P. K. H. Lee, "The roles of the outdoors and occupants in contributing to a potential pan-microbiome of the built environment: a review," *Microbiome*, vol. 4, no. 1, pp. 21–15, 2016.

[56] T. Thomas, L. Moitinho-Silva, M. Lurgi et al., "Diversity, structure and convergent evolution of the global sponge microbiome," *Nature Communications*, vol. 7, 2016.

[57] J. M. Maritz, S. A. Sullivan, R. J. Prill, E. Aksoy, P. Scheid, and J. M. Carlton, "Filthy lucre: a metagenomic pilot study of microbes found on circulating currency in New York City," *PLoS One*, vol. 12, no. 4, pp. e0175527–e0175516, 2017.

[58] H. M. Bik, J. M. Maritz, A. Luong, H. Shin, M. G. Dominguez-Bello, and J. M. Carlton, "Microbial community patterns associated with automated teller machine keypads in New York City," *mSphere*, vol. 1, no. 6, pp. e00226–e00216, 2016.

[59] R. I. Adams, A. C. Bateman, H. M. Bik, and J. F. Meadow, "Microbiota of the indoor environment: a meta-analysis," *Microbiome*, vol. 3, no. 1, p. 49, 2015.

[60] S. G. Tringe, T. Zhang, X. Liu et al., "The airborne metagenome in an indoor urban environment," *PLoS One*, vol. 3, no. 4, article e1862, 2008.

[61] E. Afshinnekoo, C. Meydan, S. Chowdhury et al., "Geospatial resolution of human and bacterial diversity with city-scale metagenomics," *Cell Systems*, vol. 1, no. 1, pp. 72–87, 2015.

[62] The MetaSUB International Consortium, "The Metagenomics and Metadesign of the Subways and Urban Biomes (MetaSUB) International Consortium inaugural meeting report," *Microbiome*, vol. 4, no. 1, p. 24, 2016.

[63] N. A. Bokulich, Z. T. Lewis, K. Boundy-Mills, and D. A. Mills, "A new perspective on microbial landscapes within food production," *Current Opinion in Biotechnology*, vol. 37, pp. 182–189, 2016.

[64] J. M. Lang, D. A. Coil, R. Y. Neches et al., "A microbial survey of the International Space Station (ISS)," *Peer J*, vol. 5, article e4029, 2017.

[65] National Academies of Sciences, *Engineering and MNA of ED on E and PSH and MDD on E and LSB on I and the CEB on ES and T. Microbiomes of the Built Environment*, National Academies Press, Washington, D.C. USA, 2017.

[66] J. S. Griffin, N. Lu, N. Sangwan et al., "Microbial diversity in an intensively managed landscape is structured by landscape connectivity," *FEMS Microbiology Ecology*, vol. 93, no. 10, pp. 1–12, 2017.

[67] G. Mhuireach, B. R. Johnson, A. E. Altrichter et al., "Urban greenness influences airborne bacterial community composition," *Science of The Total Environment*, vol. 571, pp. 680–687, 2016.

[68] A. T. Reese, A. Savage, E. Youngsteadt et al., "Urban stress is associated with variation in microbial species composition - but not richness - in Manhattan," *The ISME Journal*, vol. 10, no. 3, pp. 751–760, 2016.

[69] J. A. Gilbert, "How do we make indoor environments and healthcare settings healthier?," *Microbial Biotechnology*, vol. 10, no. 1, pp. 11–13, 2017.

[70] B. Stephens, "What have we learned about the microbiomes of indoor environments?," *mSystems*, vol. 1, no. 4, 2016.

[71] S. R. Gill, M. Pop, R. T. DeBoy et al., "Metagenomic analysis of the human distal gut microbiome," *Science*, vol. 312, no. 5778, pp. 1355–1359, 2006.

[72] J. W. Arnold, J. Roach, and M. A. Azcarate-Peril, "Emerging technologies for gut microbiome research," *Trends in Microbiology*, vol. 24, no. 11, pp. 887–901, 2016.

[73] The Integrative HMP, (iHMP) Research Network consortium, "The integrative human microbiome project: dynamic analysis of microbiome-host omics profiles during periods of human health and disease," *Cell Host & Microbe*, vol. 16, no. 3, pp. 276–289, 2014.

[74] The Human Microbiome Project Consortium, "Structure, function and diversity of the healthy human microbiome," *Nature*, vol. 486, no. 7402, pp. 207–214, 2012.

[75] MetaHIT Consortium, J. Qin, R. Li et al., "ARTICLES A human gut microbial gene catalogue established by metagenomic sequencing," *Nature*, vol. 464, no. 7285, pp. 59–65, 2010.

[76] F. Bäckhed, R. E. Ley, J. L. Sonnenburg, D. A. Peterson, and J. I. Gordon, "Host-bacterial mutualism in the human intestine," *Science*, vol. 307, no. 5717, pp. 1915–1920, 2005.

[77] P. C. Watts, K. R. Buley, S. Sanderson, W. Boardman, C. Ciofi, and R. Gibson, "Parthenogenesis in Komodo dragons," *Nature*, vol. 444, no. 7122, pp. 1021-1022, 2006.

[78] R. E. Ley, P. J. Turnbaugh, S. Klein, and J. I. Gordon, "Microbial ecology: human gut microbes associated with obesity," *Nature*, vol. 444, no. 7122, pp. 1022-1023, 2006.

[79] P. J. Turnbaugh, R. E. Ley, M. A. Mahowald, V. Magrini, E. R. Mardis, and J. I. Gordon, "An obesity-associated gut microbiome with increased capacity for energy harvest," *Nature*, vol. 444, no. 7122, pp. 1027–1031, 2006.

[80] K. Forslund, S. Sunagawa, J. R. Kultima et al., "Country-specific antibiotic use practices impact the human gut resistome," *Genome*, vol. 23, no. 7, pp. 1163–1169, 2013.

[81] K. Forslund, S. Sunagawa, L. P. Coelho, and P. Bork, "Metagenomic insights into the human gut resistome and the forces that shape it," *BioEssays*, vol. 36, no. 3, pp. 316–329, 2014.

[82] T. S. Ghosh, S. S. Gupta, G. B. Nair, and S. S. Mande, "In silico analysis of antibiotic resistance genes in the gut microflora of individuals from diverse geographies and age-groups," *PLoS One*, vol. 8, no. 12, article e83823, 2013.

[83] F. Raymond, A. A. Ouameur, M. Déraspe et al., "The initial state of the human gut microbiome determines its reshaping by antibiotics," *The ISME Journal*, vol. 10, no. 3, pp. 707–720, 2016.

[84] R. K. Singh, H.-W. Chang, D. Yan et al., "Influence of diet on the gut microbiome and implications for human health," *Journal of Translational Medicine*, vol. 15, no. 1, pp. 73–17, 2017.

[85] A. Lundin, C. M. Bok, L. Aronsson et al., "Gut flora, Toll-like receptors and nuclear receptors: a tripartite communication that tunes innate immunity in large intestine," *Cellular Microbiology*, vol. 10, no. 5, pp. 1093–1103, 2008.

[86] Y. K. Lee and S. K. Mazmanian, "Has the microbiota played a critical role in the evolution of the adaptive immune system?," *Science*, vol. 330, no. 6012, pp. 1768–1773, 2010.

[87] Y. Belkaid and T. W. Hand, "Role of the microbiota in immunity and inflammation," *Cell*, vol. 157, no. 1, pp. 121–141, 2014.

[88] M. C. Noverr and G. B. Huffnagle, "Does the microbiota regulate immune responses outside the gut?," *Trends in Microbiology*, vol. 12, no. 12, pp. 562–568, 2004.

[89] J. R. Allegretti, J. R. Korzenik, and M. J. Hamilton, "Fecal microbiota transplantation via colonoscopy for recurrent C. difficile infection," *Journal of Visualized Experiments*, no. 94, article e52154, 2014.

[90] S. Gupta, E. Allen-Vercoe, and E. O. Petrof, "Fecal microbiota transplantation: in perspective," *Therapeutic Advances in Gastroenterology*, vol. 9, no. 2, pp. 229–239, 2015.

[91] P. Seksik, L. Rigottier-Gois, G. Gramet et al., "Alterations of the dominant faecal bacterial groups in patients with Crohn's disease of the colon," *Gut*, vol. 52, no. 2, pp. 237–242, 2003.

[92] J. U. Scher, A. Sczesnak, R. S. Longman et al., "Expansion of intestinal *Prevotella copri* correlates with enhanced susceptibility to arthritis," *eLife*, vol. 2, 2013.

[93] L. Zhao, "The gut microbiota and obesity: from correlation to causality," *Nature Reviews Microbiology*, vol. 11, no. 9, pp. 639–647, 2013.

[94] C. Menni, M. A. Jackson, T. Pallister, C. J. Steves, T. D. Spector, and A. M. Valdes, "Gut microbiome diversity and high-fibre intake are related to lower long-term weight gain," *International Journal of Obesity*, vol. 41, no. 7, pp. 1099–1105, 2017.

[95] M. Knip and H. Siljander, "The role of the intestinal microbiota in type 1 diabetes mellitus," *Nature Reviews Endocrinology*, vol. 12, no. 3, pp. 154–167, 2016.

[96] J. L. Han and H. L. Lin, "Intestinal microbiota and type 2 diabetes: from mechanism insights to therapeutic perspective," *World Journal of Gastroenterology*, vol. 20, no. 47, pp. 17737–17745, 2014.

[97] J. Qin, Y. Li, Z. Cai et al., "A metagenome-wide association study of gut microbiota in type 2 diabetes," *Nature*, vol. 490, no. 7418, pp. 55–60, 2012.

[98] C. M. Velicer, S. R. Heckbert, J. W. Lampe, J. D. Potter, C. A. Robertson, and S. H. Taplin, "Antibiotic use in relation to the risk of breast cancer," *JAMA*, vol. 291, no. 7, pp. 827–835, 2004.

[99] H. T Sørensen, M. V. Skriver, S. Friis, J. K. McLaughlin, W. J. Blot, and J. A. Baron, "Use of antibiotics and risk of breast cancer: a population-based case–control study," *British Journal of Cancer*, vol. 92, no. 3, pp. 594–596, 2005.

[100] Z. Wang, E. Klipfell, B. J. Bennett et al., "Gut flora metabolism of phosphatidylcholine promotes cardiovascular disease," *Nature*, vol. 472, no. 7341, pp. 57–63, 2011.

[101] J. Yu, Q. Feng, S. H. Wong et al., "Metagenomic analysis of faecal microbiome as a tool towards targeted non-invasive biomarkers for colorectal cancer," *Gut*, vol. 66, no. 1, pp. 70–78, 2016.

[102] Q. Liang, J. Chiu, Y. Chen et al., "Fecal bacteria act as novel biomarkers for noninvasive diagnosis of colorectal cancer," *Clinical Cancer Research*, vol. 23, no. 8, pp. 2061–2070, 2017.

[103] V. Pascal, M. Pozuelo, N. Borruel et al., "A microbial signature for Crohn's disease," *Gut*, vol. 66, no. 5, pp. 813–822, 2017.

[104] R. Loomba, V. Seguritan, W. Li et al., "Gut microbiome-based metagenomic signature for non-invasive detection of advanced fibrosis in human nonalcoholic fatty liver disease," *Cell Metabolism*, vol. 25, no. 5, pp. 1054–1062.e5, 2017.

[105] S. J. Yaung, L. Deng, N. Li et al., "Improving microbial fitness in the mammalian gut by in vivo temporal functional metagenomics," *Molecular Systems Biology*, vol. 11, no. 3, p. 788, 2015.

[106] L. Fernández-Arrojo, M. E. Guazzaroni, N. López-Cortés, A. Beloqui, and M. Ferrer, "Metagenomic era for biocatalyst identification," *Current Opinion in Biotechnology*, vol. 21, no. 6, pp. 725–733, 2010.

[107] M. Ferrer, A. Ghazi, A. Beloqui et al., "Functional metagenomics unveils a multifunctional glycosyl hydrolase from the family 43 catalysing the breakdown of plant polymers in the calf rumen," *PLoS One*, vol. 7, no. 6, article e38134, 2012.

[108] M. V. Del Pozo, L. Fernández-Arrojo, J. Gil-Martínez et al., "Microbial β-glucosidases from cow rumen metagenome enhance the saccharification of lignocellulose in combination with commercial cellulase cocktail," *Biotechnology for Biofuels*, vol. 5, no. 1, pp. 73–13, 2012.

[109] C. Thompson, W. Beys-Da-Silva, L. Santi et al., "A potential source for cellulolytic enzyme discovery and environmental aspects revealed through metagenomics of Brazilian mangroves," *AMB Express*, vol. 3, no. 1, p. 65, 2013.

[110] C. Yang, Y. Xia, H. Qu et al., "Discovery of new cellulases from the metagenome by a metagenomics-guided strategy," *Biotechnology for Biofuels*, vol. 9, no. 1, p. 138, 2016.

[111] S. Courtois, C. M. Cappellano, M. Ball et al., "Recombinant environmental libraries provide access to microbial diversity for drug discovery from natural products," *Applied and Environmental Microbiology*, vol. 69, no. 1, pp. 49–55, 2003.

[112] M. Ferrer, O. V. Golyshina, T. N. Chernikova et al., "Novel hydrolase diversity retrieved from a metagenome library of bovine rumen microflora," *Environmental Microbiology*, vol. 7, no. 12, pp. 1996–2010, 2005.

[113] F. O. Aylward, K. E. Burnum, J. J. Scott et al., "Metagenomic and metaproteomic insights into bacterial communities in leaf-cutter ant fungus gardens," *The ISME Journal*, vol. 6, no. 9, pp. 1688–1701, 2012.

[114] G. Suen, J. J. Scott, F. O. Aylward, and C. R. Currie, "The microbiome of leaf-cutter ant fungus gardens," in *Handbook of Molecular Microbial Ecology II: Metagenomics in Different Habitats*, F. J. de Bruijn, Ed., John Wiley & Sons, Inc., Hoboken, NJ, USA, 2011.

[115] F. O. Aylward, K. E. Burnum-Johnson, S. G. Tringe et al., "Leucoagaricus gongylophorus produces diverse enzymes for the degradation of recalcitrant plant polymers in leaf-cutter ant fungus gardens," *Applied and Environmental Microbiology*, vol. 79, no. 12, pp. 3770–3778, 2013.

[116] S. Mirete, M. R. Mora-Ruiz, M. Lamprecht-Grandío, C. G. de Figueras, R. Rosselló-Móra, and J. E. González-Pastor, "Salt resistance genes revealed by functional metagenomics from brines and moderate-salinity rhizosphere within a hypersaline environment," *Frontiers in Microbiology*, vol. 6, 2015.

[117] S. Mirete, C. G. De Figueras, and J. E. González-Pastor, "Novel nickel resistance genes from the rhizosphere metagenome of plants adapted to acid mine drainage," *Applied and Environmental Microbiology*, vol. 73, no. 19, pp. 6001–6011, 2007.

[118] M. E. Guazzaroni, V. Morgante, S. Mirete, and J. E. González-Pastor, "Novel acid resistance genes from the metagenome of the Tinto River, an extremely acidic

environment," *Environmental Microbiology*, vol. 15, no. 4, pp. 1088–1102, 2013.

[119] D. R. Garza, M. C. van Verk, M. A. Huynen, and B. E. Dutilh, "Towards predicting the environmental metabolome from metagenomics with a mechanistic model," *Nature Microbiology*, vol. 3, no. 4, pp. 456–460, 2018.

[120] A. P. Alivisatos, M. J. Blaser, E. L. Brodie et al., "A unified initiative to harness Earth's microbiomes," *Science*, vol. 350, no. 6260, pp. 507-508, 2015.

[121] L. R. Thompson, J. G. Sanders, D. McDonald et al., "A communal catalogue reveals Earth's multiscale microbial diversity," *Nature*, vol. 551, no. 7681, pp. 457–463, 2017.

[122] C. Schmidt, "Living in a microbial world," *Nature Biotechnology*, vol. 35, no. 5, pp. 401–403, 2017.

[123] R. Sender, S. Fuchs, and R. Milo, "Revised estimates for the number of human and bacteria cells in the body," *PLoS Biology*, vol. 14, no. 8, pp. e1002533–e1002514, 2016.

[124] J. F. Meadow, A. E. Altrichter, A. C. Bateman et al., "Humans differ in their personal microbial cloud," *Peer J*, vol. 3, article e1258, 2015.

[125] J. W. Craig, F.-Y. Chang, J. H. Kim, S. C. Obiajulu, and S. F. Brady, "Expanding small-molecule functional metagenomics through parallel screening of broad-hostrRange cosmid environmental DNA libraries in diverse proteobacteria," *Applied and Environmental Microbiology*, vol. 76, no. 5, pp. 1633–1641, 2010.

[126] H. Nagayama, T. Sugawara, R. Endo et al., "Isolation of oxygenase genes for indigo-forming activity from an artificially polluted soil metagenome by functional screening using Pseudomonas putida strains as hosts," *Applied Microbiology and Biotechnology*, vol. 99, no. 10, pp. 4453–4470, 2015.

[127] B. Leis, A. Angelov, M. Mientus et al., "Identification of novel esterase-active enzymes from hot environments by use of the host bacterium Thermus thermophilus," *Frontiers in Microbiology*, vol. 6, 2015.

[128] A. Angelov, M. Mientus, S. Liebl, and W. Liebl, "A two-host fosmid system for functional screening of (meta) genomic libraries from extreme thermophiles," *Systematic and Applied Microbiology*, vol. 32, no. 3, pp. 177–185, 2009.

[129] T. Aakvik, K. F. Degnes, R. Dahlsrud et al., "A plasmid RK2-based broad-host-range cloning vector useful for transfer of metagenomic libraries to a variety of bacterial species," *FEMS Microbiology Letters*, vol. 296, no. 2, pp. 149–158, 2009.

[130] N. I. Johns, A. L. C. Gomes, S. S. Yim et al., "Metagenomic mining of regulatory elements enables programmable species-selective gene expression," *Nature Methods*, vol. 15, no. 5, pp. 323–329, 2018.

[131] R. Silva-Rocha, E. Martínez-García, B. Calles et al., "The standard European vector architecture (SEVA): a coherent platform for the analysis and deployment of complex prokaryotic phenotypes," *Nucleic Acids Research*, vol. 41, no. D1, pp. D666–D675, 2013.

[132] L. D. F. Alves, R. Silva-Rocha, and M.-E. Guazzaroni, "Enhancing metagenomic approaches through synthetic biology," in *Functional Metagenomics: Tools and Applications María-Eugenia Guazzaroni*, T. C. Charles, M. R. Liles, and A. Sessitsch, Eds., pp. 1–14, Springer International Publishing, Berlin, 1st edition, 2017.

[133] T. Uchiyama and K. Miyazaki, "Product-induced gene expression, a product-responsive reporter assay used to screen metagenomic libraries for enzyme-encoding genes," *Applied and Environmental Microbiology*, vol. 76, no. 21, pp. 7029–7035, 2010.

[134] H. J. Genee, A. P. Bali, S. D. Petersen et al., "Functional mining of transporters using synthetic selections," *Nature Chemical Biology*, vol. 12, no. 12, pp. 1015–1022, 2016.

[135] C. A. Westmann, L. d. F. Alves, R. Silva-Rocha, and M.-E. Guazzaroni, "Mining novel constitutive promoter elements in soil metagenomic libraries in Escherichia coli," *Frontiers in Microbiology*, vol. 9, 2018.

[136] Z. Fang, T. Li, Q. Wang et al., "A bacterial laccase from marine microbial metagenome exhibiting chloride tolerance and dye decolorization ability," *Applied Microbiology and Biotechnology*, vol. 89, no. 4, pp. 1103–1110, 2011.

[137] A. Ono, R. Miyazaki, M. Sota, Y. Ohtsubo, Y. Nagata, and M. Tsuda, "Isolation and characterization of naphthalene-catabolic genes and plasmids from oil-contaminated soil by using two cultivation-independent approaches," *Applied Microbiology and Biotechnology*, vol. 74, no. 2, pp. 501–510, 2007.

[138] S. Sulaiman, S. Yamato, E. Kanaya et al., "Isolation of a novel cutinase homolog with polyethylene terephthalate-degrading activity from leaf-branch compost by using a metagenomic approach," *Applied and Environmental Microbiology*, vol. 78, no. 5, pp. 1556–1562, 2012.

[139] C. C. Silva, H. Hayden, T. Sawbridge et al., "Identification of genes and pathways related to phenol degradation in metagenomic libraries from petroleum refinery wastewater," *PLoS One*, vol. 8, no. 4, pp. e61811–e61811, 2013.

[140] A. Tchigvintsev, H. Tran, A. Popovic et al., "The environment shapes microbial enzymes: five cold-active and salt-resistant carboxylesterases from marine metagenomes," *Applied Microbiology and Biotechnology*, vol. 99, no. 5, pp. 2165–2178, 2015.

[141] H. C. Rees, S. Grant, B. Jones, W. D. Grant, and S. Heaphy, "Detecting cellulase and esterase enzyme activities encoded by novel genes present in environmental DNA libraries," *Extremophiles*, vol. 7, no. 5, pp. 415–421, 2003.

[142] R. Garg, R. Srivastava, V. Brahma, L. Verma, S. Karthikeyan, and G. Sahni, "Biochemical and structural characterization of a novel halotolerant cellulase from soil metagenome," *Scientific Reports*, vol. 6, no. 1, pp. 1–15, 2016.

[143] C. Schröder, S. Elleuche, S. Blank, and G. Antranikian, "Characterization of a heat-active archaeal β-glucosidase from a hydrothermal spring metagenome," *Enzyme and Microbial Technology*, vol. 57, pp. 48–54, 2014.

[144] S. Thies, S. C. Rausch, F. Kovacic et al., "Metagenomic discovery of novel enzymes and biosurfactants in a slaughterhouse biofilm microbial community," *Scientific Reports*, vol. 6, no. 1, pp. 1–12, 2016.

[145] V. Morgante, S. Mirete, C. G. de Figueras, M. Postigo Cacho, and J. E. González-Pastor, "Exploring the diversity of arsenic resistance genes from acid mine drainage microorganisms," *Environmental Microbiology*, vol. 17, no. 6, pp. 1910–1925, 2015.

[146] Y. Wang, Y. Chen, Q. Zhou et al., "A culture-independent approach to unravel uncultured bacteria and functional genes in a complex microbial community," *PLoS One*, vol. 7, no. 10, article e47530, 2012.

[147] A. Zaprasis, Y. J. Liu, S. J. Liu, H. L. Drake, and M. A. Horn, "Abundance of novel and diverse tfdA-like genes, encoding putative phenoxyalkanoic acid herbicide-degrading

dioxygenases, in soil," *Applied and Environmental Microbiology*, vol. 76, no. 1, pp. 119–128, 2009.

[148] C. M. Rath, B. Janto, J. Earl et al., "Meta-omic characterization of the marine invertebrate microbial consortium that produces the chemotherapeutic natural product ET-743," *ACS Chemical Biology*, vol. 6, no. 11, pp. 1244–1256, 2011.

[149] M. N. IA, C. L. Tiong, C. Minor et al., "Expression and isolation of antimicrobial small molecules from soil DNA libraries," *Journal of Molecular Microbiology and Biotechnology*, vol. 3, no. 2, pp. 301–308, 2001.

12

Shiftwork-Mediated Disruptions of Circadian Rhythms and Sleep Homeostasis cause Serious Health Problems

Suliman Khan[1] **Pengfei Duan**,[2] **Lunguang Yao**,[2] **and Hongwei Hou**[1]

[1]*The Key Laboratory of Aquatic Biodiversity and Conservation of Chinese Academy of Sciences, Institute of Hydrobiology, Chinese Academy of Sciences, Wuhan, Hubei 430072, China*

[2]*Collaborative Innovation Center of Water and Security for Water Source Region of Mid-Line of South-to-North Diversion Project, College of Agricultural Engineering, Nanyang Normal University, Nanyang, Henan, China*

Correspondence should be addressed to Lunguang Yao; lunguangyao@163.com and Hongwei Hou; houhw@ihb.ac.cn

Academic Editor: Zhining Wen

Shiftwork became common during the last few decades with the growing demands of human life. Despite the social inactivity and irregularity in habits, working in continuous irregular shifts causes serious health issues including sleep disorders, psychiatric disorders, cancer, and metabolic disorders. These health problems arise due to the disruption in circadian clock system, which is associated with alterations in genetic expressions. Alteration in clock controlling genes further affects genes linked with disorders including major depression disorder, bipolar disorder, phase delay and phase advance sleep syndromes, breast cancer, and colon cancer. A diverse research work is needed focusing on broad spectrum changes caused by jet lag in brain and neuronal system. This review is an attempt to motivate the researchers to conduct advanced studies in this area to identify the risk factors and mechanisms. Its goal is extended to make the shift workers aware about the risks associated with shiftwork.

1. Introduction

Fast growing needs demand doing work in recurring periods other than the traditional diurnal periods. The rotations in working shifts disrupt natural sleep-wake cycle and eating patterns [1], which in turn cause serious health problems by affecting mental health and work effectiveness [2]. This disruption alters circadian rhythms [3] and neuronal functions to cause neuronal disorders [2]. Shiftwork for a long period increases the risk of fatigue, aggression, sleep disorders, metabolic disorders, mental abnormalities, and death [2, 4–9]. Shiftwork directly affects alertness [10, 11], causes depression, and promotes anxiety [12].

Working repeatedly during night shifts affects hormonal system and disrupts hormonal secretions and their control factors [13–16]. This altered hormonal profile increases the risk of breast cancer, prostate cancer, gastrointestinal abnormalities, cardiovascular diseases, and reproductive aberrations [17–19]. Alterations in physiological, behavioral, and psychological mechanisms [5] further develop abnormalities associated with peptic ulcer, diabetes type II, and rheumatoid arthritis [20–22]. The measurement of shiftwork is considered complex as it requires several parameters including sleep quality, fatigue level, types of sleep problems, use of stimulants, and sleepiness [2, 23].

Shiftwork has gained huge importance which is inevitable in modern world. This situation alerts researchers to focus on shiftwork and related abnormalities. Without knowing the genetic mechanisms and identifying factors related to shiftwork-promoted health problems, development of prevention and curing strategies may not be achievable. Suitable model organism, monitoring systems, and mimicking the shiftwork in lab conditions are major challenges in studying shiftwork to investigate related disorders. These challenges make the identification genes, majorly involved in shiftwork-related abnormalities, difficult. In this review, we have focused on major health problems that are linked with shiftwork directly or indirectly.

2. Shiftwork Dysregulates Circadian Rhythms by Affecting Clock Genes

Circadian rhythm mainly controls the daily wake and sleep cycle and regulates physiological processes including hormone secretion, body temperature, feeding behavior, cell cycle progression, and drug, glucose, and xenobiotic metabolism. Its disruption through environmental and genetic means causes aberrations in physiological processes. Clock genes with the effects of oscillators and endocrine and neural signals regulate circadian rhythm [24] which may be disrupted apparently by shiftwork. Circadian clock, through appropriate physiological activities in relevance to time, controls the circadian rhythms [4]. Shiftwork (chronic jet lag) suppresses the expression levels of core clock genes, including Per1 and Per2 in suprachiasmatic nuclei (SCN) and MT1 melatonin and glucocorticoid receptors in the liver. It further causes delay in acrophases of circadian expression of Per1, Per2, BMAL1, and Dbp in liver. Besides clock genes, expression levels of some cell cycle-related genes including c-Myc and p53 are also altered [25]. Vasopressin (V1a and V1b) receptors (expressed in SCN neurons) promote shiftwork effect in combination with core clock genes. Individuals lacking these receptors are normally resistant to jet lag/shiftwork effects [26].

Circadian oscillator period is determined around 24 hours genetically and adjusted by synchronizers such as LD cycle. Circadian rhythms synchronized to day time work and night time sleep require phase adjustment in altered routines by central and peripheral oscillators. This phase adjustment in certain cases disrupts the normal organization and sequence of the clock. Clock genes behave differently during phase shifts [1, 27]. Shift workers undergo circadian dysrhythmia that adversely effects mental health. This further leads to circadian rhythm disorder causing aberrations in neurogenesis and spatial cognition [28].

Many aspects of circadian rhythmicity can be modulated by serotonergic agents that indicate that serotonin is involved in the regulation of circadian rhythm. Serotonin transporter (5-HTT) control serotonin reuptake depending on serotonin transporter gene (*SLC6A4*) promoter region (5-HTTLPR) [29]. Significant associations between shiftwork and S variant of the SLC6A4 promoter and 5-HT and 5-HIAA contents of platelet can help in investigating the circadian rhythm-related mechanisms imposed by shiftwork [29]. The effect of shiftwork on circadian rhythm and sleep may cause the metabolic dysregulation and depressions by affecting the genetic pathways. The affect may lead to the disruption of other functioning systems and cause relative disorders either through direct or indirect genetic interaction in continuous/discontinuous forms.

3. Shiftwork Disrupts Normal Sleep and Behavior

Sleep consists of two repeated cyclic patterns "nonrapid eye movement and rapid eye movement" [30]. It is controlled genetically with the influence of environmental factors [31, 32]. Adenosine, a sleep-promoting molecule [33], mediates wake-promoting effects of caffeine by acting on adenosine receptors antagonistically [34]. GABA (gamma-aminobutyric acid) promotes sleep whereas dopamine, acetylcholine, norepinephrine, and histamine promote wakefulness [35]. The cyclic guanosine monophosphate (cGMP) kinase [36, 37], regulatory subunit of Shaker [38], Sleepless (sss) gene [39], and CLK and CYC proteins [40] are key players in sleep regulation. All these genes and regulators may get disrupted in shift workers which will lead to abnormal sleep and other serious health conditions.

Shiftwork due to the rotation of working schedules and light/dark disturbance directly affects sleep (as shown in Figure 1) and may cause health abnormalities including insomnia, sleep apnea, periodic leg movements, restless leg syndrome (RLS), and sleep-wake state dissociation disorders such as rapid eye movement (REM) and narcolepsy, sleep behavior disorder, and sleep walking [34]. Furthermore, the shiftwork-induced deficits in sleep homeostasis and circadian rhythms may lead to different psychiatric disorders and affect SUR2 gene, which was found involved in energy metabolism and aetiology of cardiomyopathies [41]. Shiftwork may disrupt daily patterns of human physiology controlled by circadian rhythms and sleep, including regulation of energy patterns expenditure [42, 43] and glucose metabolism [42].

Shiftwork may negatively affect genes and factors involved in sleep disorders such that a SNP marker [44] and chemokine (C-C motif) receptor 3 (*CCR3*) susceptibility gene [45] associated with narcolepsy, MEIS1 locus, and neuronal nitric oxide synthase (NOS1) and BTBD9 associated with RLS [46–48]. Serine to glycine mutation in *PER2* and mutation in casein kinase Iδ gene (*CSNK1D)* develop familial advanced sleep phase syndrome (FASPS) [34, 35] whereas point mutation in *PRNP* causes *fatal familial insomnia (FFI)* [35]. The disrupted functions of the above genes through shiftwork may lead to related disorders. Shiftwork can alter the functionality of sleep-promoting immune genes [49], glucocorticoids and NSAIDs [50], protein NF-kB (upregulated during sleep deprivation) [34], TAK1 (TGF-b-activated kinase) and *Sik3* (control sleep behaviors), and *Nalcn* gene (involved in REMS) [51]. Widening its effectiveness, shiftwork promotes acute coronary syndrome [52], menstrual disturbances, and abnormal insulin levels [53] and increases sleepiness and alcohol consumption [54]. It also lifts sleep deprivation that ultimately leads to loss in cognitive, physical, and metabolic consequences. These alterations can possibly end up with developing cardiovascular morbidity and mortality [29].

4. Shiftwork Dysregulates Metabolic Process and Develops Related Disorders

Social jet lag is a key player in developing metabolic syndrome by inducing changes in cholesterol levels and disrupting normal food processes. It was observed in an experiment that social jet lag potentiated body weight gain by increasing overconsumption of cafeteria food. As a result, it promoted high insulin and dyslipidemia indicating the risk of metabolic syndrome [55]. A chronic shift in light/dark (LD) cycles induces obesity, increases body weight and

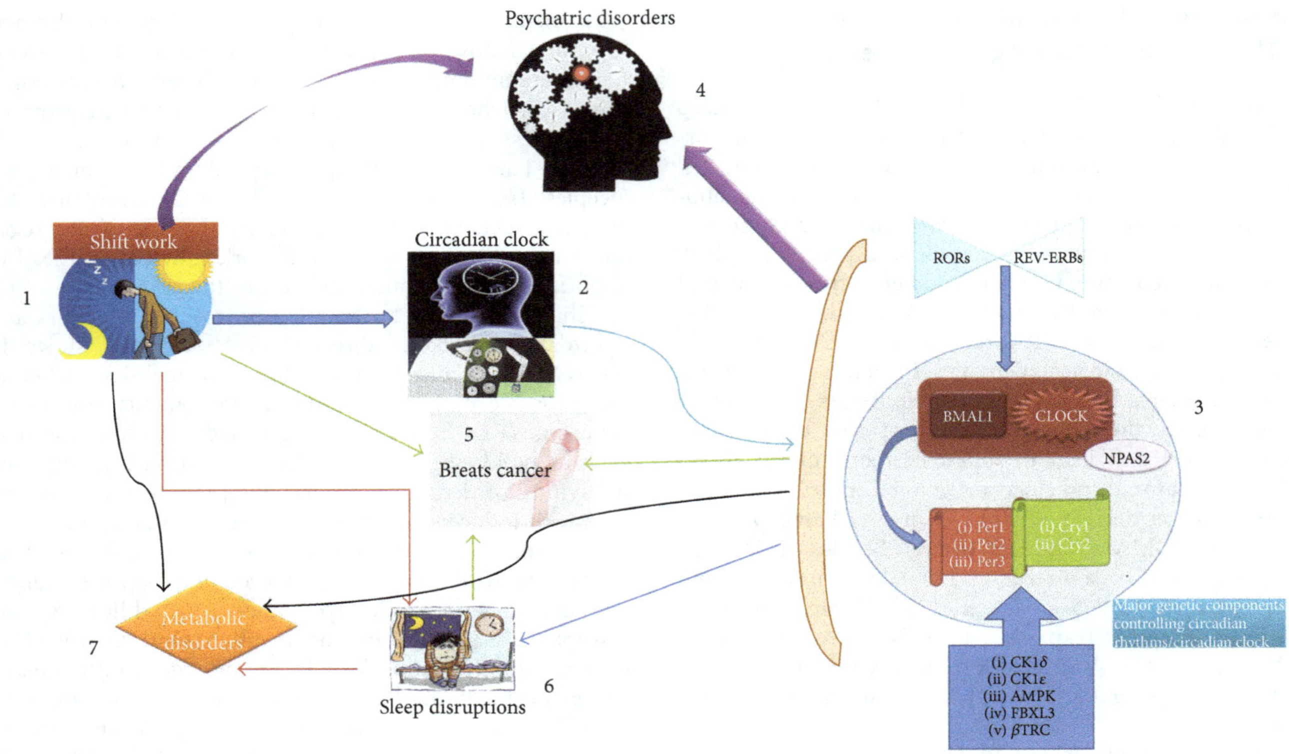

FIGURE 1: The abnormalities associated with shiftwork. (1) Shiftwork primarily affects the circadian clock and leads to several disruptions and disorders. (2) Disruption in circadian clock further affects the circadian system and alters important gene expressions which play an important role in maintaining normal body functions. (3) BMAL1 and CLOCK genes are the key factors that control circadian rhythms. Any alteration in these groups of genes further alters the genetic expression of genes (Per 1–3 and Cry 1/2) involved in clock maintenance. Other factors and proteins which play a major role in circadian rhythm are RORs, REV-ERBs, CK1δ, CK1ε, AMPK, FBXL3 and βTRC. (4) Shiftwork either directly or through circadian system alterations may cause severe psychiatric disorders including major depression, anxiety, and mood disorders. (5) Due to exposure to light irregularly, shiftwork enhances the chances of breast cancer. Breast cancer development is promoted by sleep disruptions, circadian system imbalances, and dietary conditions. (6) Sleep disorders and disrupted sleep is another condition developed by shiftwork. (7) This disruption in sleep affects the metabolism and causes metabolic disorders including obesity.

glucose intolerance, and accumulates more fat in white adipose tissues. It changes the expression profiles of metabolic genes in liver [56].

According to timed sleep restriction (TSR) study, sleep timing directly affects clock-controlled genes including ClockD19, BMAL1, Per1, Rev-erbα and Dbp, and circadian machinery. This alteration affects metabolic processes including carbohydrate and lipid utilization in liver. BMAL1, Per1, and Dbp were upregulated during early light phase and Rev-erbα during mid-light phase. Carbohydrate regulators affected were insulin receptor substrate 2 (Irs2), glucose-responsive fork head box O1 (Foxo1), and glucose transporter 2 (Slc2a2). These regulators control the functions of pyruvate carboxylase, the pyruvate transporter Slc16a7, pyruvate dehydrogenase kinase 4, the glycerol transporter aquaporin, fructose-2,6-biphosphatase 1, and liver pyruvate kinase. These alterations in carbohydrate regulators further affect glycerol kinase and glycerol phosphate dehydrogenase 2, glutamic-pyruvate transaminase, and regulators of glycerol biosynthesis [4]. This dysregulation may induce critical alterations in genetic system to develop metabolic disorders.

The presence of metabolic process controlling genes in rhythmic transcriptome indicates that alteration in circadian rhythm causes disruption in metabolic process [4, 57–59]. CLOCK, BMAL, and PER2 were found associated with obesity by affecting the metabolic process [60, 61]. Shiftwork brings about changes in appetitive hormones through circadian misalignment which causes reduction in leptin level that ultimately leads to weight gain. Shift workers are considered at higher risk of type II diabetes [42, 62].

5. Shiftwork-Related Health Risks

5.1. Shiftwork Is Associated with Mortality. Quick return occurs due to rotations in between shifts, causing major accidents [22] that normally lead to death. Shiftwork increases the risk of several disorders including mental disorders and sleep-related disorders, which may lead to death. In an experiment related to jet lag effects on transgenic aged rats, it was found that mortality light scheduled rotations at 6 hours phase advance, each week. The survival rate of rats exposed to shifted light schedule was found lesser (47%) than the survival rate (83%) of rats exposed to normal light (12L/12D) scheduled rotations. Although jet lag has no mortal effect on younger mice, it alters the behavior and circadian rhythms which further causes serious health issues by affecting brain and liver [7].

5.2. Shiftwork Induces Mental and Related Disorders. The role of SLC6A4 (serotonin transporter gene) in shift workers was confirmed to be related with time period. The proportion of short allele becomes higher than large allele if the period is more than 5 years. The effectiveness of shiftwork becomes higher in case of internal desynchronization of circadian rhythm [2, 63].

Endocrine imbalance in depression and psychosis is the hyperactivity of the HPA axis [64]. Glucocorticoid functions as an immunosuppressant and interacts with melatonin in a form of suppressive effect on the production of ACTH-induced glucocorticoid [65]. Its resistance results in hypercortisolemia in psychiatric patients and increased pituitary volume in depressed patients [66, 67]. Shiftwork causes desynchronization in circadian rhythm which in turn leads to reduced NK activity and weakening cellular immunity [68], disruptions in norepinephrine, melatonin, and serotonin production [69, 70] which may induce anxiety, depression, hypercortisolemia, and psychosis.

5.3. Shiftwork Disorder and Associated Factors. Shiftwork disorder is characterized by insomnia and excessive sleepiness which develops due to working schedules overlapping sleeping time [71]. According to a study, shiftwork disorder was found to occur frequently in males as compared to females, working during the night. Approximately, 9% of the night shift workers reported severe shiftwork disorder while one-third of the total had mild symptoms [71]. Shiftwork sleep disorder changes the behavior and sleeping periods permanently which could possibly be recovered through advanced therapies only. Treatment with modafinil, an effective compound against narcolepsy and obstructive sleep apnea, was found ineffective [6], suggesting that the mechanisms involved in shiftwork sleep disorder are not related to sleep apnea and narcolepsy. Molecular level studies with a broad range of patients, to investigate mechanisms, are needed to find out the main alterations in the system to prevent and cure the conditions properly.

The high number of nights and age were found associated with shiftwork disorder [22], whereas genotypes were found to be associated with morningness and eveningness. Allele 3111C is associated with extreme eveningness and shiftwork with semidominant influence on sleep phases without having an obvious influence on morning/evening preferences. 3111C/C homozygotes are associated with delayed shift of sleep [72] where melatonin can phase-shift the circadian clock being chronobiotic and a sleep promoting agent [73].

5.4. Shiftwork Sleep Disorder. Shiftwork sleep disorder is a condition defined by excessive sleepiness or insomnia accompanied by total sleep time reduction. It is 10–38% prevalent in shift workers [5, 43]. The disturbance in general is the sleep-wake cycle distortion of extrinsic origin. This disorder is related to the night and early morning timings. Reduction in alertness and performance along with the linkage to higher rates of comorbidity with GI disorders [67] make SWSD a severe and attention-requiring condition. Sleepiness during the night and insomnia during the day become more severe with continuous shiftwork for longer periods. Depression, ulcers, and sleepiness-related accidents are the ultimate risks associated with shiftwork sleep disorder. This sleepiness behavior during night is similar to the day time sleepiness in people with narcolepsy [6].

5.5. Advanced and Delayed Sleep Phase Syndrome. Advanced sleep phase syndrome (ASPS) is characterized by 3-4 hours advanced awakening times and sleep onsets relative to normal times. It is inherited in autosomal dominant mode, caused by circadian cycle irregularity as circadian clock genes are key players in its development. Missense mutations S662G (occurs in phosphorylation site within CSNK1-binding domain of PER2) and CSNK1D are involved in ASPS [58].

Delayed sleep phase syndrome (DSPS) is characterized by chronic inability to fall asleep and awaken at normal timings [58]. Circadian system genes are majorly involved in this dysregulated sleep behavior. Significant associations with T3111C polymorphism in the 3′UTR of CLOCK and the association of a SNP in the 5′UTR of PER2 with morning preference have been reported in DSPS [72, 74]. According to a study, amino-acid substitution S408N in the CSKN1E gene might protect the body form DSPS and non-24-hour sleep-wake syndrome development [58]. Both the shorter and longer VNTR (variable number tandem repeat) alleles were found (PER3-4/4) associated with DSPS [75]. Per3 gene is involved in delayed sleep phase syndrome and extreme diurnal preference.

Mutation in hPER2 has a notable association with advanced sleep phase syndrome, whereas some haplotypes of hPER3 have shown an association with delayed sleep phase syndrome. These conditions arise because of the alterations into casein kinase Iε (CKIε) phosphorylation of the target clock proteins, affecting morningness and eveningness. Polymorphisms in 3′ flanking region of the human clock homolog (3111T/C, hClock) is associated with eveningness and morningness such that 3111T allele has lower evening preference than 3111C allele [72]. As shiftwork has direct impacts on sleep, it may affect the normal functioning of genes to cause one of the two sleep abnormalities ASPS and DSPS.

5.6. Circadian Rhythm Sleep Disorder. Circadian rhythm sleep disorder refers to the abnormalities brought about by circadian misalignment due to variations in sleep-wake pattern. The shift in sleep-wake time is commonly caused by shiftwork, jet lag, light exposure, and insufficient sleep period. Circadian misalignment alters neuroendocrine physiology, impairs glucose tolerance, and reduces insulin sensitivity [42, 62, 76, 77]. Sleep deprivation problem rises with the incompatibility between circadian rhythms and working periods [78]. Both circadian alterations and sleep deprivations lead to fatigue, impairments in vigilance and attention, sleepiness [79], sleep deficiency, impaired physiological function, and aberrant behavior [76, 80–82]. A subset of insomnias including non-24-hour sleep-wake syndrome is also linked with circadian rhythm sleep disorder [81]. Exposure to bright light in shiftwork and working in consecutive shifts if maintained for longer time may change the system at the genetic level. This alteration will further disrupt the normal functions of gene-related behavior, sleep, and circadian system. Such modifications will ultimately lead to health-harming conditions.

6. Shiftwork-Related Alterations in Nervous System to Develop Psychiatric Disorders

Adversely affected nervous system by working in continuously rotating shifts may accelerate the rate of psychiatric disorder occurrence (Figure 1). Being major causes of disabilities [83], psychiatric disorders including bipolar disorder (BD), schizophrenia (SZ), and major depression disorder (MDD) impose enormous medical burdens [84–86]. The genetic alteration and uncontrolled expression of genes primarily causing the aforementioned disorders [83, 87–89] are associated with irregular continuous changing of shifts during work. Night shift work contributed toward several psychiatric disorders through circadian misalignment, sleep deprivation, and light-induced melatonin suppression [73]. Disrupted-in-schizophrenia-1 (DISC1) is an important genetic factor in serious mental disorders including SZ, BD, and MDD [90]. We will further discuss the previously mentioned psychiatric disorders one by one to provide detailed information.

6.1. Bipolar Disorder. BD is considered one of the severely disabling disorders. Distinctive distortions of emotion regulation make BD a severe psychiatric condition. It mainly affects emotional and social behavior with light effect on perception and thought. Being a multifactorial disorder, its risk is influenced by genetic and environmental factors [91]. Genetic and pathophysiological factors involved in the development of BD are largely unknown [92]. BD increases the risk of schizophrenia and major depression disorder which is an indication for shared genetic basis between these disorders. Although heritability has been proven, more studies are needed to investigate the genetic mechanisms [93, 94]. In BD, important genetic variants that could be affected by shiftwork either directly or indirectly are Del, ANK3 rs1938526, COMT Val158Met, DAOA, BRD1/ZBED4, BDNF Val66Met, BRD1, ASMT, CAMTA1, CCDC132, CHES1, DGKH, DRD4, HTR1A C1019G, SLC6A4, 5-HTTLPR PARD3B, PDLIM5, STin2 VNTR, KLHL3, LYPD5, MAOA T941G, MTHFR A1298C and C677T, TPH1 STAB1, HTR3B, and WNK2 [95]. They are expressed in brain and associated with CREB (cyclic adenosine monophosphate response element-binding protein). *KCNH7* R394H (rs78247304) mutation is linked to BD. *ANK3*, *CACNA1C*, an intron variant of *CACNA1C* (rs79398153), and a missense mutation of *ANK3* (N2643S) were confirmed being involved in BD [95, 96]. All these mutations are possibly enhanced in people working in several shifts or exposing to irregular light (jet lag).

6.2. Major Depression Disorder. MDD, a leading cause of loss in work productivity, is considered a fatal disorder. It is considered one of the most prevalent disorders that affect females more than males [97, 98]. Genetic and environmental risk factors mainly contribute to cause MDD. It is a neuroprogressive disorder in which each persisting episode increases impairment in function and sensitivity for upcoming episodes. A decrease in the GR messenger ribonucleic acid levels in hippocampus, frontal cortex, and amygdala has been observed in the patients suffering from MDD. Susceptibility to further episodes is increased by repeated illness which causes the permanent alteration in the normal functions of neurons [99] and genes [100] including MTHFR C677T and 5-HTTLPR [101]. Some of the important genetic variants associated with MDD that could be affected by shiftwork are APOE, SLC6A4, ACE, GNB3, HS6ST3, HTR1A, LHFPL2, PDE11A, DISC1, MAOA, SLC6A3 (DAT1), SLC25A21, VGLL4 BDNF, P2RX7, TPH2, PDE9A, and GRIK3 [95, 100]. Neutral amino acid transporter (SLC6A15) is a susceptibility gene for MDD [102].

7. Shiftwork Affects Expressions of Oncogenes to Develop Cancer

Malignancies are majorly developed by mitigation in pineal hormone melatonin by bright light at night [1]. Reduced production of melatonin, phase shift, and sleep disruption, caused by exposure to light at night, might be the possible mechanisms that cause cancer and related disorders [103]. Shiftwork causes cancer (Figure 1) via altering the clock-controlled gene expression that regulate tumor suppression [1, 104]. Per1 negatively affects the growth of tumor cell and per2 functions as tumor repressor [105]. Long-term shiftwork negatively affects IFNγ (interferon gamma) [106] The association of shiftwork-affected immune system with sleep deprivation increases inflammatory markers thereby causing malignancies and metabolic and cardiovascular disorders [1]. Circadian rhythm disruption by shiftwork or bright light exposure at night increases the rate of cancer and decreases the nocturnal rise in melatonin [19].

7.1. Breast Cancer. The disrupted circadian rhythm or circadian clock through shiftwork effects the development of breast cancer (Figure 2) [107]. Cell proliferation and apoptosis is regulated by approximately 7% of clock-controlled genes, including myelocytomatosis viral oncogene human recombinant (C-Myc), murine double minute oncogene (Mdm-2), and growth arrest and DNA damage-inducible alpha protein (Gadd 45a) and those encoding p53, caspases, and cyclins. Per1 is reduced in cancer tissue and its inhibition blunts apoptosis, whereas Per2 represses tumor in breast cancer and induces estradiol (E2) in mammary cells. Coexpression of Per2 with cytochrome inhibits growth of breast cancer cells. Deficiency of Per2 causes deregulation of Myc, cyclin A, cyclin D1, Mdm-2, and Gadd45a, while its dysfunctionality impairs apoptosis mediated by p53 by activating c-Myc signaling pathway. Overexpressing and downregulating Per1 or Per2 inhibits and accelerates growth

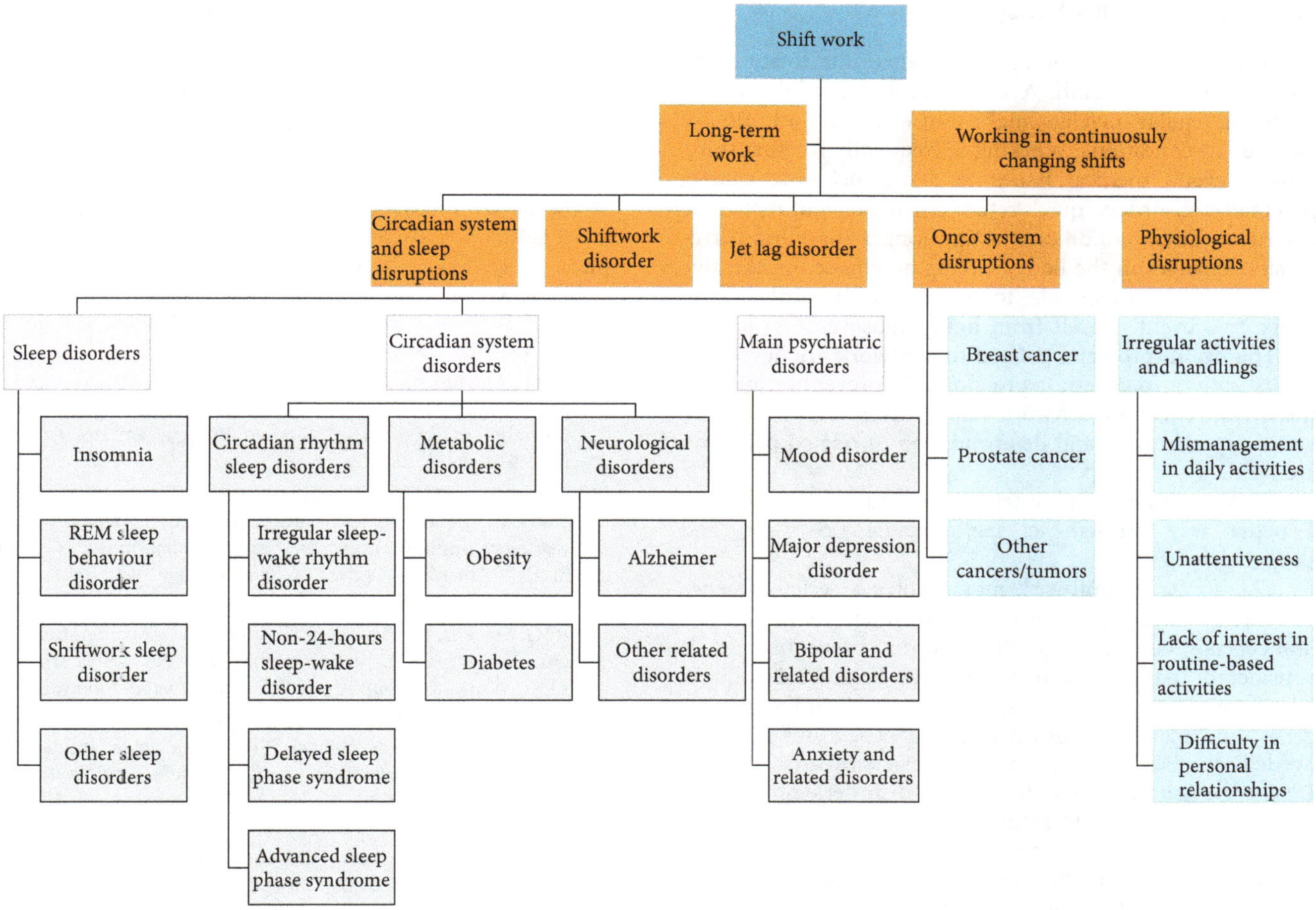

FIGURE 2: All disorders that could possibly be caused by shiftwork.

of cancer cells, respectively [1]. Telomere shortening is correlated with consecutive night shift work for a long time where it increases breast cancer risk [108].

7.2. Colonic Cancer. Per1- and Per2-mediated tumor-repressing effect is more specific to early light and early dark phase. Per2 mutations and deregulation favor development and increases cell proliferation, respectively [1]. Altered light schedules develop mutations in PER1, PER2, and PER3 which in turn promote colonic adenoma and colonic cancer [19, 109].

7.3. Prostate Cancer. Shiftwork-mediated sleep disruption is associated with elevated prostate-specific antigen (PSA), indicating an increased risk of developing prostate cancer [110]. Disrupted circadian rhythms through jet lag inhibits p53, enhances Myc expression, and induces tumorigenesis in prostate tissues targeted by endocrine [1, 111, 112].

7.4. Ovarian Cancer. Functional analysis suggested that variation, in circadian genes including *BMAL1*, *CRY2*, *CSNK1E*, *NPAS2*, *PER3*, *REV1*, and downstream transcription factors *KLF10* and *SENP3* through disruption of hormonal pathways or changes in light/dark schedules, is associated with ovarian cancer. Silencing the expression of *BMAL1* activates p53 to prevent cell cycle arrest which indicates that *BMAL1* gene may regulate the p53 tumor suppressor pathway. Per2 inhibition, reduces estrogen receptor α (ERα) response to E2 by overexpression or enhances E2 activation [112, 113]. Its lower expression along with lower expression of *BMAL1* and *CRY1* promotes lower survival of cells in ovarian cancer [114].

7.5. Lung Cancer. Not only systemic but also somatic disruption of circadian rhythms mediated by jet lag alone affects c-Myc and enhances cell proliferation. Per2 and BMAL1 lose the ability of inhibiting tumor progression and hence promotes lung cancer [115].

7.6. Shiftwork Tolerance. The effects of shiftwork regarding tolerance and responsiveness are concerned with certain factors. The youngest and oldest individuals are affected more than individuals with middle age [116]. Females due to low tolerance develop more problems, like sleep disruptions, higher risk of mortality, disability, fatigue, and obesity, while males show more cognitive disturbances [117, 118]. In case of circadian rhythms, a low score on morningness [119] and languidness and a high score on flexibility [120] are associated positively with shiftwork tolerance.

8. Conclusion and Prospects

Shiftwork has gained central importance due to its detrimental effects on health. A large population of the world, including regular travelers, night shift workers, continuously faces jet lag conditions. Working in frequently rotating shifts causes several medical issues to arise, including cancer, psychiatric disorders, quick return accidents, and metabolic disorders. These conditions lasting longer may bring irreversible changes in the body, leaving no choice for affective recovery. It is impossible to avoid working in rotating shifts or prevent oneself from light exposure.

The health problems related to shiftwork are developed by disruptions in genetic expressions. To prevent or mitigate the adverse effects of shiftwork-related disorders, unveiling of genetic mechanisms and determination of related pathways are needed.

These relations and interactions could be studied in a better way through jet lag, circadian rhythms, and sleep behaviors.

The effects of shiftwork may involve a series of genes and factors majorly involved in circadian rhythm, sleep homeostasis, and the specific disorder prevention. It is considered that circadian genes have noticeable impacts on cancer controlling genes, psychiatric disorder causing genes, and metabolic disorder-related genes. But it is still needed to be investigated whether shiftwork directly affects the genes related to certain disorders or it elongates its impact via targeting other genes such as clock genes. The proper targeted medications are possible to be developed only if the mechanisms and factors causing and promoting the shiftwork effects to cause disorders have been identified.

Conflicts of Interest

The authors declare that there is no conflict of interest regarding the publication of this paper.

Acknowledgments

This work was supported by the National Natural Science Foundation of China (Grant no. 31372381) and the Henan Scientific and Technological Innovation Team Fund (17454). The authors are thankful to the Chinese Academy of Science and The World Academy of Science (CAS-TWAS) scholarship program.

References

[1] E. L. Haus and M. H. Smolensky, "Shift work and cancer risk: potential mechanistic roles of circadian disruption, light at night, and sleep deprivation," *Sleep Medicine Reviews*, vol. 17, no. 4, pp. 273–284, 2013.

[2] I. B. Saksvik, B. Bjorvatn, H. Hetland, G. M. Sandal, and S. Pallesen, "Individual differences in tolerance to shift work – a systematic review," *Sleep Medicine Reviews*, vol. 15, no. 4, pp. 221–235, 2011.

[3] R. Salgado-Delgado, M. Angeles-Castellanos, N. Saderi, R. M. Buijs, and C. Escobar, "Food intake during the normal activity phase prevents obesity and circadian desynchrony in a rat model of night work," *Endocrinology*, vol. 151, no. 3, pp. 1019–1029, 2010.

[4] J. L. Barclay, J. Husse, B. Bode et al., "Circadian desynchrony promotes metabolic disruption in a mouse model of shiftwork," *PLoS One*, vol. 7, no. 5, article e37150, 2012.

[5] E. Flo, S. Pallesen, N. Magerøy et al., "Shift work disorder in nurses – assessment, prevalence and related health problems," *PLoS One*, vol. 7, no. 4, article e33981, 2012.

[6] C. A. Czeisler, J. K. Walsh, T. Roth et al., "Modafinil for excessive sleepiness associated with shift-work sleep disorder," *The New England Journal of Medicine*, vol. 353, no. 5, pp. 476–486, 2005.

[7] A. J. Davidson, M. T. Sellix, J. Daniel, S. Yamazaki, M. Menaker, and G. D. Block, "Chronic jet-lag increases mortality in aged mice," *Current Biology*, vol. 16, no. 21, pp. R914–R916, 2006.

[8] T. Kubo, K. Ozasa, K. Mikami et al., "Prospective cohort study of the risk of prostate cancer among rotating-shift workers: findings from the Japan Collaborative Cohort Study," *American Journal of Epidemiology*, vol. 164, no. 6, pp. 549–555, 2006.

[9] R. G. Stevens, J. Hansen, G. Costa et al., "Considerations of circadian impact for defining 'shift work' in cancer studies: IARC Working Group Report," *Occupational & Environmental Medicine*, vol. 68, no. 2, pp. 154–162, 2010.

[10] M. Sallinen and G. Kecklund, "Shift work, sleep, and sleepiness - differences between shift schedules and systems," *Scandinavian Journal of Work, Environment & Health*, vol. 36, no. 2, pp. 121–133, 2010.

[11] A.-C. Bara and S. Arber, "Working shifts and mental health – findings from the British household panel survey (1995–2005)," *Scandinavian Journal of Work, Environment & Health*, vol. 35, no. 5, pp. 361–367, 2009.

[12] M. Härmä and G. Kecklund, "Shift work and health – how to proceed?," *Scandinavian Journal of Work, Environment & Health*, vol. 36, no. 2, pp. 81–84, 2010.

[13] J.-U. Rehman, K. Brismar, U. Holmback, T. Akerstedt, and J. Axelsson, "Sleeping during the day: effects on the 24-h patterns of IGF-binding protein 1, insulin, glucose, cortisol, and growth hormone," *European Journal of Endocrinology*, vol. 163, no. 3, pp. 383–390, 2010.

[14] J. Arendt, "Shift work: coping with the biological clock," *Occupational Medicine*, vol. 60, no. 1, pp. 10–20, 2010.

[15] C. A. Crispim, J. Waterhouse, A. R. Dâmaso et al., "Hormonal appetite control is altered by shift work: a preliminary study," *Metabolism*, vol. 60, no. 12, pp. 1726–1735, 2011.

[16] E. Van Cauter and K. L. Knutson, "Sleep and the epidemic of obesity in children and adults," *European Journal of Endocrinology*, vol. 159, Supplement 1, pp. S59–S66, 2008.

[17] J. L. Marino, V. L. Holt, C. Chen, and S. Davis, "Shift work, hCLOCK T3111C polymorphism, and endometriosis risk," *Epidemiology*, vol. 19, no. 3, pp. 477–484, 2008.

[18] Y. Suwazono, M. Dochi, K. Sakata et al., "A longitudinal study on the effect of shift work on weight gain in male Japanese workers," *Obesity*, vol. 16, no. 8, pp. 1887–1893, 2008.

[19] S. Davis and D. K. Mirick, "Circadian disruption, shift work and the risk of cancer: a summary of the evidence and studies in Seattle," *Cancer Causes & Control*, vol. 17, no. 4, pp. 539–545, 2006.

[20] A. Knutsson and H. Bøggild, "Gastrointestinal disorders among shift workers," *Scandinavian Journal of Work, Environment & Health*, vol. 36, no. 2, pp. 85–95, 2010.

[21] S. Puttonen, T. Oksanen, J. Vahtera et al., "Is shift work a risk factor for rheumatoid arthritis? The Finnish Public Sector study," *Annals of the Rheumatic Diseases*, vol. 69, no. 4, pp. 679-680, 2010.

[22] M. F. Eldevik, E. Flo, B. E. Moen, S. Pallesen, and B. Bjorvatn, "Insomnia, excessive sleepiness, excessive fatigue, anxiety, depression and shift work disorder in nurses having less than 11 hours in-between shifts," *PLoS One*, vol. 8, no. 8, article e70882, 2013.

[23] R. Tamagawa, B. Lobb, and R. Booth, "Tolerance of shift work," *Applied Ergonomics*, vol. 38, no. 5, pp. 635–642, 2007.

[24] U. Albrecht, "Timing to perfection: the biology of central and peripheral circadian clocks," *Neuron*, vol. 74, no. 2, pp. 246–260, 2012.

[25] A. Iwamoto, M. Kawai, M. Furuse, and S. Yasuo, "Effects of chronic jet lag on the central and peripheral circadian clocks in CBA/N mice," *Chronobiology International*, vol. 31, no. 2, pp. 189–198, 2014.

[26] Y. Yamaguchi, T. Suzuki, Y. Mizoro et al., "Mice genetically deficient in vasopressin V1a and V1b receptors are resistant to jet lag," *Science*, vol. 342, no. 6154, pp. 85–90, 2013.

[27] W. Nakamura, S. Yamazaki, N. N. Takasu, K. Mishima, and G. D. Block, "Differential response of *Period 1* expression within the suprachiasmatic nucleus," *The Journal of Neuroscience*, vol. 25, no. 23, pp. 5481–5487, 2005.

[28] E. Abdullah, A. Idris, and A. Saparon, "PAPR reduction using SCS-SLM technique in STFBC MIMO-OFDM," *Journal of Engineering and Applied Science*, vol. 12, pp. 3218–3221, 2017.

[29] S. Sookoian, C. Gemma, T. Fernández Gianotti et al., "Effects of rotating shift work on biomarkers of metabolic syndrome and inflammation," *Journal of Internal Medicine*, vol. 261, no. 3, pp. 285–292, 2007.

[30] C. A. Landis, "Physiological and behavioral aspects of normal sleep," in *Sleep Disorders and Sleep Promotion in Nursing Practice*, N. Redeker and G. McEnany, Eds., pp. 1–18, Springer Publishing Company, LLC, New York, NY, USA, 2011.

[31] U. Ambrosius, S. Lietzenmaier, R. Wehrle et al., "Heritability of sleep electroencephalogram," *Biological Psychiatry*, vol. 64, no. 4, pp. 344–348, 2008.

[32] L. De Gennaro, C. Marzano, F. Fratello et al., "The electroencephalographic fingerprint of sleep is genetically determined: a twin study," *Annals of Neurology*, vol. 64, no. 4, pp. 455–460, 2008.

[33] T. Bjorness and R. Greene, "Adenosine and sleep," *Current Neuropharmacology*, vol. 7, no. 3, pp. 238–245, 2009.

[34] A. Sehgal and E. Mignot, "Genetics of sleep and sleep disorders," *Cell*, vol. 146, no. 2, pp. 194–207, 2011.

[35] C. Cirelli, "The genetic and molecular regulation of sleep: from fruit flies to humans," *Nature Reviews Neuroscience*, vol. 10, no. 8, pp. 549–560, 2009.

[36] D. M. Raizen, J. E. Zimmerman, M. H. Maycock et al., "Lethargus is a *Caenorhabditis elegans* sleep-like state," *Nature*, vol. 451, no. 7178, pp. 569–572, 2008.

[37] S. Langmesser, P. Franken, S. Feil, Y. Emmenegger, U. Albrecht, and R. Feil, "cGMP-dependent protein kinase type I is implicated in the regulation of the timing and quality of sleep and wakefulness," *PLoS One*, vol. 4, no. 1, article e4238, 2009.

[38] D. Bushey, R. Huber, G. Tononi, and C. Cirelli, "*Drosophila hyperkinetic* mutants have reduced sleep and impaired memory," *The Journal of Neuroscience*, vol. 27, no. 20, pp. 5384–5393, 2007.

[39] K. Koh, W. J. Joiner, M. N. Wu, Z. Yue, C. J. Smith, and A. Sehgal, "Identification of SLEEPLESS, a sleep-promoting factor," *Science*, vol. 321, no. 5887, pp. 372–376, 2008.

[40] A. C. Keene, E. R. Duboué, D. M. McDonald et al., "*Clock* and *cycle* limit starvation-induced sleep loss in *Drosophila*," *Current Biology*, vol. 20, no. 13, pp. 1209–1215, 2010.

[41] K. V. Allebrandt, N. Amin, B. Müller-Myhsok et al., "A K_{ATP} channel gene effect on sleep duration: from genome-wide association studies to function in *Drosophila*," *Molecular Psychiatry*, vol. 18, no. 1, pp. 122–132, 2011.

[42] C. M. Depner, E. R. Stothard, and K. P. Wright, "Metabolic consequences of sleep and circadian disorders," *Current Diabetes Reports*, vol. 14, no. 7, p. 507, 2014.

[43] R. R. Markwald, E. L. Melanson, M. R. Smith et al., "Impact of insufficient sleep on total daily energy expenditure, food intake, and weight gain," *Proceedings of the National Academy of Sciences of the United States of America*, vol. 110, no. 14, pp. 5695–5700, 2013.

[44] T. Miyagawa, M. Kawashima, N. Nishida et al., "Variant between *CPT1B* and *CHKB* associated with susceptibility to narcolepsy," *Nature Genetics*, vol. 40, no. 11, pp. 1324–1328, 2008.

[45] H. Toyoda, Y. Honda, S. Tanaka et al., "Narcolepsy susceptibility gene *CCR3* modulates sleep-wake patterns in mice," *PLos One*, vol. 12, no. 11, article e0187888, 2017.

[46] J. Winkelmann, D. Czamara, B. Schormair et al., "Correction: genome-wide association study identifies novel restless legs syndrome susceptibility loci on 2p14 and 16q12.1," *PLoS Genetics*, vol. 7, 2011.

[47] H. Stefansson, D. B. Rye, A. Hicks et al., "A genetic risk factor for periodic limb movements in sleep," *The New England Journal of Medicine*, vol. 357, no. 7, pp. 639–647, 2007.

[48] J. Winkelmann, P. Lichtner, B. Schormair et al., "Variants in the neuronal nitric oxide synthase (nNOS, NOS1) gene are associated with restless legs syndrome," *Movement Disorders*, vol. 23, no. 3, pp. 350–358, 2008.

[49] L. Imeri and M. R. Opp, "How (and why) the immune system makes us sleep," *Nature Reviews Neuroscience*, vol. 10, no. 3, pp. 199–210, 2009.

[50] J. Rihel, D. A. Prober, A. Arvanites et al., "Zebrafish behavioral profiling links drugs to biological targets and rest/wake regulation," *Science*, vol. 327, no. 5963, pp. 348–351, 2010.

[51] H. Funato, C. Miyoshi, T. Fujiyama et al., "Forward-genetics analysis of sleep in randomly mutagenized mice," *Nature*, vol. 539, no. 7629, pp. 378–383, 2016.

[52] L. K. Barger, S. M. W. Rajaratnam, C. P. Cannon et al., "Short sleep duration, obstructive sleep apnea, shiftwork, and the risk of adverse cardiovascular events in patients after an acute coronary syndrome," *Journal of the American Heart Association*, vol. 6, no. 10, article e006959, 2017.

[53] A. J. R. Lim, Z. Huang, S. E. Chua, M. S. Kramer, and E. L. Yong, "Sleep duration, exercise, shift work and polycystic ovarian syndrome-related outcomes in a healthy population: a cross-sectional study," *PLoS One*, vol. 11, no. 11, article e0167048, 2016.

[54] S. Lee, H.-R. Kim, J. Byun, and T. Jang, "Sleepiness while driving and shiftwork patterns among Korean bus drivers," *Annals of Occupational and Environmental Medicine*, vol. 29, no. 1, 2017.

[55] E. Espitia-Bautista, M. Velasco-Ramos, I. Osnaya-Ramírez, M. Ángeles-Castellanos, R. M. Buijs, and C. Escobar, "Social jet-lag potentiates obesity and metabolic syndrome when combined with cafeteria diet in rats," *Metabolism*, vol. 72, pp. 83–93, 2017.

[56] H. Oike, M. Sakurai, K. Ippoushi, and M. Kobori, "Time-fixed feeding prevents obesity induced by chronic advances of light/dark cycles in mouse models of jet-lag/shift work," *Biochemical and Biophysical Research Communications*, vol. 465, no. 3, pp. 556–561, 2015.

[57] J. Bass and J. S. Takahashi, "Circadian integration of metabolism and energetics," *Science*, vol. 330, no. 6009, pp. 1349–1354, 2010.

[58] J. S. Takahashi, H.-K. Hong, C. H. Ko, and E. L. McDearmon, "The genetics of mammalian circadian order and disorder: implications for physiology and disease," *Nature Reviews Genetics*, vol. 9, no. 10, pp. 764–775, 2008.

[59] L. Zhang, B. Y. Chung, B. C. Lear et al., "$DN1_p$ circadian neurons coordinate acute light and PDF inputs to produce robust daily behavior in *Drosophila*," *Current Biology*, vol. 20, no. 7, pp. 591–599, 2010.

[60] K. A. Lamia, K.-F. Storch, and C. J. Weitz, "Physiological significance of a peripheral tissue circadian clock," *Proceedings of the National Academy of Sciences of the United States of America*, vol. 105, no. 39, pp. 15172–15177, 2008.

[61] F. W. Turek, C. Joshu, A. Kohsaka et al., "Obesity and metabolic syndrome in circadian *Clock* mutant mice," *Science*, vol. 308, no. 5724, pp. 1043–1045, 2005.

[62] F. A. J. L. Scheer, M. F. Hilton, C. S. Mantzoros, and S. A. Shea, "Adverse metabolic and cardiovascular consequences of circadian misalignment," *Proceedings of the National Academy of Sciences of the United States of America*, vol. 106, no. 11, pp. 4453–4458, 2009.

[63] A. Reinberg and I. Ashkenazi, "Internal desynchronization of circadian rhythms and tolerance to shift work," *Chronobiology International*, vol. 25, no. 4, pp. 625–643, 2008.

[64] C. M. Pariante, "Risk factors for development of depression and psychosis," *Annals of the New York Academy of Sciences*, vol. 1179, no. 1, pp. 144–152, 2009.

[65] C. Campino, F. Valenzuela, E. Arteaga et al., "Melatonin reduces cortisol response to ACTH in humans," *Revista Médica de Chile*, vol. 136, no. 11, pp. 1390–1397, 2008.

[66] C. M. Pariante, "Pituitary volume in psychosis: the first review of the evidence," *Journal of Psychopharmacology*, vol. 22, 2 Supplement, pp. 76–81, 2008.

[67] M. Vogel, T. Braungardt, W. Meyer, and W. Schneider, "The effects of shift work on physical and mental health," *Journal of Neural Transmission*, vol. 119, no. 10, pp. 1121–1132, 2012.

[68] P. Boscolo, M. Di Gioacchino, M. Reale, R. Muraro, and L. Di Giampaolo, "Work stress and innate immune response," *International Journal of Immunopathology and Pharmacology*, vol. 24, pp. 51S–54S, 2011.

[69] S. R. Pandi-Perumal, V. Srinivasan, D. W. Spence, and D. P. Cardinali, "Role of the melatonin system in the control of sleep," *CNS Drugs*, vol. 21, no. 12, pp. 995–1018, 2007.

[70] S. A. Rahman, S. Marcu, L. Kayumov, and C. M. Shapiro, "Altered sleep architecture and higher incidence of subsyndromal depression in low endogenous melatonin secretors," *European Archives of Psychiatry and Clinical Neuroscience*, vol. 260, no. 4, pp. 327–335, 2010.

[71] L. Di Milia, S. Waage, S. Pallesen, and B. Bjorvatn, "Shift work disorder in a random population sample – prevalence and comorbidities," *PLoS One*, vol. 8, no. 1, article e55306, 2013.

[72] K. Mishima, T. Tozawa, K. Satoh, H. Saitoh, and Y. Mishima, "The 3111T/C polymorphism of *hClock* is associated with evening preference and delayed sleep timing in a Japanese population sample," *American Journal of Medical Genetics Part B: Neuropsychiatric Genetics*, vol. 133B, no. 1, pp. 101–104, 2005.

[73] M. R. Smith and C. I. Eastman, "Shift work: health, performance and safety problems, traditional countermeasures, and innovative management strategies to reduce circadian misalignment," *Nature and Science of Sleep*, vol. 4, pp. 111–132, 2012.

[74] J. D. Carpen, S. N. Archer, D. J. Skene, M. Smits, and M. Schantz, "A single-nucleotide polymorphism in the 5′ -untranslated region of the *hPER2* gene is associated with diurnal preference," *Journal of Sleep Research*, vol. 14, no. 3, pp. 293–297, 2005.

[75] D. S. Pereira, S. Tufik, F. M. Louzada et al., "Association of the length polymorphism in the human Per3 gene with the delayed sleep-phase syndrome: does latitude have an influence upon it?," *Sleep*, vol. 28, no. 1, pp. 29–32, 2005.

[76] R. R. Markwald and K. P. Wright Jr., "Circadian misalignment and sleep disruption in shift work: Implications for fatigue and risk of weight gain and obesity," in *Sleep Loss and Obesity*, Springer, New York, NY, USA, 2012.

[77] O. M. Buxton, S. W. Cain, S. P. O'Connor et al., "Adverse metabolic consequences in humans of prolonged sleep restriction combined with circadian disruption," *Science Translational Medicine*, vol. 4, no. 129, article 129ra43, 2012.

[78] C. L. Drake and K. P. Wright, "Shift work, shift-work disorder, and jet lag," *Principles and Practice of Sleep Medicine*, vol. 1, pp. 784–798, 2011.

[79] J. S. Ruggiero and N. S. Redeker, "Effects of napping on sleepiness and sleep-related performance deficits in night-shift workers: a systematic review," *Biological Research for Nursing*, vol. 16, no. 2, pp. 134–142, 2014.

[80] B. Marcheva, K. M. Ramsey, E. D. Buhr et al., "Disruption of the clock components CLOCK and BMAL1 leads to hypoinsulinaemia and diabetes," *Nature*, vol. 466, no. 7306, pp. 627–631, 2010.

[81] R. L. Sack, D. Auckley, R. R. Auger et al., "Circadian rhythm sleep disorders: part I, basic principles, shift work and jet lag disorders," *Sleep*, vol. 30, no. 11, pp. 1460–1483, 2007.

[82] K. P. Wright Jr., R. K. Bogan, and J. K. Wyatt, "Shift work and the assessment and management of shift work disorder (SWD)," *Sleep Medicine Reviews*, vol. 17, no. 1, pp. 41–54, 2013.

[83] Y. Zhao and F. X. Castellanos, "Annual research review: discovery science strategies in studies of the pathophysiology of child and adolescent psychiatric disorders - promises and limitations," *The Journal of Child Psychology and Psychiatry*, vol. 57, no. 3, pp. 421–439, 2016.

[84] Z. Hawi, T. D. R. Cummins, J. Tong et al., "The molecular genetic architecture of attention deficit hyperactivity disorder," *Molecular Psychiatry*, vol. 20, no. 3, pp. 289–297, 2015.

[85] S. S. Jeste and D. H. Geschwind, "Disentangling the heterogeneity of autism spectrum disorder through genetic findings," *Nature Reviews Neurology*, vol. 10, no. 2, pp. 74–81, 2014.

[86] M. P. Milham, "Open neuroscience solutions for the connectome-wide association era," *Neuron*, vol. 73, no. 2, pp. 214–218, 2012.

[87] R. T. de Sousa, A. A. Loch, A. F. Carvalho et al., "Genetic studies on the tripartite glutamate synapse in the pathophysiology and therapeutics of mood disorders," *Neuropsychopharmacology*, vol. 42, no. 4, pp. 787–800, 2017.

[88] M. Burmeister, M. G. McInnis, and S. Zöllner, "Psychiatric genetics: progress amid controversy," *Nature Reviews Genetics*, vol. 9, no. 7, pp. 527–540, 2008.

[89] A. B. Amstadter, H. H. Maes, C. M. Sheerin, J. M. Myers, and K. S. Kendler, "The relationship between genetic and environmental influences on resilience and on common internalizing and externalizing psychiatric disorders," *Social Psychiatry and Psychiatric Epidemiology*, vol. 51, no. 5, pp. 669–678, 2016.

[90] N. Sawamura, T. Ando, Y. Maruyama et al., "Nuclear DISC1 regulates CRE-mediated gene transcription and sleep homeostasis in the fruit fly," *Molecular Psychiatry*, vol. 13, no. 12, pp. 1138–1148, 2008.

[91] T. Kieseppä, T. Partonen, J. Haukka, J. Kaprio, and J. Lönnqvist, "High concordance of bipolar I disorder in a nationwide sample of twins," *The American Journal of Psychiatry*, vol. 161, no. 10, pp. 1814–1821, 2004.

[92] T. W. Mühleisen, M. Leber, T. G. Schulze et al., "Genome-wide association study reveals two new risk loci for bipolar disorder," *Nature Communications*, vol. 5, 2014.

[93] N. Craddock and P. Sklar, "Genetics of bipolar disorder: successful start to a long journey," *Trends in Genetics*, vol. 25, no. 2, pp. 99–105, 2009.

[94] Psychiatric GWAS Consortium Bipolar Disorder Working Group, "Large-scale genome-wide association analysis of bipolar disorder identifies a new susceptibility locus near *ODZ4*," *Nature Genetics*, vol. 43, pp. 977–983, 2012.

[95] J. M. Gatt, K. L. O. Burton, L. M. Williams, and P. R. Schofield, "Specific and common genes implicated across major mental disorders: a review of meta-analysis studies," *Journal of Psychiatric Research*, vol. 60, pp. 1–13, 2015.

[96] T. Kato, "Whole genome/exome sequencing in mood and psychotic disorders," *Psychiatry and Clinical Neurosciences*, vol. 69, no. 2, pp. 65–76, 2015.

[97] R. C. Kessler, W. T. Chiu, O. Demler, K. R. Merikangas, and E. E. Walters, "Prevalence, severity, and comorbidity of 12-month DSM-IV disorders in the national comorbidity survey replication," *Archives of General Psychiatry*, vol. 62, no. 6, pp. 617–627, 2005.

[98] K. Koido, T. Traks, R. Balõtšev et al., "Associations between *LSAMP* gene polymorphisms and major depressive disorder and panic disorder," *Translational Psychiatry*, vol. 2, no. 8, article e152, 2012.

[99] S. Moylan, M. Maes, N. R. Wray, and M. Berk, "The neuroprogressive nature of major depressive disorder: pathways to disease evolution and resistance, and therapeutic implications," *Molecular Psychiatry*, vol. 18, no. 5, pp. 595–606, 2013.

[100] D. F. Levinson, "The genetics of depression: a review," *Biological Psychiatry*, vol. 60, no. 2, pp. 84–92, 2006.

[101] S. López-León, A. C. J. W. Janssens, A. M. González-Zuloeta Ladd et al., "Meta-analyses of genetic studies on major depressive disorder," *Molecular Psychiatry*, vol. 13, no. 8, pp. 772–785, 2008.

[102] M. A. Kohli, S. Lucae, P. G. Saemann et al., "The neuronal transporter gene *SLC6A15* confers risk to major depression," *Neuron*, vol. 70, no. 2, pp. 252–265, 2011.

[103] L. Fritschi, D. C. Glass, J. S. Heyworth et al., "Hypotheses for mechanisms linking shiftwork and cancer," *Medical Hypotheses*, vol. 77, no. 3, pp. 430–436, 2011.

[104] S. Khan, M. W. Ullah, R. Siddique et al., "Role of recombinant DNA technology to improve life," *International Journal of Genomics*, vol. 2016, Article ID 2405954, 14 pages, 2016.

[105] S. Xiang, S. B. Coffelt, L. Mao, L. Yuan, Q. Cheng, and S. M. Hill, "Period-2: a tumor suppressor gene in breast cancer," *Journal of Circadian Rhythms*, vol. 6, no. 1, p. 4, 2008.

[106] Y. Zhu, R. G. Stevens, A. E. Hoffman et al., "Epigenetic impact of long-term shiftwork: pilot evidence from circadian genes and whole-genome methylation analysis," *Chronobiology International*, vol. 28, no. 10, pp. 852–861, 2011.

[107] E. Filipski, P. F. Innominato, M. Wu et al., "Effects of light and food schedules on liver and tumor molecular clocks in mice," *Journal of the National Cancer Institute*, vol. 97, no. 7, pp. 507–517, 2005.

[108] J. Samulin Erdem, H. Ø. Notø, Ø. Skare et al., "Mechanisms of breast cancer risk in shift workers: association of telomere shortening with the duration and intensity of night work," *Cancer Medicine*, vol. 6, no. 8, pp. 1988–1997, 2017.

[109] P. A. Wood, X. Yang, A. Taber et al., "*Period 2* mutation accelerates $Apc^{Min/+}$ tumorigenesis," *Molecular Cancer Research*, vol. 6, no. 11, pp. 1786–1793, 2008.

[110] E. E. Flynn-Evans, L. Mucci, R. G. Stevens, and S. W. Lockley, "Shiftwork and prostate-specific antigen in the national health and nutrition examination survey," *Journal of the National Cancer Institute*, vol. 105, no. 17, pp. 1292–1297, 2013.

[111] S. Lee, L. A. Donehower, A. J. Herron, D. D. Moore, and L. Fu, "Disrupting circadian homeostasis of sympathetic signaling promotes tumor development in mice," *PLoS One*, vol. 5, no. 6, article e10995, 2010.

[112] X. Yang, P. A. Wood, E.-Y. Oh, J. Du-Quiton, C. M. Ansell, and W. J. M. Hrushesky, "Down regulation of circadian clock gene period 2 accelerates breast cancer growth by altering its daily growth rhythm," *Breast Cancer Research and Treatment*, vol. 117, no. 2, pp. 423–431, 2009.

[113] S. Gery, N. Komatsu, L. Baldjyan, A. Yu, D. Koo, and H. P. Koeffler, "The circadian gene per1 plays an important role in cell growth and DNA damage control in human cancer cells," *Molecular Cell*, vol. 22, no. 3, pp. 375–382, 2006.

[114] S. B. Reeder, H. H. Hu, and C. B. Sirlin, "Proton density fat-fraction: a standardized MR-based biomarker of tissue fat concentration," *Journal of Magnetic Resonance Imaging*, vol. 36, no. 5, pp. 1011–1014, 2012.

[115] T. Papagiannakopoulos, M. R. Bauer, S. M. Davidson et al., "Circadian rhythm disruption promotes lung tumorigenesis," *Cell Metabolism*, vol. 24, no. 2, pp. 324–331, 2016.

[116] A. Blanch, B. Torrelles, A. Aluja, and J. A. Salinas, "Age and lost working days as a result of an occupational accident: a study in a shiftwork rotation system," *Safety Science*, vol. 47, no. 10, pp. 1359–1363, 2009.

[117] H. Admi, O. Tzischinsky, R. Epstein, P. Herer, and P. Lavie, "Shift work in nursing: is it really a risk factor for nurses'

health and patients' safety?," *Nursing Economics*, vol. 26, no. 4, pp. 250–257, 2008.

[118] F. Tüchsen, K. B. Christensen, and T. Lund, "Shift work and sickness absence," *Occupational Medicine*, vol. 58, no. 4, pp. 302–304, 2008.

[119] M. Takahashi, T. Tanigawa, N. Tachibana et al., "Modifying effects of perceived adaptation to shift work on health, wellbeing, and alertness on the job among nuclear power plant operators," *Industrial Health*, vol. 43, no. 1, pp. 171–178, 2005.

[120] L. Di Milia, P. A. Smith, and S. Folkard, "A validation of the revised circadian type inventory in a working sample," *Personality and Individual Differences*, vol. 39, no. 7, pp. 1293–1305, 2005.

13

Transcriptome and Expression Profiling Analysis of Recalcitrant Tea (*Camellia sinensis* L.) Seeds Sensitive to Dehydration

Xiaofang Jin,[1,2] Dandan Liu,[1] Linlong Ma,[2] Ziming Gong,[2] Dan Cao,[2] Yanli Liu,[2] Yeyun Li,[1] and Changjun Jiang[1]

[1]*State Key Laboratory of Tea Plant Biology and Utilization, Anhui Agricultural University, Hefei, Anhui 230036, China*
[2]*Fruit and Tea Research Institute, Hubei Academy of Agricultural Sciences, Wuhan, Hubei 430064, China*

Correspondence should be addressed to Yeyun Li; lyy@ahau.edu.cn and Changjun Jiang; jiangcj@ahau.edu.cn

Academic Editor: Wilfred van IJcken

The tea plant (*Camellia sinensis* (L.) O. Kuntze) is an economically important woody perennial nonalcoholic health beverage crop. Tea seeds are categorized as recalcitrant and are sensitive to dehydration treatment. However, the molecular basis of this phenomenon has not been investigated. Thus, we analyzed the genome-wide expression profiles of three dehydration stages using RNA-Seq and digital gene expression (DGE) technologies. We performed de novo assembly and obtained a total of 91,925 nonredundant unigenes, of which 58,472 were extensively annotated. By a hierarchical clustering of differentially expressed genes (DEGs), we found that 8929 DEGs were downregulated and 5875 DEGs were upregulated during dehydration treatment. A series of genes related to ABA biosynthesis and signal transduction, transcription factor, antioxidant enzyme, LEA protein, and proline metabolism that have been reported to function in dehydration process were found to be downregulated. Additionally, the expression profiles of 12 selected genes related to tea seed dehydration treatment were confirmed by qRT-PCR analysis. To our knowledge, this is the first genome-wide study elucidating the possible molecular mechanisms of sensitivity of recalcitrant tea seeds to dehydration. The results obtained in this study contribute to the preservation of tea seeds as genetic resources and can also be used to explore the mechanism of dehydration sensitivity of other recalcitrant seeds.

1. Introduction

The tea plant (*Camellia sinensis* (L.) O. Kuntze) is native to southwestern China and is an important economic woody perennial crop that is mainly grown in tropical and subtropical regions [1]. Tea is a widely consumed beverage in the world owing to its health benefits [2]. Currently, tea genetic resources are conventionally preserved in germplasm repositories, but are susceptible to potential losses due to diseases, pests, climatic hazards, and so on. Therefore, it is important to preserve tea genetic resources by seed storage for future experimentation. However, many studies have showed that tea seeds are recalcitrant [3–5].

To develop methods to preserve recalcitrant seeds, it is essential to understand the mechanism of sensitivity of these seeds to dehydration treatment. When recalcitrant seeds sense dehydration stress, a series of changes occur, including resetting of the cellular framework, alteration of the structure, composition and function of the plasma membrane, and accumulation of reactive oxygen species (ROS) [6–10]. Although we have already made some progress in elucidating the mechanism of dehydration sensitivity of some recalcitrant seeds, most of it, especially the molecular mechanism, remains unknown. Meanwhile, recalcitrant seeds of different species display a wide range of variability in sensitivity to dehydration [4, 11].

Only a few studies have been conducted to explore the potential mechanisms contributing to dehydration sensitivity of recalcitrant tea seeds. A previous study showed that when tea seeds were subjected to dehydration treatment, there was a dramatic induction of H_2O_2 accumulation in tea embryos. Such high levels of ROS were found to be detrimental to seed viability [7]. Some studies have elucidated the molecular mechanism of the response of the tea plant to dehydration stress and identified a series of dehydration-responsive genes [12, 13]. However, the molecular mechanisms of response to dehydration stress in tea plant are tissue-specific, and little is known about this mechanism in tea seeds, especially at the genome-wide transcriptional level.

Recently, rapid advances in next-generation sequencing technologies have proved to be a cost-effective method for transcriptomic studies [14, 15]. Therefore, the purpose of this study was to identify dehydration-responsive genes and elaborate on the mechanisms of dehydration sensitivity in recalcitrant tea seeds by RNA-Seq and DGE studies. The results obtained in this study not only allow us a deeper insight into the molecular mechanisms of sensitivity of recalcitrant tea seeds to dehydration, but will also be useful for preserving tea genetic resources by seed storage for a long time.

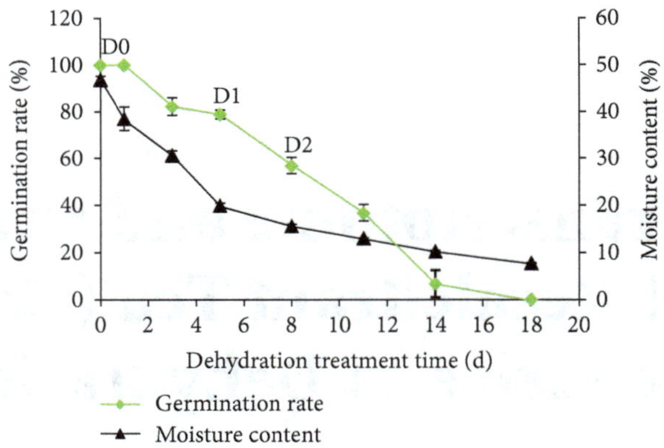

FIGURE 1: Changes in germination rate and moisture content expressed as fresh weight of recalcitrant tea seeds during dehydration treatment.

2. Materials and Methods

2.1. Plant Materials and Seed Dehydration Treatments. Mature seeds with brown pericarp were collected from the tea plant cultivar "*C. sinensis* cv. Echa 1" grown in Wuhan, Central China (114°07′E, 30°18′N). Of these, only healthy seeds were selected for use in our experiments. The pericarps of the seeds were removed and placed in an electrothermal blowing dry box (25°C) to allow dehydration for 0, 1, 3, 5, 8, 11, 14, and 18 d. The moisture contents of the seeds were determined gravimetrically by oven drying at 103°C for 17 h [16]. Three replicates of 30 seeds each were used to determine the seed moisture content measured on a fresh mass basis.

2.2. Seed Germination Tests. Seed germination test was carried out in an artificial climate chest (25°C, 500 lx, 16 h light/8 h dark) after placing the seeds in 15 cm diameter petri dishes containing wet filter paper [5, 7]. At each sampling point, three replicates of 50 seeds each were used for the germination test. Germination was defined as the appearance of a radical of at least 5 mm in length [5]. Based on seed germination rates, samples at three stages, that is, fresh seeds (CK, D0) and partially dehydrated (D1 and D2), were selected for RNA-Seq and DGE analyses. Three biological replicates of 50 seed embryos each were quickly frozen in liquid nitrogen and stored at −80°C for RNA extraction.

2.3. RNA Extraction. RNAprep pure Plant Kit (Tiangen, Beijing, China) was used for RNA extraction. The concentration, integrity, and purity of RNA were assessed using Agilent 2100 Bioanalyzer (Agilent Technologies, Santa Clara, CA, USA) and NanoDrop ND-2000 spectrophotometer (NanoDrop Technologies, Wilmington, DE, USA). All RNA samples (including a pooled RNA sample for transcriptome sequencing and RNA preparations from D0, D1, and D2 samples for DGE sequencing) with A260/A280 ratios ranging from 1.9 to 2.2 and RNA integrity number (RIN) values greater than 8 were chosen to construct cDNA libraries and quantitative real-time PCR (qRT-PCR) analyses.

2.4. Library Construction and Quality Control. Following DNase I treatment, mRNA was enriched by oligo(dT) magnetic beads. The mRNA mixed with fragmentation buffer was sheared into short fragments, which were used as templates for cDNA synthesis. Double-stranded cDNA was purified using magnetic beads and resuspended in EB buffer for end repair and addition of poly-(A) tails. Finally, sequencing adapters were ligated to the fragments, and suitable fragments were selected as templates for PCR amplification. During the quality control step, Agilent 2100 Bioanaylzer was used to check the quality and to quantify the sample libraries. The libraries were sequenced using Illumina HiSeq™ 2500 (Illumina, San Diego, USA) to generate raw data. These data are available at the NCBI sequence read archive (SRA) under the accession number SRP096975. Raw reads were preprocessed to remove reads with adapters, reads containing more than 10% unknown bases, and low-quality reads (>50% of the bases with a quality score of ≤5).

2.5. De Novo Assembly and Functional Annotation. The filtered reads were de novo assembled into nonredundant unigenes by the Trinity software [17] and TGICL software [18]. To obtain information on the functional annotation, all unigenes were used to search against the nonredundant database (NR), the nucleotide database (NT), the Swiss-Prot protein sequence database (Swiss-Prot), the cluster of orthologous group database (COG), the Kyoto encyclopedia of genes and genomes database (KEGG) by the BLAST algorithm, and the InterPro protein families database (InterPro) with InterProScan5 (E value $\leq 10^{-5}$). Proteins with highest sequence similarity to given unigenes were retrieved. Based on NR annotations, gene ontology (GO) annotations were assigned to unigenes using Blast2GO software [19].

2.6. Protein Coding Region Prediction. To predict protein coding sequences (CDSs), unigenes were first aligned by BLASTx (E value $\leq 10^{-5}$) to protein databases in the priority order of NR, Swiss-Prot, KEGG, and COG. Unigenes aligned to a higher priority database were not aligned to a subsequent lower priority database. The coding region sequences of unigenes were determined based on proteins with highest ranks in the BLAST results. These sequences were then translated into amino acid sequences using the standard codon table. When unigenes could not be found in any of the above

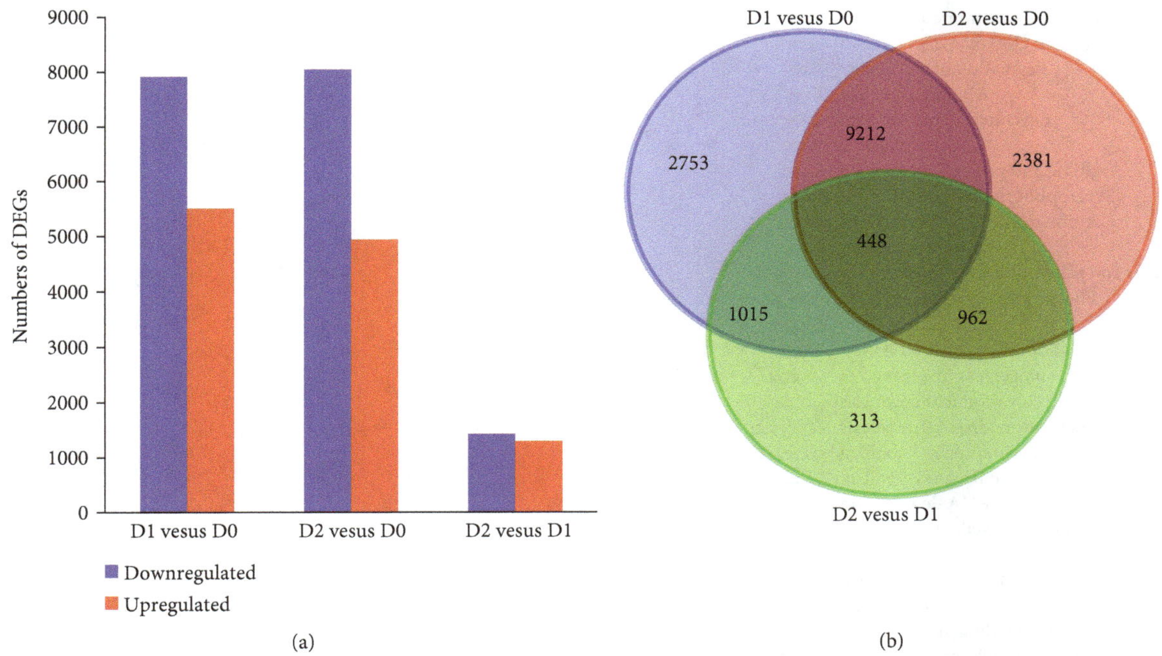

FIGURE 2: Overview of significantly differentially expressed genes (DEGs) in recalcitrant tea seeds during dehydration treatment. (a) The numbers of downregulated and upregulated DEGs. (b) Venn diagram for analysis of DEGs.

databases, ESTScan [20] was used to decipher their nucleotide and amino acid sequences.

2.7. Read Mapping to the Reference Transcriptome and Differential Expression Analysis. Clean reads and count number of DGE libraries were assessed and summarized using custom BioPerl scripts, and the reads were mapped back onto the assembled transcriptome generated by RNA-Seq. The expression levels of the unigenes were quantified by RSEM software [21] and calculated as FPKM [22]. Significantly differentially expressed unigenes with false discovery rate (FDR) $\leq 10^{-3}$, E values $\leq 10^{-5}$, and $|\log 2\,\text{ratio}| \geq 1$ were identified among D0, D1, and D2 libraries. The cluster analysis of DEGs was performed by using Cluster 3.0 software [23]. To further clarify the biological functions of DEGs, GO and KEGG pathway enrichment analyses were conducted by the Cytoscape software [24] and KOBAS software [25], respectively.

2.8. Heatmap and Transcription Factor Analysis. The heatmap was plotted using OmicShare tools, a free online platform for data analysis (http://www.omic-share.com/tools). Transcription factors (TFs) were identified and classified based on the PlantTFDB 3.0 database [26].

2.9. Quantitative Real-Time PCR Validation. To validate the expression profiles observed in RNA-Seq data, 12 significantly differentially expressed genes (DEGs) were selected for qRT-PCR analyses using SYBR Premix Ex Taq™ II Kit (Takara, Dalian, China) in a Bio-Rad CFX96 real-time PCR system (Bio-Rad, CA, USA). Glyceraldehyde-3-phosphate dehydrogenase (GAPDH) gene was used as an internal control. The primers used in qRT-PCR analyses are listed in Table S1. The relative expression values were calculated by using the $2^{-\Delta\Delta Ct}$ method [27]. Three biological replicates were performed for each experiment.

3. Results and Discussion

3.1. Germination Changes of Tea Seeds during Dehydration Treatment. In this study, we found that tea seeds at the time of harvest had a moisture content of 46.8% (D0, 0d) and a germination rate of 100% (Figure 1). With continued dehydration, the germination rate of the seeds decreased gradually. When dehydrated to 19.9% (D1, 5d), that is, less than half of the initial moisture content, the germination rate was nearly 80%. However, further dehydration to 15.6% (D2, 8d) sharply reduced the germination rate to 56.7%. When the seeds were further dehydrated, reducing the moisture content to 7.7%, the seeds permanently lost their viability. These results demonstrated that tea seeds are highly sensitive to dehydration treatment, maintaining high viability at 20% moisture content, with a moisture content of about 16% leading to a serious injury. Therefore, to evaluate the transcriptomic response of recalcitrant tea seeds during dehydration treatment, we selected D0, D1, and D2 for RNA-Seq and DGE studies.

3.2. Analysis of the Transcriptome. RNAs isolated from D0, D1, and D2 samples were mixed in equal amounts to construct a broad cDNA library using the Illumina HiSeq 2500 genome analyzer. Overviews of sequencing and de novo assembly results (Table S2), a total of 100,628,270 clean reads were obtained. These high-quality trimmed reads were then assembled de novo into 91,925 nonredundant unigenes with an average length of 854 bp and an N50

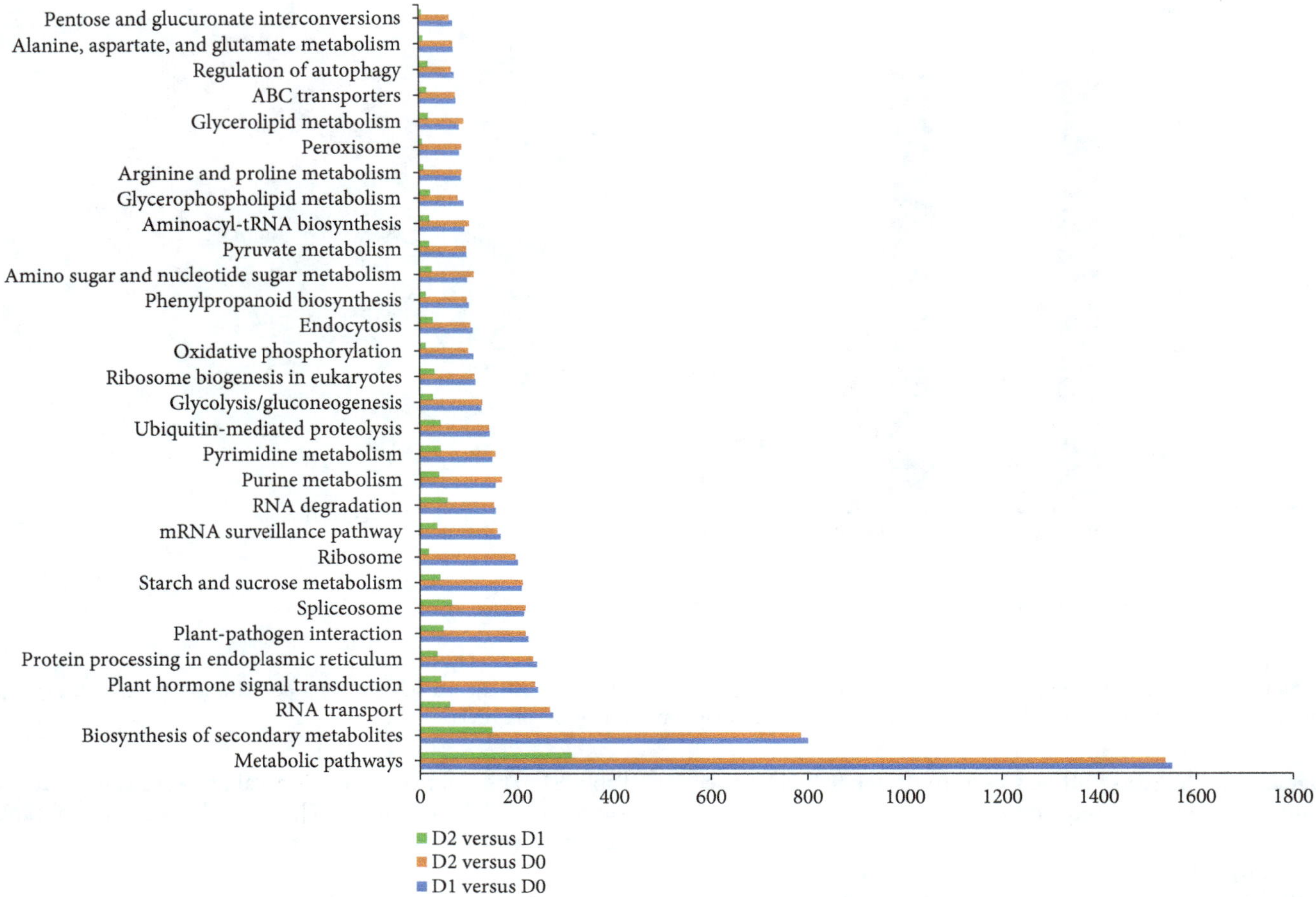

FIGURE 3: 30 main KEGG pathways based on significantly differentially expressed genes (DEGs) in recalcitrant tea seeds during dehydration treatment. The x-axis indicates the number of DEGs in each pathway, and the y-axis indicates KEGG pathway categories.

length of 1480 bp. All unigenes were longer than 300 bp, and 28,176 (30.65%) of them were longer than 1000 bp (Figure S1). A total of 48,214 (52.45%), 51,917 (56.48%), 32,028 (34.84%), 17,055 (18.55%), 33,807 (36.78%), 9535 (10.37%), and 28,088 (30.56%) unigenes had significant hits (E value $\leq 10^{-5}$) in NR, NT, Swiss-Prot, COG, InterPro, GO, and KEGG, respectively. About half of all nonredundant unigenes had significant homology with genes in NR and NT. Of the 91,925 high-quality unique sequences, 48,630 (52.90%) unigenes significantly matched known proteins in at least one of the five databases, and 12,678 (13.79%) unigenes showed similarity to proteins in all five databases (Figure S2). Based on the BLASTx protein database searches described above, we could identify 47,449 unigenes containing CDSs with an average length of 806 bp and an N50 length of 1242 bp. Of the unigenes with CDSs, 29,012 (61.14%) were longer than 500 bp, 15,153 (31.94%) were longer than 1000 bp, and 3810 were longer than 2000 bp (Figure S3). Using the ESTScan program, we were further able to assign another 2851 unigene CDSs that could not be aligned to the above databases.

3.3. DGE Analysis among the Three Stages of Dehydration. To elucidate the molecular mechanisms of the response of recalcitrant tea seeds to dehydration treatment, DGE analysis was performed to determine DEGs. Approximately 12.9 million clean reads were obtained in each library. Gene annotations were carried out by mapping clean reads to the 91,925 nonredundant unigenes from the transcriptome. About 87.3–88.4% reads in all DGE libraries were mapped to the global transcriptome, suggesting that the transcriptome was a reliable reference. Upon screening of the DEGs, we observed that the level of gene expression changed with the duration of the dehydration treatment process. As shown in the histogram, the proportion of downregulated DEGs was higher than that of the upregulated DEGs among all three dehydration stages (Figure 2(a)). In the D1 versus D0 library, we observed a total of 7917 downregulated DEGs and 5511 upregulated DEGs; in the D2 versus D0 library, we found a total 8056 downregulated DEGs and 4947 upregulated DEGs; in the D2 versus D1 library, we found a total of 1438 downregulated DEGs and 1300 upregulated DEGs. We then made a hierarchical clustering to determine the profiles of the DEGs among the three seed dehydration stages (Figure S4). The results revealed that 482 DEGs were initially downregulated in D1 and were then upregulated in D2, while the opposite trend was observed for 525 DEGs. However, 8929 DEGs and 5875

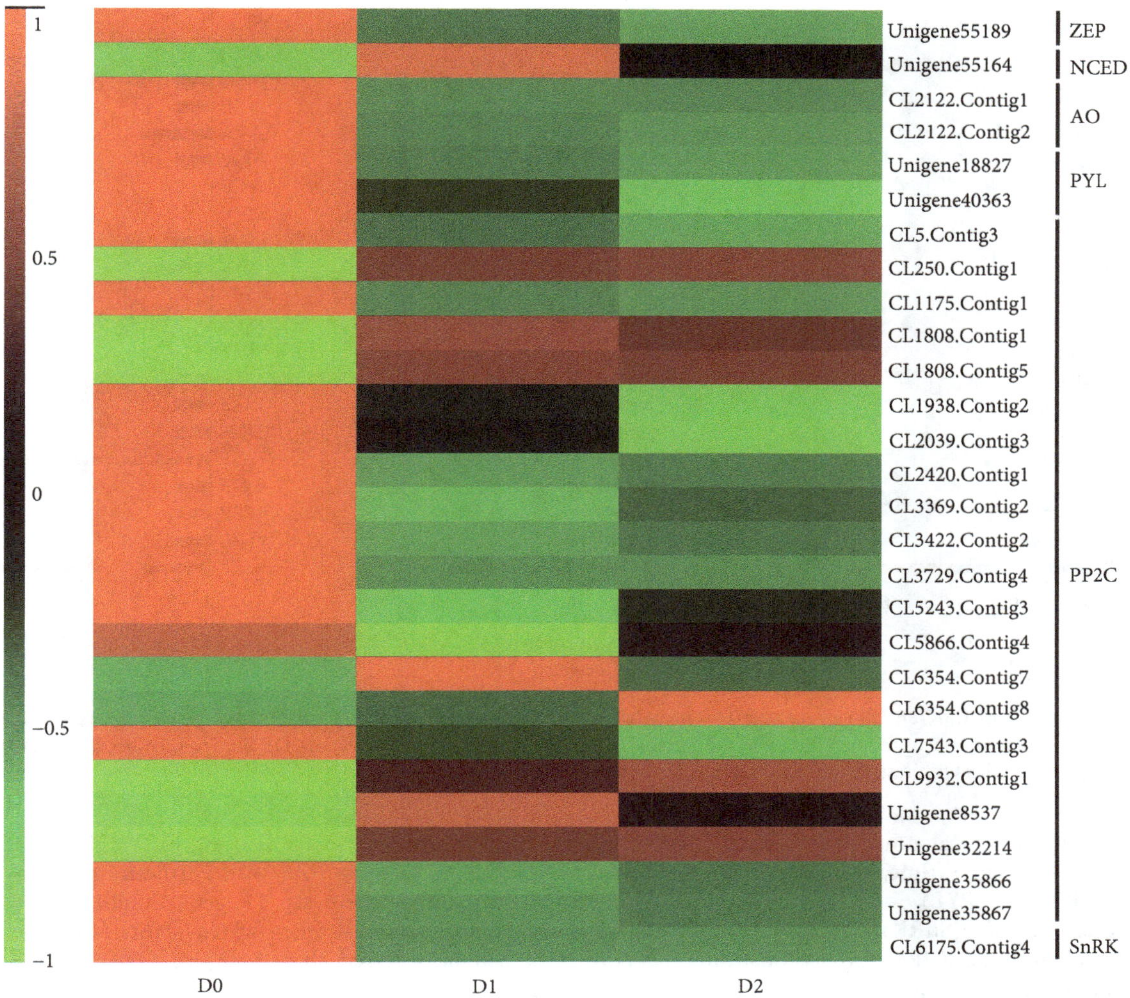

FIGURE 4: Heatmap representing relative expression levels of differentially expressed genes (DEGs) related to ABA biosynthesis and signal transduction. The numbers in the scale bar show the Z-score standardization of each DEG.

DEGs were downregulated and upregulated, respectively, in all dehydration treatment stages. Concurrently, by Venn diagram analysis, we found that only 448 DEGs overlapped among the three comparisons (Figure 2(b)).

3.4. GO and KEGG Pathway Enrichment Analyses of DEGs. Upon GO enrichment analysis, we found that several crucial biological processes, such as carbohydrate metabolic process, amino acid metabolic process, and coenzyme metabolic process, were significantly ($p \leq 0.05$) enriched (Table S3). With KEGG pathway enrichment analysis [28], 127, 127, and 113 KEGG pathways were identified in D1 versus D0, D2 versus D0, and D2 versus D1 libraries, respectively. Among these KEGG pathways, the largest groups were metabolic pathways, biosynthesis of secondary metabolites, RNA transport, and plant hormone signal transduction (Figure 3). Further, we found that a total of 14, 17, and 7 KEGG pathways were significantly ($p \leq 0.05$) enriched in D1 versus D0, D2 versus D0, and D2 versus D1 libraries, respectively (Table S4). Several key pathways, such as arginine and proline metabolism (ko00330), fatty acid biosynthesis (ko00061), inositol phosphate metabolism (ko00562), carotenoid biosynthesis (ko00906), and basal transcription factors (ko03022), were involved and function in the response to dehydration treatment of recalcitrant tea seeds. These results imply that genes involved in these biological processes and metabolic pathways may play crucial roles in the dehydration sensitivity of recalcitrant tea seeds.

3.5. Abscisic Acid Biosynthesis and Signal Transduction Genes Responding to Seed Dehydration Treatment. Abscisic acid (ABA) is an important hormone that plays vital roles in adaptive responses to various environmental stresses [29] and has been also verified as a vital factor in seed dehydration tolerance [30, 31]. In our study, one gene encoding zeaxanthin epoxidase (ZEP), one gene encoding 9-cis-epoxycarotenoid dioxygenase (NCED), and two genes encoding aldehyde oxidase (AO) related to the ABA biosynthesis pathway were identified to be differentially expressed (Figure 4 and Table S5). Although *NCED* was upregulated, the downregulation of *ZEP* and *AO* might restrain ABA

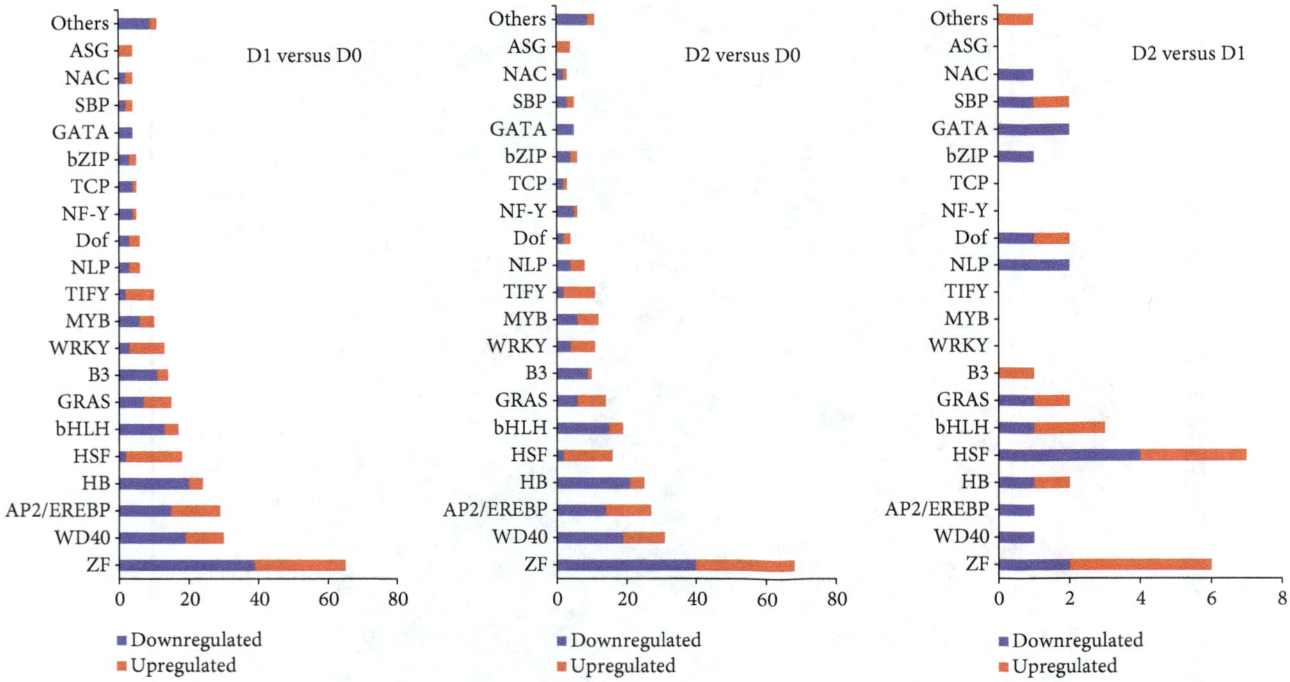

FIGURE 5: Distribution of differentially expressed TFs in recalcitrant tea seeds during dehydration treatment. The histograms show the number of downregulated and upregulated TFs in D1 versus D0, D2 versus D0, and D2 versus D1 libraries, respectively. Details are not shown for TF gene families with less than four unigenes.

biosynthesis during recalcitrant tea seed dehydration treatment. At the same time, we identified two genes encoding PYR1-like (PYL), one gene encoding SNF1-related protein kinase (SnRK), and 21 genes encoding protein phosphatase 2C (PP2C) that constitute the core regulatory network of ABA signaling pathway (Figure 4 and Table S5) and can activate a series of transcription factors (e.g., bZIP) to cope with dehydration stress [32]. Interestingly, we found that most or all of these three genes (*PYL*, *PP2C*, and *SnRK*) were also downregulated. Thus, we speculate that the repression of genes encoding ABA synthesis and signaling is likely to lead to the dehydration sensitivity of recalcitrant tea seeds.

3.6. Transcription Factors Responding to Seed Dehydration Treatment. Many studies have demonstrated that transcription factors play important functions in tolerance to abiotic stresses [33–35]. In our study, a total of 299 (171 downregulated and 128 upregulated), 299 (174 downregulated and 125 upregulated), and 34 (19 downregulated and 15 upregulated) genes encoding putative transcription factors were identified in D1 versus D0, D2 versus D0, and D2 versus D1 libraries, respectively ($|\log 2 \text{ ratio}| \geq 1$), with the number of downregulated TFs being higher. These TFs could be divided into 29 gene families based on their putative DNA-binding and kinase domains (Figure 5); most of the members of the 13 TF gene families (zinc finger protein (ZF), WD40, AP2/EREBP, HB, bHLH, B3, MYB, NLP, NF-Y, TCP, bZIP, GATA, and NAC) were downregulated. Among these, at least six TF gene families have been reported to be linked to positive regulation of the responses to dehydration stress and comprise AP2/EREBP [36], bZIP [37], HB [38], MYB [39], NAC [35], and ZF [40]. Furthermore, it has also been reported that these six TF gene families are involved in drought stress resistance of tea plants [13]. Consequently, it can be inferred that these six TF gene families might play vital roles in the response to dehydration sensitivity of recalcitrant tea seeds. In addition, a number of previously published reports showed that HSF and WRKY gene family members can improve the tolerance to drought stress in tea plants [13, 41, 42]. In the present study, we found most of the members of the HSF and WRKY gene families to be upregulated (Figure 5), and we speculated that the upregulated genes of these two TF gene families might help increase the tolerance to dehydration treatment in recalcitrant tea seeds.

3.7. Antioxidant Enzymes and Osmoprotectant Metabolism Genes Responding to Seed Dehydration Treatment. During dehydration treatment of recalcitrant seeds, the plasma membrane is believed to be a primary site of injury by ROS (e.g., hydrogen peroxide (H_2O_2) and superoxide anion ($^{\bullet}O_2^{-}$)), resulting in loss of viability [16, 43]. Several studies showed that antioxidant system comprising of enzymatic and nonenzymatic (e.g., ascorbic acid (AsA) and glutathione (GSH)) components plays an important role in scavenging excess of ROS [7, 9, 44]. The enzymatic antioxidants include catalase (CAT), glutathione peroxidase (GPx), peroxidase (POD), superoxide dismutase (SOD), and enzymes of AsA-GSH cycle (such as ascorbate peroxidase (APX), dehydroascorbate reductase (DHAR), and glutathione reductase (GR)). In our transcriptome analysis, 2 gene encoding

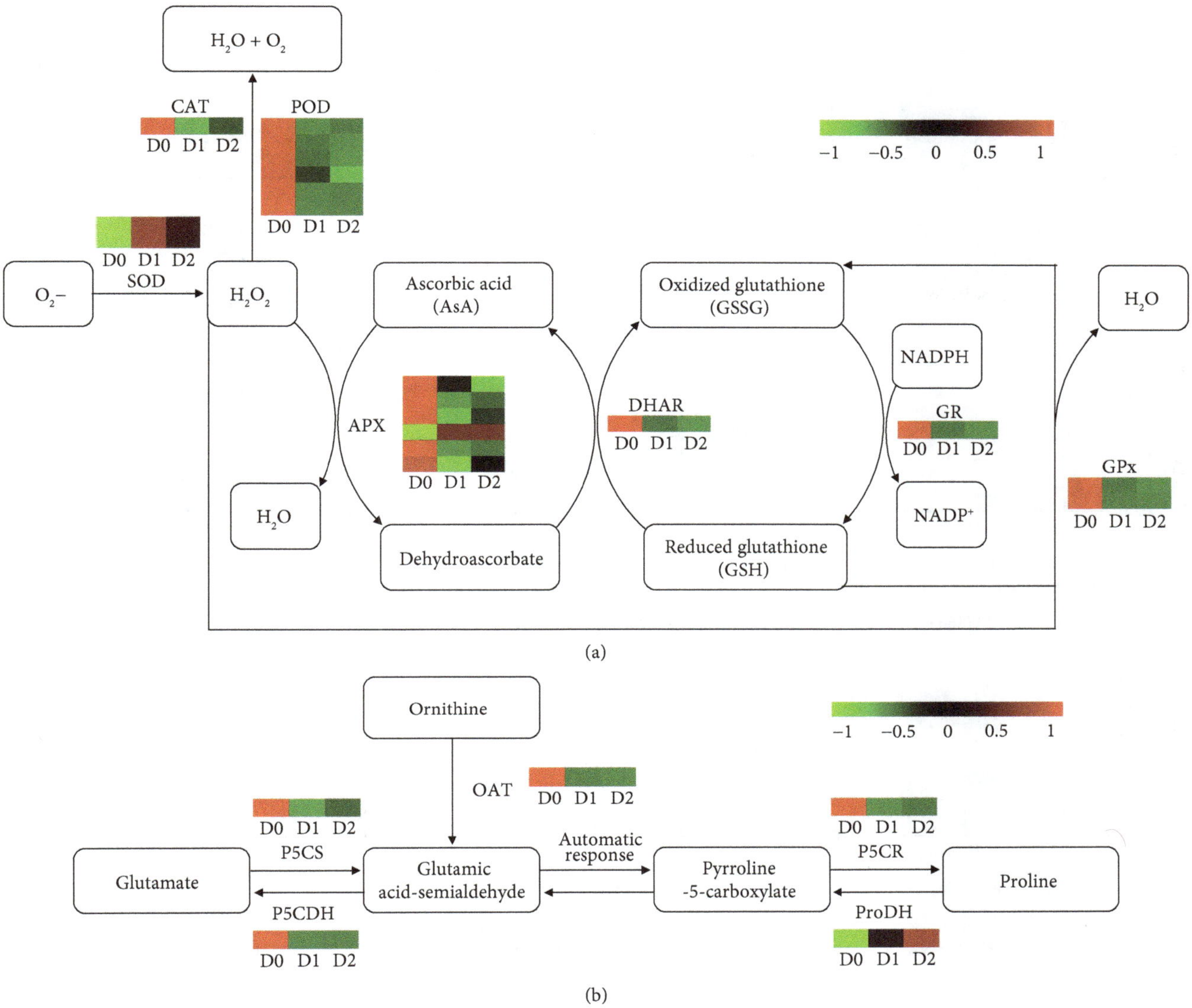

FIGURE 6: Heatmap representing relative expression levels of differentially expressed genes (DEGs) related to antioxidant enzymes (a) and proline metabolism (b). The numbers in the scale bar show the Z-score standardization of each DEG.

SOD, 1 gene encoding CAT, 6 gene encoding POD, 6 gene encoding APX, 1 gene encoding DHAR, 1 gene encoding GR, and 2 gene encoding GPx were identified (Figure 6(a) and Table S6). Most or all members of the six genes (APX, CAT, DHAR, GPx, GR, and POD) were downregulated, while the opposite trend was observed only for SOD gene. Although all members of *SOD* were upregulated, it only can scavenge $^{\bullet}O_2^-$, thereby resulting in the production of H_2O_2 [45, 46]. Therefore, we speculate that the dehydration sensitivity of recalcitrant tea seeds can be related to an accumulation of ROS due to decreased activities of these antioxidant enzymes.

Many previous studies showed that an accumulation of LEA proteins is correlated with gaining seed dehydration tolerance [8, 47]. In our study, all members of LEA proteins were found to be downregulated (Table S6). Therefore, we can speculate that reduced expression of genes encoding LEA proteins results in dehydration-intolerant tea seeds. In addition, it was also reported that accumulation of proline as an osmoprotectant has a positive influence in response to drought stress in tea plant [13]. In this study, one gene encoding pyrroline-5-carboxylate synthetase (P5CS), one gene encoding ornithine aminotransferase (OAT), one gene encoding pyrroline-5-carboxylate reductase (P5CR), one gene encoding proline dehydrogenase (ProDH), and one gene encoding pyrroline-5-carboxylate dehydrogenase (P5CDH) related to proline metabolism were also identified (Figure 6(b) and Table S6). All members of *P5CS*, *OAT*, *P5CR*, and *P5CDH* were downregulated, while those of *ProDH* were upregulated. Therefore, we inferred that proline may not play an important protective role in the response to dehydration treatment of recalcitrant tea seeds.

3.8. qRT-PCR Validation of DEGs. To validate the reliability of our RNA-Seq results, 12 DEGs related to seed dehydration treatment were selected for qRT-PCR analysis. These DEGs

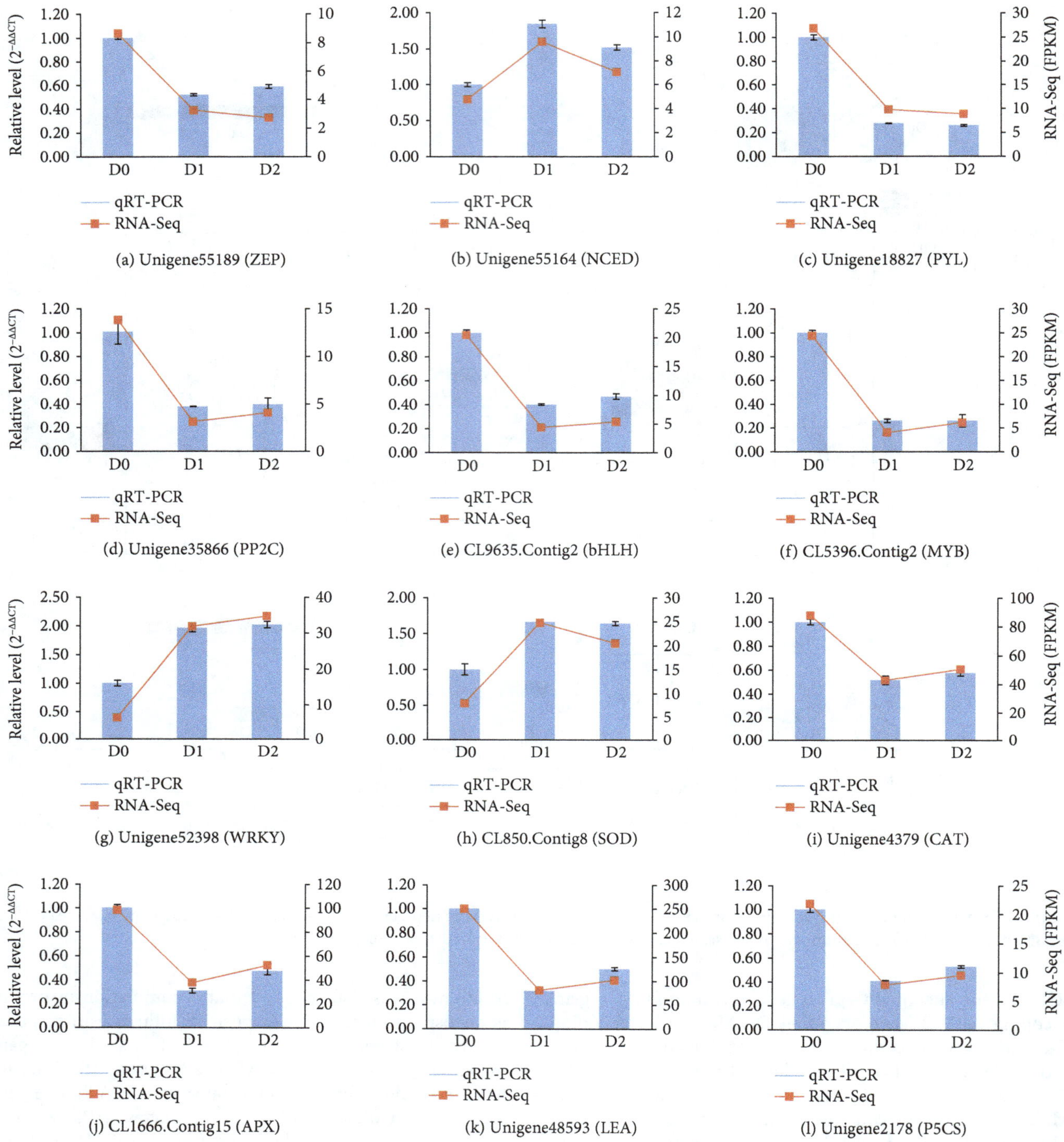

Figure 7: qRT-PCR validation of 12 selected differentially expressed genes (DEGs) responding to tea seed dehydration treatment. Validation of RNA-Seq results using qRT-PCR; GAPDH gene was chosen as the reference gene.

are involved in ABA biosynthesis and signal transduction, antioxidant enzyme, LEA protein, and so on. The results of qRT-PCR revealed that the expression trends of each of these DEGs were similar to that from the RNA-Seq results ($R^2 = 0.85$). Detailed comparisons between qRT-PCR and RNA-Seq results are shown in Figure 7.

4. Conclusion

In conclusion, to the best of our knowledge, this study is the first exploration of the global transcriptome profiles of dehydration sensitivity of recalcitrant tea seeds using RNA-Seq and DGE technologies. In our study, a total of 91,925 nonredundant unigenes were generated by de novo assembly and were extensively annotated, which were then used as the reference database for identification of DEGs. Many candidate DEGs involved in the response to dehydration treatment in recalcitrant tea seeds were identified. A series of genes related to ABA biosynthesis and signal transduction, antioxidant enzyme, LEA protein, and so on reported to function in dehydration process were downregulated, and these genes might play vital roles in the dehydration sensitivity of recalcitrant tea seeds. Therefore, our study provides insights into the molecular mechanisms of dehydration sensitivity in recalcitrant tea seeds. This knowledge is not only useful for the preservation of seeds as tea genetic resources for a long time, but also can be used to explore the mechanism of sensitivity to dehydration of other recalcitrant seeds.

Conflicts of Interest

The authors declare that they have no conflict of interest.

Authors' Contributions

Xiaofang Jin performed the experiments, analyzed the data, and wrote the paper; Dandan Liu, Ziming Gong, and Yanli Liu contributed the materials; Linlong Ma and Dan Cao carried out the seed germination tests; Yeyun Li and Changjun Jiang designed the experiments and revised the manuscript.

Acknowledgments

This work was supported by the National Natural Science Foundation of China (31270729) and the Agricultural Science and Technology Innovation Project of Hubei Province (2016-620-000-001-032).

Supplementary Materials

Supplementary 1. Figure S1: length distribution of unigenes.

Supplementary 2. Figure S2: venn diagram showing a search results against five databases.

Supplementary 3. Figure S3: CDSs predicted by BLASTx and ESTScan.

Supplementary 4. Figure S4: heatmap of differentially expressed genes (DEGs).

Supplementary 5. Table S1: primers used for qRT-PCR analysis of 12 genes.

Supplementary 6. Table S2: overviews of sequencing and de novo assembly.

Supplementary 7. Table S3: significantly enriched GO terms of DEGs.

Supplementary 8. Table S4: significantly enriched KEGG pathways of DEGs.

Supplementary 9. Table S5: DEGs related to ABA biosynthesis and signal transduction.

Supplementary 10. Table S6: DEGs related to antioxidant enzymes and osmoprotectant metabolism.

References

[1] L. Chen, Q. K. Gao, D. M. Chen, and C. J. Xu, "The use of RAPD markers for detecting genetic diversity, relationship and molecular identification of Chinese elite tea genetic resources [*Camellia sinensis* (L.) O. Kuntze] preserved in a tea germplasm repository," *Biodiversity & Conservation*, vol. 14, no. 6, pp. 1433–1444, 2005.

[2] M. Iriti, S. Vitalini, G. Fico, and F. Faoro, "Neuroprotective herbs and foods from different traditional medicines and diets," *Molecules*, vol. 15, no. 5, pp. 3517–3555, 2010.

[3] P. Berjak, C. W. Vertucci, and N. W. Pammenter, "Effects of developmental status and dehydration rate on characteristics of water and desiccation-sensitivity in recalcitrant seeds of *Camellia sinensis*," *Seed Science Research*, vol. 3, no. 3, pp. 155–166, 1993.

[4] K. P. S. Chandel, R. Chaudhury, J. Radhamani, and S. K. Malik, "Desiccation and freezing sensitivity in recalcitrant seeds of tea, cocoa and jackfruit," *Annals of Botany*, vol. 76, no. 5, pp. 443–450, 1995.

[5] Z. X. Chen, Q. Y. Lan, L. Zheng, Y. X. Bao, and Y. L. Luo, "Effects of dehydration and temperature on seed viability and antioxidative enzymes activities on three kinds of cultivars of *Camellia sinensis*," *Brazilian Journal of Botany*, vol. 38, no. 3, pp. 497–504, 2015.

[6] P. Berjak and N. W. Pammenter, "Implications of the lack of desiccation tolerance in recalcitrant seeds," *Frontiers in Plant Science*, vol. 4, p. 478, 2013.

[7] Q. Chen, L. Yang, P. Ahmad, X. Wan, and X. Hu, "Proteomic profiling and redox status alteration of recalcitrant tea (*Camellia sinensis*) seed in response to desiccation," *Planta*, vol. 233, no. 3, pp. 583–592, 2011.

[8] J. Delahaie, M. Hundertmark, J. Bove, O. Leprince, H. Rogniaux, and J. Buitink, "LEA polypeptide profiling of recalcitrant and orthodox legume seeds reveals ABI3-regulated LEA protein abundance linked to desiccation tolerance," *Journal of Experimental Botany*, vol. 64, no. 14, pp. 4559–4573, 2013.

[9] A. Francini, L. Galleschi, F. Saviozzi et al., "Enzymatic and non-enzymatic protective mechanisms in recalcitrant seeds of *Araucaria bidwillii* subjected to desiccation," *Plant Physiology and Biochemistry*, vol. 44, no. 10, pp. 556–563, 2006.

[10] V. Greggains, W. E. Finch-Savage, N. M. Atherton, and P. Berjak, "Viability loss and free radical processes during desiccation of recalcitrant *Avicennia marina* seeds," *Seed Science Research*, vol. 11, no. 3, pp. 235–242, 2001.

[11] R. Umarani, E. K. Aadhavan, and M. M. Faisal, "Understanding poor storage potential of recalcitrant seeds," *Current Science*, vol. 108, no. 11, pp. 2023–2034, 2015.

[12] S. Gupta, R. Bharalee, P. Bhorali et al., "Molecular analysis of drought tolerance in tea by cDNA-AFLP based transcript profiling," *Molecular Biotechnology*, vol. 53, no. 3, pp. 237–248, 2013.

[13] S. C. Liu, J. Q. Jin, J. Q. Ma et al., "Transcriptomic analysis of tea plant responding to drought stress and recovery," *PLoS One*, vol. 11, no. 1, article e0147306, 2016.

[14] B. J. Haas and M. C. Zody, "Advancing RNA-Seq analysis," *Nature Biotechnology*, vol. 28, no. 5, pp. 421–423, 2010.

[15] Z. Wang, M. Gerstein, and M. Snyder, "RNA-Seq: a revolutionary tool for transcriptomics," *Nature Reviews Genetics*, vol. 10, no. 1, pp. 57–63, 2009.

[16] T. Roach, R. P. Beckett, F. V. Minibayeva et al., "Extracellular superoxide production, viability and redox poise in response to desiccation in recalcitrant *Castanea sativa* seeds," *Plant, Cell & Environment*, vol. 33, no. 1, pp. 59–75, 2010.

[17] M. G. Grabherr, B. J. Haas, M. Yassour et al., "Full-length transcriptome assembly from RNA-Seq data without a reference genome," *Nature Biotechnology*, vol. 29, no. 7, pp. 644–652, 2011.

[18] G. Pertea, X. Huang, F. Liang et al., "TIGR gene indices clustering tools (TGICL): a software system for fast clustering of large EST datasets," *Bioinformatics*, vol. 19, no. 5, pp. 651-652, 2003.

[19] A. Conesa, S. Götz, J. M. García-Gómez, J. Terol, M. Talón, and M. Robles, "Blast2GO: a universal tool for annotation, visualization and analysis in functional genomics research," *Bioinformatics*, vol. 21, no. 18, pp. 3674–3676, 2005.

[20] C. Iseli, C. V. Jongeneel, and P. Bucher, "ESTScan: a program for detecting, evaluating, and reconstructing potential coding regions in EST sequences," in *Proceedings: International Conference on Intelligent Systems for Molecular Biology*, pp. 138–148, Heidelberg, Germany, August 1999.

[21] B. Li and C. N. Dewey, "RSEM: accurate transcript quantification from RNA-Seq data with or without a reference genome," *BMC Bioinformatics*, vol. 12, no. 1, p. 323, 2011.

[22] A. Mortazavi, B. A. Williams, K. McCue, L. Schaeffer, and B. Wold, "Mapping and quantifying mammalian transcriptomes by RNA-Seq," *Nature Methods*, vol. 5, no. 7, pp. 621–628, 2008.

[23] M. B. Eisen, P. T. Spellman, P. O. Brown, and D. Botstein, "Cluster analysis and display of genome-wide expression patterns," *Proceedings of the National Academy of Sciences of the United States of America*, vol. 95, no. 25, pp. 14863–14868, 1998.

[24] S. Maere, K. Heymans, and M. Kuiper, "*BiNGO*: a Cytoscape plugin to assess overrepresentation of gene ontology categories in biological networks," *Bioinformatics*, vol. 21, no. 16, pp. 3448-3449, 2005.

[25] X. Mao, T. Cai, J. G. Olyarchuk, and L. Wei, "Automated genome annotation and pathway identification using the KEGG Orthology (KO) as a controlled vocabulary," *Bioinformatics*, vol. 21, no. 19, pp. 3787–3793, 2005.

[26] J. Jin, H. Zhang, L. Kong, G. Gao, and J. Luo, "PlantTFDB 3.0: a portal for the functional and evolutionary study of plant transcription factors," *Nucleic Acids Research*, vol. 42, no. D1, pp. D1182–D1187, 2014.

[27] K. J. Livak and T. D. Schmittgen, "Analysis of relative gene expression data using real-time quantitative PCR and the $2^{-\Delta\Delta C}{}_{T}$ method," *Methods*, vol. 25, no. 4, pp. 402–408, 2001.

[28] M. Kanehisa, S. Goto, Y. Sato, M. Furumichi, and M. Tanabe, "KEGG for integration and interpretation of large-scale molecular data sets," *Nucleic Acids Research*, vol. 40, no. D1, pp. D109–D114, 2012.

[29] F. Hauser, R. Waadt, and J. I. Schroeder, "Evolution of abscisic acid synthesis and signaling mechanisms," *Current Biology*, vol. 21, no. 9, pp. R346–R355, 2011.

[30] A. Khandelwal, S. H. Cho, H. Marella et al., "Role of ABA and ABI3 in desiccation tolerance," *Science*, vol. 327, no. 5965, p. 546, 2010.

[31] S. H. Son, V. R. Chitnis, A. Liu, F. Gao, T.-N. Nguyen, and B. T. Ayele, "Abscisic acid metabolic genes of wheat (*Triticum aestivum* L.): identification and insights into their functionality in seed dormancy and dehydration tolerance," *Planta*, vol. 244, no. 2, pp. 429–447, 2016.

[32] T. Umezawa, K. Nakashima, T. Miyakawa et al., "Molecular basis of the core regulatory network in ABA responses: sensing, signaling and transport," *Plant & Cell Physiology*, vol. 51, no. 11, pp. 1821–1839, 2010.

[33] K. C. Babitha, R. S. Vemanna, K. N. Nataraja, and M. Udayakumar, "Overexpression of *EcbHLH57* transcription factor from *Eleusine coracana L.* in tobacco confers tolerance to salt, oxidative and drought stress," *PLoS One*, vol. 10, no. 9, article e0137098, 2015.

[34] H. Li, W. Huang, Z. W. Liu, Y. X. Wang, and J. Zhuang, "Transcriptome-based analysis of Dof family transcription factors and their responses to abiotic stress in tea plant (*Camellia sinensis*)," *International Journal of Genomics*, vol. 2016, Article ID 5614142, 15 pages, 2016.

[35] Z. Y. Xu, S. Y. Kim, D. Y. Hyeon et al., "The *Arabidopsis* NAC transcription factor ANAC096 cooperates with bZIP-type transcription factors in dehydration and osmotic stress responses," *The Plant Cell*, vol. 25, no. 11, pp. 4708–4724, 2013.

[36] G. Mallikarjuna, K. Mallikarjuna, M. K. Reddy, and T. Kaul, "Expression of *OsDREB2A* transcription factor confers enhanced dehydration and salt stress tolerance in rice (*Oryza sativa* L.)," *Biotechnology Letters*, vol. 33, no. 8, pp. 1689–1697, 2011.

[37] X. S. Huang, J. H. Liu, and X. J. Chen, "Overexpression of *PtrABF* gene, a bZIP transcription factor isolated from *Poncirus trifoliata*, enhances dehydration and drought tolerance in tobacco via scavenging ROS and modulating expression of stress-responsive genes," *BMC Plant Biology*, vol. 10, no. 1, p. 230, 2010.

[38] V. Belamkar, N. T. Weeks, A. K. Bharti, A. D. Farmer, M. A. Graham, and S. B. Cannon, "Comprehensive characterization and RNA-Seq profiling of the HD-Zip transcription factor family in soybean (*Glycine max*) during dehydration and salt stress," *BMC Genomics*, vol. 15, no. 1, p. 950, 2014.

[39] P. R. Shingote, P. G. Kawar, M. C. Pagariya, R. S. Kuhikar, A. S. Thorat, and K. H. Babu, "SoMYB18, a sugarcane MYB transcription factor improves salt and dehydration tolerance in tobacco," *Acta Physiologiae Plantarum*, vol. 37, no. 10, p. 217, 2015.

[40] Y. J. Jung, I. H. Lee, I. S. Nou, K. D. Lee, A. M. Rashotte, and K. K. Kang, "*BrRZFP1* a *Brassica rapa* C3HC4-type RING zinc finger protein involved in cold, salt and dehydration stress," *Plant Biology*, vol. 15, no. 2, pp. 274–283, 2013.

[41] Z. W. Liu, Z. J. Wu, X. H. Li et al., "Identification, classification, and expression profiles of heat shock transcription factors in tea plant (*Camellia sinensis*) under temperature stress," *Gene*, vol. 576, no. 1, pp. 52–59, 2016.

[42] Z. J. Wu, X. H. Li, Z. W. Liu, H. Li, Y. X. Wang, and J. Zhuang, "Transcriptome-wide identification of *Camellia sinensis* WRKY transcription factors in response to temperature

stress," *Molecular Genetics and Genomics*, vol. 291, no. 1, pp. 255–269, 2016.

[43] S. Pukacka, M. Malec, and E. Ratajczak, "ROS production and antioxidative system activity in embryonic axes of *Quercus robur* seeds under different desiccation rate conditions," *Acta Physiologiae Plantarum*, vol. 33, no. 6, pp. 2219–2227, 2011.

[44] K. Apel and H. Hirt, "Reactive oxygen species: metabolism, oxidative stress, and signal transduction," *Annual Review of Plant Biology*, vol. 55, no. 1, pp. 373–399, 2004.

[45] R. K. Sairam and A. Tyagi, "Physiology and molecular biology of salinity stress tolerance in plants," *Current Science*, vol. 86, no. 3, pp. 407–421, 2004.

[46] H. Wang, T. Wang, and I. Ahmad, "Involvement of phosphate supplies in different transcriptional regulation pathway of *Oryza sativa* L.'s antioxidative system in response to arsenite and cadmium stress," *Ecotoxicology*, vol. 24, no. 6, pp. 1259–1268, 2015.

[47] E. M. Kalemba and S. Pukacka, "Association of protective proteins with dehydration and desiccation of *orthodox* and *recalcitrant* category seeds of three *Acer* genus species," *Journal of Plant Growth Regulation*, vol. 31, no. 3, pp. 351–362, 2012.

14

Lingguizhugan Decoction Protects against High-Fat-Diet-Induced Nonalcoholic Fatty Liver Disease by Alleviating Oxidative Stress and Activating Cholesterol Secretion

Lili Yang,[1] Weili Lin,[2,3] Colleen A. Nugent,[4] Shijun Hao,[1] Haiyan Song,[1] Tao Liu,[1] and Peiyong Zheng[1]

[1]*Institute of Digestive Diseases, Longhua Hospital, China-Canada Center of Research for Digestive Diseases (ccCRDD), Shanghai University of Traditional Chinese Medicine, Shanghai 200032, China*
[2]*Key Laboratory of Computational Biology, CAS-MPG Partner Institute for Computational Biology, Shanghai Institutes for Biological Sciences, Chinese Academy of Sciences, Shanghai 200032, China*
[3]*University of Chinese Academy of Sciences, Beijing 100049, China*
[4]*Digestive Diseases and Nutrition Center, Women and Children's Hospital of Buffalo, Department of Pediatrics, The State University of New York at Buffalo, Buffalo, NY 14214, USA*

Correspondence should be addressed to Tao Liu; lh2488@126.com and Peiyong Zheng; zpychina@sina.com

Academic Editor: Peijian He

Background. Nonalcoholic fatty liver disease (NAFLD) has become a leading cause of liver transplantation. *Lingguizhugan* decoction (LGZG), a classical Chinese herbal formula, has beneficial effects on NAFLD animal models. Our study examined the impact of LGZG on hepatic global transcriptome of high-fat-diet-induced NAFLD rats. *Methods.* Three groups of Wistar rats were included: normal, NAFLD model, and LGZG-treated NAFLD groups. Four weeks for the treatment, liver tissues were harvested for RNA sequencing. Differentially expressed genes (DEGs) and enriched pathways were detected on hepatic global transcriptome profile. Real-time PCR validated the regulatory patterns of LGZG on NAFLD rats. *Results.* DEGs between the NAFLD model and normal groups indicated the elevated peroxisome proliferator-activated receptor (PPAR) and hedgehog signaling pathways in NAFLD rats. In bile secretion pathway, genes involved in cholesterol secretion were activated by LGZG treatment. Increased expression of antioxidant *OSIGN1* and decreased expression of genes (*AHR*, *IRF2BP2*, and *RASGEF1B*) that induce oxidative stress and inflammation were observed in NAFLD rats treated with LGZG. The regulatory patterns of LGZG treatment on these oxidative stress-related genes were confirmed by real-time PCR. *Conclusion.* Our study revealed a "two-hits-targeting" mechanism of LGZG in the treatment for NAFLD: alleviating oxidative stress and activating cholesterol secretion.

1. Introduction

Nonalcoholic fatty liver disease (NAFLD) encompasses a spectrum of pathological conditions, including simple steatosis, nonalcoholic steatohepatitis (NASH), fibrosis, and cirrhosis. NAFLD has been increasingly prevalent worldwide and has become a leading cause of liver transplantation, along with increasing obesity rate and metabolic syndrome [1]. According to the "two-hits" hypothesis of NAFLD pathogenesis, the first hit is represented by lipid accumulation in the hepatocytes, after which oxidative stress leads to severe NASH [2]. Additionally, recent studies demonstrated that other risk factors also contribute to the development of severe NASH, including altered gut microbiota [3, 4], endogenous alcohol metabolism [5–7], and endoplasmic reticulum stress [8]. Currently, many potential targets for the treatment of NAFLD are identified, including lipid metabolism, oxidative stress, inflammation, fibrosis, and altered gut microbiota [9]. However, clinical trials exclusively aimed at only one of these targets at a time and achieved limited effects. For example, vitamin E therapy targets the "second hit"—oxidative stress, and pioglitazone therapy targets insulin resistance. Both of these two therapies decrease serum AST/ALT and reduce lobular inflammation. They, however, have no impact on insulin resistance, portal inflammation, or liver fibrosis and did not obtain satisfactory sustained results [10].

Given NAFLD is a consequence of multiple risk factors, especially the major "two hits"—lipid accumulation and oxidative stress. A therapeutic strategy that targets both of "two hits" or several hits simultaneously could be more effective than one single therapeutic target. Traditional Chinese medicine (TCM) has been increasingly applied as the potential anti-NAFLD drugs and exhibited multipronged preventive and therapeutic effects [9, 11, 12]. LGZG is a classic TCM formula, which is a mixture of four herbs: *Poria*, *Ramulus Cinnamomi*, *Atractylodes macrocephala Koidz*, and *Radix Glycyrrhizae*. LGZG has a long time of clinical practice in the treatment of chronic congestive heart failure. In recent years, LGZG has been effectively used to treat obesity and hyperlipidemia [13] and our previous study showed that LGZG has a benefit in treating NAFLD [14].

TCM is an ancient medical practice system which emphasizes the integrity of the entire human body, and it usually exerts therapeutic effects via multiple targets or pathways. Recently, high-throughput omics technologies, especially transcriptomics RNA sequencing, have been increasingly applied in TCM research and revealed multipronged therapeutic mechanisms of TCM [11, 15]. Herein, to better understand the therapeutic mechanisms of LGZG, we examined the effects of LGZG treatment on the hepatic global transcriptome of HFD-induced NAFLD rats. Combined with real-time PCR validation, we identified a "two-hits-targeting" mechanism for LGZG in the treatment of NAFLD.

2. Materials and Methods

2.1. Experimental Animals and Treatment. This study was reviewed and approved by the Animal Experiment Ethics Committee of Shanghai University of TCM and carried out in accordance with their recommendations. A total of 24 male Wistar rats with weights of 130 g ± 10 g in specific pathogen free (SPF) grade were purchased from Shanghai Si-Lai-Ke Experimental Animal Ltd. (Shanghai, China). Animals were randomly divided into three groups (eight rats per group): normal, NAFLD model, and LGZG-treated NAFLD groups. Rats in the normal group were fed with a standard diet. Rats in the NAFLD model and LGZG-treated NAFLD groups were fed with HFD, which consists of 10% lard oil, 2% cholesterol, and 88% standard chow. Rats in the LGZG-treated NAFLD group received a dosage of 10 mL/kg/d (pure solution) via drinking freely. The dosage was 6 g crude medical material per kilogram body weight, approximately seven times of the standard dosage in practice, on the basis of the dose-equivalence equation between rats and humans [16]. Treatment lasted for four weeks. After a 12-hour fast, the animals were sacrificed under pentobarbital sodium (2%, 5.5 mL/kg) anesthesia. Liver tissues were harvested for subsequent analysis.

2.2. Drug Preparation. LGZG decoction is comprised of four Chinese herbs: *Poria* (20 g), *Ramulus Cinnamomi* (15 g), *Rhizoma Atractylodis Macrocephalae* (15 g), and *Radix Glycyrrhizae* (10 g). The dosage is determined according to the test book of "*The Hndouts of JinguiYaoyue*." All herbs were purchased from Longhua Hospital affiliated to Shanghai University of TCM. LGZG decoction was made according to conventional TCM decocting methods [14]. Briefly, all herbs were boiled with 500 mL water after 30 min of soaking. After 20 min, the liquid was transferred by filtration as a first dose of medicine. The remaining of filtration was boiled after adding 400 mL water, and then liquid was transferred by filtration to make a second dose. Two doses were mixed to form 100 mL (pure solution) final decoction. The quality of LGZG was controlled with high-performance liquid chromatography (HPLC). HPLC-grade reagents were purchased from Burdick & Jackson. An Agilent 1100 HPLC system consisting of a G1354A pump, a G1313A autosampler, and a UV/VIS Photodiode Array G1315B detector was used for all analyses (Figure S1).

2.3. RNA Sequencing Analysis. Liver tissue from the right lobe was collected after four weeks of LGZG treatment as described above. Total RNA was isolated with NanoPhotometer spectrophotometer (IMPLEN, CA, USA) and qualified on the Qubit RNA Assay (Qubit 2.0 Fluorometer, Life Technologies, CA, USA). The RNA libraries were sequenced on IlluminaHiseq 4000 platform with paired-end 150 base pair long reads. Clean data were obtained from raw data by removing reads containing adapter, N base, and low-quality reads with NGS QC Toolkit (version: 2.3.3). Clean data were mapped to the reference genome of R.norvegicus6.0 and estimated for gene expression level using TopHat2 and cufflinks (version: 2.2.1).

2.4. Real-Time PCR. *OSGIN1*, *IRF2BP2*, *AHR*, and *RASGEF1B* mRNA levels were determined by real-time PCR. Primers were designed with the primer premier 5.0 software (Table S1). Total RNA of liver tissues was extracted with TRIzol reagent (Invitrogen, USA). The concentration of RNA was measured with NanoDrop 2000 (Thermo Scientific, USA). Quantitative measurement was performed with the Premix Ex Taq kit (TakaRa) according to the manufacturer's instructions on Applied Biosystems StepOne Plus Sequence Detection System. The real-time cycler conditions were as follows: first denatured at 95°C for 30 s and then amplified with 40 cycles (each cycle was denaturated at 90°C for 5 s and annealing/extension at 60°C for 30 min). Product purity was determined by dissociation curve analysis. Gene expression was quantified relative to the values of the control group after adjusting for β-actin by the $2^{-\Delta\Delta CT}$ method [17].

2.5. Statistical Analysis. Differentially expressed genes (DEGs) were identified by cuffdiff (version: 2.2.1), with a Q value < 0.05 and an absolute value of log2 fold change > 0.58. Pathway enrichment analysis was performed with clusterProfiler R package (version: 2.4.3) with P value < 0.05. Data were denoted as mean ± standard deviation (SD). In biochemical analysis, statistical analysis was performed with

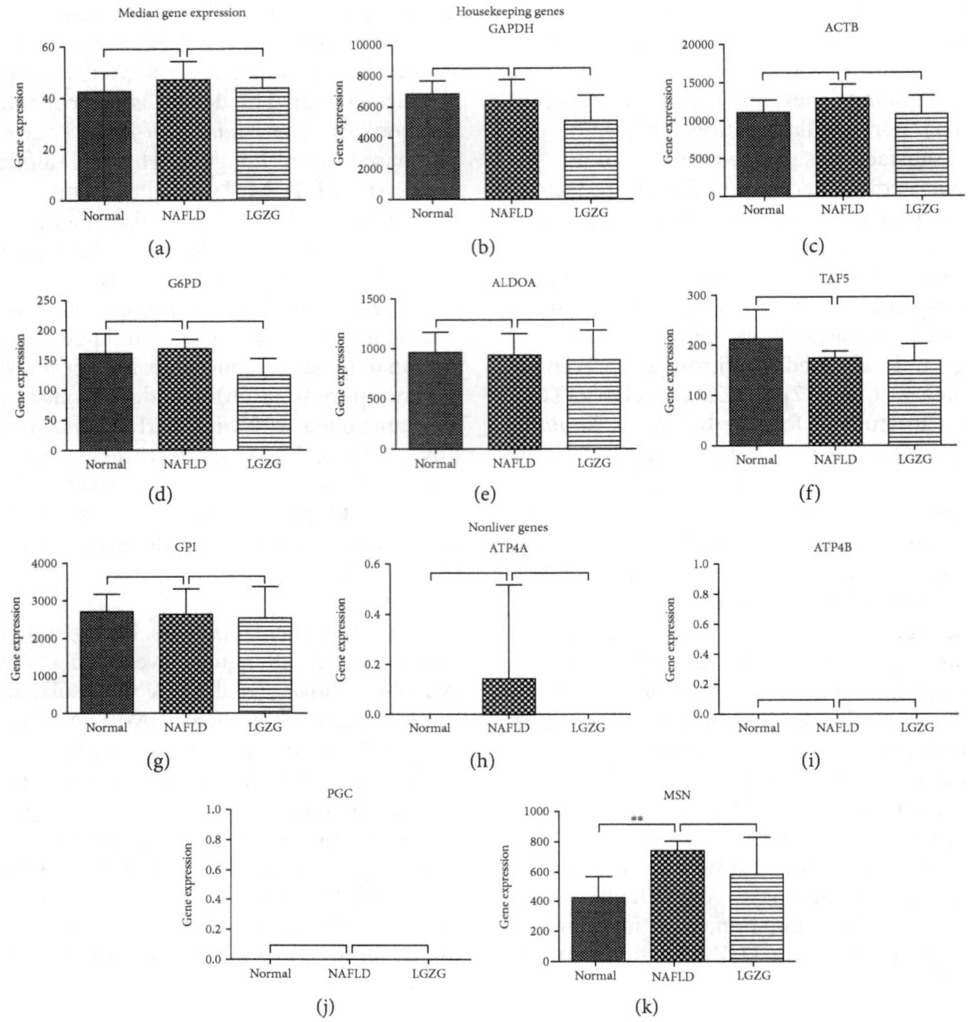

Figure 1: Median gene expression (a) was similar among all study groups. Housekeeping genes *GAPDH* (b), *ACTB* (c), *G6PD* (d), *ALDOA* (e), *TAF* (f), and *GPI* (g) and nonliver genes *ATP4A* (h), *ATP4B* (i), *PGC* (j), and *MSN* (k) exhibited similar expression among all study groups. **Q value < 0.01 calculated from cuffdiff.

one-way analysis of variance (ANOVA) and Dunnett's test. In RT-PCR validation, statistical analysis was performed with Mann–Whitney test with a two-tailed distribution. P values < 0.05 were considered statistically significant. Statistical analysis was performed in R 3.2.3 software.

3. Results

3.1. Genes and Pathways Associated with NAFLD Pathology. Expression levels of about 22,077 genes of rat livers were quantified on a global RNA sequencing. Similar median gene expression and expression levels for housekeeping genes and nonliver genes are indicative of the qualified RNA-seq data in the present study (Figure 1). A total of 931 genes were differentially expressed between the NAFLD model and normal groups, including 494 upregulated and 437 downregulated genes. These 931 DEGs were regarded as NAFLD-regulated genes. The top 40 regulated genes (20 most upregulated and 20 most downregulated) were depicted in Figure 2. These genes are known to be associated with lipid metabolism, including triglyceride metabolic process (*PCSK9*), steroid biosynthesis (*SQLE*), glycerolipid metabolism (*MGLL*), ether lipid metabolism (*PLA2G7*), and phosphatidylinositol signaling system (*IP6K1*). Additionally, the chemokine *CXCL13*, ubiquitin *UBD*, and somatomedin *SBSPON* play roles in inflammatory responses, and their expressions were elevated in NAFLD rats (Figure 2).

Further, 76 KEGG pathways were enriched with NAFLD-regulated genes (Table S2). The top 30 enriched KEGG pathways were shown in Figure 3. In addition to several well-known pathways that are related to NAFLD pathology (including fatty acid elongation, AMPK signaling pathway, and NF-kappa B signaling pathway), peroxisome proliferator-activated receptor (PPAR) and hedgehog (Hh) signaling pathways (Table S2) are of special interest because of their essential roles in hepatic fibrosis of NAFLD progression. As to individual genes of PPAR signaling pathway (Figure 4(a)), NAFLD rats exhibited increased expression of one of the nuclear receptors—*PPARG*, which regulates

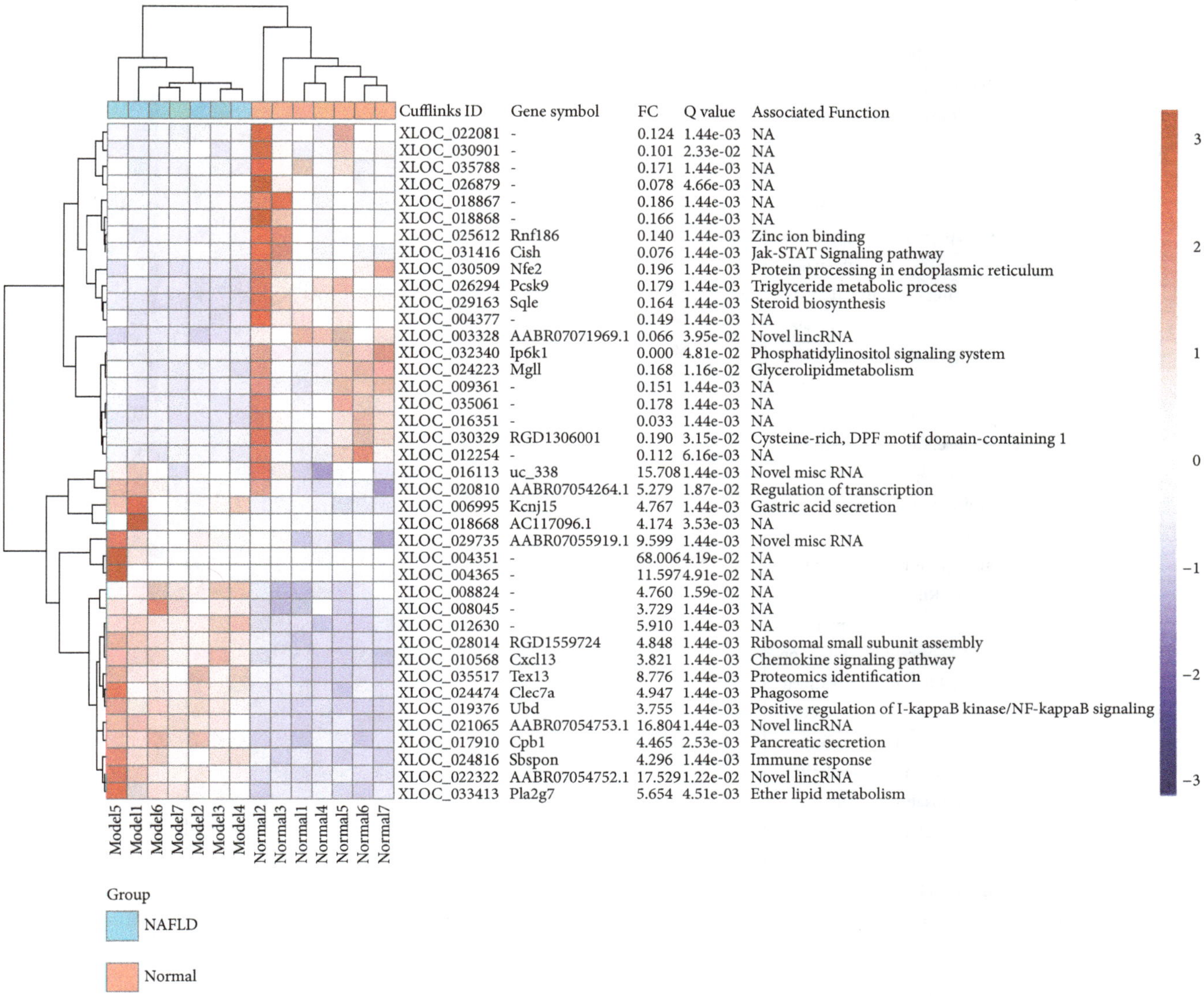

FIGURE 2: Differential expression of top 40 NAFLD-regulated genes. The samples are broadly divided into two groups, normal and NAFLD model groups. The color scale shown at the top right illustrates the relative expression level of the genes across all samples. FC, fold change between NAFLD model and normal groups.

lipogenesis and cholesterol metabolism and represents a good candidate gene for NAFLD [18]. Consistently, increased expression of stearoyl-CoA desaturase (*SCD1* and *SCD2*) was observed in NAFLD rat livers, which are regulated by *PPARG* and the rate-limiting enzymes in lipid biosynthesis [19, 20]. *CYP7A1*, a rate-limiting enzyme in cholesterol metabolism, was upregulated in NAFLD rats. However, some target genes of *PPARG* were downregulated in NAFLD rats. And these decreased expressions of the target genes resulted in impaired fatty acid transport (*DBI* and *FABP1*) and oxidation (*CYP4A1*), which contributed to fatty acid accumulation in the NAFLD liver.

G protein-coupled transmembrane receptor Smoothened (*SMO*) is one of important components of canonical Hh signaling pathway. Our results showed upregulation of *SMO* in NAFLD rats, compared with normal rats. A trend of decreased expression of *SMO* was shown in LGZG-treated rats. In addition, *ADRBK2* (upstream regulator of *SMO*) and *CCND2* (Hh downstream target gene) were also significantly elevated in NAFLD rats (Figure 4(b)).

3.2. Genes and Pathways Mediating the Therapeutic Effects of LGZG. Compared with the NAFLD model group, elevated expression of 110 genes and decreased expression of 89 genes were observed in the livers of the LGZG-treated NAFLD rats. These total 199 genes were considered as LGZG-regulated genes. Figure 5 showed the top 40 LGZG-regulated genes (20 most upregulated and 20 most downregulated). The LGZG-treated NAFLD rats exhibited reduced expression of *INSIG1* and *LPIN1*, which indicated the decreased cholesterol biosynthesis and triglyceride accumulation in the liver. Pathway enrichment analysis was performed with LGZG-regulated genes and unearthed 29 enriched pathways (Figure 6, Table S3). In bile secretion pathway, genes required

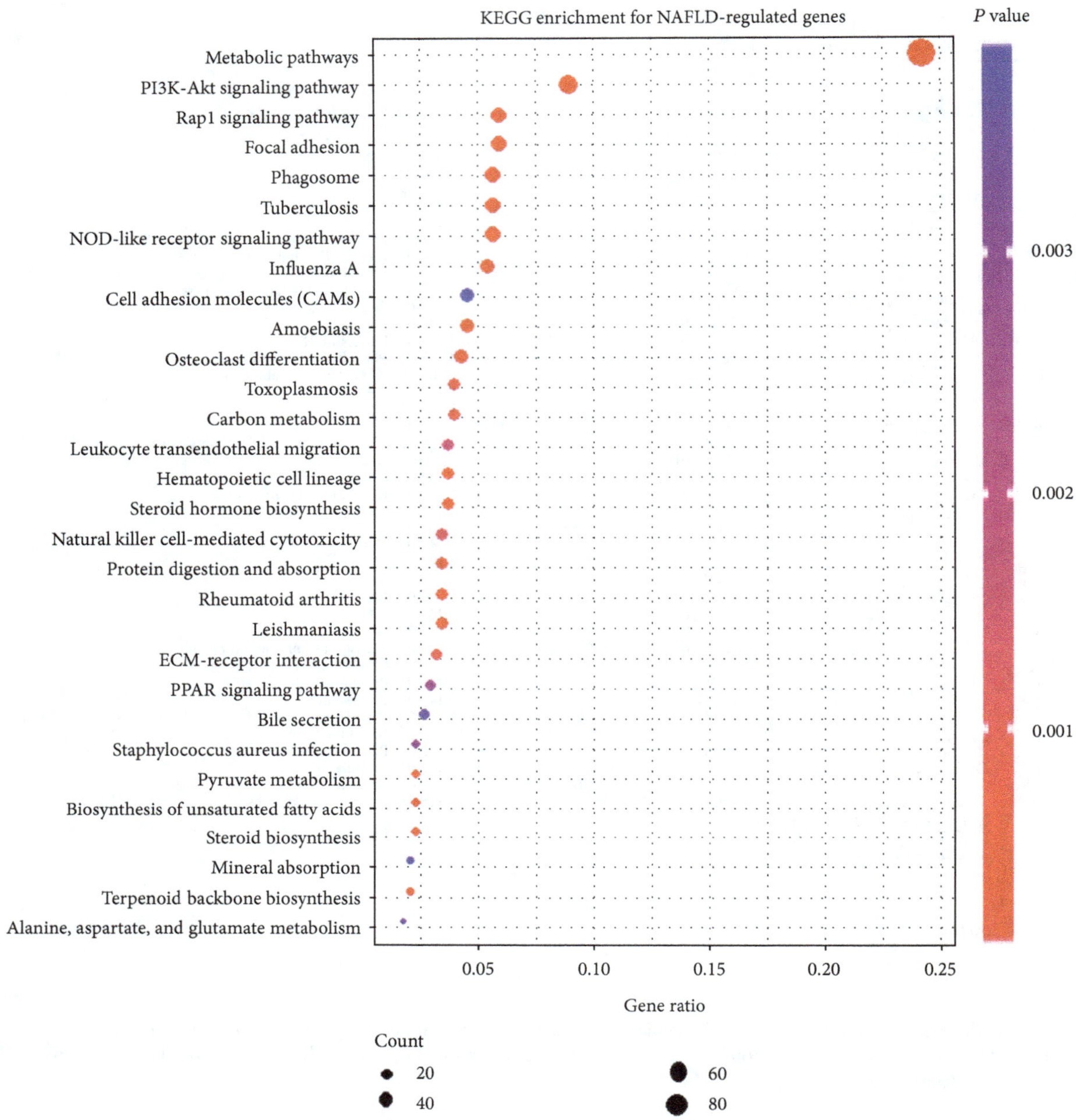

FIGURE 3: Scatter plot of the top 30 enriched KEGG pathways for NAFLD-regulated genes. The x-axis represents the ratio of NAFLD-regulated gene and all gene numbers annotated in this KEGG pathway. The y-axis is enriched KEGG pathways. P value was calculated from hypergeometric test. A smaller P value indicates higher significance ($P < 0.05$).

for cholesterol secretion were elevated, including *ABCG8*, *ABCG5*, and *NCEH1*. Some metabolism and signaling pathways were also enriched, including fatty acid metabolism, Jak-STAT, and FoxO signaling pathways.

Furthermore, four (*OSGIN1*, *AHR*, *IRF2BP2*, and *RASGEF1B*) were shown to be strongly regulated upon LGZG therapy, which may play potential roles in NAFLD physiopathology (Figure 7, Table 1). These four genes also showed differential expression changes in NAFLD rats, compared with normal rats. Also, the abnormal expression change in NAFLD rats was significantly reversed by the LGZG therapy. Thus, they were considered as the important target genes of LGZG.

3.3. Validation of LGZG-Regulated Genes with Real-Time PCR. Expression levels of four LGZG-regulated genes were validated with real-time PCR, including *OSGIN1*, *AHR*, *IRF2BP2*, and *RASGEF1B* (Figure 8). The real-time PCR results confirmed that the expression of *OSGIN1* was reversely elevated, and the expression levels of *IRF2BP2*, *AHR*, and *RASGEF1B* were inversely suppressed under the LGZG treatment.

4. Discussion

LGZG is a classic TCM formula that has been effectively used to treat obesity and hyperlipidemia. Recently, it has exhibited

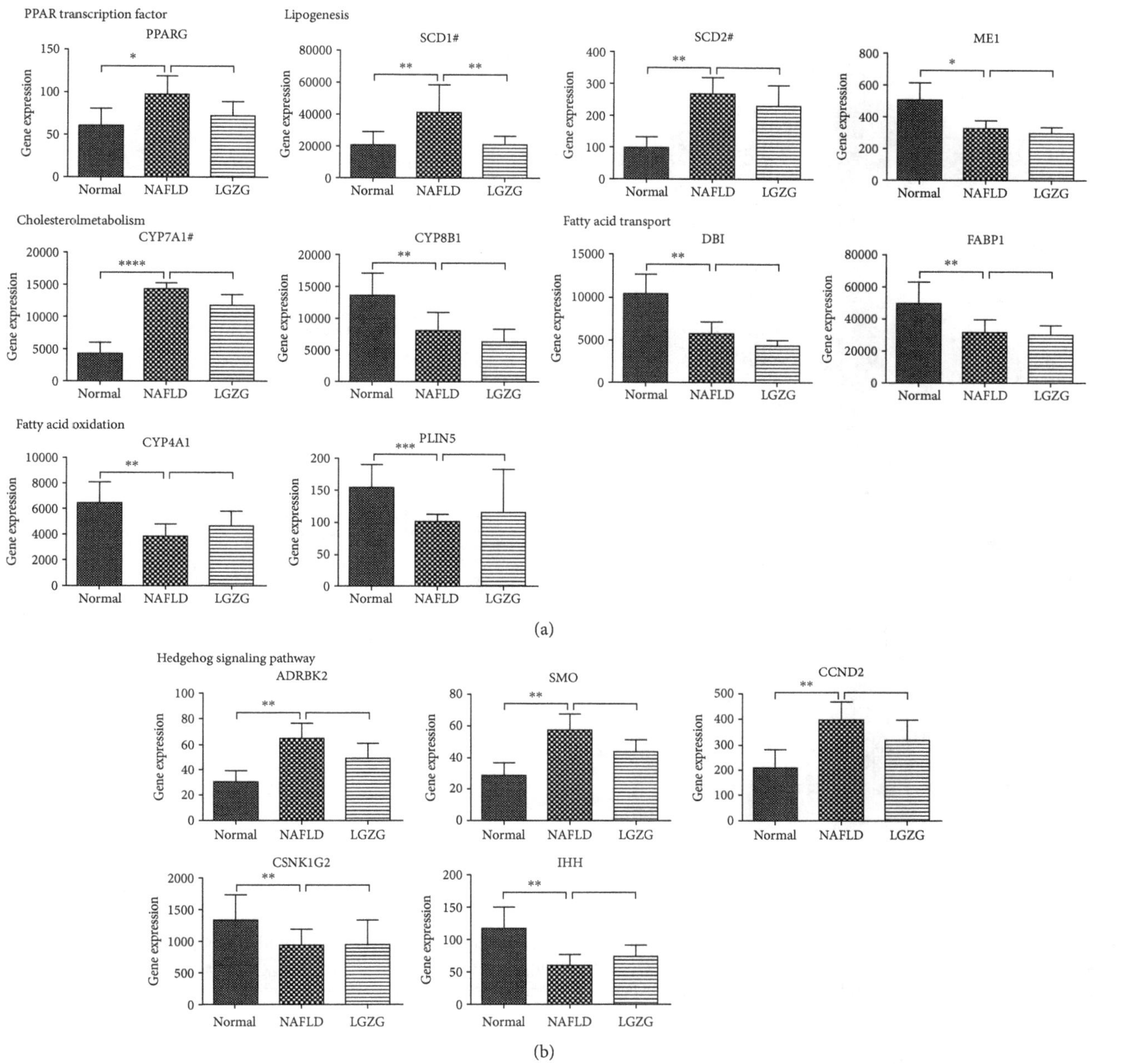

FIGURE 4: Differential gene expression of (a) PPAR and (b) hedgehog signaling pathways in a NAFLD rat model. $^{*}Q$ value < 0.05; $^{**}Q$ value < 0.01; $^{***}Q$ value < 0.001; $^{****}Q$ value < 0.0001 calculated from cuffdiff. $^{\#}$Rate-limiting enzyme.

potent effects on HFD-induced NAFLD [14]. In this study, we examined the comprehensive effects of LGZG on the hepatic global gene expression profile in NAFLD progression. Many of the changes in expression profile reflected decreased hepatic cholesterol, oxidative stress, and inflammation. Subsequently, the elevated gene expression in PPAR and Hh signaling pathways of NAFLD rats provided an assurance of RNA-seq dataset quality. Two potential mechanisms under the efficacy of LGZG for NAFLD were identified: (i) alleviated oxidative stress and (ii) promoted cholesterol secretion to reduce hepatic cholesterol accumulation (Figure 9). These observations indicated a "two-hits-targeting" [2, 21] mechanism for LGZG in the treatment of NAFLD.

4.1. Characteristics of HFD-Induced NAFLD. NAFLD physiopathology-associated pathways were significantly enriched with NAFLD-regulated genes. Among these pathways, activation of PPAR and Hh signaling pathways is of special interest because their deregulation contributes to liver damage and metabolic syndrome [22–27]. Hh signaling is significantly upregulated in NASH, compared with the normal healthy liver [28]. Recent study showed that activation of *SMO* could induce Hh-responsive hepatocytes in NAFLD [28], which lends strong support to our results (Figure 4(b)). Notably, multiple studies showed that in different rodent models of diet-induced NASH, pharmacological inhibition of *SMO* (vismodebig or LDE225) can deactivate Hh signaling pathway and consistently improve liver inflammation and

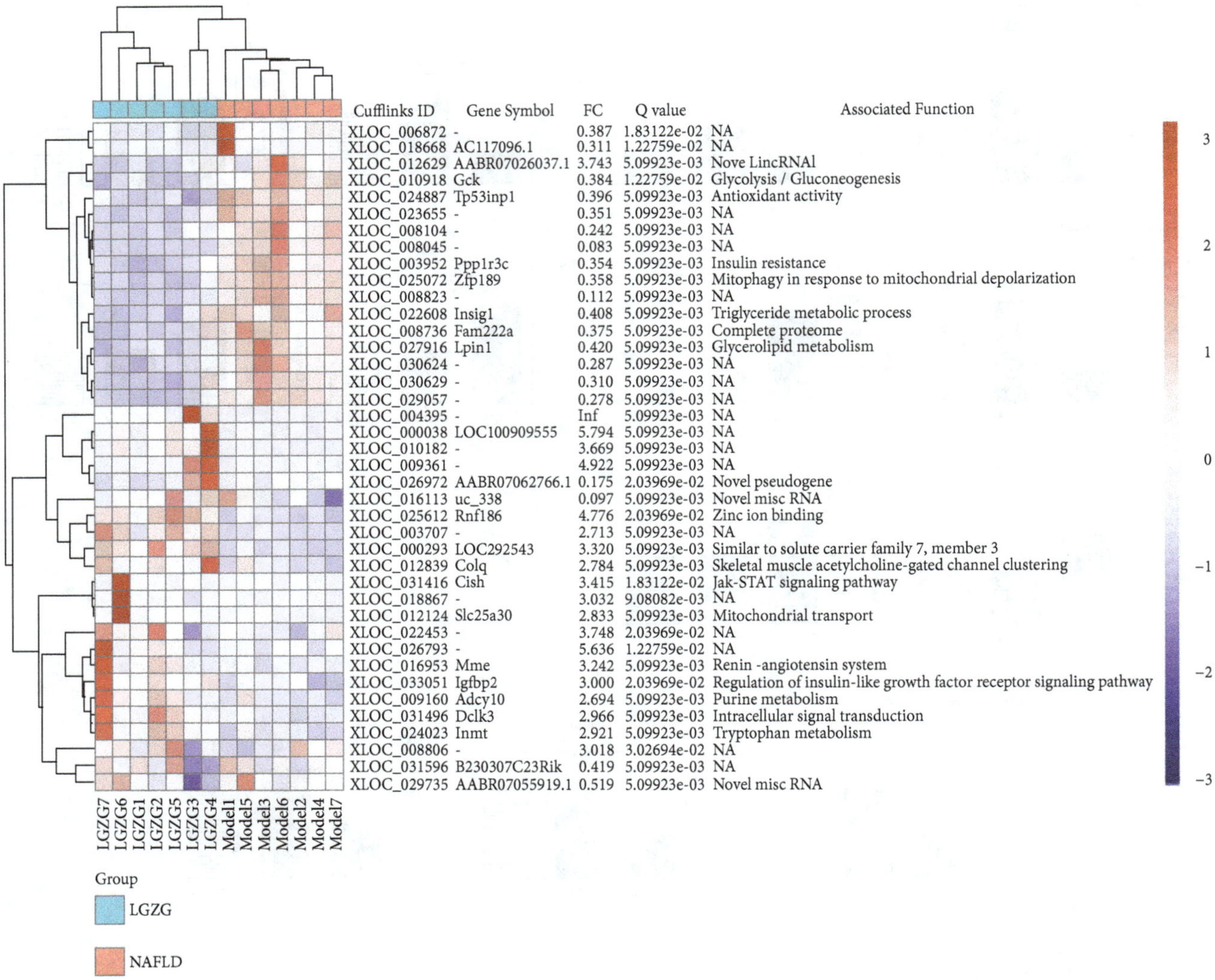

FIGURE 5: Differential expression of top 40 LGZG-regulated genes. The samples of the NAFLD model group formed a clade and were separated from the samples of the LGZG-treated NAFLD groups. The color scale shown at the top right illustrates the relative expression level of the genes across all samples. FC, fold change between LGZG-treated NAFLD and NAFLD model groups.

fibrosis [25, 28–30]. Despite no significant difference was observed between NAFLD and LGZG treatment groups for *SMO*, a trend of decreased expression was shown in LGZG-treated NAFLD rats. In addition, lipid metabolism-related genes were markedly altered in NAFLD rats, which contributed to hepatic TG accumulation. Genes and pathways associated with NAFLD pathology were the manifestation of the RNA-seq dataset quality.

4.2. Alleviation of Oxidative Stress. Significant expression changes of four LGZG-regulated genes (*OSGIN1*, *AHR*, *IRF2BP2*, and *RASGEF1B*) exhibited alleviation of oxidative stress upon LGZG treatment, compared with the NAFLD group. As an antioxidant, *OSGIN1* is a cell growth inhibitor to resist oxidative stress [31]. Liu et al. reported that the mRNA level of *OSGIN1* was reduced in HCC specimens, and in HCC pateints, the inhibition of *OSGIN1* was related to shorter overall and disease-free survival times [32], suggesting that the significant upregulation of *OSGIN1* is critical for antioxidant response. *AHR* is identified to induce cellular oxidative stress and increase lipid peroxidation in NAFLD [33–36], and activation of *AHR* has pleotropic effects on steatosis of NAFLD [37, 38].

RASGEF1B and *IRF2BP2* were involved in inflammatory responses and were dramatically inhibited by the LGZG treatment. *RASGEF1B* was identified as a Ras-associated guanine nucleotide exchange factor and upregulated in macrophages stimulated with bacterial lipopolysaccharides (LPS) [39]. LPS is currently considered one of the major "hits" in NAFLD pathogenesis and progression [40]. Our current study showed significant upregulation of *RASGEF1B* in NAFLD model rats and reverse inhibition of *RASGEF1B* by LGZG treatment. Increased expression of *RASGEF1B* may be a strong defense response to LPS in NAFLD progression, and its restoration indicated the alleviation of inflammatory responses. *IRF2BP2* acts as a negative regulator of

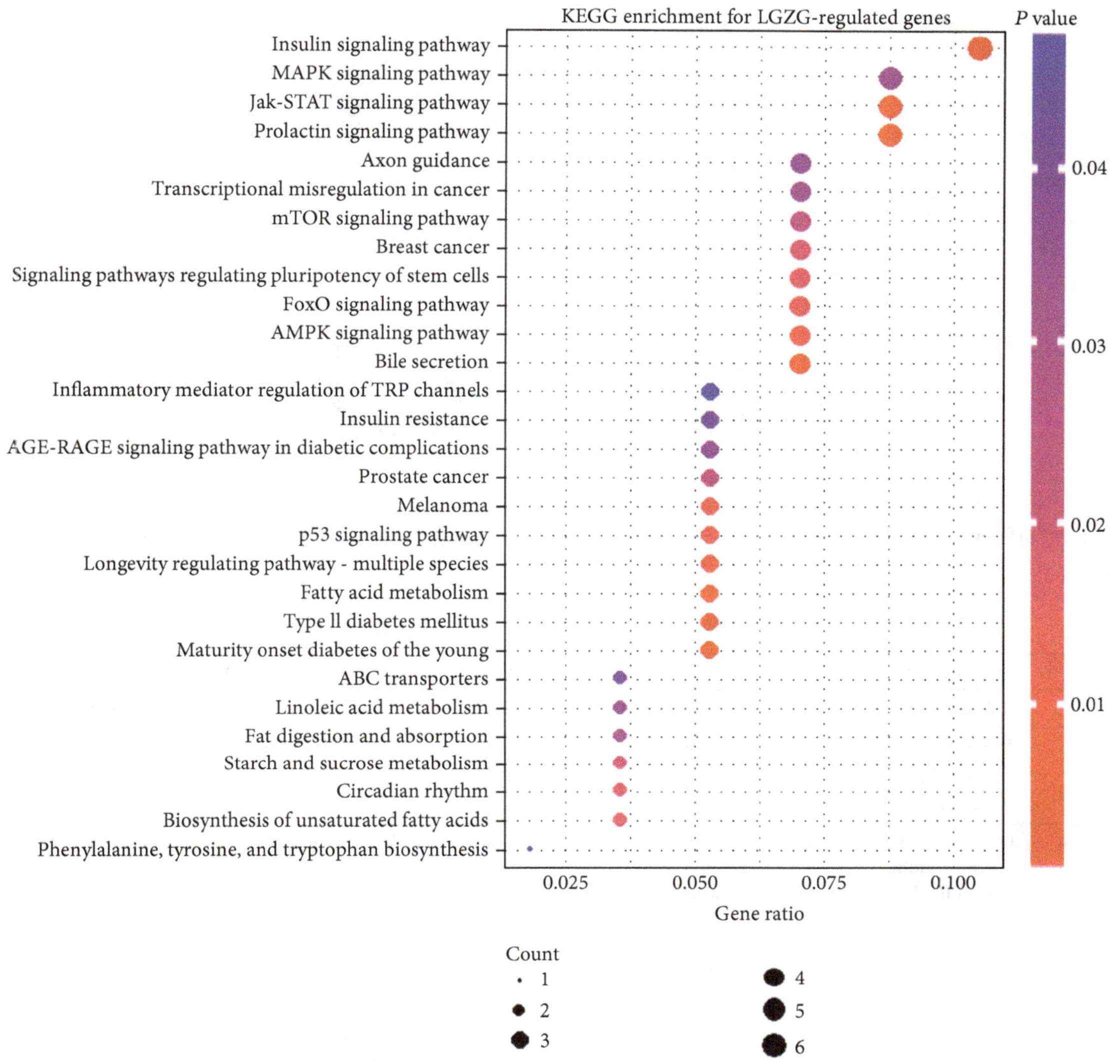

Figure 6: Scatter plot of the enriched KEGG pathways for LGZG-regulated genes. The x-axis represents the ratio of LGZG-regulated gene numbers annotated in this KEGG pathway to all gene numbers annotated in this KEGG pathway. The y-axis is enriched KEGG pathways. P value was calculated from hypergeometric test. A smaller P value indicates higher significance ($P < 0.05$).

the nuclear factor of activated T cell (NFAT) transcription factor [41]. Restoration of *IRF2BP2* by LGZG treatment also suggested the alleviation of inflammatory responses in NAFLD. Continuous oxidative stress may lead to chronic inflammation. Thus, the alleviation of inflammatory responses also indicates the reduced oxidative stress upon LGZG treatment.

4.3. Activation of Cholesterol Secretion. Liver lipid accumulation is the first "hit" in the pathogenesis of NAFLD, and its removal is a desired intervention for NAFLD. Cholesterol is one type of important lipid in the liver, which has been shown as an emerging factor involved in the development of many metabolic diseases [8]. Free cholesterol stores in the liver in the formation of cholesterol esters. In the cholesterol secretion pathway, the upregulation of neutral cholesteryl ester hydrolase 1 (*NCEH1*) can accelerate the transformation of cholesterol esters into free cholesterol. *ABCG5* and *ABCG8* are two half-transporters that dimerize to create a cholesterol transporter, and their activation promotes the excretion of hepatic cholesterol [42]. In addition, *NCEH1*, *ABCG5*, and *ABCG8* are reported to be drug targets of pioglitazone [43] and ezetimibe [42] in the therapy of human gallbladder cholesterolosis and HFD-induced fatty liver.

5. Conclusion

Based on transcriptome analysis and experimental validation, our study examined the comprehensive effects of LGZG on hepatic global gene expression profile in HFD-induced NAFLD rats. Compared with normal rats, our data revealed significant upregulation of PPAR and Hh signaling pathways in NAFLD rats, which are known to be involved in NAFLD pathology and thus provide an assurance of the quality of our RNA-seq dataset. Of particular interest, NAFLD rats

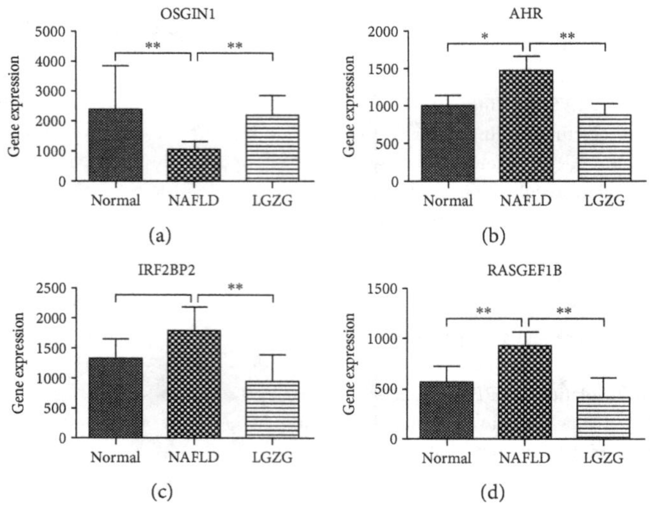

FIGURE 7: Differential gene expression of LGZG-regulated genes. *Q value < 0.05; **Q value < 0.01 calculated from cuffdiff.

TABLE 1: Important LGZG-regulated genes.

Gene	Q value*	FC#	Q value$	FC+	Associated function
OSGIN1	1.44e-03	0.43	5.10e-03	2.31	Oxidoreductase activity
IRF2BP2	6.62e-02	1.43	5.10e-03	0.52	Interferon regulatory factor 2-binding protein 1 & 2
AHR	2.20e-02	1.37	5.10e-03	0.67	Blood vessel development
RASGEF1B	1.44e-03	1.63	5.10e-03	0.46	Ras guanyl-nucleotide exchange factor activity

*Q value of comparison between the NAFLD model and normal groups, calculated from cuffdiff. #Fold change of comparison between the NAFLD model and normal groups, calculated from cuffdiff. $Q value of comparison between the LGZG-treated NAFLD and NAFLD model groups, calculated from cuffdiff. +Fold change of comparison between the LGZG-treated NAFLD and NAFLD model groups, calculated from cuffdiff.

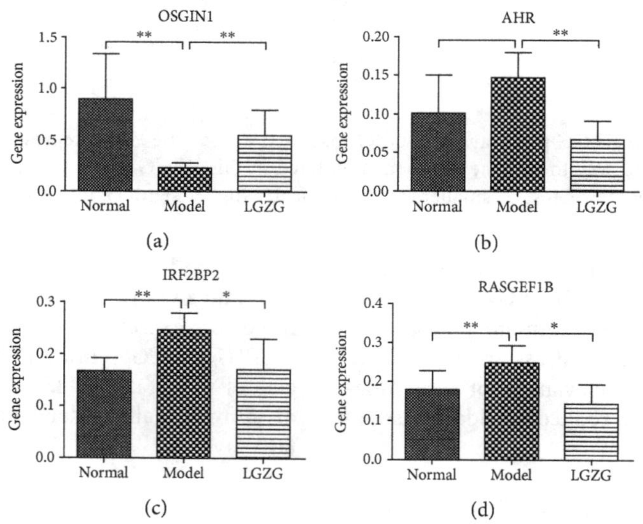

FIGURE 8: Differential gene expression of LGZG-regulated genes was examined by real-time PCR. Plotted values are the mean ± SD of mRNA expression levels in the livers of normal rats (normal), NAFLD model rats, and NAFLD rats treated with LGZG decoction (LGZG). $N = 6$ per group. *$P < 0.05$; **$P < 0.01$ Mann–Whitney test.

with LGZG treatment exhibited elevated expression of antioxidant and suppressed expression of prooxidant and proinflammatory genes in oxidative stress. Additionally, genes required for cholesterol secretion were increased by the LGZG treatment. These findings supported a "two-hits-targeting" mechanism for LGZG in the treatment of NAFLD: alleviating oxidative stress and activating cholesterol secretion.

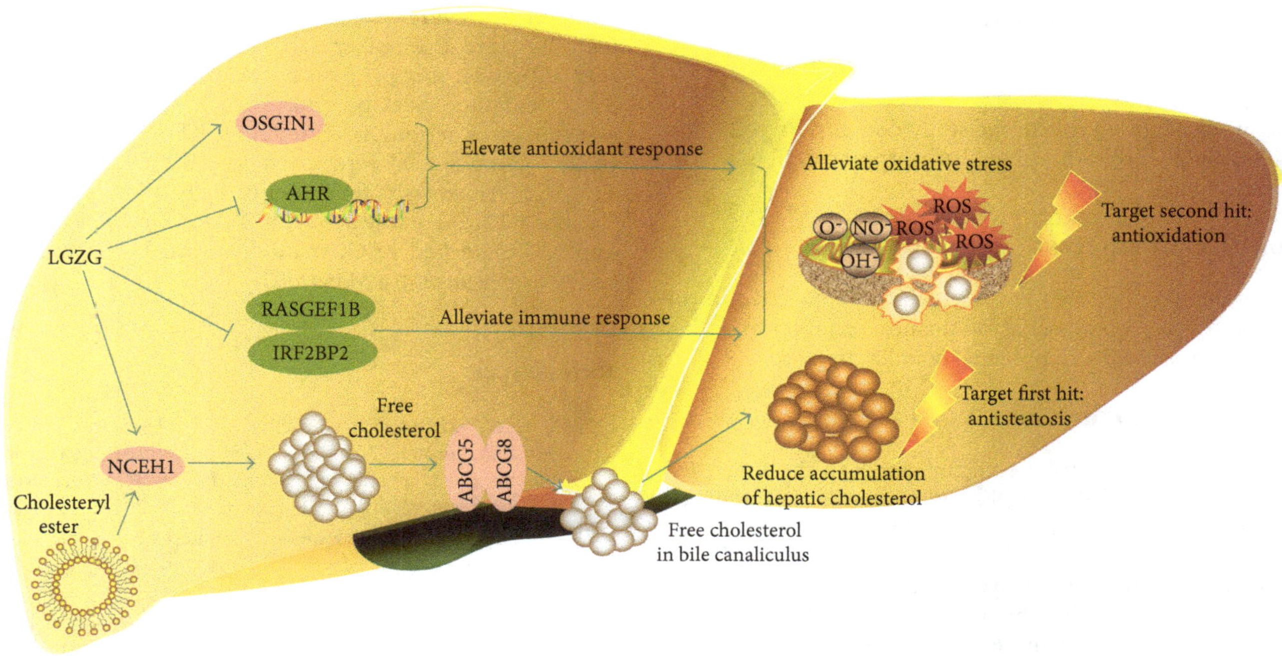

FIGURE 9: Summary of the "two-hits-targeting" therapeutic mechanisms of LGZG for NAFLD. Data presented in this study supports two independent mechanisms: (i) LGZG activates antioxidant (*OSGIN1*) and suppresses prooxidant and proinflammatory genes (*AHR*, *IRF2BP2*, and *RASGEF1B*), causing alleviated oxidative stress, and (ii) LGZG induces cholesterol secretion genes in the liver, leading to reduced accumulation of liver lipid. These mechanisms reflect that LGZG treatment prominately impeded classic "two hits" of NASH physiopathclogy. Red genes indicate increased expression; green genes indicate decreased expression. Normal arrows indicate activation. Arrows with a vertical line at the end indicate suppression.

Abbreviations

NAFLD: Nonalcoholic fatty liver disease
NASH: Nonalcoholic steatohepatitis
LGZG: Lingguizhugan
EBP/BW: Epididymal fat pad/body weight ratio
SPF: Specific pathogen free
HFD: High-fat diet
HPLC: High-performance liquid chromatography
H&E: Hematoxylin and eosin
ALT: Alanine aminotransferase
AST: Aspartate aminotransferase
TG: Triglyceride
TC: Total cholesterol
VLDL: Very-low-density lipoprotein
RT-PCR: Real-time PCR
DEG: Differentially expressed genes
SD: Standard deviation
ANOVA: One-way analysis of variance
CV: Central vein
PCSK9: Proprotein convertase subtilisin/kexin type 9
SQLE: Squalene monooxygenase
MGLL: Acylglycerol lipase
PLA2G7: Platelet-activating factor acetylhydrolase
IP6K1: Inositol hexakisphosphate 5-kinase
CXCL13: C-X-C motif chemokine 13
UBD: Ubiquitin D
SBSPON: Somatomedin B and thrombospondin type 1 domain containing
PPAR: Peroxisome proliferator-activated receptor
PPARG: Peroxisome proliferator-activated receptor gamma
SCD: Stearoyl-CoA desaturase
SCD2: Stearoyl-Coenzyme A desaturase 2
CYP7A1: Cholesterol 7alpha-monooxygenase
DBI: Diazepam-binding inhibitor (GABA receptor modulator, acyl-CoA-binding protein)
FABP1: Fatty acid binding protein 1, liver
CYP4A1: Alkane 1-monooxygenase
SMO: Smoothened
Hh: Hedgehog
ADRBK2: Beta-adrenergic receptor kinase
CCND2: Cyclin D2
INSIG1: Insulin-induced gene 1
LPIN1: Phosphatidate phosphatase LPIN
ABCG8: ATP-binding cassette, subfamily G (WHITE), member 8 (sterolin 2)
ABCG5: ATP-binding cassette, subfamily G (WHITE), member 5 (sterolin 1)
NCEH1: Neutral cholesterol ester hydrolase 1
OSGIN1: Oxidative stress-induced growth inhibitor 1
AHR: Aryl hydrocarbon receptor
IRF2BP2: Interferon regulatory factor 2 binding protein 2
RASGEF1B: RasGEF domain family, member 1B

HCC: Hepatocellular carcinoma.

Disclosure

The funders had no role in the study design, data collection and analysis, decision to publish, or preparation of the manuscript.

Conflicts of Interest

The authors declare that the research was conducted in the absence of any commercial or financial relationships that could be construed as a potential conflict of interest.

Authors' Contributions

Each author has contributed significantly to the submitted work. Tao Liu and Peiyong Zheng conceived and designed the experiments. Lili Yang and Weili Lin performed the experiments. Lili Yang, Weili Lin, Colleen A. Nugent, Shijun Hao, and Haiyan Song analyzed the data. Lili Yang, Weili Lin, Colleen A. Nugent, Shijun Hao, Haiyan Song, and Tao Liu wrote the paper. All authors read and approved the final manuscript. Lili Yang and Weili Lin contributed equally to this work.

Acknowledgments

This work was supported by the National Natural Science Foundation of China 81620108030 and Natural Science Foundation of Shanghai 15ZR1441900.

Supplementary Materials

Figure S1. Quality control of Lingguizhugan Decoction under High Performance Liquid Chromatography (HPLC). A. Reference substance; B. Lingguizhugan Decoction 1. glycyrrhizic acid. Standardization of extract by HPLC glycyrrhizic acid was purchased from Shanghai R&D Centre for standardization of Chinese Medicines (Shanghai, China), and the purity was higher than 98.0%;. HPLC-grade reagents were purchased from Burdick & Jackson. An Agilent 1100 HPLC system consisting of a G1354A pump, a G1313A auto-sampler, and a UV/VIS Photodiode Array G1315B Detector was used for all analyses. Chromatographic separations were carried out on an Merck C18 hibar column (4.6 mm × 250mm, 5 μm) with methanol:acetonitrile : water :acetid acid(as solvent) (15:35:45:0.9,v/v) in the mobile phase at a flow rate of 0.8 mL/min at 25°C for 40 min. Twenty μL (after a high speed centrifugation) of the sample was injected, and the signals were detected at 254 nm with UV detection. The HPLC fingerprint of Lingguizhugan decoction revealed the major peaks (glycyrrhizic acid) at 254 nm. The content of glycyrrhizic acid were 0.9548%. Figure S2. Hierarchical clustering of replicate samples based on correlation matrix in three groups: (A) Normal group; (B) NAFLD model group; (C) LGZG-treated group. Each column and row represents samples in each group. Each cell of the heatmap reflects the distance of two samples. Distance was measured with Euclidean metric. The legend of distance value is shown on the upper right. Figure S2A and figure S2B revealed that s12 and s24 were the outlier samples of normal group and NAFLD model group, respectively. Figure S2C showed no outlier samples, but the data in s37 was missed during assessment of serum ALT and AST. Thus, s37 in LGZG-treated group was also removed in the RNA-seq analysis. Table S1. Primer pairs for real-time PCR analysis. Table S2. Enriched KEGG pathways with NAFLD regulated genes. Table S3. Enriched KEGG pathways with LGZG regulated genes. *(Supplementary Materials)*

References

[1] A. Wree, L. Broderick, A. Canbay, H. M. Hoffman, and A. E. Feldstein, "From NAFLD to NASH to cirrhosis–new insights into disease mechanisms," *Nature Reviews Gastroenterology & Hepatology*, vol. 10, no. 11, pp. 627–636, 2013.

[2] C. P. Day and O. F. James, "Steatohepatitis: a tale of two "hits"?," *Gastroenterology*, vol. 114, no. 4, pp. 842–845, 1998.

[3] L. Zhu, S. S. Baker, C. Gill et al., "Characterization of gut microbiomes in nonalcoholic steatohepatitis (NASH) patients: a connection between endogenous alcohol and NASH," *Hepatology*, vol. 57, no. 2, pp. 601–609, 2013.

[4] L. Zhu, R. D. Baker, R. Zhu, and S. S. Baker, "Gut microbiota produce alcohol and contribute to NAFLD," *Gut*, vol. 65, no. 7, article 1232, 2016.

[5] Y. Zhang, S. S. Baker, R. D. Baker, R. Zhu, and L. Zhu, "Systematic analysis of the gene expression in the livers of nonalcoholic steatohepatitis: implications on potential biomarkers and molecular pathological mechanism," *PLoS One*, vol. 7, no. 12, article e51131, 2012.

[6] R. Zhu, S. S. Baker, C. A. Moylan et al., "Systematic transcriptome analysis reveals elevated expression of alcohol-metabolizing genes in NAFLD livers," *The Journal of Pathology*, vol. 238, no. 4, pp. 531–542, 2016.

[7] S. S. Baker, R. D. Baker, W. Liu, N. J. Nowak, and L. Zhu, "Role of alcohol metabolism in non-alcoholic steatohepatitis," *PLoS One*, vol. 5, no. 3, article e9570, 2010.

[8] E. Sozen and N. K. Ozer, "Impact of high cholesterol and endoplasmic reticulum stress on metabolic diseases: an updated mini-review," *Redox Biology*, vol. 12, pp. 456–461, 2017.

[9] L. Zhu, R. D. Baker, and S. S. Baker, "Editorial: from multiple hits to multiple therapeutic targets of non-alcoholic fatty liver disease," *Current Drug Targets*, vol. 16, no. 12, pp. 1272-1273, 2015.

[10] A. J. Sanyal, N. Chalasani, K. V. Kowdley et al., "Pioglitazone, vitamin E, or placebo for nonalcoholic steatohepatitis," *New England Journal of Medicine*, vol. 362, no. 18, pp. 1675–1685, 2010.

[11] Q. Feng, W. Liu, S. S. Baker et al., "Multi-targeting therapeutic mechanisms of the Chinese herbal medicine QHD in the treatment of non-alcoholic fatty liver disease," *Oncotarget*, vol. 8, no. 17, pp. 27820–27838, 2017.

[12] J.-Y. Xu, L. Zhang, Z.-P. Li, and G. Ji, "Natural products on nonalcoholic fatty liver disease," *Current Drug Targets*, vol. 16, no. 12, pp. 1347–1355, 2015.

[13] D. Chen, B. Ke, Y. Huang et al., "Effects of the modified Linggui Zhugan Decoction () combined with short-term very low calorie diets on glycemic control in newly diagnosed type

2 diabetics," *Journal of Traditional Chinese Medicine*, vol. 31, no. 3, pp. 185–188, 2011.

[14] T. Liu, L.-L. Yang, L. Zou et al., "Chinese medicine formula Lingguizhugan decoction improves beta-oxidation and metabolism of fatty acid in high-fat-diet-induced rat model of fatty liver disease," *Evidence-Based Complementary and Alternative Medicine*, vol. 2013, Article ID 429738, pp. 1-9, 2013.

[15] Y. Wang, W. Lin, C. Li et al., "Multipronged therapeutic effects of Chinese herbal medicine Qishenyiqi in the treatment of acute myocardial infarction," *Frontiers in Pharmacology*, vol. 8, p. 98, 2017.

[16] C. X. Chen, *Pharmacology of Traditional Chinese Medicine*, Shanghai Scientific and Technical Publishers, Shanghai, China, 2006.

[17] K. J. Livak and T. D. Schmittgen, "Analysis of relative gene expression data using real-time quantitative PCR and the $2^{-\Delta\Delta C}{}_{T}$ method," *Methods*, vol. 25, no. 4, pp. 402–408, 2001.

[18] S. Gawrieh, M. C. Marion, R. Komorowski et al., "Genetic variation in the peroxisome proliferator activated receptor-gamma gene is associated with histologically advanced NAFLD," *Digestive Diseases and Sciences*, vol. 57, no. 4, pp. 952–957, 2011.

[19] Y. C. Kim and J. M. Ntambi, "Regulation of stearoyl-CoA desaturase genes: role in cellular metabolism and preadipocyte differentiation," *Biochemical and Biophysical Research Communications*, vol. 266, no. 1, pp. 1–4, 1999.

[20] J. M. Ntambi, "Regulation of stearoyl-CoA desaturase by polyunsaturated fatty acids and cholesterol," *Journal of Lipid Research*, vol. 40, no. 9, pp. 1549–1558, 1999.

[21] P. Marzuillo, A. Grandone, L. Perrone, and E. M. del Giudice, "Understanding the pathophysiological mechanisms in the pediatric non-alcoholic fatty liver disease: the role of genetics," *World Journal of Hepatology*, vol. 7, no. 11, article 1439, 1443 pages, 2015.

[22] A. M. Hamed, O. A. El-Kharashi, S. S. Boctor, and L. F. Abd-Elaziz, "Potential involvement of PPAR α activation in diminishing the hepatoprotective effect of fenofibrate in NAFLD: accuracy of non-invasive panel in determining the stage of liver fibrosis in rats," *Biomedicine & Pharmacotherapy*, vol. 85, pp. 68–78, 2017.

[23] A. Federico, M. Dallio, M. Masarone, M. Persico, and C. Loguercio, "The epidemiology of non-alcoholic fatty liver disease and its connection with cardiovascular disease: role of endothelial dysfunction," *European Review for Medical and Pharmacological Sciences*, vol. 20, no. 22, pp. 4731–4741, 2016.

[24] K. H. Liss and B. N. Finck, "PPARs and nonalcoholic fatty liver disease," *Biochimie*, vol. 136, pp. 65–74, 2017.

[25] M. Verdelho Machado and A. M. Diehl, "Role of hedgehog signaling pathway in NASH," *International Journal of Molecular Sciences*, vol. 17, no. 6, 2016.

[26] M. V. Machado, G. A. Michelotti, G. Xie et al., "Mouse models of diet-induced nonalcoholic steatohepatitis reproduce the heterogeneity of the human disease," *PLoS One*, vol. 10, no. 5, article e0127991, 2015.

[27] C. D. Guy, A. Suzuki, M. Zdanowicz et al., "Hedgehog pathway activation parallels histologic severity of injury and fibrosis in human nonalcoholic fatty liver disease," *Hepatology*, vol. 55, no. 6, pp. 1711–1721, 2012.

[28] H. Kwon, K. Song, C. Han et al., "Inhibition of hedgehog signaling ameliorates hepatic inflammation in mice with nonalcoholic fatty liver disease," *Hepatology*, vol. 63, no. 4, pp. 1155–1169, 2015.

[29] P. Hirsova, S. H. Ibrahim, S. F. Bronk, H. Yagita, and G. J. Gores, "Vismodegib suppresses TRAIL-mediated liver injury in a mouse model of nonalcoholic steatohepatitis," *PLoS One*, vol. 8, no. 7, article e70599, 2013.

[30] W. K. Syn, Y. Jung, A. Omenetti et al., "Hedgehog-mediated epithelial-to-mesenchymal transition and fibrogenic repair in nonalcoholic fatty liver disease," *Gastroenterology*, vol. 137, no. 4, pp. 1478–1488.e8, 2009.

[31] H. Huynh, C. Ng, C. Ong, K. Lim, and T. M. Chan, "Cloning and characterization of a novel pregnancy-induced growth inhibitor in mammary gland," *Endocrinology*, vol. 142, no. 8, pp. 3607–3615, 2001.

[32] M. Liu, Y. Li, L. Chen et al., "Allele-specific imbalance of oxidative stress-induced growth inhibitor 1 associates with progression of hepatocellular carcinoma," *Gastroenterology*, vol. 146, no. 4, pp. 1084–1096.e5, 2014.

[33] J. He, B. Hu, X. Shi et al., "Activation of the aryl hydrocarbon receptor sensitizes mice to nonalcoholic steatohepatitis by deactivating mitochondrial sirtuin deacetylase Sirt3," *Molecular and cellular biology*, vol. 33, no. 10, pp. 2047–2055, 2013.

[34] T. P. Dalton, A. Puga, and H. G. Shertzer, "Induction of cellular oxidative stress by aryl hydrocarbon receptor activation," *Chemico-Biological Interactions*, vol. 141, no. 1-2, pp. 77–95, 2002.

[35] J. H. Lee, T. Wada, M. Febbraio et al., "A novel role for the dioxin receptor in fatty acid metabolism and hepatic steatosis," *Gastroenterology*, vol. 139, no. 2, pp. 653–663, 2010.

[36] H. Lu, W. Cui, and C. D. Klaassen, "Nrf2 protects against 2, 3, 7, 8-tetrachlorodibenzo-p-dioxin (TCDD)-induced oxidative injury and steatohepatitis," *Toxicology and Applied Pharmacology*, vol. 256, no. 2, pp. 122–135, 2011.

[37] J. He, J. H. Lee, M. Febbraio, and W. Xie, "The emerging roles of fatty acid translocase/CD36 and the aryl hydrocarbon receptor in fatty liver disease," *Experimental Biology and Medicine*, vol. 236, no. 10, pp. 1116–1121, 2011.

[38] B. Wahlang, M. Song, J. I. Beier et al., "Evaluation of Aroclor 1260 exposure in a mouse model of diet-induced obesity and non-alcoholic fatty liver disease," *Toxicology and Applied Pharmacology*, vol. 279, no. 3, pp. 380–390, 2014.

[39] L. R. Ferreira, E. F. Abrantes, C. V. Rodrigues et al., "Identification and characterization of a novel mouse gene encoding a Ras-associated guanine nucleotide exchange factor: expression in macrophages and myocarditis elicited by Trypanosoma cruzi parasites," *Journal of Leukocyte Biology*, vol. 72, no. 6, pp. 1215–1227, 2002.

[40] S. Ceccarelli, N. Panera, M. Mina et al., "LPS-induced TNF-α factor mediates pro-inflammatory and pro-fibrogenic pattern in non-alcoholic fatty liver disease," *Oncotarget*, vol. 6, no. 39, pp. 41434–41452, 2015.

[41] M. D. Keller, R. Pandey, D. Li et al., "Mutation in IRF2BP2 is responsible for a familial form of common variable immunodeficiency disorder," *Journal of Allergy and Clinical Immunology*, vol. 138, no. 2, pp. 544–550.e4, 2016.

[42] J. S. Bae, J. M. Park, J. Lee et al., "Amelioration of non-alcoholic fatty liver disease with NPC1L1-targeted IgY or n-3 polyunsaturated fatty acids in mice," *Metabolism*, vol. 66, pp. 32–44, 2017.

[43] J. M. Wang, D. Wang, Y. Y. Tan, G. Zhao, and Z. L. Ji, "Pioglitazone reduces lipid droplets in cholesterolosis of the gallbladder by increasing ABCA1 and NCEH1 expression," *Molecular and Cellular Biochemistry*, vol. 399, no. 1-2, pp. 7–15, 2015.

15

The Endophytic Bacterial Microbiota Associated with Sweet Sorghum (*Sorghum bicolor*) is Modulated by the Application of Chemical N Fertilizer to the Field

Cintia Mareque,[1] Thais Freitas da Silva,[2] Renata Estebanez Vollú,[2] Martín Beracochea,[1] Lucy Seldin,[2] and Federico Battistoni[1]

[1]*Departamento de Bioquímica y Genómica Microbianas, Instituto de Investigaciones Biológica Clemente Estable, Avenida Italia 3318, Montevideo 11600, Uruguay*
[2]*Laboratório de Genética Microbiana, Instituto de Microbiologia Paulo de Góes, Universidade Federal do Rio de Janeiro (UFRJ), Rio de Janeiro, Brazil*

Correspondence should be addressed to Federico Battistoni; fbattistoni@iibce.edu.uy

Academic Editor: Rafael Silva-Rocha

Sweet sorghum (*Sorghum bicolor*) is a multipurpose crop used as a feedstock to produce bioethanol, sugar, energy, and animal feed. However, it requires high levels of N fertilizer application to achieve the optimal growth, which causes environmental degradation. Bacterial endophytes, which live inside plant tissues, play a key role in the health and productivity of their host. This particular community may be influenced by different agronomical practices. The aim of the work was to evaluate the effects of N fertilization on the structure, diversity, abundance, and composition of endophytic and diazotrophic bacterial community associated with field-grown sweet sorghum. PCR-DGGE, quantitative PCR, and high-throughput sequencing were performed based on the amplification of *rrs* and *nifH* genes. The level of N fertilization affected the structure and abundance but not the diversity of the endophytic bacterial communities associated with sweet sorghum plants. This effect was pronounced in the roots of both bacterial communities analyzed and may depend on the physiological state of the plants. Specific bacterial classes and genera increased or decreased when the fertilizer was applied. The data obtained here contribute to a better understanding on the effects of agronomical practices on the microbiota associated with this important crop, with the aim to improve its sustainability.

1. Introduction

Globally, sweet sorghum (*Sorghum bicolor*) is the fourth most important cereal and is a multipurpose crop that is used in grain, forage, syrup, fodder, and bioethanol production [1]. Microbial communities play a crucial role in ecosystems, and particularly plant microbiomes can modulate the growth, health, productivity, C-sequestration, and phytoremediation of plants and play a key role in global biogeochemical cycles [2].

Bacterial endophytes are defined as bacteria that can be detected at a particular moment within the internal tissues of an apparently healthy host plant [3]. Many endophytes are likely to have positive effects on their hosts, with the best examples being N_2-fixing bacteria [4], but the potential of applying endophytic bacteria as inoculants is underexplored [5–7]. Several members of the phyla Proteobacteria, Firmicutes, Bacteriodetes, and Actinobacteria have been isolated as endophytes. These bacteria have been recognized to have profoundly favourable impacts on plant growth by producing phytohormones, synthesizing fungicidal and/or bactericidal substances, enhancing the availability of minerals, possessing phosphate-solubilizing activity, and providing nitrogen to plants [4]. In addition, endophytic bacteria are an effective agent for stimulating plant secondary metabolism and for improving or producing functional components [4, 8]. These features mentioned stress the potential of the endophytic bacteria to be used as a bioinoculant of agronomical important crops, with the aim to develop more sustainable production systems [9].

The structure and diversity of the endophytic communities have been shown to be potentially influenced by several factors, such as the plant species and genotype, agricultural practices, and environmental conditions [10–12]. To better understand and manipulate the contribution of endophytic bacteria to plants when used as bioinoculants, it is crucial to decipher the community structure, diversity, composition, metabolic processes, adaptability, and beneficial features of the microbiome associated with target crops. Plants have been shown to harbour an enormous diversity of bacteria, including diazotrophs [4, 13, 14]. Therefore, a better understanding of the endophytic bacteria microbiota of host plants may help elucidate their role within their hosts, moving toward the development of more sustainable agronomical practices.

The availability of nitrogen often limits crop productivity, and chemical nitrogen fertilization is widely used to increase the yield of agronomical crops. However, this practice has a high environmental and economic cost, since crop plants are able to use only 50% of the applied fertilizer, while the rest is lost from the plant-soil system through gaseous emissions, runoff erosion, and leaching [15, 16]. The environmental impacts of this loss range from greenhouse effects, ozone layer damage, and acid rain to changes in the global N cycle and nitrate pollution of surface and ground water [17]. In addition, the application of N fertilizer could also inhibit the N_2-fixation process by diazotrophic bacteria in the soil or associated with the plants [18].

These problems emphasize the urgent need for new technologies based on plant growth-promoting bacteria (PGPB) to help achieve more sustainable agricultural production systems.

Previously, part of the cultivable community associated with sweet sorghum (cv. M81E) was described. Isolates from this community showed several plant growth-promoting (PGP) traits, and some of them were described as plant growth promoters of this sweet sorghum cultivar [19].

The aim of the present study was to evaluate the effects of N fertilization on the structure, diversity, abundance, and composition of endophytic and diazotrophic bacterial communities associated with sweet sorghum in fields. To study these communities, DNA fingerprinting, quantitative PCR, and high-throughput sequencing techniques were used.

2. Materials and Methods

2.1. Plant Sampling and DNA Extraction. Sweet sorghum plants were sampled from two different fields with contrasting N fertilization levels: 0 and 100 kg N ha^{-1}, at the cropping region Bella Union-Artigas, Uruguay (30°37′56.0″S 57°21′ 18.0″W). The soil physicochemical characteristics were analyzed as follows: pH 6.2, 42% sand, 25% silt, 33% clay, 1.51% organic matter, and 0.12% total N.

At the laboratory, each collected plant (three plants from each fertilization level) was independently divided into roots and upper and lower stems, which were used for total microbial DNA extraction. The total microbial community DNA was extracted from 1.5 g of each plant fraction using the PowerSoil® DNA Isolation Kit (MO BIO Laboratories, Inc. USA) and purified with the Kit Wizard® DNA Clean-Up System (Promega, USA). Prior to the bacterial DNA extraction, the stem samples were disinfected with 70% EtOH and the epidermis was peeled off with a sterile scalpel. In the case of the roots, the rhizospheric soil was removed by vortexing for 10 min in 0.9% NaCl, the material was disinfected for 10 min in 70% EtOH, surface-sterilized 30 min in 4% sodium hypochlorite, and then rinsed with sterile deionized water. Finally, the root samples were sonicated for 15 min and vortexed for 1 min. Subsequently, the ends of the material were removed with a sterile scalpel and discarded, and sterility tests were conducted on the remainder of the tissue on TSA (trypticase soy agar) plates. DNA preparations were visualized after electrophoresis in a 1% (*w*/*v*) agarose gel in 1x TBE buffer to access their integrity, stained with GoodView (Beijing SBS; Genetech), and stored at −20°C prior to PCR amplification.

2.2. Nested PCR Amplification of the nifH and 16S rRNA Coding Genes. For DGGE analyses, the 16S rRNA and *nifH* gene sequences from stem and root samples were amplified by triplicate, using nested PCR as previously described [20–22].

For 16S rRNA coding gene amplification, the first PCR was carried out with the forward primer 799F and the reverse primer 1492R (Table S1) [20] generating a product of approximately 700 bp. The second step was carried out with the forward primer F968 containing a GC clamp and the reverse primer R1401 (Table S1) [23] yielding a product of 433 bp. For both reactions, the PCR mixture contained 1.0 μl of template DNA (16 ng μl^{-1}), 2x *GoTaq*® Reaction Buffer, 2.1 mM $MgCl_2$, 0.4 mM dNTPs, 0.2 μM of each primer, and 2 U *Taq* polymerase, in a final reaction volume of 25 μl. In the second PCR, 1 μl of the first PCR product was used as a template. The PCR conditions were as follows: 35 cycles consisting of denaturing at 94°C for 20 sec, annealing at 53°C for 40 sec for the first PCR and 48°C for 90 sec for the second PCR, and primer extension for 40 sec for the first PCR and 90 sec for the second PCR at 72°C with a final extension at 72°C for 10 min.

DGGE analyses based on the 16S rRNA gene, of the Proteobacteria, Actinobacteria, and Firmicutes phyla, were performed using specific primers for each group (Table S1). The reaction mixtures and the PCR conditions were as previously reported [20, 23–27], while the conditions for the second PCR were the same as mentioned above.

For the *nifH* gene amplification, the first PCR was carried out with the forward primer FGPH19 and the reverse primer POLR (Table S1) [21, 22], generating a product of 429 bp. The second PCR was carried out with the forward primer POLF containing a GC clamp and the reverse primer AQER (Table S1 [21]), yielding a product of 320 bp. The PCR reaction mixture contained 1 μl of template DNA (16 ng μl^{-1}), 10 μl 5x *GoTaq*® Reaction Buffer, 2.1 mM $MgCl_2$, 0.8 mM dNTPs, 0.2 μM of both sets of primers, and 1.25 U *Taq* polymerase, in a final reaction volume of 50 μl. For the second reaction, 1 μl of the first PCR product was used as a template

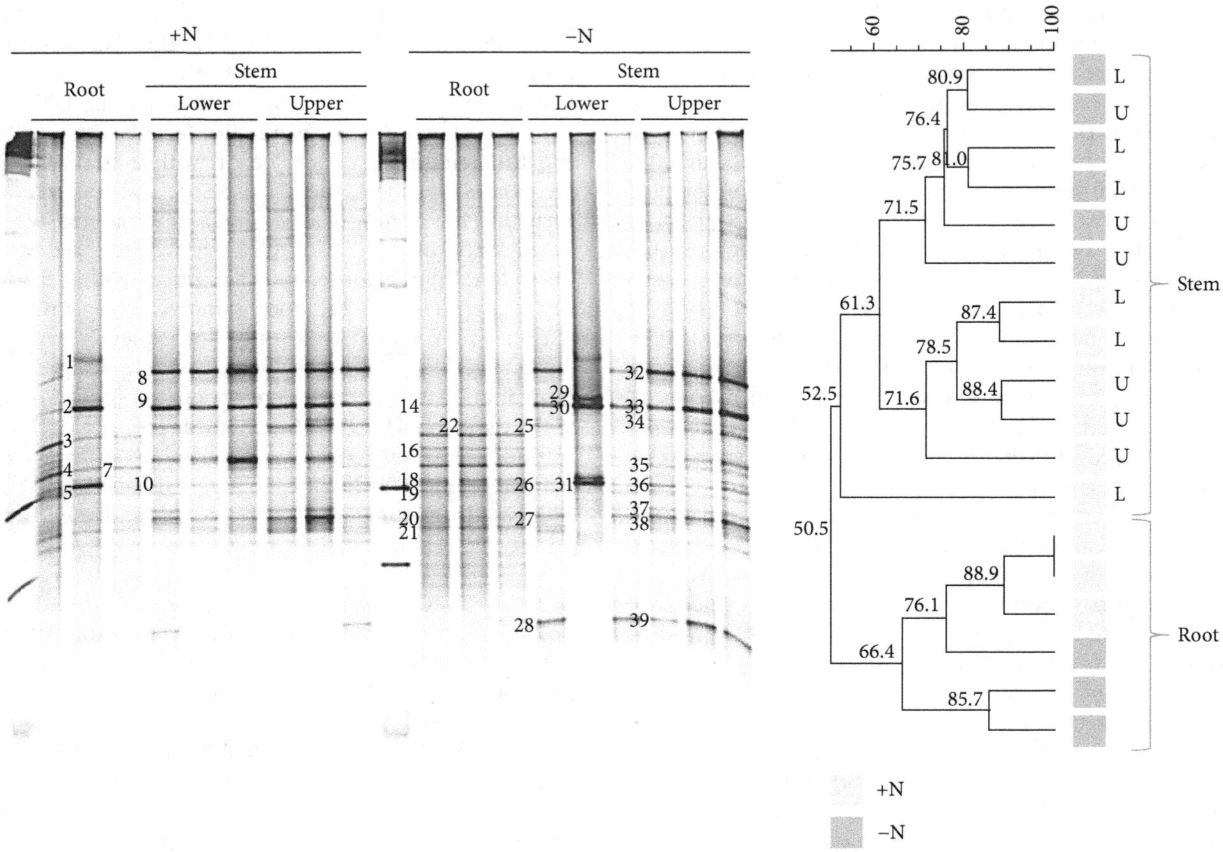

(a)

(b)

FIGURE 1: (a) Denaturing gradient gel electrophoresis (DGGE) fingerprints of 16S rRNA gene fragments amplified from endophytic DNA templates isolated from sweet sorghum plants grown under high (+) and low (−) N fertilization levels (100 and 0 kg N ha^{-1}, respectively). Numbers indicate the excised bands from which sequences were determined. (b) Dendrogram obtained using the unweighted pair group method with mathematical averages and DICE similarity coefficients. Grey and black squares: high and low nitrogen fertilization levels, respectively. L and U: lower and upper stem parts, respectively.

for the next reaction. The PCR conditions were as follows: 30 cycles consisting of denaturing at 94°C for 1 min, annealing at 50°C for 1 min for the first PCR and 55°C for the second PCR, and primer extension for 2 min at 72°C with a final extension at 72°C for 5 min. The amplification products obtained were analyzed on 1% (*w/v*) agarose gels via electrophoresis in TAE buffer (20 mM Tris-acetate, 0.5 mM EDTA; pH 9) and stained with GoodView (Beijing SBS; Genetech).

2.3. Denaturing Gradient Gel Electrophoresis (DGGE). PCR product (20 μl) was loaded onto 8% (*w/v*) polyacrylamide gels, 1 mm thick, in 1x TAE buffer (20 mM Tris-acetate, 0.5 mM EDTA; pH 9), with a 40–65% urea and formamide denaturant gradient for the study of Alphaproteobacteria, Betaproteobacteria, and Firmicutes, 45–75% for Actinobacteria, and 20–70% for the diazotrophic community. The electrophoresis was run at 60 V for 16 hours using the D-Code system from Bio-Rad Laboratories. The gels were stained for 30 min with 1x SYBR® Green (Invitrogen™) and the bands visualized digitalized using Storm™ (GE Healthcare). Selected bands (numbered in Figures 1 and 2) were excised from the gels with a sterile scalpel, reamplified, and sequenced. For the latter, the bands were eluted at 4°C overnight in 50 μl of Milli-Q water, and 1 μl of each supernatant was used as a PCR template. The reamplification reaction conditions were the same as used for the second cycle described above using the primers F968-GC and R1401 for the 16S rRNA; or the primers POLF-GC and AQER for the *nifH* gene. PCR products obtained were sent for sequencing to Macrogen Inc., Korea. Forward and reverse sequences obtained were assembled using the DNA Baser Sequence Assembler v3.x 302 (2010) (http://www.DnaBaser.com). Nucleotide sequences obtained were identified by BLASTn analyses [28], using the GenBank database from the National Center for Biotechnology Information. All sequences obtained in this study were deposited in the GenBank database under the accession numbers: KY062497-KY062552.

2.4. Bioinformatics and Statistical Analysis. The dendrograms and the binary matrix based on the digitalized image of the DGGE gels were constructed with the UPGMA algorithm with mathematical averages and Dice similarity coefficients using the GelCompar II 6.5 software (Applied Maths NV). The Shannon-Wiener, Simpson, and Chao-1 alpha indices

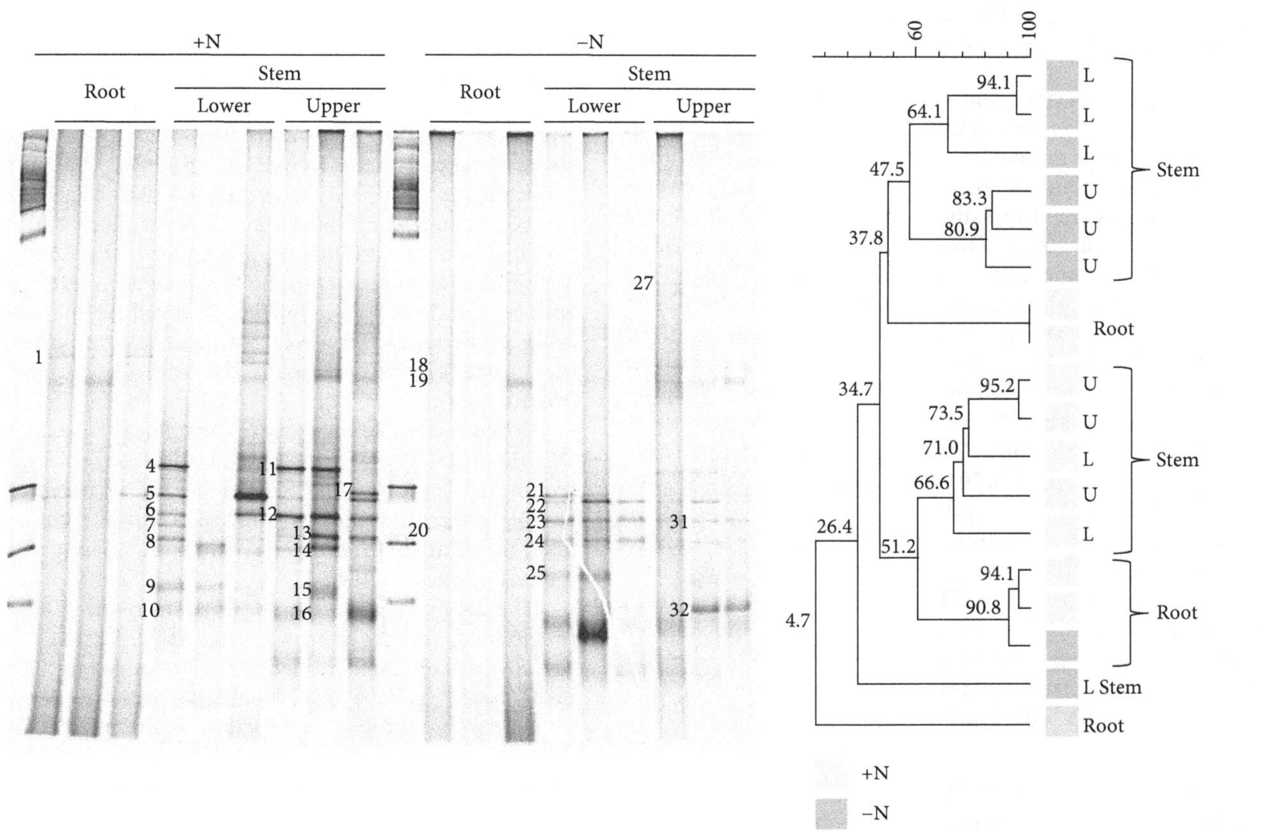

(a) (b)

Figure 2: (a) Denaturing gradient gel electrophoresis (DGGE) fingerprints of *nifH* gene fragments amplified from endophytic DNA templates isolated from sweet sorghum plants grown under high (+) and low (−) N fertilization levels (100 and 0 kg N ha^{-1}, respectively). Numbers indicate the excised bands from which sequences were determined. (b) Dendrogram obtained using the unweighted pair group method with mathematical averages and DICE similarity coefficients. Grey and black squares: high and low nitrogen fertilization levels, respectively. L and U: lower and upper stem parts, respectively.

were calculated from the band patterns using the densitometry curves and then exported into a quantitative numeric matrix, relative to the band surface.

ANOVA test was performed using the Fisher LSD post hoc test at a significance level of $P < 0.05$. All the statistic analyses were performed in InfoStat programme [29].

2.5. Quantification of the Bacterial Community by Quantitative PCR (qPCR). The abundances of 16S rRNA and *nifH* genes were quantified by real-time PCR (qPCR) using the primers 6S-27F/338R and POLF/POLR (Table S1), respectively [21, 22, 30]. The qPCR reaction was performed in a CFX96 Touch Real-Time PCR (Bio-Rad) equipment, and all measurements were performed using the SYBR Green approach. The PCR mixture was 12.5 μl of the iQSYBR Green Supermix (Bio-Rad), 1 μM of each primer, and 4 to 25 ng of DNA template, within a total volume of 25 μl. The qPCR cycle for the 16S rRNA coding gene consisted of a denaturation step for 10 min at 95°C, 40 cycles for 15 s at 95°C, 30 s at 58°C, and 30 s at 72°C. The qPCR cycle for the *nifH* gene consisted of a denaturation step for 5 min at 95°C and 40 cycles for 10 s at 95°C, for 10 s at 59°C, and 30 s at 72°C. Product specificity was confirmed by melting curve analysis (58–95°C, 0.5°C per read, 5 s hold) and visualization in agarose gels, which showed specific product bands at the expected size of 180 bp for the 16S rRNA gene and 360 bp for the *nifH* gene.

For both genes, three replicates in duplicate were used. For the standard curve, triplicates were employed for every run using a known number of each gene from the genome of *Herbaspirillum seropedicae* SmR1 [31], from 6.62×10^1 to 6.62×10^5 copies of the *nifH* gene and from 1.99×10^3 to 1.99×10^6 copies of the 16S rRNA encoding gene.

For the standard curve, mass concentrations of standard DNA were converted into copy concentrations using the following equation [32]:

$$\mathrm{DNA(copy)} = \frac{6.02 \times 10^{23}\,(\mathrm{copy/mol}) \times \mathrm{DNA\ amount\ (g)}}{\mathrm{DNA\ length(bp)} \times 660(\mathrm{g/mol/bp})}. \tag{1}$$

For statistical analyses, an ANOVA test was performed using the InfoStat programme and in those circumstances where significant differences were confirmed, the means were compared using the Tukey test with a $P < 0.05$ [29].

2.6. Ion Torrent® High-Throughput Sequencing of the Bacterial 16S rRNA Coding Gene

2.6.1. Sample Collection and DNA Extraction. For bacterial DNA extraction, same plant samples used in the DGGE experiments were employed. In this case, stems from four sweet sorghum plants were sampled and pooled from two different fields with contrasting N fertilization levels (0 and 100 kg N ha^{-1}). Bacterial DNA extraction was performed following the protocol previously described [33], with the following modifications: stems were peeled with a sterilized scalpel, and 50 g of the inner stem tissues were homogenized in 300 ml of Milli-Q water. Bacterial DNA extraction from the enriched fraction was obtained using a CTAB bacterial DNA isolation method (Joint Genome Institute protocols/ http://1ofdmq2n8tc36m6i46scovo2e.wpengine.netdna-cdn.com/wp-content/uploads/2014/02/JGIBacterial-DNA-isolation-CTAB-Protocol-2012.pdf).

The purity of the extracted DNA was checked with the NanoDrop ND-1000 spectrophotometer (NanoDrop Technologies, Wilmington, DE, USA) (260/280 nm ratio), and it was quantified using an Agilent 2100 Bioanalyzer (Agilent Technologies, Santa Clara, CA, USA). The integrity of the DNA was also confirmed by electrophoresis in a 0.8% agarose gel with 1x TAE buffer.

2.6.2. Sequencing and Data Analysis. PCR amplification of the V6 region of the 16S rRNA gene sequences was carried out using a pool of six forward and reverse degenerate primers each (Table S2) [34]. High-throughput sequencing of the amplicons was conducted using the Ion Torrent Personal Genome Machine (PGM) platform at the Genomic Department of the IIBCE (Uruguay). Raw sequencing reads were checked using the following quality criteria: (i) polyclonal reads with Ion Torrent Suite (5.03) were discarded, (ii) reads were trimmed to 90 bp and shorter ones were discarded, (iii) reads with an expected error rate equal to or greater than 1.0 using Usearch (Edgar, 2010) were also discarded, and (iv) reads that match with the *Sorghum bicolor* complete genome [35] using Bowtie 2 [36] were discarded. Downstream analysis followed the pipeline described by Pylro et al. [37]. A custom set of bash scripts was constructed to automatize the pipeline; these are available on the site https://github.com/mberacochea/sorghum-bicolor-M81E-16S. All sequences obtained in this study were deposited under the NCBI accession number PRJNA352426.

3. Results

3.1. Endophytic Bacterial Communities Associated with Different Tissues of Sweet Sorghum Plants. DNA was recovered from all the sweet sorghum samples (roots and stems) of cv. M81E grown in the field under different N fertilization levels (0 and 100 kg N ha^{-1}). All DNA samples were used as templates for PCR amplification using primers based on general 16S rRNA coding genes, on specific 16S rRNA genes for Alpha-, Beta-, and Gammaproteobacteria, Actinobacteria, Firmicutes, and on the *nifH* gene (Table S1). DNA fragments of the expected size obtained using each set of primers were resolved by DGGE.

The results from the DGGE analysis of the total endophytic bacterial community based on the 16S rRNA gene amplicon are shown in Figure 1(a). The UPGMA-assisted cluster analysis of the DGGE gels revealed the endophytic communities clustered according to the organ analyzed (roots and stem) with 50.5% similarity (Figure 1(b)). Moreover, within the aforementioned organs, the communities are grouped with respect to the N fertilization level analyzed, with 61.3 and 66.4% similarities observed for the stems and roots, respectively. However, no distinct community structuring was observed between the lower and upper stems (Figure 1(b)).

When the whole endophytic community was analyzed by DGGE, alpha diversity indices (Simpson 1-D, Chao-1, and Shannon H) showed that the higher values were obtained for the root communities at the low N fertilization level, but no significant differences were observed among the treatments analyzed (Table S3).

The DNA bands were retrieved from the DGGE gels, reamplified, and sequenced (Figure 1(a)). BLASTn analyses revealed that all the identified genera belonged to the phylum Proteobacteria (Table S2), which was primarily represented by genera from the classes Betaproteobacteria (*Duganella*, *Aquabacterium*, *Bordetella*, and *Massilia*) and Gammaproteobacteria (*Pantoea*, *Salmonella*, *Klebsiella*, *Kosakonia*, *Pseudomonas*, *Serratia*, and *Stenotrophomonas*).

Interestingly, when the phyla Proteobacteria (classes Alpha, Beta, and Gamma), Firmicutes, and Actinobacteria were analyzed using specific primers based on the 16S rRNA gene amplicon (Table S1), the UPGMA-assisted cluster analysis of the DGGE gels revealed that the structure of those communities did not cluster according to the plant organ (roots and stem) or to the fertilization level analyzed (0 and 100 kg N ha^{-1}) (data not shown). Selected bands retrieved from each DGGE gel are shown in Table S4.

Analysis of the Alphaproteobacteria DNA bands amplified using the specific primers and retrieved from the DGGE gels showed that only 45% of the sequences were related to the expected genera, such as *Agrobacterium*, *Ancylobacter*, *Brevundimonas*, and *Pleomorphomonas* (Table S4). In contrast, analysis of the Betaproteobacteria bands isolated from the DGGE gels, which were amplified using the specific primers, revealed that 100% of the sequences were related to the expected genera. From the Betaproteobacteria DGGE gel, only the genera *Massilia* and *Methyloversatilis* were identified (Table S4). Despite the specificity of the primers, none of the selected bands sequenced from the Gammaproteobacteria DGGE gels was assigned to a genus belonging to this class. By contrast, the identities of the bands retrieved from the Actinobacteria DGGE gels showed that 100% of the genera identified were related to the expected phyla, including *Curtobacterium*, *Microbacterium*, *Nocardia*, *and Sediminihabitans* (Table S4). Similarly, the sequence identities of the bands retrieved from the Firmicutes DGGE gel were also 100% related to the expected phyla, including *Bacillus*, *Macrococcus*, *Staphylococcus*, and *Exiguobacterium* (Table S4).

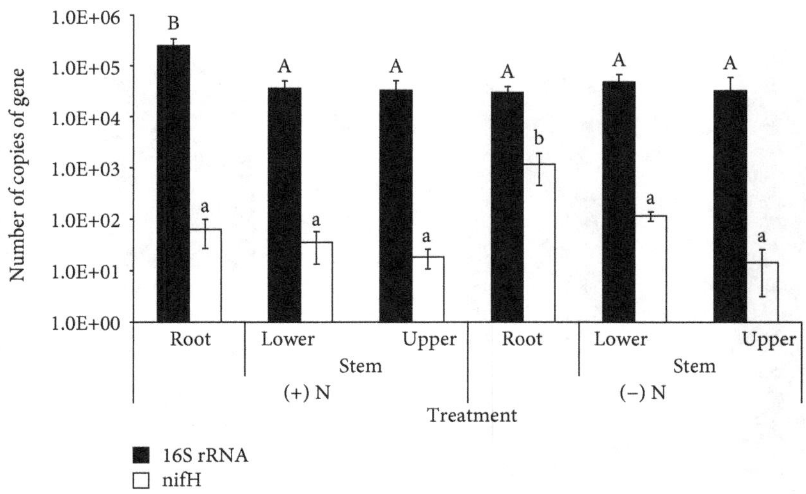

FIGURE 3: Quantification of 16S rRNA and *nifH* genes copies in samples taken from different organs of sweet sorghum plants (cv. M81E) grown in the field under high (+) and low (−) N fertilization (100 and 0 kg N ha^{-1}, respectively). Means within two treatments that have the same letter are not significantly different by Tukey test with a $P < 0.05$.

In contrast to the 16S rRNA-based data, analysis of the endophytic-diazotrophic bacterial community based on DGGE gels with the *nifH* gene amplicons showed that the stems produced a greater number of bands than the roots in both treatments analyzed (Figure 2(a)). In addition, the UPGMA-assisted cluster analysis of the endophytic-diazotrophic community structure showed that this community was clustered according to treatment (0 and 100 kg N ha^{-1}) that exhibited 34.7% similarity, except for a single root replicate. Moreover, within these treatments, the communities were grouped according to the organs analyzed, with 51.2 and 37.8%, respectively. In particular, under low N fertilization conditions, the stem community separated into two groups (lower and upper stem samples) that exhibited 47.5% similarity (Figure 2(b)).

The diversity of diazotrophic bacterial communities was also evaluated based on the DGGE gels obtained. In this case, higher alpha index values were obtained in the stems from the high N fertilization condition. However, in all the treatments analyzed, no significant differences were observed (Table S5). Selected bands from the DGGE gels in which the diazotrophic community was analyzed were excised and reamplified, and the products were sequenced (Figure 2(a)). BLASTn analyses revealed that all the bands were closely related to *nifH* genes of Gammaproteobacteria (71%), Betaproteobacteria, and Cyanobacteria (4%), while 21% were related to unculturable bacteria (Table S5). From the last group, the first hit from the BLASTn analysis that matched to a culturable strain was also taken into account. In this case, all the sequences were closely associated with *nifH* genes from Gammaproteobacteria members (Table S5).

3.2. Quantification of the Endophytic and Diazotrophic-Endophytic Bacterial Communities. The abundances of the bacterial endophytic and endophytic-diazotrophic communities were assessed by qPCR using the 16S rRNA and *nifH* genes. The standard curves for the 16S rRNA gene amplification showed a linear correlation (R^2) of 0.99, corresponding to a PCR efficiency of 90%; for the *nifH* gene, the linear correlation was 0.98, corresponding to a PCR efficiency of 110%. The 16S rRNA gene abundance varied from 3.0×10^4 to 2.5×10^5, while for the *nifH* gene, the number of copies varied from 1.9×10^1 to 1.2×10^3 (Figure 3).

The number of copies of the 16S rRNA gene in the roots of plants grown under high N fertilization (+N) conditions was significantly higher than in all other conditions analyzed (Figure 3). By contrast, the number of *nifH* gene copies varied markedly in the roots and stems (lower and upper) of plants grown under low N fertilization (−N) conditions, while no differences were observed within the plants grown under +N conditions (Figure 3). Indeed, under −N conditions, the abundance of the *nifH* genes was significantly higher in the roots than in the other organs analyzed.

3.3. Bacterial Community Composition. After quality control, the number of retained reads was 41,736 and 44,890 for the DNA samples from plants grown under +N and −N conditions, respectively. Rarefaction curves were used to assess OTU richness and showed that an asymptote was reached for both treatments analyzed, with a higher number of OTUs observed in the +N treatment (Figure S1).

The relative abundance of the microbial clades at two taxonomic levels (phylum and class) is summarized in Figure 4. At the phylum level, OTUs related to Proteobacteria and Firmicutes dominated both treatments analyzed, with relative abundances of 75/65 and 18/27% in each treatment (+/−N), respectively. In addition, Actinobacteria and Bacteroidetes OTUs accounted for approximately 2% of the relative bacterial abundances in both treatments.

An analysis at the class level showed that in the +N treatment, Gammaproteobacteria was the most abundant class (45%), followed by Betaproteobacteria (21%), Bacilli (18%), Alphaproteobacteria (8.9%), and Actinobacteria (2%). The most abundant class in the −N treatment was also

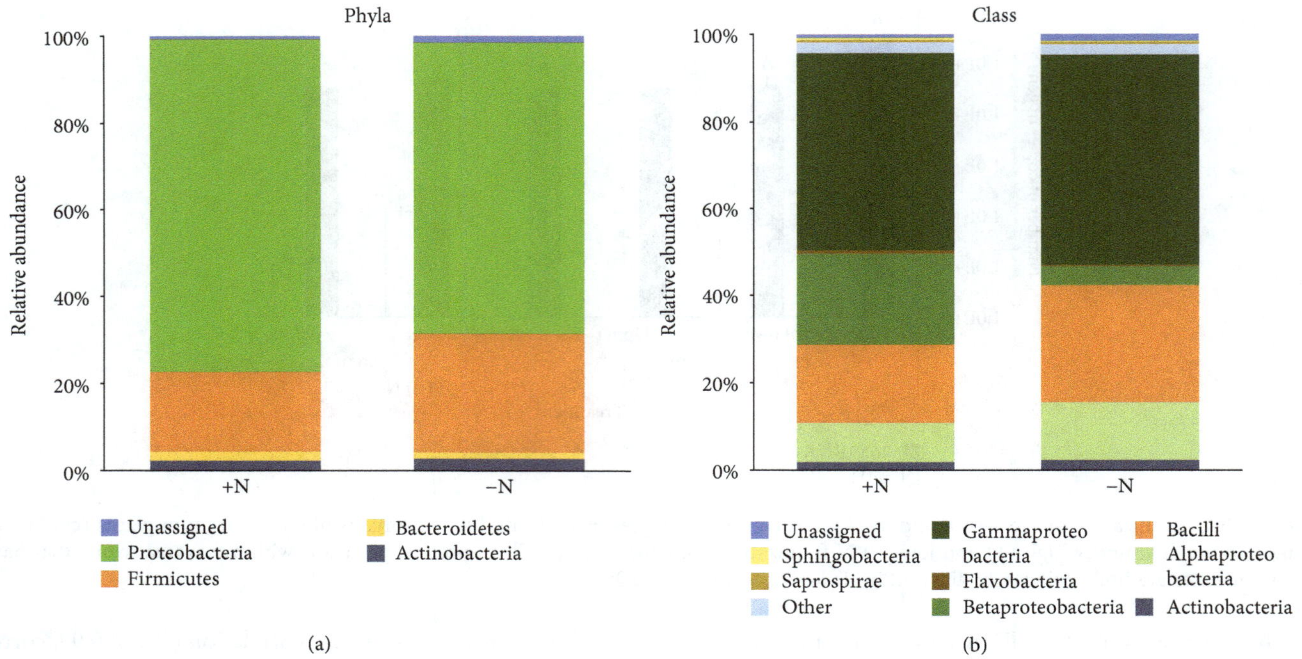

FIGURE 4: Taxonomic composition of the 16S rRNA samples associated with sweet sorghum plants (cv. M81E) grown in the field under different N fertilization levels (+/−N). Relative abundance (over 0.5%) of the bacteria at the level of (a) phylum and (b) class.

Gammaproteobacteria (48%), followed by Bacilli (27%), Alphaproteobacteria (13%), Betaproteobacteria (4.2%), and Actinobacteria (2.4%) (Figure 4).

Analysis of OTUs exhibiting a 10-fold change in abundance between treatments showed that OTUs that were most affected by the N fertilization treatment were members of the Beta- and Gammaproteobacteria, as well as Firmicutes. Within the Betaproteobacteria class, the OTUs associated with the genus *Herbaspirillum* increased from 0.1 to 5.0% after the N fertilization treatment. For the Gammaproteobacteria, large changes were observed in OTUs related to the genus *Erwinia*, showing a decrease from 21.6 to 1.1%, while *Pseudomonas* increased from 1.3 to 17.9% after the N fertilization treatment. Finally, within Firmicutes, OTUs related to the genus *Bacillus* decreased from 14.8 to 8.6% in response to fertilization with 100 kg N ha^{-1}.

4. Discussion

Understanding the effects of agronomical practices, e.g., chemical fertilization, on plant microbiota is necessary to optimize plant-microbiota communities with the aim of improving agronomical sustainability.

In this work, different culture-independent methods were used to evaluate the impact of chemical N fertilization on the endophytic and diazotrophic-endophytic communities associated with the commercial sweet sorghum cv. M81E grown under field conditions. These methodologies are powerful tools that have greatly contributed to identifying the microbial composition and diversity in a wide range of ecosystems, including the interior of plant tissues [38–41].

Our results demonstrated that N fertilization in the field influenced the structure but not the diversity of the endophytic bacterial community within each organ analyzed (roots and stems). This result agrees with a previous study, in which the structures of sorghum stem and root communities were shown to be significantly different [42]. Moreover, the effects of chemical fertilization on the endophytic bacterial community structure were reported for several grasses, such as *Dactylis glomerata*, *Festuca rubra*, and *Lolium perenne* [43, 44]. Different selective processes act together during the recruitment of bacteria that finally colonize the internal tissues of plants [8]. Thus, the combination of N fertilization and the specific plant organ features (e.g., high sugar content) could be the primary factors that influenced the bacterial endophyte structure observed in this study.

Additionally, we observed that the bacterial abundance in the stems of both treatments analyzed was statistically the same and was not influenced by the N fertilization treatment. Nevertheless, the bacterial abundance on roots was increased in response to N fertilization, a case where the diversity was lower but not significantly so. Because root tissues are the primary entry point for bacteria, N fertilization may directly affect the physiological state of roots and the bacteria in the vicinity that can effectively infect the root internal tissues as a consequence [39, 45].

With respect to the bacteria identified, the results showed that the 16S rRNA gene sequences from bands retrieved from DGGE were all related to Proteobacteria, as was previously reported for arable sweet sorghum [46]. Moreover, high-throughput sequencing analysis based on the 16S rRNA gene from the potential endophytic bacterial community also showed that most of the OTUs obtained from both treatments were also related to Proteobacteria. These results agree with those obtained by Maropola et al. [42], who observed that Proteobacteria, in addition to Firmicutes and

Actinobacteria, were the most dominant phyla in both communities analyzed (roots and stems) of sweet sorghum plants.

Interestingly, when Proteobacteria (Alpha and Beta), Actinobacteria, and Firmicutes phyla were analyzed by DGGE, several genera were identified, in agreement with previous a study in which the culturable endophytic community associated with sweet sorghum was analyzed [19, 47, 48]. Moreover, from the DGGE analysis, we showed that the genera *Massilia* from the class Betaproteobacteria, as well as *Bacillus* from the phylum Firmicutes, were well represented. Species of both genera are common inhabitants of the inner tissues of various species of plants, where they play an important role in plant protection and growth promotion [4, 8, 14].

With respect to the effect of the chemical fertilization on the composition of the bacterial community, it was interesting to observe that the Betaproteobacteria, as well as members of the class Bacilli, had large shifts in the number of OTUs present after the application of the N fertilizer. OTUs that increased in abundance in the presence of N fertilizer were associated with the genera *Pseudomonas* and *Herbaspirillum*, while the OTUs that decreased were associated with the genera *Bacillus* and *Erwinia*. The association of representatives of these genera with different plants (including sweet sorghum) has previously been described [14]. Our results support the hypothesis that the physiological state of the plant modulates the bacterial microbiota composition recruiting specific bacteria, a phenomenon that may play a key role in promoting plant health and growth [8, 39, 45]. Further experiments are needed to determine the specific functions of the identified bacterial genera in the microbiota of sweet sorghum.

In addition, we observed that the level of N fertilization was the primary factor affecting the structure of the diazotrophic-endophytic bacterial community, but it did not significantly affect its diversity. In addition, the abundances of the *nifH* gene in the internal tissues of both analyzed treatments were not significantly different, except for the roots of plants grown without N fertilization. In the latter case, the abundance of the diazotrophic-endophytic bacterial community was higher and less diverse than that in the roots of plants grown without N fertilization. These results also support the hypothesis that under certain conditions, a specific endophytic community is recruited from a pool of opportunistic bacteria present in the soil and that those most competitive can infect and survive within the plant tissue environment [8, 42, 46]. In this study, the absence of N fertilization may have prompted plants to recruit diazotrophic bacteria, which may contribute to the plant growth promotion via the BNF process. This is supported by the observation that positive BNF was detected in the roots of sweet sorghum plants by an acetylene reduction assay (ARA) [48]. Moreover, our results are consistent with previous studies in which the *nifH* gene abundance of sweet sorghum, maize, and rice treated with different levels of nitrogen fertilizer was studied [49–51]. However, it should be noted that because the present study was based on the extraction of total DNA from the roots and stems of sorghum plants, we cannot assume that the *nifH* genes were actually active.

The identities of the *nifH* amplicon sequences retrieved from the DGGE gels belonged to the phyla Cyanobacteria and Proteobacteria. Within the Proteobacteria, as was observed in the 16S rRNA gene analysis, the most abundant class detected was the Gammaproteobacteria. Similar results were obtained when the diversity of the *nifH* gene pools of sweet sorghum was studied, but in this case, the most abundant classes affected by the chemical N fertilization were Alpha- and Betaproteobacteria [49]. Is interesting to note that within the Gammaproteobacteria, most of the sequences were from the genera *Enterobacter* and *Klebsiella*, which are well described as diazotrophic-endophytic plant growth promoters [52, 53] and as being associated with sweet sorghum plants [19, 48]. These results stress the role that these genera may play as PGP diazotrophs in the microbiota associated with sweet sorghum plants.

5. Conclusions

The results obtained in our study showed that the application of N fertilizer affected the structure, abundance, and composition of the endophytic bacterial communities associated with sweet sorghum plants. This effect was pronounced in the roots of both bacterial communities analyzed and may have depended of the physiological state of the plants. Moreover, specific bacterial classes and genera increased or decreased when the fertilizer was applied. The data obtained in this study contribute to a better understanding of the effects of different agronomical practices on the microbiota associated with this important crop, which may help improve its sustainability.

Abbreviations

DGGE: Denaturing gradient gel electrophoresis
PGPB: Plant growth-promoting bacteria
BNF: Biological nitrogen fixation.

Conflicts of Interest

The authors declare that there is no conflict of interest regarding the publication of this paper.

Acknowledgments

This work was supported by grants from the Sectorial Energy Fund (Project FSE_2011_1_5911) of the Uruguayan National Agency for Innovation and Research (Agencia Nacional de Investigación e Innovación—ANII) and the Uruguayan Program for the Development of the Basic Sciences (Programa de Desarrollo de las Ciencias Básicas—-PEDECIBA). The authors are very grateful to Ing. Agr. Fernando Hackembruch from the Agriculture Department of the Alcoholes Uruguay S.A. (ALUR S.A.) for the plant material supplied and Dr. Claudia Piccini for her critical reading and inputs to the manuscript.

Supplementary Materials

Table S1: primers used for PCR amplification. Table S2: primers used for Ion Torrent pyrosequencing analysis. Table S3: alpha diversity indices. Statistical analysis of the total endophytic and diazotrophic-endophytic bacterial community associated with sweet sorghum cv. M81E, grown under different N fertilization levels (+/−N). Table S4: identification by NCBI BLASTn of 16S rRNA sequences retrieved from DGGE bands. Table S5: identification by NCBI BLASTx of nifH sequences retrieved from DGGE bands. Figure S1: rarefactions curves of sweet sorghum samples treated with high and low N fertilization levels (+N and −N, respectively). *(Supplementary Materials)*

References

[1] A. Almodares and M. R. Hadi, "Production of bioethanol from sweet sorghum: a review," *African Journal of Agricultural Research*, vol. 4, pp. 772–780, 2009.

[2] R. Mendes, P. Garbeva, and J. M. Raaijmakers, "The rhizosphere microbiome: significance of plant beneficial, plant pathogenic, and human pathogenic microorganisms," *FEMS Microbiology Reviews*, vol. 37, no. 5, pp. 634–663, 2013.

[3] B. J. E. Schulz and C. J. C. Boyle, "What are endophytes?," in *Microbial Root Endophytes*, B. Schulz, C. Boyle, and T. N. Sieber, Eds., Springer, Berlin, Heidelberg, 2006.

[4] G. Santoyo, G. Moreno-Hagelsieb, M. del Carmen Orozco-Mosqueda, and B. R. Glick, "Plant growth-promoting bacterial endophytes," *Microbiological Research*, vol. 183, pp. 92–99, 2016.

[5] D. Egamberdieva, S. J. Wirth, V. V. Shurigin, A. Hashem, and E. F. Abd_Allah, "Endophytic bacteria improve plant growth, symbiotic performance of chickpea (*Cicer arietinum* L.) and induce suppression of root rot caused by *Fusarium solani* under salt stress," *Frontiers in Microbiology*, vol. 8, pp. 1–13, 2017.

[6] R. D. Lally, P. Galbally, A. S. Moreira et al., "Application of endophytic *Pseudomonas fluorescens* and a bacterial consortium to *Brassica napus* can increase plant height and biomass under greenhouse and field conditions," *Frontiers in Plant Science*, vol. 8, pp. 1–10, 2017.

[7] Y. Liu, L. Cao, H. Tan, and R. Zhang, "Surface display of ACC deaminase on endophytic *Enterobacteriaceae* strains to increase saline resistance of host rice sprouts by regulating plant ethylene synthesis," *Microbial Cell Factories*, vol. 16, no. 1, pp. 214–219, 2017.

[8] P. R. Hardoim, L. S. van Overbeek, and J. D. Elsas, "Properties of bacterial endophytes and their proposed role in plant growth," *Trends in Microbiology*, vol. 16, no. 10, pp. 463–471, 2008.

[9] A. V. Sturz, B. R. Christie, and J. Nowak, "Bacterial endophytes: potential role in developing sustainable systems of crop production," *Critical Reviews in Plant Sciences*, vol. 19, no. 1, pp. 1–30, 2000.

[10] P. G. Dennis, A. J. Miller, and P. R. Hirsch, "Are root exudates more important than other sources of rhizodeposits in structuring rhizosphere bacterial communities?," *FEMS Microbiology Ecology*, vol. 72, no. 3, pp. 313–327, 2010.

[11] P. R. Hardoim, L. S. van Overbeek, G. Berg et al., "The hidden world within plants: ecological and evolutionary considerations for defining functioning of microbial endophytes," *Microbiology and Molecular Biology Reviews*, vol. 79, no. 3, pp. 293–320, 2015.

[12] K. Ulrich, A. Ulrich, and D. Ewald, "Diversity of endophytic bacterial communities in poplar grown under field conditions," *FEMS Microbiology Ecology*, vol. 63, no. 2, pp. 169–180, 2008.

[13] M. R. R. Coelho, I. E. Marriel, S. N. Jenkins, C. V. Lanyon, L. Seldin, and A. G. O'Donnell, "Molecular detection and quantification of *nifH* gene sequences in the rhizosphere of sorghum (*Sorghum bicolor*) sown with two levels of nitrogen fertilizer," *Applied Soil Ecology*, vol. 42, no. 1, pp. 48–53, 2009.

[14] M. Rosenblueth and E. Martínez-Romero, "Bacterial endophytes and their interactions with hosts," *Molecular Plant-Microbe Interactions*, vol. 19, no. 8, pp. 827–837, 2006.

[15] D. Tilman, "The greening of the green revolution," *Nature*, vol. 396, no. 6708, pp. 211-212, 1998.

[16] P. M. Vitousek, J. D. Aber, R. W. Howarth et al., "Human alteration of the global nitrogen cycle: sources and consequences," *Ecological Applications*, vol. 7, no. 3, pp. 737–750, 1997.

[17] A. O. Adesemoye and J. W. Kloepper, "Plant-microbes interactions in enhanced fertilizer-use efficiency," *Applied Microbiology and Biotechnology*, vol. 85, no. 1, pp. 1–12, 2009.

[18] M. M. Roper and V. V. S. R. Gupta, "Enhancing non-symbiotic N_2 fixation in agriculture," *The Open Agriculture Journal*, vol. 10, no. 1, pp. 7–27, 2016.

[19] C. Mareque, C. Taulé, M. Beracochea, and F. Battistoni, "Isolation, characterization and plant growth promotion effects of putative bacterial endophytes associated with sweet sorghum (*Sorghum bicolor* (L) Moench)," *Annales de Microbiologie*, vol. 65, no. 2, pp. 1057–1067, 2015.

[20] M. K. Chelius and E. W. Triplett, "The diversity of archaea and bacteria in association with the roots of Zea mays L," *Microbial Ecology*, vol. 41, no. 3, pp. 252–263, 2001.

[21] F. Poly, L. Ranjard, S. Nazaret, F. Gourbiere, and L. J. Monrozier, "Comparison of *nifH* gene pools in soils and soil microenvironments with contrasting properties," *Applied and Environmental Microbiology*, vol. 67, no. 5, pp. 2255–2262, 2001.

[22] P. Simonet, M. C. Grosjean, A. K. Misra, S. Nazaret, B. Cournoyer, and P. Normand, "Frankia genus-specific characterization by polymerase chain reaction," *Applied and Environmental Microbiology*, vol. 57, no. 11, pp. 3278–3286, 1991.

[23] H. Heuer, M. Krsek, P. Baker, K. Smalla, and E. M. Wellington, "Analysis of Actinomycete communities by specific amplification of genes encoding 16S rRNA and gel-electrophoretic separation in denaturing gradients," *Applied and Environmental Microbiology*, vol. 63, no. 8, pp. 3233-3241, 1997.

[24] J. K. Brons and J. D. Van Elsas, "Analysis of bacterial communities in soil by use of denaturing gradient gel electrophoresis and clone libraries, as influenced by different reverse primers," *Applied and Environmental Microbiology*, vol. 74, no. 9, pp. 2717–2727, 2008.

[25] P. Garbeva, J. A. Van Veen, and J. D. Van Elsas, "Predominant *Bacillus* spp. in agricultural soil under different management regimes detected via PCR-DGGE," *Microbial Ecology*, vol. 45, no. 3, pp. 302–316, 2003.

[26] N. C. M. Gomes, H. Heuer, J. Schönfeld, R. Costa, L. Mendonça-Hagler, and K. Smalla, "Bacterial diversity of the rhizosphere of maize (Zea mays) grown in tropical soil studied by temperature gradient gel electrophoresis," *Plant and Soil*, vol. 232, no. 1/2, pp. 167–180, 2001.

[27] H. Heuer, K. Hartung, G. Wieland, I. Kramer, and K. Smalla, "Polynucleotide probes that target a hypervariable region of 16S rRNA genes to identify bacterial isolates corresponding to bands of community fingerprints," *Applied and Environmental Microbiology*, vol. 65, no. 3, pp. 1045–1049, 1999.

[28] S. F. Altschul, W. Gish, W. Miller, E. W. Myers, and D. J. Lipman, "Basic local alignment search tool," *Journal of Molecular Biology*, vol. 215, no. 3, pp. 403–410, 1990.

[29] J. A. Di Rienzo, F. Casanoves, M. G. Balzarini, L. Gonzalez, M. Tablada, and C. W. Robledo, *InfoStat Versión*, Grupo InfoStat, FCA, Universidad Nacional de Córdoba, Argentina, 2016, http://www.infostat.com.ar.

[30] D. Bulgari, P. Casati, F. A. Quaglino, and P. A. Bianco, "Endophytic bacterial community of grapevine leaves influenced by sampling date and phytoplasma infection process," *BMC Microbiology*, vol. 14, no. 1, p. 198, 2014.

[31] T. Pellizzaro Pereira, F. P. do Amaral, P. Dall'Asta, F. C. A. Brod, and A. C. M. Arisi, "Real-time PCR quantification of the plant growth promoting bacteria *Herbaspirillum seropedicae* strain SmR1 in maize roots," *Molecular Biotechnology*, vol. 56, pp. 660–670, 2014.

[32] J. A. Whelan, N. B. Russell, and M. A. Whelan, "A method for the absolute quantification of cDNA using real-time PCR," *Journal of Immunological Methods*, vol. 278, no. 1-2, pp. 261–269, 2003.

[33] H. X. Wang, Z. L. Geng, Y. Zeng, and Y. M. Shen, "Enriching plant microbiota for a metagenomic library construction," *Environmental Microbiology*, vol. 10, no. 10, pp. 2684–2691, 2008.

[34] S. Jünemann, K. Prior, R. Szczepanowski et al., "Bacterial community shift in treated periodontitis patients revealed by ion torrent 16S rRNA gene amplicon sequencing," *PLoS One*, vol. 7, no. 8, article e41606, 2012.

[35] A. H. Paterson, J. E. Bowers, R. Bruggmann et al., "The Sorghum bicolor genome and the diversification of grasses," *Nature*, vol. 457, no. 7229, pp. 551–556, 2009.

[36] B. Langmead and S. L. Salzberg, "Fast gapped-read alignment with Bowtie 2," *Nature Methods*, vol. 9, no. 4, pp. 357–359, 2012.

[37] V. S. Pylro, L. F. W. Roesch, D. K. Morais, I. M. Clark, P. R. Hirsch, and M. R. Tótola, "Data analysis for 16S microbial profiling from different benchtop sequencing platforms," *Journal of Microbiological Methods*, vol. 107, pp. 30–37, 2014.

[38] A. Campisano, L. Antonielli, M. Pancher, S. Yousaf, M. Pindo, and I. Pertot, "Bacterial endophytic communities in the grapevine depend on pest management," *PLoS One*, vol. 9, no. 11, article e112763, 2014.

[39] T. F. da Silva, R. E. Vollú, J. M. Marques, J. F. Salles, and L. Seldin, "The bacterial community associated with rose-scented geranium (*Pelargonium graveolens*) leaves responds to anthracnose symptoms," *Plant and Soil*, vol. 414, no. 1-2, pp. 69–79, 2017.

[40] D. Fischer, B. Pfitzner, M. Schmid et al., "Molecular characterisation of the diazotrophic bacterial community in uninoculated and inoculated field-grown sugarcane (*Saccharum* sp.)," *Plant and Soil*, vol. 356, no. 1-2, pp. 83–99, 2012.

[41] A. Sessitsch, P. Hardoim, J. Döring et al., "Functional characteristics of an endophyte community colonizing rice roots as revealed by metagenomic analysis," *Molecular Plant-Microbe Interactions*, vol. 25, no. 1, pp. 28–36, 2012.

[42] M. K. A. Maropola, J.-B. Ramond, and M. Trindade, "Impact of metagenomic DNA extraction procedures on the identifiable endophytic bacterial diversity in *Sorghum bicolor* (L. Moench)," *Journal of Microbiological Methods*, vol. 112, pp. 104–117, 2015.

[43] F. Wemheuer, K. Kaiser, P. Karlovsky, R. Daniel, S. Vidal, and B. Wemheuer, "Bacterial endophyte communities of three agricultural important grass species differ in their response towards management regimes," *Scientific Reports*, vol. 7, no. 1, article 40914, 2017.

[44] F. Wemheuer, B. Wemheuer, D. Kretzschmar et al., "Impact of grassland management regimes on bacterial endophyte diversity differs with grass species," *Letters in Applied Microbiology*, vol. 62, no. 4, pp. 323–329, 2016.

[45] L. C. Carvalhais, P. G. Dennis, B. Fan et al., "Linking plant nutritional status to plant-microbe interactions," *PLoS One*, vol. 8, no. 7, article e68555, 2013.

[46] J.-B. Ramond, F. Tshabuse, C. W. Bopda, D. A. Cowan, and M. I. Tuffin, "Evidence of variability in the structure and recruitment of rhizospheric and endophytic bacterial communities associated with arable sweet sorghum (*Sorghum bicolor (L)* Moench)," *Plant and Soil*, vol. 372, no. 1-2, pp. 265–278, 2013.

[47] J. L. Grönemeyer, C. S. Burbano, T. Hurek, and B. Reinhold-Hurek, "Isolation and characterization of root-associated bacteria from agricultural crops in the Kavango region of Namibia," *Plant and Soil*, vol. 356, no. 1-2, pp. 67–82, 2012.

[48] W. L. Pedersen, K. Chakrabarty, R. V. Klucas, and A. K. Vidaver, "Nitrogen fixation (acetylene reduction) associated with roots of winter wheat and sorghum in Nebraska," *Applied and Environmental Microbiology*, vol. 35, no. 1, pp. 129–135, 1978.

[49] M. R. R. Coelho, M. de Vos, N. P. Carneiro, I. E. Marriel, E. Paiva, and L. Seldin, "Diversity of *nifH* gene pools in the rhizosphere of two cultivars of sorghum (*Sorghum bicolor*) treated with contrasting levels of nitrogen fertilizer," *FEMS Microbiology Letters*, vol. 279, no. 1, pp. 15–22, 2008.

[50] A. Rodríguez-Blanco, M. Sicardi, and L. Frioni, "Plant genotype and nitrogen fertilization effects on abundance and diversity of diazotrophic bacteria associated with maize (*Zea mays* L.)," *Biology and Fertility of Soils*, vol. 51, no. 3, pp. 391–402, 2015.

[51] Z. Tan, T. Hurek, and B. Reinhold-Hurek, "Effect of N-fertilization, plant genotype and environmental conditions on *nifH* gene pools in roots of rice," *Environmental Microbiology*, vol. 5, no. 10, pp. 1009–1015, 2003.

[52] A. L. Iniguez, Y. Dong, and E. W. Triplett, "Nitrogen fixation in wheat provided by Klebsiella pneumoniae 342," *Molecular Plant-Microbe Interactions*, vol. 17, no. 10, pp. 1078–1085, 2004.

[53] C. Taulé, A. Castillo, S. Villar, F. Olivares, and F. Battistoni, "Endophytic colonization of sugarcane (*Saccharum officinarum*) by the novel diazotrophs *Shinella sp.* UYSO24 and *Enterobacter sp.* UYSO10," *Plant and Soil*, vol. 403, no. 1-2, pp. 403–418, 2016.

16

Transposable Elements in the Organization and Diversification of the Genome of *Aegilops speltoides* Tausch (Poaceae, Triticeae)

Olga Raskina

Institute of Evolution University of Haifa, Aba-Hushi Avenue 199, Mount Carmel, Haifa 498838, Israel

Correspondence should be addressed to Olga Raskina; olga@evo.haifa.ac.il

Academic Editor: Graziano Pesole

Repetitive DNA—specifically, transposable elements (TEs)—is a prevailing genomic fraction in cereals that underlies extensive genome reshuffling and intraspecific diversification in the wild. Although large amounts of data have been accumulated, the effect of TEs on the genome architecture and functioning is not fully understood. Here, plant genome organization was addressed by means of cloning and sequencing TE fragments of different types, which compose the largest portion of the *Aegilops speltoides* genome. Individual genotypes were analyzed cytogenetically using the cloned TE fragments as the DNA probes for fluorescence in situ hybridization (FISH). The obtained TE sequences of the Ty1-*copia*, Ty3-*gypsy*, LINE, and CACTA superfamilies showed the relatedness of the *Ae. speltoides* genome to the Triticeae tribe and similarities to evolutionarily distant species. A significant number of clones consisted of intercalated fragments of TEs of various types, in which Fatima (Ty3-*gypsy*) sequences predominated. At the chromosomal level, different TE clones demonstrated sequence-specific patterning, emphasizing the effect of the TE fraction on the *Ae. speltoides* genome architecture and intraspecific diversification. Altogether, the obtained data highlight the current species-specific organization and patterning of the mobile element fraction and point to ancient evolutionary events in the genome of *Ae. speltoides*.

1. Introduction

Repetitive DNA—specifically, transposable elements (TEs)—constitutes at least 45% of the human genome, wherein the fraction of long interspersed nucleotide element (LINE) retrotransposons is 17% [1]. In plants, TEs comprise up to 80% of the genomes, with prevailing long terminal repeat (LTR) families of Ty1-*copia* and Ty3-*gypsy* retrotransposons [2, 3], which vary extensively in their sequence motifs and abundances, even between closely related species [4, 5]. Mobile elements move to new sites in the genome either through an RNA intermediate via a copy-and-paste mechanism (retrotransposons of Class I) or directly through a cut-and-paste mechanism (transposons of Class II) [1, 2, 6], generating the basis for genetic variability in somatic and generative tissues and resulting in intraspecific variations [7, 8]. TEs modify the host genome via insertional mutagenesis, affect both the expression of neighboring genes and translation, and contribute to new gene generation [9–12]. TE mobilization, especially under conditions of environmental stress and/or hybridization, causes prompt karyotype changes that accompany speciation [13–15].

Many epigenetically silent copies and fragments of TEs accumulate in the genome as an integral part of heterochromatin [10, 16], and the methylation and epigenetic remodeling of heterochromatin-specific repeats have been involved in the siRNA-mediated transcriptional silencing of full-length, transpositionally competent TEs [17, 18]. At the cytological level, heterochromatic DNA is traced as condensed chromatin blocks throughout the cell cycle, except during replication in the late S-phase [19]; the replication of euchromatic gene-rich DNA occurs earlier in the S-phase. Nuclear chromatin organization and dynamics are associated with genome functioning; during cell differentiation, gene replication and expression timing can change due to repositioning in the nuclei and chromatin remodeling [20]. Regardless of whether high polymorphism is present, the heterochromatin pattern is an integral chromosome- and species-specific characteristic. In the wild, ongoing chromosomal rearrangements lead to considerable changes in the numbers, sizes, and positions of highly repetitive DNA clusters and underlie the divergence of natural populations [21].

Despite the large amount of accumulated data, the significance of the complex repetitive DNA fraction in the eukaryotic genome restructuring and functioning is still not completely understood. Here, plant genome organization is addressed in terms of the genomic content and chromosomal patterns of different TE types. The present study was conducted using wild diploid ($2n = 2x = 14$) predominantly cross-pollinated, but self-compatible, goatgrass, *Aegilops speltoides* Tausch (sect. Sitopsis, Triticeae), which is considered the G- and B-genome ancestor of wild and cultivated allopolyploid wheats [22]. In natural panmictic populations, *Ae. speltoides* is presented with two morphs, ssp. *ligustica* (dominant) and ssp. *aucheri* (recessive) [23, 24]. In addition, plants with intermediate *ligustica*/*aucheri* phenotypes have been revealed in natural populations, suggesting genetic changes in the linked group of genes encoding the spike morphology [25]. *Ae. speltoides* has a large genome of 5.5–5.8 pg/1C [26], comprising an extraordinary number of TEs, especially LTR retrotransposons. In the wild, *Ae. speltoides* possesses a wide spectrum of chromosomal rearrangements and extranumerary B chromosomes [25, 27]. The next-generation 454 sequencing of the individual genotype showed that the predominant Ty1-*copia* superfamily exceeds 12% of the identified TEs in the *Ae. speltoides* genome and the Ty3-*gypsy* family Fatima makes up to 7% of the TEs [4]. Mobile elements are considered perpetual rebuilders of the *Ae. speltoides* genome, especially in stressful environments, and they are likely recruited into evolutionarily significant events, leading to population divergence and speciation at the diploid level and via allopolyploidy [28–30]. In the present study, the composition of the *Ae. speltoides* genome was investigated by means of cloning and sequencing of fragments of different TE types, which compose the largest portion of the genome and form the major fraction of heterochromatin. The data obtained from TE fragments' cloning and sequencing revealed different types of TEs, which showed the relatedness of *Ae. speltoides* to the Triticeae tribe and similarities to evolutionarily distant species. At the chromosomal level, different TE clones demonstrated sequence-specific patterning, highlighting the effect of the TE fraction on the *Ae. speltoides* genome architecture and intraspecific diversification.

2. Materials and Methods

2.1. Plant Material. Original plants of *Ae. speltoides* from contrasting allopatric populations from Cankiri (Turkey; PI 573448, USDA), Ankara (Turkey; PI 573452, USDA), and Katzir (Israel; 2.93, Institute of Evolution University of Haifa), line TS43 from Giv'at Koach (Israel; TS43, Weizmann Institute of Science), and artificial F_1–F_2 intraspecific hybrids of these plants [25] were analyzed using fluorescence in situ hybridization (FISH). The genotype F_1_K5/A2 carries one extranumerical B chromosome ($2n = 2x = 14 + B$) inherited from the K5 maternal genome. This set of plants was used to evaluate and characterize the TE chromosomal patterns under the native and artificial heterozygosity of the *Ae. speltoides*.

2.2. DNA Isolation and Polymerase Chain Reaction (PCR) Amplification of Reverse Transcriptase (RT) Gene Sequences of Ty1-copia, Ty3-gypsy, and LINE Retroelements. Genomic DNA was isolated from the young leaves of the individual TS43 genotype using the CTAB method [31]. Degenerate oligonucleotide primers were used for PCR amplification from the genomic DNA of conserved regions of the RT genes of the Ty1-*copia* [32], Ty3-*gypsy* [33], and LINE [34] retroelements. PCR amplifications were conducted in 25 μl reaction volumes containing 12.5 μl of DreamTaq™ Green PCR Master Mix (2x; Fermentas), 150–200 ng of genomic DNA from TS43 leaves, and each degenerate primer in a final concentration of 2 μM. For PCR amplification of the individual TE clones, 0.5–1.0 ng of plasmid DNA was used as the template and standard T7 and SP6 primers for the pGEM®-T Vector (Promega, USA) were employed in a final concentration of 0.5 μM. The PCR conditions were as follows: an initial denaturation for 4 min at 94°C, 35 cycles of amplification (30 s at 94°C, 1 min at 50°C, 1 min at 72°C), and a final elongation of 10 min at 72°C. The PCR-amplified fragments were purified using the HiYield Gel/PCR DNA Fragments Extraction Kit (RBC Bioscience, Taiwan).

2.3. The Cloning and Sequencing of PCR-Amplified TE Fragments. Total purified PCR-amplified products were ligated into a pGEM®-T Easy Vector and transformed into *Escherichia coli* JM109 blue competent cells according to the standard manufacturer's protocol (pGEM-T Easy Vector System II; Promega, USA). Recombinant clones were isolated using the PureYield™ Plasmid Miniprep System (Promega, USA), screened for inserts by PCR, and, following digestion with *Eco*RI (Supplementary Figure S1), sequenced with universal T7 primer. The obtained sequences of TE fragments, 34 in total, were deposited in the National Center for Biotechnology Information (NCBI) GenBank database with the accession numbers KY404239 (Aesp1) to KY404272 (Aesp34) (Supplementary Table S1).

The selected clones were PCR amplified using plasmid DNA as the template. They were used as the DNA probes for FISH.

2.4. Identification and Comparison of the Sequences. The sequences obtained in this study were analyzed for similarity to known sequences using the BLAST packages provided by the NCBI (https://blast.ncbi.nlm.nih.gov/Blast.cgi) [35], TRansposable Elements Platform (TREP; http://botserv2.uzh.ch/kelldata/trep-db/index.html) [36], SENSOR software (Genetic Information Research Institute (GIRI); http://www.girinst.org/) [37], and Rice Genome Annotation Project (http://rice.plantbiology.msu.edu/index.shtml) [38]. They were compared to each other using the NCBI ALIGN program and CLUSTALW software (Multiple Sequence Alignment by

CLUSTALW; Kyoto University Bioinformatics Center; http://www.genome.jp/tools/clustalw).

2.5. Cytogenetic Analysis of the TEs' Chromosomal Patterns. For the FISH experiments, cytological slides of individual anthers and seedling shoot apical meristems containing well-spread chromosomal plates were used. The chromosome spreads, DNA probe labeling, and FISH procedures were conducted as previously described [39]. Tandem repeats Spelt1 [40], pTa71 (for the localization of 45S rDNA) [41], and As5SDNAE (for the localization of 5S rDNA) [42] were used as the DNA probes for FISH. The PCR-amplified fragments were purified using the HiYield Gel/PCR DNA Fragments Extraction Kit (RBC Bioscience, Taiwan) and used as the DNA probes in the standard oligolabeling protocol as previously described [39, 43]. The DNA probes were directly labeled with Cy-3, fluorescein-12-dUTP, and ATTO-425 (Jena Bioscience, Germany). AT-specific 4′,6-diamidino-2-phenylindole (DAPI) and GC-specific chromomycin A_3 (CMA_3) fluorochromes were used for differential staining to reveal AT-enriched heterochromatin patterns and GC-enriched heterochromatic clusters in the nuclear organizer regions (NORs) on chromosomes 1 and 6 in the *Ae. speltoides* genome. The slides were examined on a Leica DMR microscope equipped with a DFC300 FX CCD camera.

3. Results

3.1. Identification and Comparative Characterization of the Sequenced TE Fragments. Three sets of nucleotide sequences were obtained via the cloning and sequencing of PCR products amplified using degenerate primers for RT of Ty1-*copia* (10 sequences, clones Aesp1 to Aesp10), Ty3-*gypsy* (13 sequences, clones Aesp11 to Aesp23), and LINE (11 sequences, clones Aesp24 to Aesp34) retrotransposons (Supplementary Table S1). PCR with degenerate primers for RT of Ty1-*copia* amplified a mix of approximately 280 bp fragments (Supplementary Figure S1). Comparing the sequences with each other showed that the homology between clones Aesp1 to Aesp6 was 95-96% at 100% coverage (Supplementary Table S1). These sequences show high similarity (83–98%, coverage of 85–99%) to RT of the WIS family of Ty1-*copia* retrotransposons, and clone Asp7 showed the highest similarity to Angela. Clone Aesp8 shows 93% homology at 95% coverage to the Ty1-*copia* element Rada reported in the storage protein activator (spa) locus in *Ae. speltoides* and the *Triticum aestivum* and *T. durum* genomes. Clone Aesp9 was identified as exhibiting similarity to the Ty1-*copia* retroelement LeojygB_RLC_Hvul_LeojygB_Hn582D21 in *Hordeum vulgare*. The Aesp10 sequence showed 88% homology to the chromosome 3B of *T. aestivum* (accession no. HG6703064) and 64–70% similarity to the RT genes of *Oryza sativa*, *Zea mays*, and *Setaria italica* (Supplementary Table S1). Thus, six extremely similar sequences of the Ty1-*copia* type—Aesp1 to Aesp6—showed high identity to the WIS retrotransposon; Aesp7 was extremely similar to Angela; and three sequences—Aesp8, Aesp9, and Aesp10—significantly differed from each other and the group of Aesp1–Aesp7,

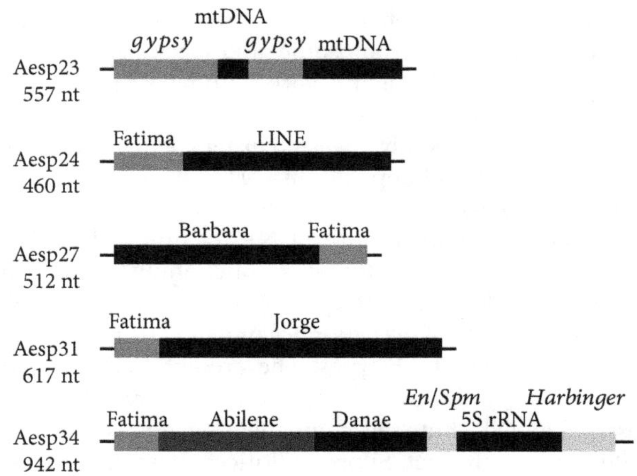

Figure 1: Structural organization of individual cloned transposable element (TE) sequences.

demonstrating homology of 64–85% to non-Triticeae species, such as rice, foxtail millet, false brome, and bamboo.

The cloning of the PCR-amplified fragments of Ty3-*gypsy* retrotransposons and following sequencing of randomly chosen colonies yielded a wide range of sequences of different lengths (Supplementary Figure S1). The analysis for similarity to known sequences showed that clones Aesp11 to Aesp16 (from 361 to 1006 nt in length) exhibited high similarity to the Fatima family of Ty3-*gypsy* retrotransposons; however, they differed from each other, except the Aesp11 and Aesp13 clones, which showed identity of 92%. In addition, the sequences Aesp11 and Aesp13 showed 93–97% similarity to the clones Gas-1 to Gas-5 of *Ae. speltoides*. Two clones, Aesp17 (505 nt) and Aesp18 (502 nt) showed significant homology to Ty3-*gypsy* retrotransposon Carmilla. Two almost identical clones, Aesp20 (997 nt) and Aesp21 (1001 nt), exhibited similarity to Ty3-*gypsy* of *Brachypodium distachyon* and 66–68% of homology at 88–89% coverage to the Ty3-*gypsy* VRN-B1 retrotransposon of allopolyploid wheat. One sequence, Aesp22 (202 nt), was identified as DNA transposon Jorge (TIR, CACTA) of Class II. Clone Aesp23 comprised two fragments showing 73–78% similarity to Ty3-*gypsy* of the chickpea (*Cicer arietinum*) and two fragments of mitochondrial DNA with high similarity to the mitochondrial genomes of *Ae. speltoides* and allopolyploid wheats (Figure 1; Supplementary Table S1). Therefore, the set of 13 clones comprised 11 sequences of Ty3-*gypsy*, wherein the Fatima family is presented by six sequences, two sequences were similar to Carmilla, one clone was similar to Nusif, and two sequences were similar to the VRN-B1 Ty3-*gypsy* retrotransposon. The Aesp23 sequence comprises fragments of Ty3-*gypsy* retroelements and mitochondrial DNA. One sequence was identified as CACTA transposon Jorge of Class II.

PCR with degenerate primers for RT of LINE elements and the following cloning and sequencing of randomly selected colonies yielded three clones, namely Aesp24 (460 nt), Aesp25 (384 nt), and Aesp26 (710 nt), comprising similar sequences to non-LTR LINE-like elements and short fragments similar to Ty3-*gypsy* Fatima

(Supplementary Table S1, Figure S1). Four extremely similar clones, Aesp27 to Aesp30 (509–512 nt) comprised 404 nt of Ty1-*copia* element Barbara and a fragment of 82 nt of Ty3-*gypsy* Fatima (Figure 1). Two almost identical sequences, Aesp31 (617 nt) and Aesp32 (605 nt), comprised fragments comparable to DNA transposon Jorge (TIR, CACTA; 90–93% identity) and 81 nt of Ty3-*gypsy* Fatima. Clone Aesp33 (491 nt) showed 87–89% similarity to the Ty3-*gypsy* Fatima family. The Aesp34 (942 nt) sequence comprised three Ty3-*gypsy* fragments similar to Fatima (84 nt), Abilene (308 nt), and Danae (190 nt), a short fragment of 47 nt of the *En/Spm* (TIR, CACTA) transposon, a fragment of 201 nt of 5S rDNA, and a fragment of 94 nt of the *Harbinger* (TIR, *Harbinger*) transposon. In turn, the part of the Aesp34 sequence (540 nt) containing the fragments of Danae, *En/Spm*, 5S rDNA, and *Harbinger* exhibited 84% homology (at 100% coverage) to the genomic scaffold of chromosome 3B of *T. aestivum* (accession no. HG670306.1) and 70% homology to the cytokinin oxidase/dehydrogenase (CKX2.5) gene (accession no. JN381556). Thus, in this set of 11 clones, only three sequences were similar to non-LTR LINE elements, while four clones comprised fragments of Ty1-*copia* Barbara and Ty3-*gypsy* Fatima, one sequence was highly similar to Fatima, and two clones comprised fragments of CACTA transposons Jorge and Fatima. One clone, Aesp34, contained fragments of three different Ty3-*gypsy* elements, two fragments of DNA transposons, and a fragment of 5S (Figure 1; Supplementary Table S1).

3.2. Chromosomal Patterning of Individual TE Clones in the Genome of Ae. speltoides. The genomes of parental plants from Katzir (K5 and K17) and Giv'at Koach (GK) were highly enriched with Spelt1 tandem repeat clusters, in contrast to plants from Ankara (A1 and A2) and Cankiri (C1), and the F_1 and F_2 genomes were heterozygous for chromosomal markers, specifically, Spelt1 clusters [25]. The species-specific Spelt1 tandem repeat typically forms clusters in the distal/terminal heterochromatic chromosomal regions in *Ae. speltoides* (Figures 2(c), 2(e), 2(g)–2(j)).

The cytogenetic analysis displayed certain features in the chromosomal patterning of the different TE sequences obtained in this study. The FISH experiments revealed a pan-chromosomal distribution of Ty1-*copia*, Ty3-*gypsy*, and LINE retrotransposons in the *Ae. speltoides* genome and TE-sequence-specific clustering in certain chromosomal positions.

The Ty1-*copia* retroelement WIS (clone Aesp2) demonstrated dispersed distribution along the chromosomes' length, forming more prominent rare clusters in distal positions on meiotic (Figure 2(a)) and somatic (Figure 2(b)) chromosomes. Extranumerical B chromosome carried a large intercalary WIS cluster and small TE clusters adjacent to 5S rDNA clusters in both arms (Figure 2(a)). The Ty1-*copia* Barbara (clone Aesp29) showed predominant intercalary clustering and significant depletion in the distal/terminal and pericentromeric chromosome regions (Figure 2(c)), in contrast to WIS. The homologs demonstrated significant similarity in the Aesp2 and Aesp29 chromosomal patterns, as shown for individual chromosomes in the small boxes (Figures 2(a)–2(c)).

The Ty3-*gypsy* retrotransposons Fatima (Aesp15), Carmilla (Aesp18), and Nusif (Aesp19; Figures 2(d)–2(g)) demonstrated TE-sequence-specific chromosome patterning and formed numerous intercalary, distal/terminal, and pericentromeric clusters of different sizes and fluorescence intensities. In particular, small terminal clusters of Spelt1 and Carmilla were observed in the long arm of chromosome 5 (Figure 2(e)); differences in homologous chromosome 5 morphology indicated heterozygosity for rearrangements in the C1 genome. These three Ty3-*gypsy* clones flanked the regions of 5S rDNA and 45S rDNA, rather than intercalating in them.

Clone Aesp23, combining fragments of *gypsy* and mitochondrial DNA, demonstrated dispersed distribution throughout all the chromosomes (Figure 2(h)). Intensive clustering was detected in the intercalary, pericentromeric, and distal/terminal chromosome positions.

Two sequences of non-LTR LINE retrotransposons, clones Aesp24 and Aesp25, were dissimilar in their chromosomal distributions. Total fluorescence of Aesp25 appeared significantly lower, and the number of clusters was considerably less compared with Aesp24 (Figures 2(i)-2(j)); however, these rare clusters were extremely distinct.

Thus, all the TE clones demonstrated sequence-specific peculiarities in their distribution, with preferential clustering in certain chromosomal positions, specifically, in the distal/terminal and pericentromeric regions corresponding to DAPI- and Giemsa-positive [44] heterochromatic blocks in the *Ae. speltoides* genome. In many cases, homologous meiotic and somatic chromosomes demonstrated similarities in TE patterning. The same clone(s) showed similar chromosomal distribution/patterning in different genotypes, and in the same genotype(s), different TE clones demonstrated TE-sequence-specific distribution.

4. Discussion

In the present study, TE fragments of different classes and families were sequenced and cytologically visualized in the *Ae. speltoides* genome. The composition and pattern of repetitive DNA largely determine distinctiveness and reflect the evolution of the species. Various types of repetitive DNA constitute the genome of *Ae. speltoides*, wherein the fraction of mobile elements is the largest. Among the sequences of Ty1-*copia* obtained in this research (Supplementary Table S1), there were six fragments of WIS (clones Aesp1 to Aesp6) and one fragment of Angela (Aesp7) belonging to the BARE1 clade, which represents the largest TE portion of the *Ae. speltoides* genome [4]. The pairwise alignment of these clones and sequences in the publicly available databases suggests enrichments of the *Ae. speltoides* genome with highly diverged elements of the *copia* superfamily. In addition, four extremely similar sequences, Aesp27 to Aesp30, were classified as belonging to the Barbara family of LTR Ty1-*copia* retrotransposons. Clone Aesp9 was identified as similar to Ty1-*copia* retroelement LeojygB in *H. vulgare*, and clone Aesp8 was extremely different from all the other sequences, showing high homology to the Ty1-*copia* element Rada reported in

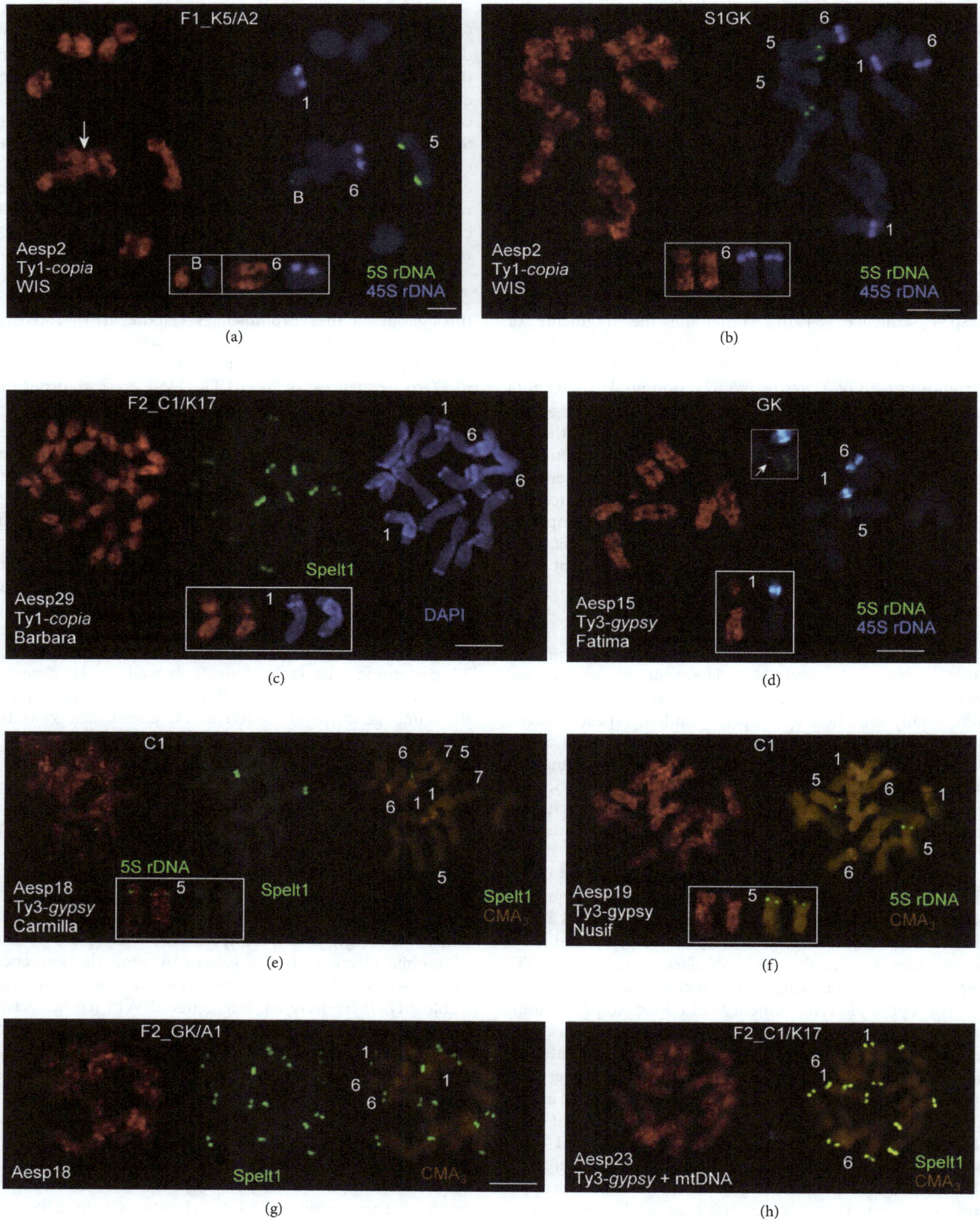

Figure 2: Continued.

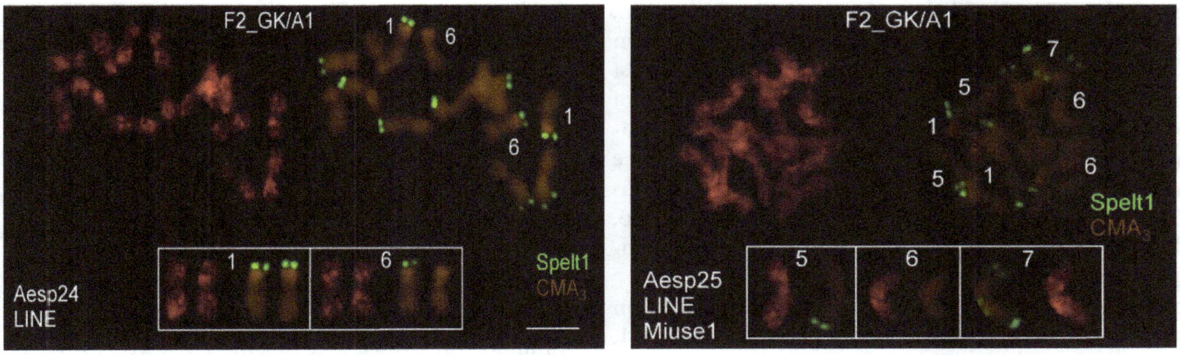

(i) (j)

Figure 2: Fluorescent in situ hybridization (FISH) with individual cloned transposable element (TE) sequences on meiotic and somatic chromosomes of *Aegilops speltoides*. (a–b) Chromosomal distribution of the clone Aesp2 (WIS, Ty1-*copia*) on meiotic and somatic chromosomes of two different genotypes. (a) Meiotic chromosomes of the hybrid genotype F_1_K5/A2 obtained in the cross ♀Katzir-5 × ♂Ankara-2 (K5/A2). In the small box, supernumerary B chromosome and chromosome 6 forming bivalents are enlarged. In the B chromosome, large intercalary TE and small TE clusters adjacent to the 5S rDNA blocks are revealed in both arms. Large TE clusters in both arms in both instances of homologous chromosome 6 are detected. Nonhomologous synapses between chromosome 6 and another bivalent is indicated by an arrow. (b) Somatic chromosomes of the genotype S1GK achieved by self-pollination of the TS43 (Giv'at Koach (GK)) plant. In the small box, instances of homologous chromosome 6 are enlarged; TE clusters are detected in the long arms. (c) Chromosomal distribution of clone Aesp29 (Barbara, Ty1-*copia*) in the hybrid genotype F_2_C1/K17 obtained by self-pollination of the F_1_C1/K17 genotype; cross-combination ♀Cankiri-1 × ♂Katzir-17 (C1/K17). Homologous chromosomes 1 are enlarged in the small box. Significant depletion in TE abundance is observed in the pericentromeric and distal/terminal chromosomal regions, including DAPI-positive AT-enriched heterochromatic and Spelt1 clusters. (d) Distribution of the clone Aesp15 (Fatima, Ty3-*gypsy*) on meiotic chromosomes in the genotype GK (line TS43). Panchromosomal TE clustering is observed; homologous chromosomes are similar in their TE patterns. In the small box, the bivalent of chromosome 1 is enlarged. The nonhomologous synapsis between chromosomes 1 and 5 is shown with an arrow in the small upper box; the ectopic chromatin fiber between bivalents is shown with a dashed line. (e) Chromosomal pattern of clone Aesp18 (Carmilla, Ty3-*gypsy*) in the somatic chromosomes of genotype C1 (Cankiri-1). Intensive clustering throughout the whole chromosome lengths is observed. In the small box, instances of homologous chromosome 5 are enlarged. (f) Patterning of clone Aesp19 (Nusif, Ty3-*gypsy*) in the somatic chromosomes of genotype C1. The homologous chromosomes show similarity in TE patterning. In the small box, instances of chromosome 5 are enlarged. Differences in the chromosome 5′s morphology give evidence for heterozygosity in terms of rearrangement in the C1 genotype. (g) Distribution of clone Aesp18 (Carmilla, Ty3-*gypsy*) in genotype F_2_GK/A1 obtained by self-pollination of the F1_GK/A1 (cross-combination ♀GK × ♂Ankara-1 (GK/A1)). The TE pattern is comparable to the pattern observed in genotype C1 (e). (h) Dispersal distribution of clone Aesp23 (Ty3-*gypsy* + mitochondrial DNA) throughout the somatic chromosomes in genotype F_2_C1/K17. (i) Chromosomal patterning of clone Aesp24 (long interspersed nucleotide element (LINE)) in the F_2_GK/A1 genotype. Large clusters are observed in the intercalary regions. Chromosomes 1 and 6 are enlarged in the small box. (j) Distribution of clone Aesp25 (LINE) in the somatic chromosomes of the F_2_GK/A1 genotype. Rear large TE clusters are revealed in some chromosomes. In the small box, individual chromosomes 5, 6, and 7 are enlarged. Scale bar = 10 μm.

the storage protein activator (spa) locus in *Ae. speltoides* and the *T. aestivum* and *T. durum* genomes. Some clones did not reveal any significant similarity to the sequences mentioned for *Ae. speltoides* in the publicly available databases but showed high homology to the RT gene of diploid and allopolyploid wheats, *H. vulgare*, *O. sativa*, *Z. mays*, *B. distachyon*, *S. italica*, *Phyllostachys edulis*, and *P. heterocycla*. Specifically, clones Aesp8, Aesp9, and Aesp10 showed significant homology to the non-Triticeae species, that is, rice, foxtail, false brome, maize, and bamboo, likely representing the evolutionarily ancient TE fraction in the *Ae. speltoides* genome.

Among the Ty3-*gypsy* fragments obtained in this study, the Fatima element was identified in 16 clones, including those containing other TE types (Aesp31/32, Aesp27–Aesp30, and Aesp34) and mitochondrial DNA (Aesp23; Figure 1). All Fatima sequences obtained in this research showed high homology to retrotransposons in Triticeae species. The highly heterogeneous Fatima family composed up to 7% of the TE fraction in the *Ae. speltoides*, which significantly exceeds the abundances of other *gypsy* families [4], and it was highly abundant in the B-genome of allopolyploid wheat [45, 46]. In addition to Fatima, two sequences, Aesp17 and Aesp18, showed significant homology to Carmilla, and the Aesp19 clone was comparable to Nusif retrotransposons in the genome of *T. aestivum*. In contrast to the high abundance in the genome Fatima elements, two similar Ty3-*gypsy*-like fragments, Aesp20 and Aesp21, showed homology only to the Vrn-B1 retrotransposon in the *T. aestivum* genome and the Ty3-*gypsy* retrotransposon in *B. distachyon*. Thus, the *gypsy*-type sequences obtained in this study demonstrated significant nucleotide diversity and different abundances in the *Ae. speltoides* genome and Triticeae species.

Unlike mammalian genomes, non-LTR LINE retrotransposons exhibit low abundance in plants and are much less represented in GenBank than the *copia* and *gypsy* superfamilies. In the *T. aestivum* genome, LINE elements were identified in the subtelomeric DNA marked by the Spelt52 tandem repeat [47], and they comprised 1.3%

of chromosome 3B [45]. Three clones, Aesp24 to Aesp26, contained sequences showing homology to the known LINE elements in the genomes of cereals. Clone Aesp24 comprised a Fatima fragment of 72 nt and sequence of 358 nt, which is comparable to the non-LTR retrotransposon in *H. vulgare* (Figure 1). The Aesp25 sequence revealed a homology of 68% to the MIUSE1 retrotransposon in *T. monococcum*. The largest sequence, Aesp26, showed 65–80% similarity to the LINE elements in allopolyploid wheat and *B. distachyon*.

In this study, the CACTA superfamily in the TE fraction of *Ae. speltoides* was represented by clones Aesp22, Aesp31, and Aesp32. These sequences showed identity of 90–97% to the DNA transposon Jorge (CACTA, TIR) in the wheat genome (Supplementary Table S1). Among the total TE content, CACTA elements composed 4.9% of chromosome 3B [45] and were identified in the subtelomeric DNA in *T. aestivum* [47]. In *Ae. speltoides*, Jorge made up 1.84% of the TE content [4].

Most sequences obtained in this research comprised TE fragments of various types (Figure 1). In addition, fragments of Fatima and mitochondrial DNA were identified in clone Aesp23; such a sequence composition points to the TEs' effect on the mitochondrial DNA invasion of the *Ae. speltoides* nuclear genome. The complex Aesp34 clone contained three fragments of *gypsy*-type elements, Fatima, Abilene, and Danae and two fragments of different DNA transposons, *En*/*Spm* and *Harbinger*, flanking a short sequence of 5S rRNA (Figure 1). Part of this sequence, that is, Danae–*En*/*Spm*–5S rRNA–*Harbinger* (540 nt in total), showed an identity of 94% to the genomic scaffold of chromosome 3B of *T. aestivum* and 70% homology to the CKX2.5 gene. The high similarity to chromosome 3B likely evidences nonhomologous recombination and/or mobile element transposition activity event(s) in the diploid B-genome progenitor, resulting in the origin of this complex fragment. The existence of such a comprehensive sequence in the genome is the potential target for illegitimate recombination, in which, specifically, the regular 5S rDNA cluster could be involved. In particular, the insertions of *En*/*Spm* were considered as a potential factor in the 5S rDNA mobility in the genome of *Ae. speltoides* [28, 39]. Furthermore, clone Aesp34 could be evidence for the existence of other evolutionarily conserved complex sequences in the genome and reflect species- and chromosome-specific repetitive DNA patterning/organization.

The fragments of TE sequences obtained in this study, on the one hand, demonstrated dispersed distribution throughout the chromosomes and, on the other, exhibited sequence-specific peculiarities in their patterns. The significant clustering of different TEs was observed in the chromosomal regions adjacent to the 5S rDNA and 45S rDNA loci and in the pericentromeric and distal/terminal positions, that is, chromosome regions corresponding to the Giemsa-positive and tandem repeat-enriched heterochromatic blocks forming species- and chromosome-specific patterns in the *Ae. speltoides* genome [25, 44]. In euchromatin, widely interspersed retrotransposons demonstrate nonrandom and TE-sequence-specific chromosomal clustering [43] (present work, Figure 2), indicating the TEs' impact on the genome architecture and diversification. However, widespread throughout the chromosomes, various TEs and nested mobile elements of different classes and families [11, 46, 48, 49] (present work) are the hotspots for illegitimate recombination (Figures 2(a) and 2(d)), provoking chromosome aberrations in both hetero- and euchromatin regions and eventually leading to genome instability. In contrast, the availability of the same/highly similar sequences in virtually any genome region could prevent the appearance of numerous deleterious DNA lesions, especially in the cases of critical double-strand breaks, as repetitive DNAs—primarily TEs—-could serve as the overabundant and ubiquitous templates for nonhomologous DNA repair [50]. Thus, mobile elements perform a dual function in the genome as the main structural fraction of chromatin and, at the same time, a platform for chromosome/genome restructuring under the influence of a variety of internal and external factors, resulting in widespread intraspecific polymorphism in the repetitive DNAs' patterns and abundances in natural populations of *Ae. speltoides* [27, 30].

5. Conclusions

In the present research, the genome constitution of wild goatgrass, *Ae. speltoides*, was explored by means of cloning and sequencing different types of mobile elements and cytogenetic analysis of individual TE sequences' chromosomal distributions. The obtained TE clones provided evidence for the enrichments of the genome with different types of TEs, which demonstrated wide nucleotide diversity among sequences of the same superfamily. In addition, widespread intercalating events resulted in complex organization in most of the obtained TE clones. Among the TEs that were found in this study, there were sequences common to Triticeae, as well as sequences showing the similarities of *Ae. speltoides* repetitive DNA fraction to distant genera and reflecting the evolutionary history of the species. Clear differences in the chromosome patterns of individual TE clones will allow these sequences to be used in future studies for understanding the chromosome/genome organization and repatterning under various internal and external impacts.

Conflicts of Interest

The author declares no conflict of interests.

Acknowledgments

The author is most grateful to Imad Shams for the useful suggestions. This work was supported by the Israel Science Foundation under grant number 723/07.

References

[1] E. S. Lander, L. M. Linton, B. Birren et al., "Initial sequencing and analysis of the human genome," *Nature*, vol. 409, no. 6822, pp. 860–921, 2001.

[2] J. L. Bennetzen, "The contributions of retroelements to plant genome organization, function and evolution," *Trends in Microbiology*, vol. 4, no. 9, pp. 347–353, 1996.

[3] C. Feschotte, N. Jiang, and S. R. Wessler, "Plant transposable elements: where genetics meets genomics," *Nature Reviews Genetics*, vol. 3, no. 5, pp. 329–341, 2002.

[4] C. P. Middleton, N. Stein, B. Keller, B. Kilian, and T. Wicker, "Comparative analysis of genome composition in Triticeae reveals strong variation in transposable element dynamics and nucleotide diversity," *The Plant Journal*, vol. 73, no. 2, pp. 347–356, 2013.

[5] P. Novák, E. Hřibová, P. Neumann, A. Koblížková, J. Doležel, and J. Macas, "Genome-wide analysis of repeat diversity across the family Musaceae," *PLoS One*, vol. 9, no. 6, article e98918, 2014.

[6] T. Wicker, F. Sabot, A. Hua-van et al., "A unified classification system for eukaryotic transposable elements," *Nature Reviews Genetics*, vol. 8, no. 12, pp. 973–982, 2007.

[7] A. R. Muotri, V. T. Chu, M. C. N. Marchetto, W. Deng, J. V. Moran, and F. H. Gage, "Somatic mosaicism in neuronal precursor cells mediated by L1 retrotransposition," *Nature*, vol. 435, no. 7044, pp. 903–910, 2005.

[8] J. K. Baillie, M. W. Barnett, K. R. Upton et al., "Somatic retrotransposition alters the genetic landscape of the human brain," *Nature*, vol. 479, no. 7374, pp. 534–537, 2011.

[9] K. Kashkush, M. Feldman, and A. A. Levy, "Transcriptional activation of retrotransposons alters the expression of adjacent genes in wheat," *Nature Genetics*, vol. 33, no. 1, pp. 102–106, 2003.

[10] Z. Lippman, A. V. Gendrel, M. Black et al., "Role of transposable elements in heterochromatin and epigenetic control," *Nature*, vol. 430, no. 6998, pp. 471–476, 2004.

[11] T. Wicker, J. P. Buchmann, and B. Keller, "Patching gaps in plant genomes results in gene movement and erosion of colinearity," *Genome Research*, vol. 20, no. 9, pp. 1229–1237, 2010.

[12] D. Lisch and J. L. Bennetzen, "Transposable element origins of epigenetic gene regulation," *Current Opinion in Plant Biology*, vol. 14, no. 2, pp. 156–161, 2011.

[13] M. A. Grandbastien, C. Audeon, E. Bonnivard et al., "Stress activation and genomic impact of Tnt1 retrotransposons in Solanaceae," *Cytogenetic and Genome Research*, vol. 110, no. 1–4, pp. 229–241, 2005.

[14] J. S. P. Heslop-Harrison and T. Schwarzacher, "Organisation of the plant genome in chromosomes," *The Plant Journal*, vol. 66, no. 1, pp. 18–33, 2011.

[15] R. J. Waugh O'Neill, M. J. O'Neill, and J. A. Marshall Graves, "Undermethylation associated with retroelement activation and chromosome remodelling in an interspecific mammalian hybrid," *Nature*, vol. 393, no. 6680, pp. 68–72, 1998.

[16] S. I. S. Grewal and D. Moazed, "Heterochromatin and epigenetic control of gene expression," *Science*, vol. 301, no. 5634, pp. 798–802, 2003.

[17] H. H. Kavi, W. Xie, H. R. Fernandez, and J. A. Birchler, "Global analysis of siRNA-mediated transcriptional gene silencing," *BioEssays*, vol. 27, no. 12, pp. 1209–1212, 2005.

[18] R. K. Slotkin and R. Martienssen, "Transposable elements and the epigenetic regulation of the genome," *Nature Reviews Genetics*, vol. 8, no. 4, pp. 272–285, 2007.

[19] P. Fransz and H. De Jong, "From nucleosome to chromosome: a dynamic organization of genetic information," *The Plant Journal*, vol. 66, no. 1, pp. 4–17, 2011.

[20] M. R. Hübner, M. A. Eckersley-Maslin, and D. L. Spector, "Chromatin organization and transcriptional regulation," *Current Opinion in Genetics & Development*, vol. 23, no. 2, pp. 89–95, 2013.

[21] D. A. Levin, *The Role of Chromosomal Change in Plant Evolution*, Oxford University Press, New York, NY, USA, 2002.

[22] B. Maestra and T. Naranjo, "Genome evolution in Triticeae-*Chromosomes Today*, E. Olmo and C. A. Redi, Eds., Birkhäuser Basel, Basel, 2000.

[23] D. Zohary and D. Imber, "Genetic dimorphism in fruit types in *Aegilops speltoides*," *Heredity*, vol. 18, no. 2, pp. 223–231, 1963.

[24] G. Kimber and M. Feldman, *Wild Wheat, an Introduction., Vol. 353*, College of Agriculture University of Missouri Special Report, Columbia, MO, USA, 1987.

[25] O. Raskina, "Genotype- and cell-specific dynamics of tandem repeat patterns in *Aegilops speltoides* Tausch (Poaceae, Triticeae)," *Cytogenetic and Genome Research*, vol. 153, no. 2, pp. 105–116, 2017.

[26] T. Eilam, Y. Anikster, E. Millet, J. Manisterski, O. Sagi-Assif, and M. Feldman, "Genome size and genome evolution in diploid Triticeae species," *Genome*, vol. 50, no. 11, pp. 1029–1037, 2007.

[27] O. Raskina, L. Brodsky, and A. Belyayev, "Tandem repeats on an eco-geographical scale: outcomes from the genome of *Aegilops speltoides*," *Chromosome Research*, vol. 19, no. 5, pp. 607–623, 2011.

[28] O. Raskina, A. Belyayev, and E. Nevo, "Quantum speciation in *Aegilops*: molecular cytogenetic evidence from rDNA cluster variability in natural populations," *Proceedings of the National Academy of Sciences of the United States of America*, vol. 101, no. 41, pp. 14818–14823, 2004.

[29] B. Yaakov, K. Meyer, S. Ben-David, and K. Kashkush, "Copy number variation of transposable elements in *Triticum–Aegilops* genus suggests evolutionary and revolutionary dynamics following allopolyploidization," *Plant Cell Reports*, vol. 32, no. 10, pp. 1615–1624, 2013.

[30] E. Hosid, L. Brodsky, R. Kalendar, O. Raskina, and A. Belyayev, "Diversity of long terminal repeat retrotransposon genome distribution in natural populations of the wild diploid wheat *Aegilops speltoides*," *Genetics*, vol. 190, no. 1, pp. 263–274, 2012.

[31] K. K. Kidwell and T. C. Osborn, Eds.J. S. Beckmann and T. C. Osborn, Eds., "Simple plant DNA isolation procedures," in *Plant Genomes: Methods for Genetic and Physical Mapping*, pp. 1–13, Springer Netherlands, Dordrecht, 1992.

[32] P. L. Vanderwiel, D. F. Voytas, and J. F. Wendel, "*Copia*-like retrotransposable element evolution in diploid and polyploid cotton (*Gossypium* L.)," *Journal of Molecular Evolution*, vol. 36, no. 5, pp. 429–447, 1993.

[33] M. D. Purugganan and S. R. Wessler, "Molecular evolution of the plant R regulatory gene family," *Genetics*, vol. 138, no. 3, pp. 849–854, 1994.

[34] S. E. Kubis, J. S. Heslop-Harrison, C. Desel, and T. Schmidt, "The genomic organization of non-LTR retrotransposons (LINEs) from three *Beta* species and five other angiosperms," *Plant Molecular Biology*, vol. 36, no. 6, pp. 821–831, 1998.

[35] S. F. Altschul, T. L. Madden, A. A. Schäffer et al., "Gapped BLAST and PSI-BLAST: a new generation of protein database search programs," *Nucleic Acids Research*, vol. 25, no. 17, pp. 3389–3402, 1997.

[36] C. Camacho, G. Coulouris, V. Avagyan et al., "BLAST+: architecture and applications," *BMC Bioinformatics*, vol. 10, no. 1, p. 421, 2009.

[37] O. Kohany, A. J. Gentles, L. Hankus, and J. Jurka, "Annotation, submission and screening of repetitive elements in Repbase: RepbaseSubmitter and Censor," *BMC Bioinformatics*, vol. 7, no. 1, p. 474, 2006.

[38] S. Ouyang, W. Zhu, J. Hamilton et al., "The TIGR Rice Genome Annotation Resource: improvements and new features," *Nucleic Acids Research*, vol. 35, Supplement 1, pp. D883–D887, 2007.

[39] O. Raskina, A. Belyayev, and E. Nevo, "Activity of the *En/Spm*-like transposons in meiosis as a base for chromosome repatterning in a small, isolated, peripheral population of *Aegilops speltoides* Tausch," *Chromosome Research*, vol. 12, no. 2, pp. 153–161, 2004.

[40] E. A. Salina, E. G. Pestsova, I. G. Adonina, and A. V. Vershinin, "Identification of a new family of tandem repeats in Triticeae genomes," *Euphytica*, vol. 100, no. 1/3, pp. 231–237, 1998.

[41] S. Taketa, H. Ando, K. Takeda, G. E. Harrison, and P. J. Heslop-Harrison, "The distribution, organization and evolution of two abundant and widespread repetitive DNA sequences in the genus *Hordeum*," *Theoretical and Applied Genetics*, vol. 100, no. 2, pp. 169–176, 2000.

[42] B. R. Baum and L. G. Bailey, "The 5S rRNA gene sequence variation in wheats and some polyploid wheat progenitors (Poaceae: Triticeae)," *Genetic Resources and Crop Evolution*, vol. 48, no. 1, pp. 35–51, 2001.

[43] A. Belyayev, O. Raskina, and E. Nevo, "Variability of the chromosomal distribution of Ty3-*grypsy* retrotransposons in the populations of two wild Triticeae species," *Cytogenetic and Genome Research*, vol. 109, no. 1–3, pp. 43–49, 2005.

[44] E. D. Badaeva, B. Friebe, and B. S. Gill, "Genome differentiation in *Aegilops*. 1. Distribution of highly repetitive DNA sequences on chromosomes of diploid species," *Genome*, vol. 39, no. 2, pp. 293–306, 1996.

[45] E. Paux, D. Roger, E. Badaeva et al., "Characterizing the composition and evolution of homoeologous genomes in hexaploid wheat through BAC-end sequencing on chromosome 3B," *The Plant Journal*, vol. 48, no. 3, pp. 463–474, 2006.

[46] E. A. Salina, E. M. Sergeeva, I. G. Adonina et al., "The impact of Ty3-*gypsy* group LTR retrotransposons *Fatima* on B-genome specificity of polyploid wheats," *BMC Plant Biology*, vol. 11, no. 1, p. 99, 2011.

[47] E. A. Salina, E. M. Sergeeva, I. G. Adonina et al., "Isolation and sequence analysis of the wheat B genome subtelomeric DNA," *BMC Genomics*, vol. 10, no. 1, p. 414, 2009.

[48] P. SanMiguel, A. Tikhonov, Y. K. Jin et al., "Nested retrotransposons in the intergenic regions of the maize genome," *Science*, vol. 274, no. 5288, pp. 765–768, 1996.

[49] V. V. Kapitonov and J. Jurka, "Molecular paleontology of transposable elements from *Arabidopsis thaliana*," *Genetica*, vol. 107, no. 1/3, pp. 27–37, 1999.

[50] A. Knoll, F. Fauser, and H. Puchta, "DNA recombination in somatic plant cells: mechanisms and evolutionary consequences," *Chromosome Research*, vol. 22, no. 2, pp. 191–201, 2014.

17

Metabolic Pathway Genes Associated with Susceptibility Genes to Coronary Artery Disease

Heng Lu,[1] Yi Chen,[1] and Linlin Li[2]

[1]*Department of Cardiovascular Surgery, Union Hospital, Fujian Medical University, Fuzhou, China*
[2]*Department of Nephrology, Union Hospital, Fujian Medical University, Fuzhou, China*

Correspondence should be addressed to Linlin Li; libb1997@sina.com

Academic Editor: Margarita Hadzopoulou-Cladaras

Coronary artery disease (CAD) is one of the leading threats to global health. Previous research has proven that metabolic pathway disorders, such as high blood lipids and diabetes, are one of the risk factors that mostly cause CAD. However, the crosstalk between metabolic pathways and CAD was mostly studied on physiology processes by analyzing a single gene function. A canonical correlation analysis was used to identify the metabolic pathways, which were integrated as a unit to coexpress with CAD susceptibility genes, and to resolve additional metabolic factors that are related to CAD. Seven pathways, including citrate cycle, ubiquinone, terpenoid quinone biosynthesis, and N-glycan biosynthesis, were identified as an integrated unit coexpressed with CAD genes. These pathways could not be revealed as a coexpressed pathway through traditional methods as each single gene has weak correlation. Furthermore, sets of genes in these pathways were candidate markers for diagnosis and detection from patients' serum.

1. Introduction

Coronary artery disease (CAD) is a leading cause of global death with several risk factors, including smoking, hypertension, diabetes, obesity, high blood lipids, and stress [1–4]. The current risk assessment of CAD is based on the determination of these factors. Moreover, avoiding these risk factors is the most efficient means of preventing CAD [5, 6]. Generally, almost all risk factors of CAD are related to metabolism in views of basic biological research. For example, high blood lipids, which play an important role in causing CAD, is regulated by a complex metabolic network that involves the biosynthesis and degradation of cholesterol, triglycerides, and lipoproteins [7–10]. Another risk factor, diabetes, is a metabolic disease related to glucose [11, 12]. The relationship between diet and risk of CAD also suggests that the metabolic pathway plays a pivotal role in the pathophysiology of CAD. Recent studies revealed that vegetarians as well as Mediterranean diet people have less cases of CAD [13–17]. In general, pathophysiology and dietetics research have determined the close link between metabolism and CAD.

In another side, molecular genetics research also accumulated several evidences which suggest that genetic factors contribute lots to the susceptibility of CAD. Although, in comprehensive studies, susceptible genes to CAD have been revealed to be involved in diverse biological pathways such as inflammation, innate immunity, and cholesterol metabolism. However, genome-wide crosstalks between susceptibility genes and metabolism pathways still remained to be characterized.

To understand the crosstalk between genes in the genome level, the transcriptome network is an efficient way to mining important interaction. The most common model for transcriptome network analysis is coexpression which calculates the correlation between genes in certain conditions. However, the traditional approach is only suitable for a single gene pair. So far, limited transcriptome network analysis works with integrated pathways which is consisted of many genes [18]. Canonical correlation analysis (CCA) is a powerful method to measure the coexpression between two sets of genes [19]. To detect the genetic regulatory variants on the Childhood Asthma Management Program, researchers used CCA to successfully detect candidate genes. The genes related to the glioma pathway were also identified by CCA from the transcriptome of glioblastoma in patients. The transcriptome plays a key role in studies on AIDS restriction genes using the CCA-determined purine metabolism pathway [20].

In this study, CCA was used to identify the coexpression between integrated metabolic pathways and CAD genes. The most significant metabolic pathways that could be a potential candidate disease marker for CAD diagnosis are discussed.

2. Materials and Methods

2.1. Datasets. Human genome expression datasets were downloaded from the website of COPRESDB (http://coxpresdb.jp/), which contains approximately 4000 experiments and expression data on 20,000 human genes. Metabolic pathway genes were downloaded from KEGG (http://www.kegg.jp), which includes 129 typical metabolic pathways with predicted genes. The CAD genes were collected from published literature. Two expression datasets were generated to include the metabolic pathway gene and CAD expression data.

2.2. CCA. CCA was performed to dissect the correlations between CAD and metabolic pathway gene expression. In concept, CCA could integrate multiple correlations into an abstract group. The correlation level between two sets of genes can be determined by scoring the relationship between the groups. In statistics, this method generates independent pairs of new variables from the original two sets of variables, namely, the canonical variable. CCA is a method processing cross-covariance matrices. If there are correlations among the variables from two vectors $\mathbf{X} = (X_1, \ldots, X_n)$ and $\mathbf{Y} = (Y_1, \ldots, Y_m)$ of random variables, CCA could detect linear combination of the X_i and Y_j with maximum correlation. As a result, the canonical variable is a linear combination of the original variables.

Vector $\mathbf{A}_i = (a_1, a_2, a_3, \ldots, a_n)$ describes the CAD gene i with different expression levels from different experiments (1 to n). Vector $\mathbf{B}_i = (b_1, b_2, b_3, \ldots, b_n)$ describes the metabolic gene i with different expression levels from different experiments (1 to n). Matrix $\mathbf{M} = (A_1, A_2, A_3, \ldots, A_n)$ is for CAD genes, and matrix $\mathbf{N}_i = (B_1, B_2, B_3, \ldots, B_n)$ is for a certain metabolic pathway i. CCA were calculated between matrix **M** and each $\mathbf{N}_i$ by following the steps in the *R* program:

(1) *zscore (M) and zscore* (N_i);

(2) r_{11} *= corr (zscore (M)),* r_{22} *= corr (zscore(*N_i*));*

(3) *[AA, BB, r, U, V, stats] = canoncorr (zscore (M), zscore* (N_i)*);*

(4) $r_{1A} = AA'^{*} r_{11}$, $r_{2B} = BB'^{*} r_{22}$;

(5) S_a *= sum* (r_{1A}.^2/length (r_{1A}), S_b *= sum* (r_{2B} ^2/length (r_{2B}));

(6) *Screen the candidate CCA using a certain threshold for r P value,* S_a*, and* S_b*. (in this study, the thresholds were 0.6, 0.00001, 0.3, and 0.15)*

S_a and S_b were used to validate the percentage of genes in total CADs or a certain metabolic pathway, which could be presented by canonical variables. A high level of S_a or S_b indicates that more genes belonging to the group are involved in the correlation. Thus, not only the correlation between gene pairs was identified but also the correlation between sets.

2.3. Software Tools. CCA was performed using the *R* platform (http://www.r-project.org/). The web-based DAVID tool (http://david.abcc.ncifcrf.gov/) was used for gene ontology enrichment analysis for the metabolic pathway. The STRING platform was used for the contraction of the PPI network.

2.4. Sample Collection. Fifty serum samples from Chinese patients (Union Hospital of Fujian Medical University) were collected and divided into two groups, namely, CAD (from CAD patients) and control (from healthy people). The blood samples were maintained at 37°C for 1 h and centrifuged at 3000 g/min for 20 min at 4°C, and the supernatant fluid is collected at −80°C until testing. All clinical research experiments were approved by the Medical Ethics Committee of Union Hospital of Fujian Medical University and performed according to the approved guidelines.

2.5. RNA Isolation, cDNA Synthesis, and q-PCR Analysis. The total RNA was extracted using TRIzol reagent (Life Technologies), following the manufacturer's instructions with slight modifications. Up to 1 μg of the total RNA was used by utilizing the First Strand cDNA Synthesis Kit (TOYOBO) to generate cDNA according to the manufacturer's instructions. q-PCR was performed using the TransStart Top Green q-PCR SuperMix (TransGen Biotech) and was conducted using the iQTM5 multicolor real-time PCR detection system (Bio-Rad).

3. Results and Discussion

3.1. Strategy. In this study, a clustering for the expression pattern of CAD and metabolic pathway genes was first created to estimate the possible gene group, which could provide an evidence for identifying the correlation among gene groups. Next, a coexpression network was constructed among individual CAD and metabolic pathway genes. This step is prepared for comparing with CCA. Then, CCA was performed to identify the canonical variables that delegate the integrated CAD and metabolic pathway genes (Figure 1). In particular, additional genes in a pathway could be regulated by the same factor. Thus, the whole pathway was suggested to be controlled by the same factor. In this study, an entire metabolic pathway was correlated with the entire CAD genes.

3.2. CAD and Metabolic Pathway Genes. A published literature shows that SNPs associated with coronary heart disease and mean arterial pressure were collected for further analysis. There are Mendelian forms of CAD that show mutations in genes that are related to low- and high-density lipoprotein metabolism or homeostasis. The heritability of fatal coronary events is estimated at 57% and 38% in monozygotic and dizygotic twin cohorts, respectively. Researchers have long recognized the familial clustering of CAD. The first linkage of

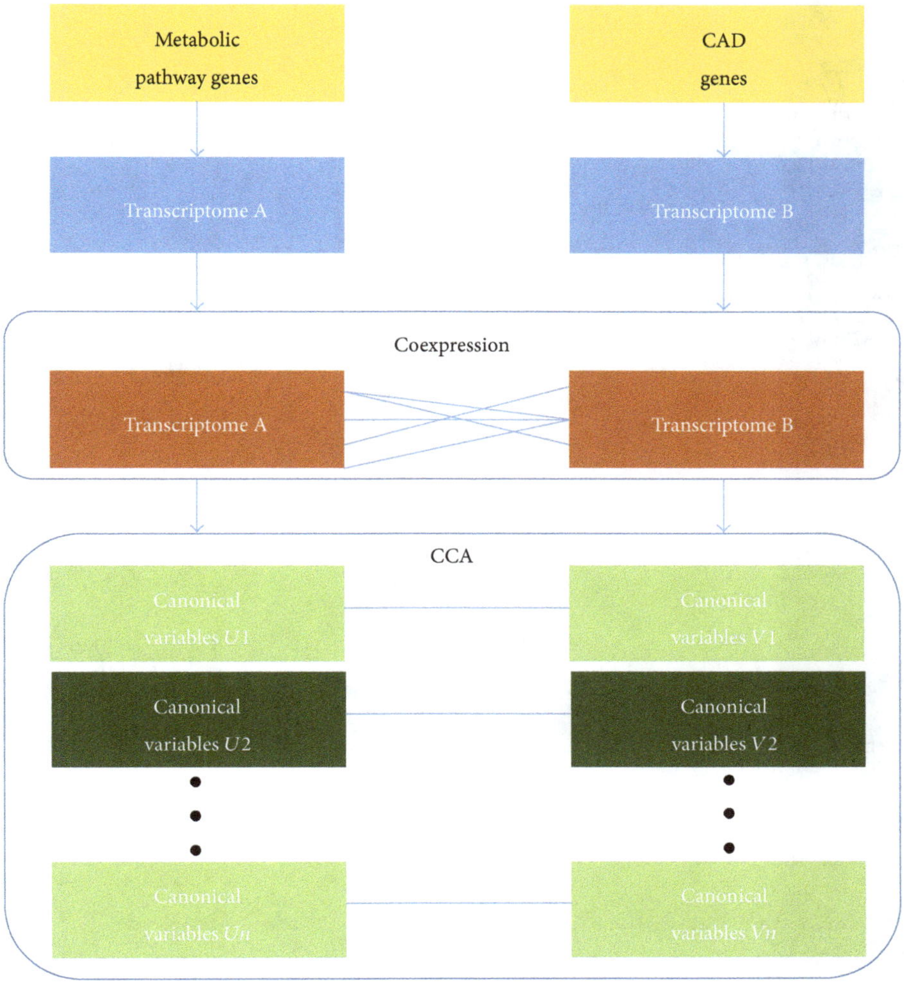

Figure 1: Strategy of study for the canonical correlation analysis. Collection of CAD genes and metabolic pathway genes; construction of the coexpression network among individual CAD and metabolic pathway genes; performance of CCA; and analysis of canonical variables.

premature CAD to loci on chromosomes 2q21.2-22 and Xq23-26 was found in Finland [21]. Subsequently, several studies from Germany, the United States, and England identified a novel chromosome locus linked with CAD [22]. Different genes, such as ALOX5AP (5-lipoxygenase-activating protein), are characterized by haplotypes that are associated with CAD. Several genome-wide association studies promoted by the improved genome sequence technology are performed on CAD. A novel replicated locus on chromosome 17 was found by genome-wide mapping of susceptibility to CAD [23]. In Indo-Mauritians, a susceptibility locus on chromosome 16p13 is revealed by a genome-wide scan for CAD [24]. Recently, a new locus on chromosome 10p11.23 for CAD is discovered by a genome-wide association study [25]. In addition, a genome-wide linkage analysis for CAD and its related risk factors is conducted [26].

Eight studies identified 373 SNPs associated with coronary heart disease-related phenotype at a 10E5 P level. These SNPs are located at several chromosomes inside coding or noncoding regions. In general, the coding regions match 120 genes involved in diverse biological processes, such as response to a stimulus, developmental, rhythmic, metabolic, and immune system processes (Figure S1, Table S1). Moreover, the expression of CAD-associated genes was controlled during the progress of the disease. Thus, the expression levels of 128 genes were collected from human transcriptome database with approximately 5000 experiments.

Then, genes that are related to metabolic pathways were collected from the KEGG database. In general, 84 metabolic pathways, including metabolism of carbohydrate, energy, lipid, nucleotide, amino acid, glycan biosynthesis, cofactors and vitamins, terpenoids and polyketides, and biosyntheses of other secondary metabolites (Table S2), cover major metabolic processes in human biology. A total of 300 gene code putative enzymes exist for these pathways according to the current knowledge on the human genome. Currently, certain metabolic pathways, such as D-arginine and D-ornithine metabolism, valine, leucine and isoleucine biosyntheses, lysine biosynthesis, and biotin and lipoic acid metabolism, contain only 1–3 genes. These pathways have a limited value for further CCA. A total of 77 pathways, such as glycolysis/gluconeogenesis, TCA cycle, and pentose phosphate

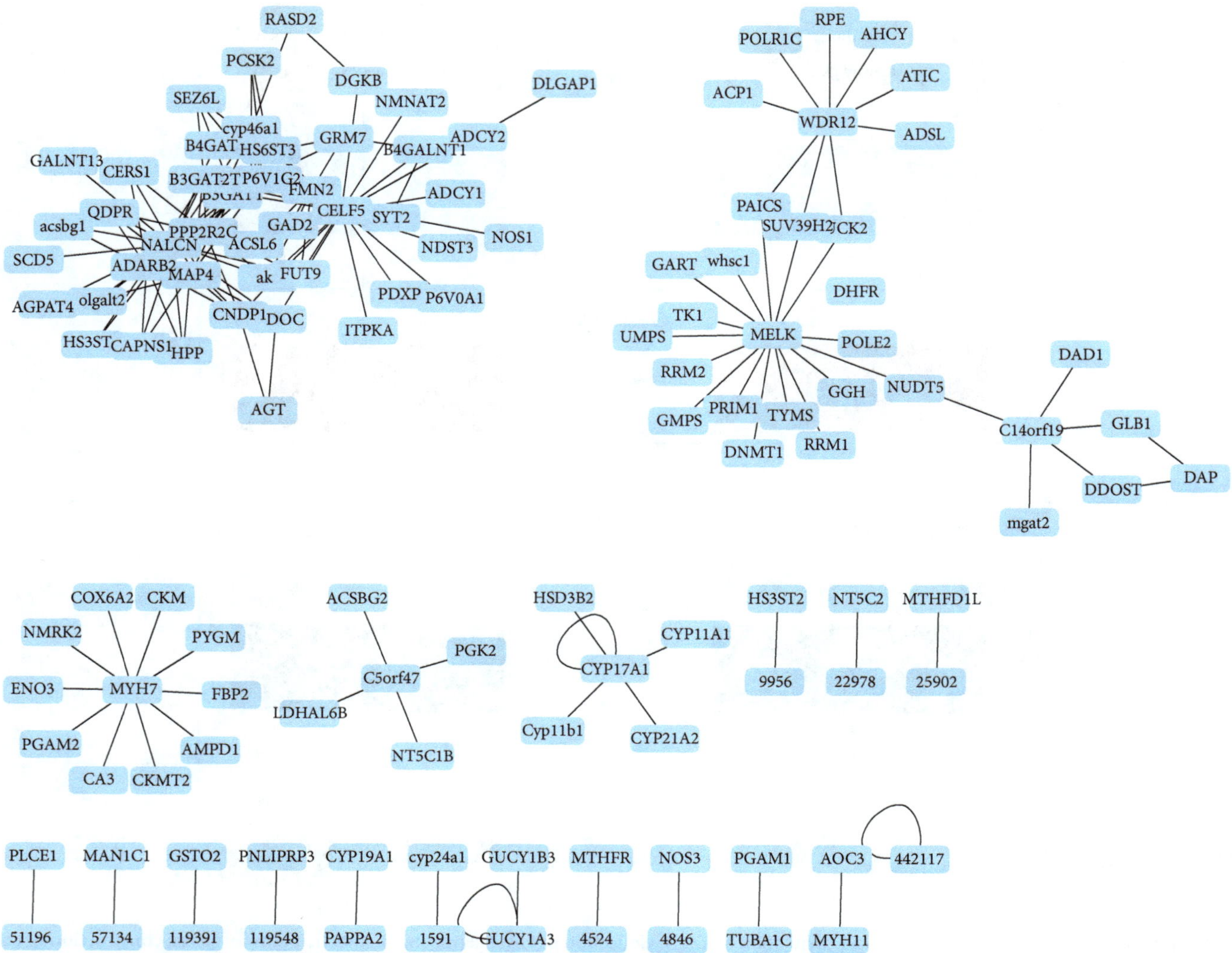

FIGURE 2: Coexpression network of CAD genes and metabolic pathway genes. The connection indicates the Pearson correlation factor $r > 0.6$.

pathway (Table S2), have more than five genes in the human genome. These pathways could be subject to CCA because we aim to estimate the correlation between the integrated CAD gene group and metabolic gene pathway, not the individual gene.

3.3. Coexpression and the PPI Network. There are huge transcriptome data related to CAD carried out by microarray and next-generation sequencing. For example, researchers establish a mouse heart transcriptomic network sensitivity to various heart diseases using microarray [27]. Cardiac hypertrophy was also studied by transcriptome microarrays to identify long noncoding RNAs [28]. A gene network was established in low-dose, radiation-affected cardiomyocytes [29]. Many transcriptional regulators in human ischemic cardiomyopathy are revealed from the gene expression network [30]. Transcriptome sequencing of different cell lines identifies several candidate genes related to coronary artery calcification [31]. Some potential candidate genes for heart failure were also discovered by transcriptome network analysis [32]. Transcriptome network analysis is also used to find divergent cardiovascular disease pathways between animal and humans [33].

A coexpression network was constructed using the Pearson correlation to determine the relationship between all CAD susceptibility and metabolic pathway genes. Numerous genes have connections with high r value ($r > 0.6$). A total of 24 CAD genes, namely, *105* (*ADARB2*), *183* (*AGT*), *1586* (*CYP17A1*), *1611* (*DAP*), *2917* (*GRM7*), *2982* (*GUCY1A3*), *4134* (*MAP4*), *4625* (*MYH7*), *4629* (*MYH11*), *5126* (*PCSK2*), *5522* (*PPP2R2C*), *9229* (*DLGAP1*), *9833* (*MELK*), *23544* (*SEZ6L*), *23551* (*RASD2*), *55017* (*C14orf119*), *55759* (*WDR12*), *56776* (*FMN2*), *60676* (*PAPPA2*), *60680* (*CELF5*), *84790* (*TUBA1C*), *127833* (*SYT2*), *133491* (*C5orf47*), and *259232* (*NALCN*), were identified in this coexpression network (Figure 2, Table S3). The network clearly shows several disconnected subnetworks. Six subnetworks, namely, 9833 (maternal embryonic leucine zipper kinase (MELK)), 55759 (WD repeat domain 12 (WDR12)), 55017 (C14orf119), 4625 (myosin, heavy chain 7, cardiac muscle (MYH7)), 133491

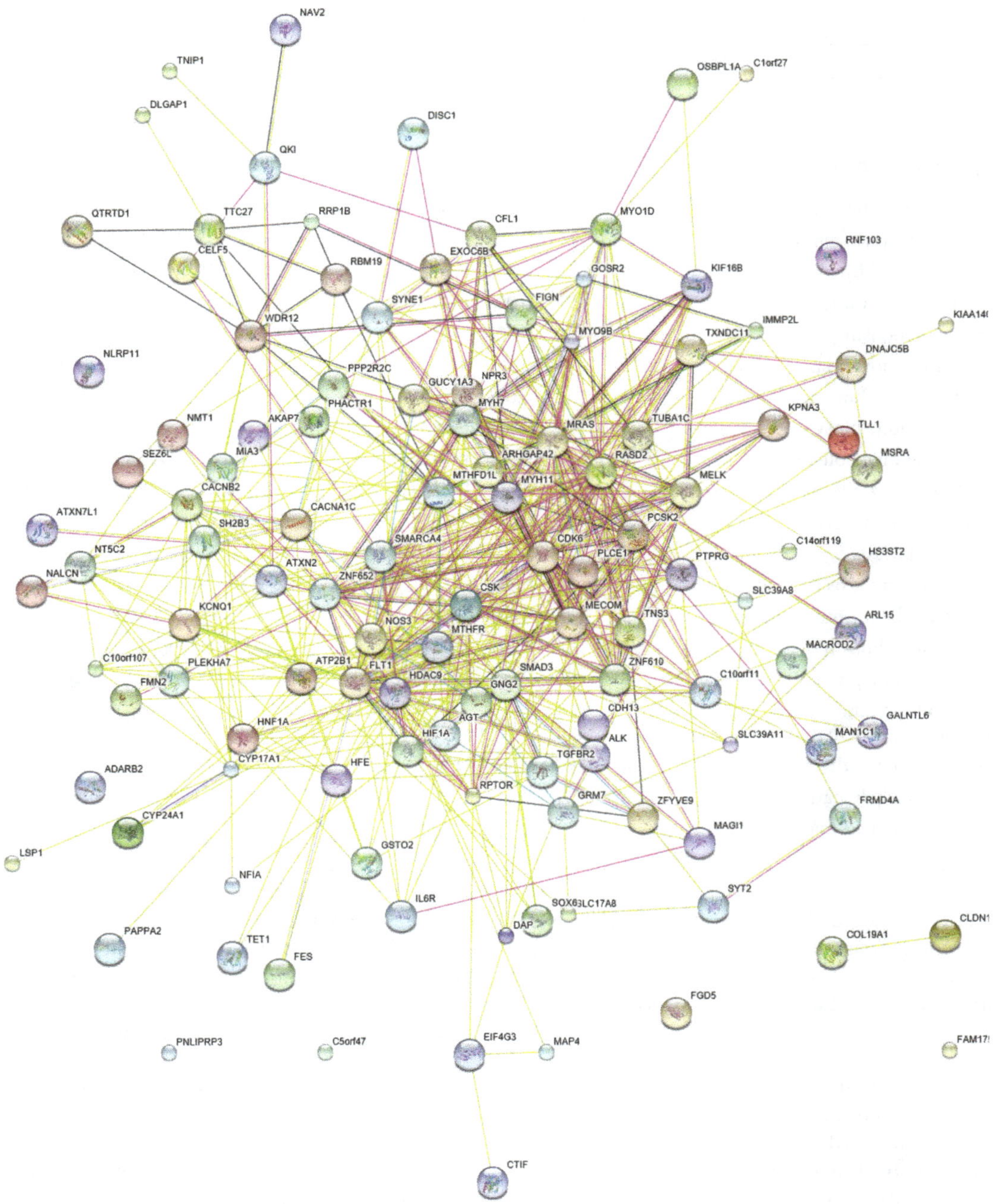

FIGURE 3: PPI network of CAD genes. Note that no metabolic pathway was enriched in the PPI network.

(C5orf47), and 1586 (cytochrome P450, family 17, subfamily A, polypeptide 1 (CYP17A1)) (Figure 2), have different genes at the center. In addition, one subnetwork has multiple genes at the center.

In this coexpression network, CAD genes were connected to 80 metabolic pathway genes (Table S4). CAD genes were coexpressed with the metabolic pathway. These metabolic pathway genes were enriched in a diverse metabolic pathway, such as purine metabolism, biosynthesis of antibiotics, pyrimidine metabolism, and glycolysis/gluconeogenesis (P value $< 10E-03$). Furthermore, several metabolic pathways with less than four genes coexpressed with CAD genes. For example, 23205, 81616, and 23305 belonging to fatty acid biosynthesis exist in this network. The coexpression network between CAD and metabolic pathway genes could not identify the connection between the integrated pathways but simply link from point to point.

Furthermore, a PPI network was constructed using CAD genes as baits with low constraints by the STRING platform (Figure 3). Several interactions between CAD genes were observed. Moreover, many interaction proteins were identified through CAD baits. However, the KEGG

Table 1: Enriched metabolic pathway in the coexpression network.

Term	Count	Pop	P value	Genes
hsa00230: purine metabolism	17	176	$8.46E-11$	5427, 9533, 2618, 6240, 5557, 11164, 6241, 26289, 270, 158, 93034, 8833, 10606, 471, 108, 107, 2983
hsa01130: biosynthesis of antibiotics	14	212	$6.52E-07$	6120, 5224, 5232, 5223, 230, 2618, 26289, 270, 158, 8789, 92483, 10606, 471, 2027
hsa00240: pyrimidine metabolism	10	104	$2.40E-06$	93,034, 5427, 9533, 7372, 7371, 6240, 5557, 7298, 6241, 7083
hsa00010: glycolysis/gluconeogenesis	7	67	$1.06E-04$	92483, 5224, 5232, 5223, 230, 2027, 8789
hsa00670: one carbon pool by folate	4	20	0.001425311	1719, 471, 2618, 7298
hsa01230: biosynthesis of amino acids	6	74	0.001495954	6120, 5224, 5232, 5223, 230, 2027
hsa01200: carbon metabolism	7	113	0.001764573	6120, 5224, 5232, 5223, 230, 2027, 8789
hsa04913: ovarian steroidogenesis	5	49	0.002269906	1583, 108, 107, 3284, 1588
hsa00330: arginine and proline metabolism	5	50	0.002446109	84735, 1158, 4842, 57571, 1160
hsa00140: steroid hormone biosynthesis	5	58	0.004206383	1583, 1589, 3284, 1588, 1584
hsa00410: beta-alanine metabolism	4	31	0.005135035	84735, 2572, 57571, 8639
hsa00514: other types of O-glycan biosyntheses	4	31	0.005135035	10690, 135152, 27087, 23127
hsa00061: fatty acid biosynthesis	3	13	0.00927769	23205, 81616, 23305
hsa00790: folate biosynthesis	3	14	0.010744279	5860, 1719, 8836
hsa04925: aldosterone synthesis and secretion	5	81	0.013544109	1583, 108, 107, 1589, 3284
hsa00983: drug metabolism—other enzymes	4	46	0.015345429	8833, 7372, 7371, 7083
hsa01212: fatty acid metabolism	4	48	0.017202919	23205, 79966, 81616, 23305
hsa04922: glucagon signaling pathway	5	99	0.026291005	92483, 5224, 5223, 5837, 108
hsa00534: glycosaminoglycan biosynthesis—heparan sulfate/heparin	3	24	0.030276378	222537, 9348, 266722
hsa03320: PPAR signaling pathway	4	67	0.040871389	23205, 79966, 81616, 23305
hsa00030: pentose phosphate pathway	3	29	0.042939357	6120, 230, 8789
hsa00760: nicotinate and nicotinamide metabolism	3	29	0.042939357	93034, 27231, 23057

enrichment analysis showed that all these genes only were enriched in adrenergic signaling in cardiomyocytes, tight junction, and glutamatergic synapse. No metabolic pathway was found while enriched KEGG is processed in the PPI network.

3.4. CCA of CAD Susceptibility and Metabolic Pathway Genes. CCA was used to resolve canonical variables, correlation coefficient, and standard deviation to determine the significant canonical correlations between CAD and metabolic pathway genes in the transcriptome. The canonical variables are a new set of statistically independent variable pairs generated from the original variables, which describe the level of gene expression in different experiments. The canonical variables, which are independent, could be linearly combined to create original variables and could be considered as a component of the original variables. We set a standard to evaluate the canonical variables, which could delegate an integrated pathway at a maximum. Thus, we isolated the canonical variables with large S_a, S_b (>0.2), r values (>0.5), and minimal P values (<0.001). The standard deviations indicated the extent of coverage of canonical variables over the entire pathway. Thus, we selected metabolic pathways with more than 10 genes in the genome for further analysis to identify the correlation among the integrated pathways or among single genes. Table 1 indicated that only seven sets of canonical variables that satisfy all our requirements were detected. These variables are related to seven metabolic pathways, including TCA cycle, ubiquinone and other terpenoid quinone biosyntheses, N-glycan biosynthesis, other glycan degradation, glycosaminoglycan degradation, and glycosylphosphatidylinositol (GPI) anchor biosynthesis and glycosphingolipid biosynthesis–ganglioseries. None of these metabolic pathways were identified by the Pearson correlation or PPI networks. The correlation coefficient factors were more than 0.9 (Figure 4(a)). S_a indicated that the standard deviation for metabolic pathways had a maximum of 0.3 Table 1. S_b indicated that the standard deviation of CAD genes varies from 0.19 to 0.93 Table 1. The results suggested that the integrated coexpression in metabolic pathways is not at high level. However, CAD genes were highly integrated to coexpress with metabolic pathways. S_b was equal to 0.93 for coexpression between CAD genes and ubiquinone and other terpenoid quinone biosyntheses, indicating that 93% of CAD genes can be delegated by canonical variables (Figure 4(a)–4(c)). The results were discussed with published literature by showing the interaction between metabolic pathways and CAD genes.

3.5. TCA Cycle. Table 1 indicates that the canonical variable represents 20% of the variability in the original expression pattern of TCA cycle pathway genes and 19% of the

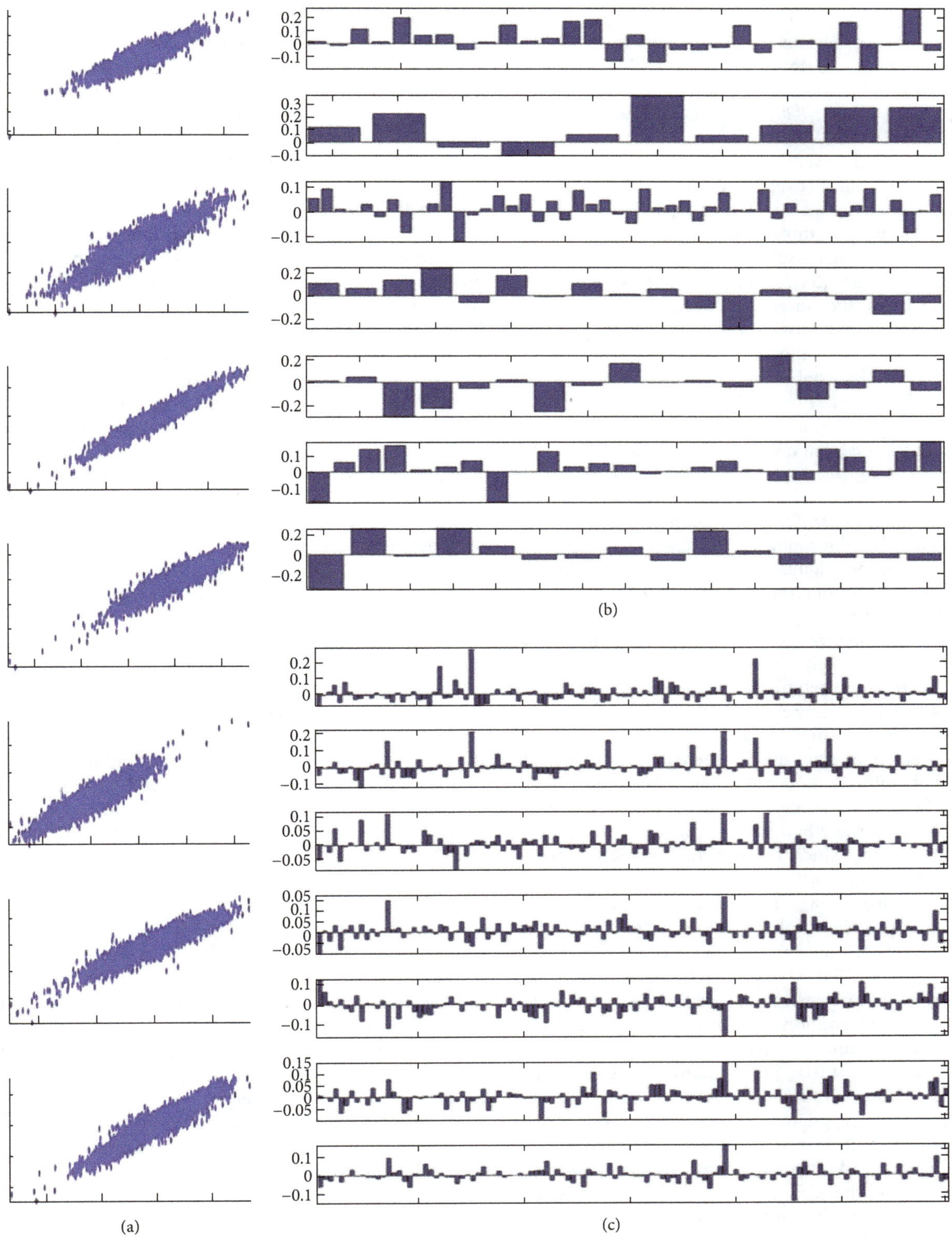

FIGURE 4: Characteristic of canonical variables indicating integrate pathway. A threshold was set to identify the CCA which cover whole pathway (S_a, $S_b > 0.2$, r values > 0.5, and P values < 0.001). (a) Scatter plots of seven CCA pairs with top r values. Note that all identified CCA were highly correlated. (b) The coefficient of metabolic pathway genes on seven canonical variables. Each metabolic pathway contains different number of genes. The genes with big values contributed more in CCA. (c) The coefficient of CAD genes on seven canonical variables. The genes with big values contributed more in CCA.

variability in the expression pattern of CAD genes. The top 6 genes with big absolute value of coefficient factor is selected as representative genes for this canonical variable (Figure 4(b), Table 1). These genes were *8803* (*SUCLA2*), *1737* (*DLAT*), *8801* (*SUCLG2*), *4191* (*MDH2*), *6392* (*SDHD*), and *4190* (*MDH1*).

In the last decades, several studies provide evidence that supports the interaction between CAD and TCA cycle. For example, the degradation of myocardial aspartate and glutamate is induced by the complete citric acid cycle, ischemia, and hypoxia. This process resulted in the production of succinate in isolated hearts [34]. Also, it has been revealed the ATP production in the absence of oxygen through this channeling which could provide cardioprotection [35]. In another case, the accumulation of TCA cycle intermediate fumarate is beneficial for stabilizing hypoxia-inducible factor 1α (HIF-1α), which is a key transcription factor that controls the response to hypoxia and myocardial protection [36]. Together, it is proposed that manipulation of the citrate cycle could increase cardioprotection.

3.6. Ubiquinone and Other Terpenoid Quinone Biosyntheses. Table 1 presents that the canonical variable comprises 30% of the variability in the original expression pattern of ubiquinone and other terpenoid quinone biosynthesis pathway genes and 93% of the variability in the expression pattern of CAD genes. The top three genes with a high absolute value of coefficient factor, namely, *27235* (*COQ2*), *84274* (*COQ5*), and *79001* (*vkorc1*), are selected to represent the genes for this canonical variable (Figure 4(b), Table 1).

The key role of ubiquinone in CAD has been accepted by several researchers extensively. Coenzyme Q10, which is a kind of ubiquinone, is supplemented to improve diastolic heart functions, especially in CAD patients. Moreover, it is the major method of treating current diseases [37]. Coenzyme Q10, which could reduce oxidative stress and induce antioxidant activity in CAD patients, is the proposed mechanism [38]. This method is helpful for low-density lipoprotein cholesterol and vitamin E. In particular, this mitochondrial coenzyme is a vital factor for producing ATP [39]. These functions of coenzyme Q10 in antioxidant and energy production imply its role in coronary revascularization. The research demonstrates a significant correlation between the risk of CAD and levels of coenzyme Q10 [40]. Many patients with cardiovascular diseases, such as CAD, cardiomyopathy, congestive heart failure, angina pectoris, and hypertension, showed a deficiency in coenzyme Q10. The deficiency of coenzyme Q10 is a risk factor for cardiovascular disorder, especially increasing the early mortality in myalgic encephalomyelitis/chronic fatigue syndrome [41]. Current studies have proposed that low plasma coenzyme Q10 should be considered a risk factor for CAD and a marker for chronic fatigue in depression [42]. Coenzyme Q10 has been regarded as an independent factor of mortality in congestive heart failure. Furthermore, coenzyme Q10 improves the immune system in vertigo and Meniere' disease-like syndrome [43]. In general, coenzyme Q10 is significant in clinical therapy and diagnosis of CAD patients.

3.7. Glycosaminoglycan Degradation. As shown in Table 1, the canonical variable represents 22% of the variability in the original expression pattern of glycosaminoglycan degradation pathway genes and 69% of the variability in the expression pattern of CAD genes. The top four genes with a high absolute value of coefficient factor, namely, *2720* (*GLB1*), *3074* (*HEXB*), *6677* (*SPAM1*), and *2799* (*GNS*), are selected to represent the genes for this canonical variable (Figure 4(b), Table 1).

Studies on the link between glycosaminoglycan and CAD were limited in the last decades. However, the clues were already revealed by research on certain kinds of glycosaminoglycan with antioxidative activity. A pilot study has been conducted with sulodexide, a glycosaminoglycan that primarily consists of heparin, which mainly causes oxidative stress [44]. Sulodexide treatment could decrease only the level of plasma 8-isoprostane but not LDL cholesterol, triglycerides, fibrinogen, and C-reactive protein. However, a significant reduction in oxidative stress in CAD patients administered with sulodexide was observed, implying the important role of plasma 8-isoprostane and glycosaminoglycan. A study showed that another long-chain glycosaminoglycan-hyaluronan could interact with versican to create matrices, which are required for arterial smooth muscle growth during the vascular disease [45]. The histopathological studies on coronary arteries revealed that abundant glycosaminoglycan is associated with percutaneous transluminal coronary angioplasty [46].

3.8. Glycosphingolipid Biosynthesis–Ganglioseries. In Table 1, the canonical variable accounts for 22% of the variability in the original expression pattern of glycosphingolipid biosynthesis–ganglioseries pathway genes and 92% of the variability in the expression pattern of CAD genes. The top three genes with a high absolute value of coefficient factor, namely, 2583 (*B4GALNT1*), 2720 (*GLB1*), and 3074 (*HEXB*), are selected to represent genes for this canonical variable (Figure 4(b), Table 1).

The role of glycosphingolipid biosynthesis–ganglioseries in CAD disease was also revealed by recent studies. Glycosphingolipids act as messengers that are integrated into the cell membrane to transduce the growth factor. High levels of cholesterol in the blood are proven to be associated with glycosphingolipid. Recent studies concluded that glycosphingolipids were involved in the major lipoprotein classes in normal and dyslipoproteinemic sera [47]. Cholesterol efflux could be inhibited by the accumulation of glycosphingolipid [48]. The cellular cholesterol homeostasis is induced by glycosphingolipid storage [49]. Studies found that arterial stiffness and atherosclerosis could be ameliorated by inhibiting glycosphingolipid synthesis in mice and rabbit models given the pivotal role of cholesterol in CAD [50].

3.9. Other Pathways without Literature. In Table 2, the canonical variable comprises 23% of the variability in the original expression pattern of N-glycan biosynthesis pathway genes and 28% of the variability in the expression pattern of CAD genes. The top 11 genes with a high absolute value of coefficient factor, namely, *4247* (*mgat2*), *4248*

Table 2: The canonical variable delegate integrates metabolic pathways.

Metabolic pathways	Gene number	r	S_a	S_b	Represent gene ID
Citrate cycle (TCA cycle)	30	0.90	0.20	0.19	8803*, 1737*, 8801*, 4191*, 6392*, 4190*
Ubiquinone and other terpenoid quinone biosyntheses	10	0.90	0.30	0.93	27235, 84274, 79001
N-glycan biosynthesis	48	0.96	0.23	0.28	4247, 4248, 85365, 1650, 11282, 79868, 57134, 8813, 4122, 146664, 29929
Other glycan degradation	17	0.92	0.27	0.62	10825, 2720, 3074, 129807, 2519
Glycosaminoglycan degradation	17	0.91	0.22	0.69	2720, 3074, 6677, 2799
Glycosylphosphatidylinositol (GPI) anchor biosynthesis	25	0.93	0.21	0.43	9091, 284098, 2822, 5281, 5279
Glycosphingolipid biosynthesis–ganglioseries	15	0.93	0.22	0.92	2583, 2720, 3074

S_a and S_b indicate the extent of canonical variables covering the whole pathway. The representative genes were selected based on the percentage of covering (*gene number in pathways S_b).

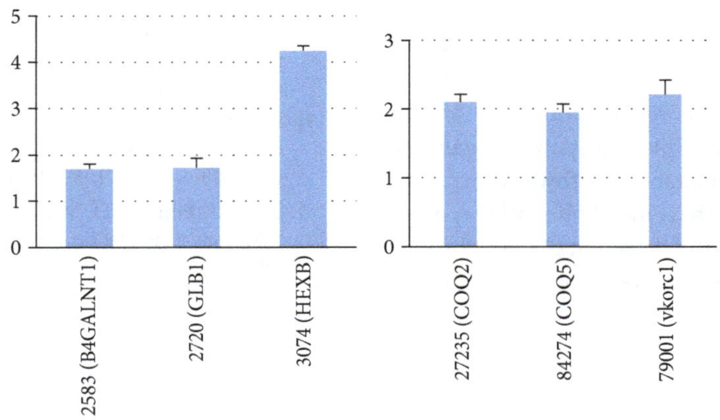

Figure 5: qRT-PCR of metabolic pathway genes in CAD patients. (a) *2583* (*B4GALNT1*), *2720* (*GLB1*), and *3074* (*HEXB*) in glycosphingolipid biosynthesis. (b) *27235* (*COQ2*), *84274* (*COQ5*), and *79001* (*vkorc1*) in ubiquinone and other terpenoid quinone biosyntheses.

(*MGAT3*), *85365* (*ALG2*), *1650* (*DDOST*), *11282* (*MGAT4B*), *79868* (*ALG13*), *57134* (*MAN1C1*), *8813* (*DPM1*), *4122* (*MAN2A2*), *146664* (*MGAT5B*), and 29929 (*ALG6*), are selected to represent the genes for this canonical variable (Figure 4(b), Table 1).

As shown in Table 1, the canonical variable explains 27% of the variability in the original expression pattern of other glycan degradation pathway genes and 62% of the variability in the expression pattern of CAD genes. The top five genes with a high absolute value of coefficient factor, namely, *10825* (*NEU3*), *2720* (*GLB1*), *074* (*HEXB*), *129807* (*NEU4*), and 2519 (*FUCA2*), are selected to represent the genes for this canonical variable (Figure 4(b), Table 1).

Table 1 also shows that the canonical variable represents 21% of the variability in the original expression pattern of the GPI anchor biosynthesis pathway genes and 43% of the variability in the expression pattern of CAD genes. The top five genes with a high absolute value of coefficient factor, namely, (*PIGQ*), *284098* (*PIGW*), *2822* (*GPLD1*), *5281* (*PIGF*), and 5279 (*PIGC*), are selected to represent the genes for this canonical variable (Figure 4(b), Table 1).

3.10. Validation of Expression Genes as Candidate Disease Markers for CAD. To explore the potential application of these metabolic pathway genes as disease markers for CAD, we used q-PCR for analyzing the expression pattern of CAD genes and its canonical correlated metabolic pathways. As more than 90% of CAD genes are associated with ubiquinone and other terpenoid quinone biosyntheses and glycosphingolipid biosyntheses–ganglioseries, we selected genes from both metabolic pathways for further analysis. Fifty serum samples were collected from the CAD patients and the control group. The expression levels of metabolic genes, namely, *2583* (*B4GALNT1*), *2720* (*GLB1*), and *3074* (*HEXB*), were determined by q-PCR. Except *3074* (*HEXB*) with 4 times induction, *2583* (*B4GALNT1*) and *2720* (*GLB1*) only increased 1.7 times in CAD patients (Figure 5(a)). The metabolic genes showed significant difference between CAD patients and the control group with a P value of less than $10E-4$. Six metabolic genes showed a significant difference between the CAD patients and the control group with a P value of less than $10E-4$. In traditional method, it is difficult to justify these genes as disease markers with less than two times induction. However, three genes together identified from CCA showed a significant P value. Thus, it suggests that a combination of genes is a potential disease marker for CAD by checking the expression levels rather than single genes. Also, genes *27235* (*COQ2*), *84274* (*COQ5*), and *79001* (*vkorc1*) in ubiquinone and other terpenoid quinone biosynthesis pathways were also determined by qRT-PCR in these samples. A similar result was found as the glycosphingolipid biosynthesis pathway. Each gene was induced only 2 times in CAD patients (Figure 5(b)). However, three genes together showed a very significant P value. In statistics, the CCA method indicated that this combination of genes could be potential disease markers.

4. Conclusion

In this study, the canonical correlation between CAD and metabolic pathways gene expressions was analyzed. The results showed that the TCA cycle, ubiquinone and other terpenoid quinone biosyntheses, N-glycan biosynthesis, other glycan degradation, glycosaminoglycan degradation, GPI anchor biosynthesis, and glycosphingolipid biosynthesis–ganglioseries are the most significant metabolic pathways correlated with CAD genes. Furthermore, these metabolic pathway genes are beneficial for the diagnosis and detection of CAD through the quantification of the expression level of patients' serum.

Conflicts of Interest

The authors declare no financial interest related to this work.

Acknowledgments

The authors would like to thank Fuzhou Shichen Biotheon Technology Co. Ltd. (http://www.fuzhoushichen.com) for the technical support.

References

[1] R. Calderon Jr., R. H. Schneider, C. N. Alexander, H. F. Myers, S. I. Nidich, and C. Haney, "Stress, stress reduction and hypercholesterolemia in African Americans: a review," *Ethnicity & disease*, vol. 9, no. 3, pp. 451–462, 1998.

[2] M. Kivimäki, S. T. Nyberg, G. D. Batty et al., "Job strain as a risk factor for coronary heart disease: a collaborative meta-analysis of individual participant data," *The Lancet*, vol. 380, no. 9852, pp. 1491–1497, 2012.

[3] R. Lozano, M. Naghavi, K. Foreman et al., "Global and regional mortality from 235 causes of death for 20 age groups in 1990 and 2010: a systematic analysis for the Global Burden of Disease Study 2010," *The Lancet*, vol. 380, no. 9859, pp. 2095–2128, 2013.

[4] J. A. Finegold, P. Asaria, and D. P. Francis, "Mortality from ischaemic heart disease by country, region, and age: statistics from World Health Organisation and United Nations," *International Journal of Cardiology*, vol. 168, no. 2, pp. 934–945, 2013.

[5] H. C. McGill, C. A. McMahan, and S. S. Gidding, "Preventing heart disease in the 21st century: implications of the Pathobiological determinants of atherosclerosis in youth (PDAY) Study," *Circulation*, vol. 117, no. 9, pp. 1216–1227, 2008.

[6] C. J. McNeal, T. Dajani, D. Wilson, A. E. Cassidy-Bushrow, J. B. Dickerson, and M. Ory, "Hypercholesterolemia in youth: opportunities and obstacles to prevent premature atherosclerotic cardiovascular disease," *Current Atherosclerosis Reports*, vol. 12, no. 1, pp. 20–28, 2010.

[7] W. B. Kannel and R. S. Vasan, "Triglycerides as vascular risk factors: new epidemiologic insights," *Current Opinion Cardiology*, vol. 24, no. 4, pp. 345–350, 2009.

[8] J. Danesh, R. Collins, and R. Peto, "Lipoprotein (a) and coronary heart disease: meta-analysis of prospective studies," *Circulation*, vol. 102, no. 10, pp. 1082–1085, 2000.

[9] B. Smolders, R. Lemmens, and V. Thijs, "Lipoprotein (a) and stroke: a meta-analysis of observational studies," *Stroke*, vol. 38, no. 6, pp. 1959–1966, 2007.

[10] P. J. Schreiner, J. D. Morrisett, A. R. Sharrett et al., "Lipoprotein[a] as a risk factor for preclinical atherosclerosis," *Arteriosclerosis, Thrombosis, and Vascular Biology*, vol. 13, no. 6, pp. 826–833, 1993.

[11] A. E. Kitabchi, G. E. Umpierrez, J. M. Miles, and J. N. Fisher, "Hyperglycemic crises in adult patients with diabetes," *Diabetes Care*, vol. 32, no. 7, pp. 1335–1343, 2009.

[12] V. S. Malik, B. M. Popkin, G. A. Bray, J. P. Despres, and F. B. Hu, "Sugar-sweetened beverages, obesity, type 2 diabetes mellitus, and cardiovascular disease risk," *Circulation*, vol. 121, no. 11, pp. 1356–1364, 2010.

[13] X. Wang, Y. Ouyang, J. Liu et al., "Fruit and vegetable consumption and mortality from all causes, cardiovascular disease, and cancer: systematic review and dose-response meta-analysis of prospective cohort studies," *BMJ*, vol. 349, article g4490, 2014.

[14] D. Li, "Effect of the vegetarian diet on non-communicable diseases," *Journal of the Science of Food and Agriculture*, vol. 94, no. 2, pp. 169–173, 2014.

[15] T. Huang, B. Yang, J. Zheng, G. Li, M. L. Wahlqvist, and D. Li, "Cardiovascular disease mortality and cancer incidence in vegetarians: a meta-analysis and systematic review," *Annals of Nutrition & Metabolism*, vol. 60, no. 4, pp. 233–240, 2012.

[16] C. Walker and B. V. Reamy, "Diets for cardiovascular disease prevention: what is the evidence?," *American Family Physician*, vol. 79, no. 7, pp. 571–578, 2009.

[17] D. E. Threapleton, D. C. Greenwood, C. E. L. Evans et al., "Dietary fibre intake and risk of cardiovascular disease: systematic review and meta-analysis," *BMJ*, vol. 347, article f6879, 2013.

[18] C. Chen, T. K. Hyun, X. Han et al., "Coexpression within integrated mitochondrial pathways reveals different networks in normal and chemically treated transcriptomes," *International Journal of Genomics*, vol. 2014, Article ID 452891, 10 pages, 2014.

[19] M. G. Naylor, X. Lin, S. T. Weiss, B. A. Raby, and C. Lange, "Using canonical correlation analysis to discover genetic regulatory variants," *PLoS One*, vol. 5, no. 5, article e10395, 2010.

[20] H. Ye, J. Yuan, Z. Wang et al., "A canonical correlation analysis of AIDS restriction genes and metabolic pathways identifies purine metabolism as a key cooperator," *Computational and Mathematical Methods in Medicine*, vol. 2016, Article ID 2460184, 10 pages, 2016.

[21] P. Pajukanta, M. Cargill, L. Viitanen et al., "Two loci on chromosomes 2 and X for premature coronary heart disease identified in early- and late-settlement populations of Finland," *American Journal of Human Genetics*, vol. 67, no. 6, pp. 1481–1493, 2000.

[22] Y. Guo, F. Wang, L. Li et al., "Genome-wide linkage analysis of large multiple multigenerational families identifies novel genetic loci for coronary artery disease," *Scientific Reports*, vol. 7, no. 1, p. 5472, 2017.

[23] M. Farrall, F. R. Green, J. F. Peden et al., "Genome-wide mapping of susceptibility to coronary artery disease identifies a novel replicated locus on chromosome 17," *PLoS Genetics*, vol. 2, no. 5, article e72, 2006.

[24] S. Francke, M. Manraj, C. Lacquemant et al., "A genome-wide scan for coronary heart disease suggests in Indo-Mauritians a susceptibility locus on chromosome 16p13 and replicates linkage with the metabolic syndrome on 3q27," *Human Molecular Genetics*, vol. 10, no. 24, pp. 2751–2765, 2001.

[25] J. Erdmann, C. Willenborg, J. Nahrstaedt et al., "Genome-wide association study identifies a new locus for coronary artery disease on chromosome 10p11. 23," *European Heart Journal*, vol. 32, no. 2, pp. 158–168, 2010.

[26] U. Broeckel, C. Hengstenberg, B. Mayer et al., "A comprehensive linkage analysis for myocardial infarction and its related risk factors," *Nature Genetics*, vol. 30, no. 2, pp. 210–214, 2002.

[27] S. E. Hong, I. Park, H. Cha et al., "Identification of mouse heart transcriptomic network sensitive to various heart diseases," *Biotechnology Journal*, vol. 3, no. 5, pp. 648–658, 2008.

[28] L. Zhang, E. A. Hamad, M. Vausort et al., "Whole transcriptome microarrays identify long non-coding RNAs associated with cardiac hypertrophy," *Genomics Data*, vol. 5, pp. 68–71, 2015.

[29] M. A. Coleman, S. P. Sasi, J. Onufrak et al., "Low-dose radiation affects cardiac physiology: gene networks and molecular signaling in cardiomyocytes," *American Journal of Physiology-Heart and Circulatory Physiology*, vol. 309, no. 11, pp. H1947–H1963, 2015.

[30] I. Herrer, E. Roselló-Lletí, A. Ortega et al., "Gene expression network analysis reveals new transcriptional regulators as novel factors in human ischemic cardiomyopathy," *BMC Medical Genomics*, vol. 8, no. 1, pp. 1–14, 2015.

[31] S. K. Sen, J. J. Barb, P. F. Cherukuri et al., "Identification of candidate genes involved in coronary artery calcification by transcriptome sequencing of cell lines," *BMC Genomics*, vol. 15, no. 1, p. 198, 2014.

[32] J. Chen, H. Y. Wang, and C. Y. Zeng, "Transcriptome network analysis of potential candidate genes for heart failure," *Genetics and Molecular Research*, vol. 12, no. 4, pp. 4687–4697, 2013.

[33] Y. Zhao, Z. Sheng, and J. Huang, "A systematic analysis of heart transcriptome highlights divergent cardiovascular disease pathways between animal models and humans," *Molecular BioSystems*, vol. 8, no. 2, pp. 504–510, 2012.

[34] G. Czibik, V. Steeples, A. Yavari, and H. Ashrafian, "Citric acid cycle intermediates in cardioprotection," *Circulation Cardiovascular Genetics*, vol. 7, no. 5, pp. 711–719, 2014.

[35] H. Taegtmeyer, "Metabolic responses to cardiac hypoxia. Increased production of succinate by rabbit papillary muscles," *Circulation Research*, vol. 43, no. 5, pp. 808–815, 1978.

[36] H. Ashrafian, G. Czibik, M. Bellahcene et al., "Fumarate is cardioprotective via activation of the Nrf2 antioxidant pathway," *Cell Metabolism*, vol. 15, no. 3, pp. 361–371, 2012.

[37] S. Balta, S. Demırkol, and T. Celik, "Coenzyme Q10 supplementation may improve diastolic heart functions especially coronary artery disease patients," *Hemodialysis International*, vol. 17, no. 3, pp. 467-468, 2013.

[38] B. J. Lee, Y. C. Huang, S. J. Chen, and P. T. Lin, "Coenzyme Q10 supplementation reduces oxidative stress and increases antioxidant enzyme activity in patients with coronary artery disease," *Nutrition*, vol. 28, no. 3, pp. 250–255, 2012.

[39] A. Kumar, H. Kaur, P. Devi, and V. Mohan, "Role of coenzyme Q10 (CoQ10) in cardiac disease, hypertension and Meniere-like syndrome," *Pharmacology & Therapeutics*, vol. 124, no. 3, pp. 259–268, 2009.

[40] B. J. Lee, C. H. Yen, H. C. Hsu, J. Y. Lin, S. Hsia, and P. T. Lin, "A significant correlation between the plasma levels of coenzyme Q10 and vitamin B-6 and a reduced risk of coronary artery disease," *Nutrition Research*, vol. 32, no. 10, pp. 751–756, 2012.

[41] M. Maes, I. Mihaylova, M. Kubera, M. Uytterhoeven, N. Vrydags, and E. Bosmans, "Coenzyme Q10 deficiency in myalgic encephalomyelitis/chronic fatigue syndrome (ME/CFS) is related to fatigue, autonomic and neurocognitive symptoms and is another risk factor explaining the early mortality in ME/CFS due to cardiovascular disorder," *Neuroendocrinology Letters*, vol. 30, no. 4, pp. 470–476, 2008.

[42] M. Maes, I. Mihaylova, M. Kubera, M. Uytterhoeven, N. Vrydags, and E. Bosmans, "Lower plasma coenzyme Q 10 in depression: a marker for treatment resistance and chronic fatigue in depression and a risk factor to cardiovascular disorder in that illness," *Neuroendocrinology Letters*, vol. 30, no. 4, pp. 462–469, 2009.

[43] R. B. Singh, M. A. Niaz, V. Rastogi, and S. S. Rastogi, "Coenzyme Q in cardiovascular disease," *The Journal of the Association of Physicians of India*, vol. 46, no. 3, pp. 299–306, 1998.

[44] M. Bilinska, J. Wolszakiewicz, M. Duda, J. Janas, A. Beresewicz, and R. Piotrowicz, "Antioxidative activity of sulodexide, a glycosaminoglycan, in patients with stable coronary artery disease: a pilot study," *Medical Science Monitor*, vol. 15, no. 12, pp. CR618–CR623, 2009.

[45] T. N. Wight and M. J. Merrilees, "Proteoglycans in atherosclerosis and restenosis," *Circulation Research*, vol. 94, no. 9, pp. 1158–1167, 2004.

[46] M. Ueda, A. E. Becker, and T. Fujimoto, "Pathological changes induced by repeated percutaneous transluminal coronary angioplasty," *Heart*, vol. 58, no. 6, pp. 635–643, 1987.

[47] G. Dawson, A. W. Kruski, and A. M. Scanu, "Distribution of glycosphingolipids in the serum lipoproteins of normal human subjects and patients with hypo-and hyperlipidemias," *Journal of Lipid Research*, vol. 17, no. 2, pp. 125–131, 1976.

[48] S. Chatterjee and P. O. Kwiterovich, "Glycosphingolipids of human plasma lipoproteins," *Lipids*, vol. 11, no. 6, pp. 462–466, 1976.

[49] E. N. Glaros, W. S. Kim, C. M. Quinn et al., "Glycosphingolipid accumulation inhibits cholesterol efflux via the ABCA1/apolipoprotein A-I pathway: 1-phenyl-2-decanoylamino-3-morpholino-1-propanol is a novel cholesterol efflux accelerator," *The Journal of Biological Chemistry*, vol. 280, no. 26, pp. 24515–24523, 2005.

[50] S. Chatterjee, D. Bedja, S. Mishra et al., "Inhibition of glycosphingolipid synthesis ameliorates atherosclerosis and arterial stiffness in apolipoprotein $E^{-/-}$ mice and rabbits fed a high-fat and -cholesterol diet," *Circulation*, vol. 129, no. 23, pp. 2403–2413, 2014.

18

Classification of Complete Proteomes of Different Organisms and Protein Sets based on their Protein Distributions in Terms of Some Key Attributes of Proteins

Hao-Bo Guo,[1] Yue Ma,[1] Gerald A. Tuskan,[2] Xiaohan Yang,[2] and Hong Guo[1]

[1]*Department of Biochemistry and Cellular and Molecular Biology, University of Tennessee, Knoxville, TN 37996, USA*
[2]*Biosciences Division, Oak Ridge National Laboratory, Oak Ridge, TN 3783, USA*

Correspondence should be addressed to Xiaohan Yang; yangx@ornl.gov and Hong Guo; hguo1@utk.edu

Academic Editor: Joshua Xu

The existence of complete genome sequences makes it important to develop different approaches for classification of large-scale data sets and to make extraction of biological insights easier. Here, we propose an approach for classification of complete proteomes/protein sets based on protein distributions on some basic attributes. We demonstrate the usefulness of this approach by determining protein distributions in terms of two attributes: protein lengths and protein intrinsic disorder contents (ID). The protein distributions based on L and ID are surveyed for representative proteome organisms and protein sets from the three domains of life. The two-dimensional maps (designated as fingerprints here) from the protein distribution densities in the LD space defined by $\ln(L)$ and ID are then constructed. The fingerprints for different organisms and protein sets are found to be distinct with each other, and they can therefore be used for comparative studies. As a test case, phylogenetic trees have been constructed based on the protein distribution densities in the fingerprints of proteomes of organisms without performing any protein sequence comparison and alignments. The phylogenetic trees generated are biologically meaningful, demonstrating that the protein distributions in the LD space may serve as unique phylogenetic signals of the organisms at the proteome level.

1. Introduction

Determination of complete genome sequences for a number of organisms has offered an unprecedented opportunity for biological community and transformed biology into a discipline that depends significantly on how to classify and interpret large-scale data sets and to extract biological insights from these data sets. The traditional ways of thinking and approaches from the pregenomic era (e.g., the sequence comparison/alignment and homology identification) are of fundamental importance in the postgenomic era. Nevertheless, new approaches based on some global features of omics data sets need to be explored in order to make classification and comparison of large-scale data sets easier. For proteomes, this may be achieved, for instance, through identification of key parameters or attributes of proteins and comparison of protein distributions within complete proteomes of different organisms or protein sets in terms of such parameters or attributes.

In this paper, we adapt this approach and use two parameters of proteins for the purpose of classifying complete proteomes of different organisms (for simplicity, proteomes) and protein sets: the length of protein amino acid (aa) sequence (protein length L hereafter) and intrinsic disorder content (protein disorder ID hereafter). It had been proposed that the protein sizes, folding rates, and many other physical properties could be associated or even determined by L [1, 2]. At the level of proteomes, previous studies have suggested that the eukaryotic proteomes may exhibit averagely longer L compared to the prokaryotic proteomes [3, 4], even though further analysis may still be necessary. The importance of intrinsically disordered proteins (IDPs) and protein regions (IDPRs) has been recognized [5–13], and it has been observed that relatively high contents of intrinsic disorders may exist for eukaryotic proteins than for prokaryotic proteins [14]. Moreover, proteins expressed in two eukaryotic organelles, chloroplasts and mitochondria, which evolved from cyanobacteria and alphaproteobacteria, respectively, seem to have a lower disorder content, on average, compared to nuclear-encoded proteins in their host eukaryotes [15]. Interestingly, it has been demonstrated that intrinsically disordered proteins are associated with a variety of human diseases [16, 17], including cancers [18, 19]. As a result, intrinsically disordered proteins have become important targets for drug design [20–25]. Thus, understanding intrinsically disordered proteins at the proteomic levels would be of considerable interest. The observations that the distributions of proteins in terms of ID and L may be different for proteomes and for different protein sets suggest that such distributions may be used to classify proteomes of different organisms or protein sets. They may also be used in the future to help understand the properties of proteomes in different disease states, as there seems to be a wide variability of predicted disorder among different diseases [26]. It is interesting to see that a recent study revealed that the overall disorder fractions are positively correlated to the size of the proteomes (estimated by the total aa numbers) and that the disorder fractions of the proteomes of large bacteria (more than 2.5 M aa) are comparable to those of eukaryotes [27].

Here we analyze the protein distributions in terms of L and ID from proteomes of different organisms across the three domains of life, collective data sets of organelles (plasmids, chloroplasts, and mitochondria), and the proteome data of two giant DNA viruses (termed giruses in literature). We noticed that the eukaryotic proteomes do not always exhibit averagely longer proteins than the prokaryotic proteomes. Our observation on protein disorder agrees well with the previous finding, that is, the average disorder contents in eukaryotic proteins are indeed higher than those in prokaryotic proteins. The two-dimensional maps (designated as fingerprints here) based on the protein distribution densities in the LD space defined by $\ln(L)$ and ID for the representative proteomes of different organisms and protein sets were constructed, and these fingerprints show distinct patterns for different organisms and protein sets. The features and relationships among the fingerprints are analyzed and compared. To test if our classification of proteomes of different organisms and protein sets proposed here is meaningful, we generated phylogenetic trees based on the protein distribution densities in the fingerprints of proteomes of different organisms without performing any protein sequence comparison and alignments. The phylogenetic trees generated in this way were found to be meaningful, as they contain important information of evolution. Thus, the proposed approach may represent a useful and simple way for proteome classification and comparison. In present study, for each protein-encoding gene locus only the prime protein has been used, therefore, the protein densities (Figure 1 and Figure S1) could be regarded as the gene densities. Moreover, using the poplar proteome as an example, it was found that the phylogenies show little difference with or without using alternative splicing proteins (Figure S3). Discussions are made concerning the possibility for extending this approach through introduction of additional attributes.

2. Results

2.1. Protein Distributions in Terms of L and ID. Here, we discuss the proteins (811,600 entries in total) from the proteomes of different organisms and protein sets listed in Table 1, with the protein lengths varying over three degrees of magnitude from 5 (*Os06g47230* of rice) to 34,350 aa (*titin* of human). For the protein length comparison, as pointed out previously [4], the median length is a better measure than the average length to avoid biases from extremely long proteins. Table 1 lists both the median and average lengths of all the proteomes and proteins from gene sets. It should be pointed out that in the present analysis, only the primary protein at each gene locus is selected. This allows a significant simplification of proteome classification. This approximation seems to be reasonable for the main purpose of this work, as there is little difference in the results for the test cases with or without using alternative splicing proteins. Table 1 shows that the eukaryotic proteomes do not *always* have averagely longer proteins than those in the prokaryotic proteomes, as previously suggested [3, 4]. For instance, the basal flowering plant *Amborella trichopoda* has a median protein length of 218, shorter than all prokaryotes (Archaea and bacteria) surveyed here. In addition, *Giardia intestinalis* in the Eukaryota domain has an even shorter median protein length of 147. The average *L*s show the same trend as the median values (Table 1).

Nevertheless, the proteins in a eukaryotic proteome do have a significantly higher intrinsic disorder in average ($41.1 \pm 6.4\%$) compared to those in a prokaryotic proteome ($15.6 \pm 6.5\%$), consistent with previous studies [14, 28]. This trend stands for the average disorder contents of all residues from the proteomes ($47.5 \pm 6.4\%$ for eukaryotes compared to $32.9 \pm 1.4\%$ for prokaryotes). Proteomes from the archaeon *N. equitans* and bacterium *Rickettsiales* have the lowest disorder content at the protein level (7.0% for *N. equitans* and 7.7% for *Rickettsiales*) for the systems examined. As the smallest known archaeon, *N. equitans* is an obligate symbiont on the other archaeon *I. hospitalis*, which is the smallest known free-living archaeon [29]. The free-living alphaproteobacterium *Rickettsiales*, on the other hand, was suggested to be a living candidate that is close to the ancient endosymbiotic alphaproteobacteria that were merged into an archaeon and eventually transferred into the mitochondria of the first eukaryotic cell [30]. These two symbiotic or presymbiotic organisms have retained more ordered proteins compared to other free-living bacteria and Archaea surveyed here.

Consistent with previous studies [15], the proteins from the mitochondrion (88,405 proteins from 6119 species) and chloroplast (80,807 proteins form 935 species) sets have relatively low disorder contents compared to the proteins

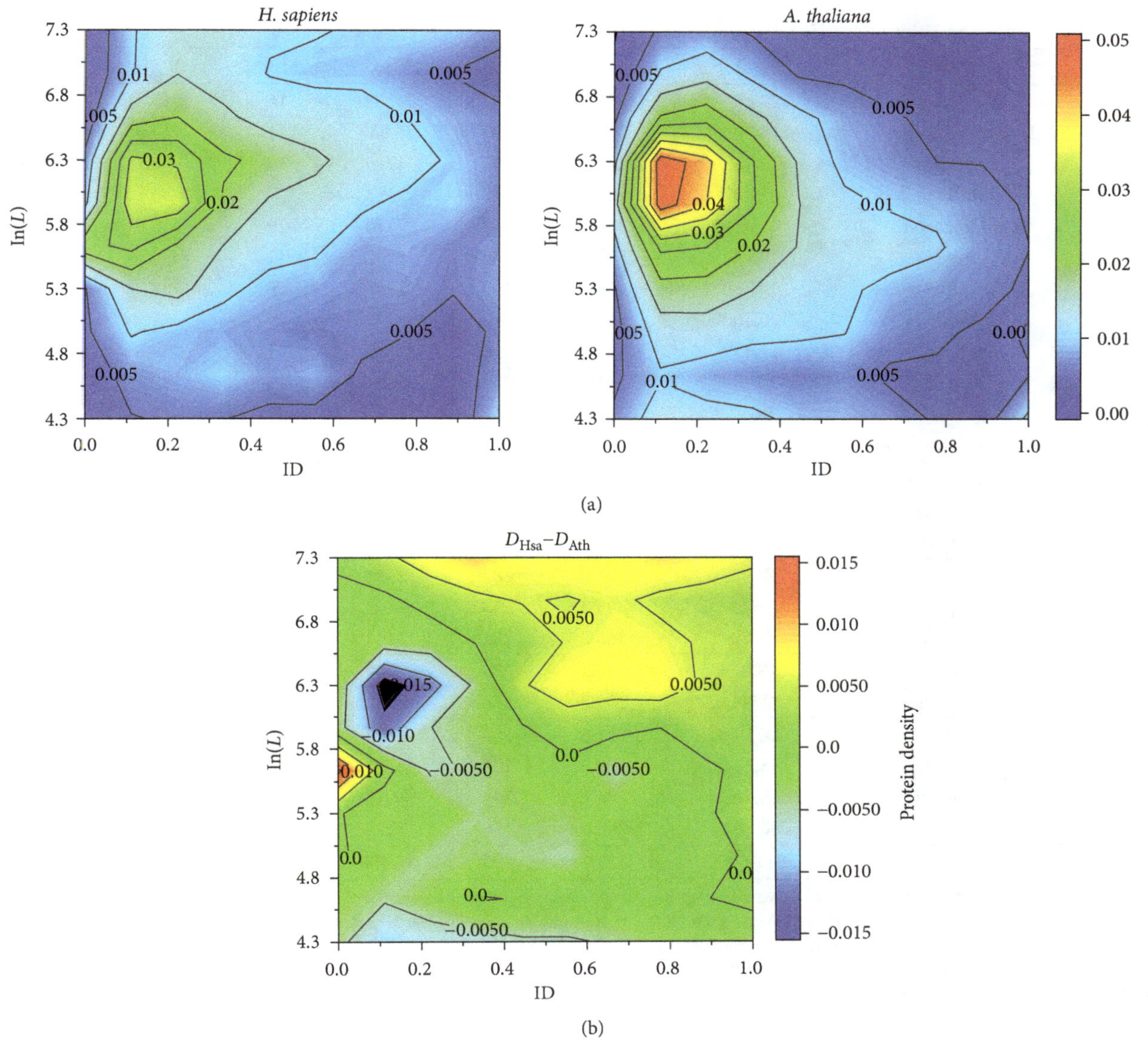

FIGURE 1: (a) Representative protein-density contour maps of (left) an animal (*H. sapiens*) and (right) a plant (*A. thaliana*) proteome. Short proteins ($\ln(L) < 4.3$ or $L < 74$) and long proteins ($\ln(L) > 7.3$ or $L > 1480$) are treated as $\ln(L) = 4.3$ and $\ln(L) = 7.3$, respectively, for statistics. (b) Differential protein density contour map between *H. sapiens* (D_{Hsa}) and *A. thaliana* (D_{Ath}) indicates that short disordered proteins are enriched in the plant proteome; and the animal proteome has more long disordered proteins.

encoded in nuclear genes of eukaryotic organisms, for example, the mitochondrial protein set has a considerably lower disorder content of 8.6% at protein level. The mitochondria have lost most of their ancestral genes either by transferring to the nucleus or by being discarded [31]. Here, we show that the mitochondrial proteins have relatively low disorder contents (i.e., highly ordered) at both the protein and the residue levels (Table 1). The genes retained in the mitochondrial genomes have been proposed preferentially to encode core proteins involved in electron transfers [32], and a colocalization of the redox regulation (CoRR) mechanism was proposed to explain why the mitochondrial and chloroplastic organelles retain their own genes, or proteins [33, 34]. Our analysis indicates that the chloroplast genes have their proteins with disorder contents close to the free-living prokaryotes, but higher than those from the symbiotic Archaeon *N. equitans* and alphaproteobacterium *Rickettsiales*, as well as the mitochondrial set (Table 1).

The proteomes of two giant DNA viruses (giruses), the Mimivirus and Pandoravirus, were also analyzed. The numbers of proteins encoded in these two giruses are comparable to the prokaryotic proteomes. The disorder content of the proteome of the Mimivirus is larger than that of the prokaryotes, but smaller than that of the eukaryotes surveyed here. However, the Pandoravirus has a proteome with disorder content close to that of the eukaryotes.

Finally, the viral and plasmid gene sets were analyzed. The viral gene set contains 237,463 genes collected from 4942 strains and the plasmid set contains 95,214 genes cultivated from 985 bacteria. Interestingly, the proteins from

TABLE 1: A summary of the proteomes and gene sets.

Domain[a]	Species	Gene number[b]	Ave[c]	Med[c]	Max[c]	Min[c]	ID_{pep}%[b,d]	ID_{res}%[b,e]
Eukaryota	*H. sapiens*	20,193	561.0	417	34,350	16	45.2	49.3
	D. melanogaster	13,700	537.2	396	22,949	11	44.3	49.0
	S. cerevisiae	5917	494.1	405	4910	16	38.1	44.6
	A. thaliana	27,407	405.2	348	5393	7	36.8	43.6
	P. trichocarpa	41,434	385.0	317	5410	29	35.5	42.6
	A. comosus	29,772	372.6	288	5407	31	39.5	45.4
	O. sativa	48,788	376.1	290	4957	5	38.0	44.5
	A. trichopoda	26,460	317.0	218	4990	29	37.5	43.9
	C. reinhardtii	17,819	732.9	498	23,859	31	54.8	61.9
	P. patens	32,400	351.9	250	5199	13	40.2	45.5
	G. intestinalis	9667	353.8	147	8161	33	35.1	41.7
	Monocercomonoides	16,780	784.6	393	14,902	49	52.7	60.1
Archaea	*Lokiarchaeum*	5348	268.4	224	3592	20	20.0	33.0
	I. hospitalis	1434	278.3	240	1392	33	20.4	34.3
	N. equitans	540	280.2	228	2197	45	7.0	30.6
Bacteria	*E. coli*	4140	316.9	282	2358	14	17.5	32.2
	S. elongatus	2612	305.3	258	1807	29	20.8	34.3
	Rickettsiales	1780	365.2	251	2243	31	7.7	32.8
Giruses	*Mimivirus*	979	356.7	289	2959	25	25.0	36.6
	Pandoravirus	2541	259.2	178	2321	26	36.4	43.5
Gene sets	*Viruses*	237,463	251.8	154	8573	9	28.0	38.8
	Plasmids	95,214	258.9	206	16,990	9	27.2	38.1
	Mitochondria	88,405	286.1	261	2640	13	8.6	20.0
	Plastids	80,807	280.0	156	5242	12	20.5	32.0
All proteins[f]		811,600	325.7	225	34,350	5	32.2	39.8

[a]Proteomes in the three domains of life; the giant DNA viruses (giruses) and collective protein sets are listed after the cellular species; [b]Total gene numbers; [c]Protein length statistics: Ave: average; Med: median; Max: maximal; Min: minimal protein lengths; [d]Percentage of the intrinsically disordered proteins in the proteome or gene set; [e]Average intrinsic disorder contents of all residues carried by the proteome or gene set; [f]All proteins studied in the present work. The protein length statistics covers all proteins in a proteome or gene set; however, the proteins with unknown sequence(s) (*X* residues) are excluded in the intrinsic disorder calculations.

these two sets yield similar trends in both length and disorder distributions.

2.2. Definition of the LD Space. Consistent with a previous report [3], exponential distributions of the protein lengths (L) in all proteomes and protein data sets have been observed. In this analysis, all proteins of a proteome or protein set have been ranked hierarchically from the shortest to the longest, and the proteins then distribute linearly on $\ln(L)$ (the natural base was used for the logarithm function in this study). Similar linear distribution trend is observed for the percentage of residues located in the IDPR, ID (Figure 2). Therefore, a two-dimensional LD space could be defined with one phase for the content of the protein intrinsic disorder, ID, and the other phase for the logarithm of the protein length, $\ln(L)$. Figure 2 exemplifies the protein distribution in the LD space of the human proteome.

2.3. Dependency of the Two Attributes for the LD Space. We defined a two-dimensional LD space with the two attributes, $\ln(L)$ and ID, and these two attributes need to be independent of each other. Therefore, we calculated the correlation coefficients (CCs) between $\ln(L)$ and ID of proteins in all proteomes and protein sets (Figure 3). Pearson's and Spearman's CCs for all proteins (811,600 entries, Table 1 and S1) are −0.101 and −0.129, respectively. The overall slight negative CC (anticorrelation) indicates that there may be a trend that shorter proteins have averagely higher disorder contents than the longer proteins. However, the anticorrelational trend does not hold for all species surveyed in this study and positive CC values were found, too, such as in the animals (human and fruit fly) and green algae *C. reinhardtii* (Table S1). The variations in the correlational trends between $\ln(L)$ and ID, therefore, may have been driven by the evolutionary processes rather than a cause-and-effect relationship. As such, the validity of the protein LD space and the related architecture of protein distributions in the LD space (i.e., the "fingerprint") should be discussed in an evolutionary framework (see below).

2.4. Architecture of Protein Distribution (Fingerprint) in the LD Space. The most thoroughly annotated animal and plant

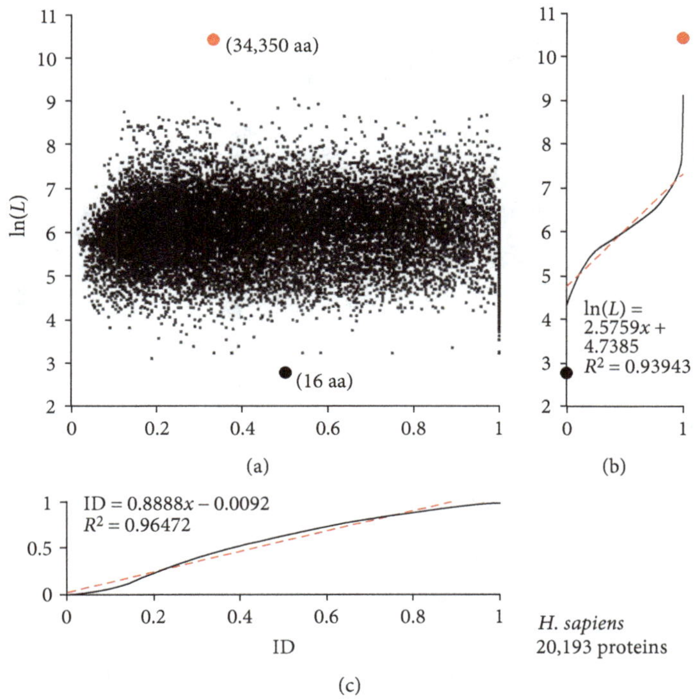

FIGURE 2: Protein distributions for the human (*H. sapiens*) proteome in the LD space defined by ln(L) (the protein length in a logarithm scale) and ID (protein intrinsic disorder contents with 1.0 corresponding to proteins with 100% residues disordered and 0.0 corresponding to proteins with 0% residues disordered). The distributions in the hierarchical scale are shown in (b) and (c), respectively (see text). Linear fittings of ln(L) and ID are shown in red dashed lines with satisfactory R^2 and hence support the linear participations shown in Table 2. The blue and red dots indicate the shortest (16 aa) and longest (34,350 aa) proteins, respectively.

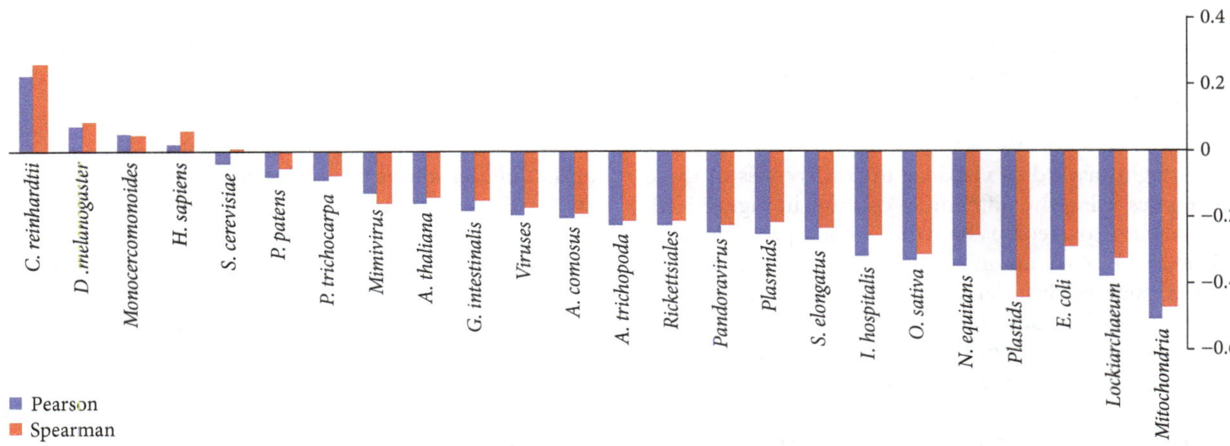

FIGURE 3: Pearson's (blue) and Spearman's (red) correlation coefficients (CCs) between ln(L) and ID of the proteins in proteomes and gene sets surveyed in the present work. All species were ranked by the Pearson's CCs from the highest positive (*C. reinhardtii*) to highest negative (mitochondrial gene set).

genomes may be those of human (*H. sapiens*) and *Arabidopsis thaliana*, respectively. Using proteomes form the two representative animal and plant, the protein distributions of proteomes in the LD space were converted to the protein-density contour maps in Figure 1(a) (see Materials and Methods). As we will show below, this approach may be useful in comparative proteomes/genomics.

At a first glance, the plant proteome has more proteins of medium lengths (~$5.7 < \ln(L) < 6.4$ or ~$300 < L < 600$) and relatively low disorder contents (ID < 0.3) whereas the animal proteome contains more long and disordered proteins (e.g., $L > 600$ and ID > 0.5). This may partly explain the slightly positive correlations between ln(L) and ID in the animal proteomes but negative correlations in the plant proteomes. The protein distribution contour maps of other proteomes and gene sets can be found in Figure S1 in Supplementary Materials and have been trimmed in the phylogenetic tree in Figure 4 (see below).

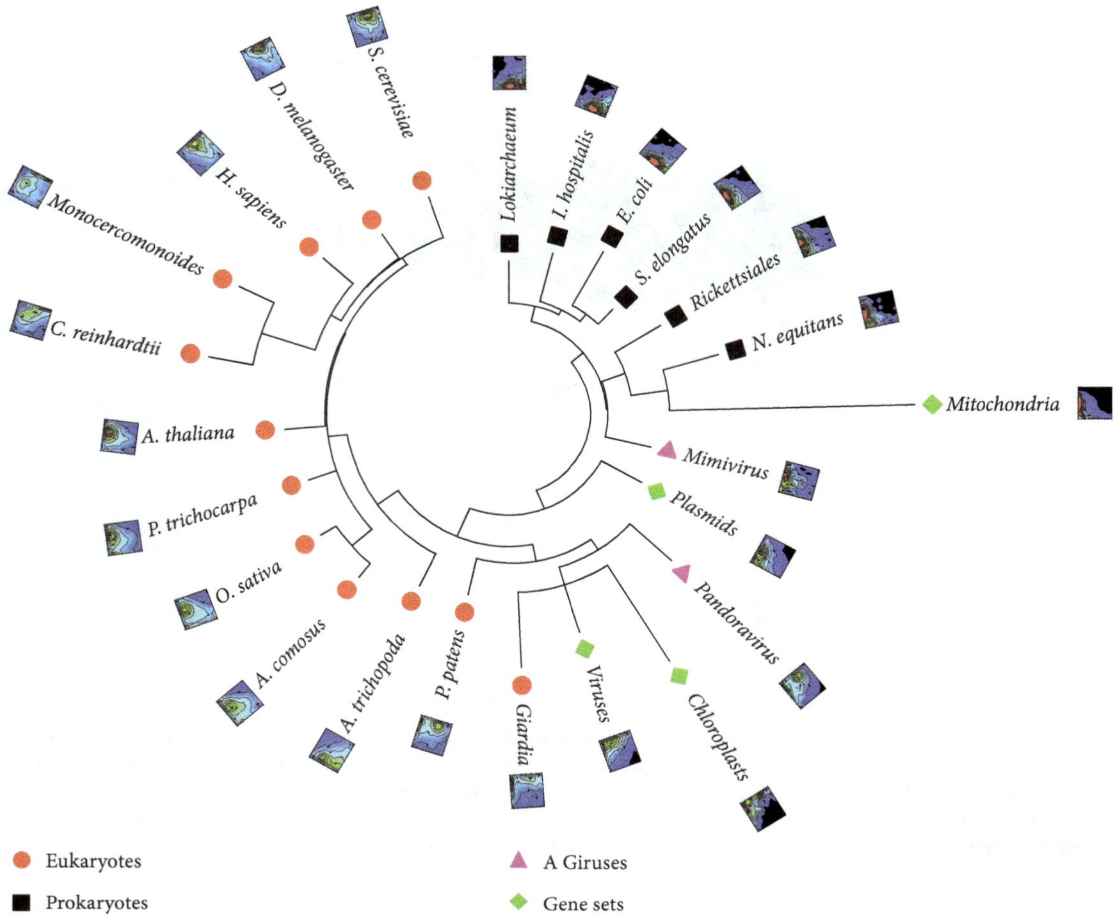

FIGURE 4: The phylogenetic tree reconstructed using the protein distribution densities on the LD space. The protein density distributions in the LD space for each species or gene set are also shown (Figure S1 in Supplementary Materials shows higher resolution figures). MEGA5 [69] was used to plot the tree.

It is straightforward to visualize the differences of these two proteomes using the differential contour in Figure 1(b). The *H. sapiens* proteome has 657 short proteins (i.e., $L < 100$ or $\ln(L) < 4.6$), among which 294 (1.5% of all proteins) are considered disordered (ID > 0.5); in the *A. thaliana* proteome, 888 (3.2%) out of 2292 short proteins are disordered. On the other hand, in the *H. sapiens* proteome, 1135 (5.6%) out of 2384 long proteins (i.e., $L \geq 1000$ or $\ln(L) \geq 6.9$) are disordered; whereas, in the *A. thaliana* proteome, 306 (1.1%) out of 1157 long proteins are disordered. Therefore, a significant difference between the animal (*H. sapiens*) and the plant (*A. thaliana*) could be recognized as that the former has more long disordered proteins, whereas the latter has more short disordered proteins. This difference shown in Figure 1(b) allows us to narrow down the protein/gene distributions related to the architectural differences between the two organisms.

A recent report also indicates that the overall disorder contents of the *A. thaliana* proteome are lower than those of the *H. sapiens* proteome [35], which was attributed that more IDP genes functioning in environmental adaptations may have been enriched in plants [35]. Based on our analysis and the apparent abundance of the short disordered proteins in *A. thaliana* compared to *H. sapiens* (Figure 1(b)), we focus on the 888 short (<100 aa, see above) IDP (sIDP) genes of *A. thaliana*. Among these genes, the GO annotations of 203 sIDPs could not be identified, that is, they may be considered among the "dark matter" of the *A. thaliana* proteome [36]. However, among the 685 annotated sIDPs (occupying 545 GO terms), only 20 (~0.2% of all sIDPs) with 32 GO annotations were included in the previous analysis showing "enrichment" of 74 GO annotations related to the environmental adaptations in *A. thaliana* compared to *H. sapiens* [35]. Based on our analysis, this enrichment might not be significant for the sIDPs. We suggest that it might be possible that in animals and other organism (e.g., the green algae *C. reinhardtii*), some of the sIDPs had been lost whereas long IDPs were enriched. Here, GO annotations of the plant genes were adopted from the plant comparative genomics database PLAZA 3.0 [37].

2.5. Phylogeny Reconstructed Based on Protein Distribution Densities in the LD Space. As the first test concerning whether our classification of proteomes and protein sets is biologically reasonable, we generated phylogenetic trees based on the protein distribution densities in the fingerprints of proteomes

TABLE 2: Intervals that partition the LD spaces into $M \times N$ blocks with $M = N = 10$.

	Number	1	2	3	4	5	6	7	8	9	10
	ln(L)	(0,4.6)	(4.6,4.9)	(4.9,5.2)	(5.2,5.5)	(5.5,5.8)	(5.8,6.1)	(6.1,6.4)	(6.4,6.7)	(6.7,7.0)	(7.0,∞)
10 × 10	L	(1100)	(101,135)	(135,182)	(182,245)	(245,331)	(331,446)	(446,602)	(602,813)	(813,1097)	(1097,∞)
	ID%	(0,0.1)	(0.1,0.2)	(0.2,0.3)	(0.3,0.4)	(0.4,0.5)	(0.5,0.6)	(0.6,0.7)	(0.7,0.8)	(0.8,0.9)	(0.9,1.0)

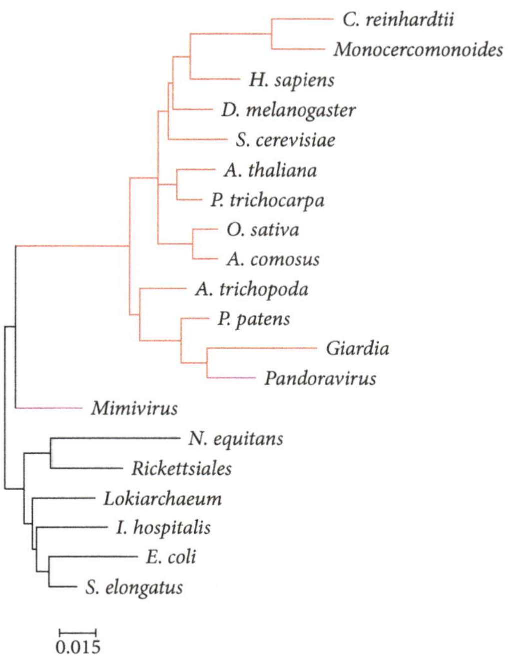

FIGURE 5: The phylogenetic tree reconstructed from the protein distributions in the LD space using $M = N = 10$ in (3) in Materials and Methods. Eukaryotes are in red, prokaryotes (bacteria and Archaea) in blue and giruses in pink branches. MEGA5 [69] was used to plot the trees. Because this tree is based on normalized protein densities (of 100 blocks in the $M = N = 10$ tree here), the branch length of the tree is relatively small with a scale bar of 0.015.

without performing any protein sequence comparison and alignments. Here, aiming to quantify the *architectural* differences among proteomes, the LD space was divided into $M \times N$ blocks and then, the distance between two species A and B was calculated using a Euclidian-type formula based on the protein distributions in all blocks (see (3) in Materials and Methods). In this *architectural*-distance calculation, no rigorous biological function annotations and/or genomic comparisons using BLAST or other protocols are required.

By dividing the LD space with $M = N = 10$ (Table 2), the distance matrices for all proteomes including those from giruses (Table 1) were calculated and converted to phylogenetic trees as shown in Figure 5. We also tested the 5 × 5 or 2 × 2 partitioning; the 10 × 10 partitioning of the LD space seems to yield relatively high accuracy (Table S2 and Figure S2 in Supplementary Materials). Nevertheless, some of the key properties are not very sensitive to the M and N values. Several interesting features have been found in the trees that we reconstructed: (1) the eukaryotes are clearly separated from the prokaryotes and (2) plants and animals are grouped together, even the eudicot plants (*A. thaliana* and *P. trichocarpa*) and monocot plants (*O. sativa* and *A. comosus*) are separated. The tree in Figure 5 correctly puts *A. trichopoda* before the other plant species and after *P. patens*. Interestingly, it is consistent with our understanding of the plants-fungi-animals phylogenetic relationships [38] and stays in the framework of the natural classification of three domains of life [39]. Based on the phylogenetic tree, the definition of the protein LD space might be considered meaningful to the proteomes, at least to those chosen in present work.

3. Discussion

To the best of our knowledge, this is the first time to classify proteomes and protein sets based on the protein distribution densities in the LD space (fingerprints), and a detailed comparison with the previous work is therefore not straightforward. Nevertheless, the survey of protein distributions in terms of each of the two attributes is consistent with the work published previously. We noticed that the eukaryotic proteomes do not always exhibit averagely longer proteins than the prokaryotic proteomes. Our observation on protein disorder agrees well with the previous finding, that is, the average disorder contents in eukaryotic proteins are indeed higher than those in prokaryotic proteins. We have also generated phylogenetic trees based on the protein distribution densities in the fingerprints of proteomes, and this allows us to make some comparisons of the results that we obtained here with the knowledge in the field and to examine the consistency and differences with earlier investigations. Such comparison may also provide certain alternative views that were generated through this unique approach.

3.1. Giant DNA Viruses and the Tree of Life. It has been in the debate over the years concerning if viruses should be included in the tree of life [40, 41] or if they are alive at all [42, 43]. The discovery of Mimivirus [44] that belongs to the nucleocytoplasmic large DNA viruses (NCLDV) and the following discoveries of other giant DNA viruses (giruses) [45], for example, the Pandoravirus with a genome size exceeding some of the cellular organisms [46], invoked questions on if a "fourth domain" should be added to the tree of life [46, 47] and potentially important roles that viruses played in eukaryogenesis [48]. Interestingly, we found that Mimivirus is located in between the Eukaryota and prokaryote (Archaea + Eubacteria) branches, that is, at the prokaryote-to-eukaryote transition zone. This is consistent with the original phylogenetic analysis inferred based on seven universally conserved protein sequences [44]. The Pandoravirus, on the other hand, is located within the

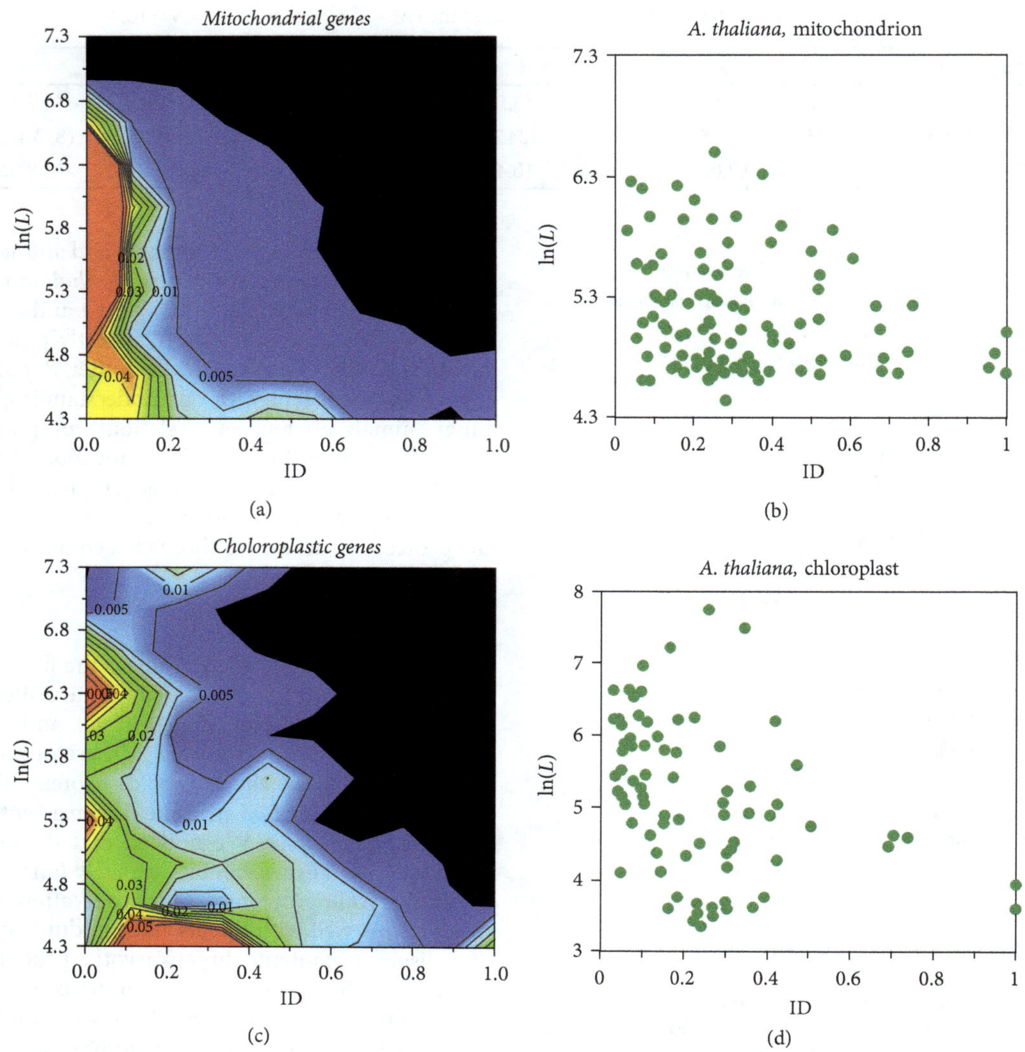

FIGURE 6: Protein distributions in the LD space for (a) the mitochondrial gene set, (b) the mitochondrial genes in *A. thaliana*, (c) the chloroplast gene set, and (d) the chloroplastic genes in *A. thaliana*.

Eukaryota branch. The vast majority (>93%) of the Pandoravirus genes exhibit no homology to anything known [46]; however, our approach puts it in the same branch of the parasite Giardia (Figure 5(c)), owing to the abundance of short proteins (both in ordered and disordered states) in these two organisms (Figure S1).

3.2. Organelles. The phylogenetic tree with the viral and organelle (mitochondria, chloroplasts, and plasmids) gene sets is shown in Figure 4 along with the fingerprints in the LD space. In this tree, the viral gene set is located in the same branch as the Pandoravirus. The plasmid gene set is located in between prokaryotic and eukaryotic branches, or more accurately, between Mimivirus and Pandoravirus. These results suggest the importance of horizontal gene transfers in eukaryogenesis carried by the viral and plasmid genes.

In Figure 4, the mitochondrial gene set sits in the same branch as the symbiont *N. equitans* and alphaproteobacterium *Rickettsiales*, owing to that majorities of the proteins in these proteomes and protein set are highly ordered (Table 1). The chloroplast set is located at the same branch as the viral gene set and *Giardia* (Figure 4). Using the full set of annotated mitochondrial genomes for 2015 species, a recent report [32] revealed that the proteins retained in the eukaryotic mitochondria are preferentially the structural cores in the electron transportation chains. Our survey with the mitochondrial proteins obtained from the NCBI database indicates that the mitochondrial proteins are mainly structurally ordered (Figure 6(a)), thereby possibly structurally and functionally conserved, too. However, using the model plant species *A. thaliana* as an example, the mitochondrial protein distribution in the LD space (Figure 6(b)) does not match that from the mitochondrial gene set (Figure 6(a)). This inconsistency may originate from a considerable amount of highly disordered proteins retained in the mitochondria. For instance, *A. thaliana* has 115 mitochondrial genes, 23 of which are IDPs (i.e., ID $\geq$ 0.5; here, ID refers to the ratio of residues). However, we found that 19 (out of 23) mitochondrial IDPs have unknown functions involved in unknown biological processes (Table S3 in Supplementary Materials), immediately raising a question on the validity of the results obtained from annotated mitochondrial genomes (Figure 6(a) in the present study and [32]). The protein distribution profile of *A. thaliana* chloroplast (Figure 6(d)) resembles that of the collective chloroplast gene set (Figure 6(c)). Only 6 out of 85 *A. thaliana* proteins are IDPs, all of which have been annotated as ribosomal proteins (Table S3).

4. Conclusion

Our two-dimensional contour maps (or proteome fingerprints) based on the protein distribution densities in the LD space show distinct patterns for different organisms and protein sets and may therefore be used for classification of proteomes and protein sets. The phylogenetic trees generated based on the protein distribution densities from the fingerprints were found to be meaningful, as they seem to contain important information of evolution. Thus, the proposed approach and its further extension may represent a useful and alternative way for proteome classification and comparison. It should be pointed out that although in the present work we used protein lengths (L) and protein intrinsic disorder contents (D) as the basic attributes, other attributes (not limited to those from proteins) may be introduced as well. One can imagine that one of the properties for the attributes would be that protein distributions in terms of the new attributes would be different for different proteomes (protein sets) so that the purpose of classification of proteomes (protein sets) can be achieved.

5. Materials and Methods

5.1. Proteomes and Gene Set. The plant proteomes in this study were downloaded from Phytozome, and the proteomes of bacteria, Archaea, and animals were downloaded from UniProt; the organelle protein sets were obtained from NCBI, at or before December 2016.

Here, we surveyed 12 eukaryotic proteomes from two animal species *Homo sapiens* [49, 50] and *Drosophila melanogaster* [51], two monocot plant species *Oryza sativa L. ssp. indica* [52] and *Ananas comosus* [53], two dicot plant species *Arabidopsis thaliana* [54] and *Populus trichocarpa* [55], the basal angiosperm *Amborella trichopoda* [56], the moss *Physcomitrella patens* [57], the fungus *Saccharomyces cerevisiae strain S288C* [58], the green algae *Chlamydomonas reinhardtii* [59], the metamonada Giardia (previously known as an Archezoa that lacks conventional mitochondrion) [60], and *Monocercomonoides sp. PA203* that completely lacks the mitochondrial or mitochondrial-derived genes [61]. We also analyzed three bacterial species *Escherichia coli K12 MG1655* [62], the cyanobacterium *Synechococcus elongatus PCC 7942* [63], and the alphaproteobacterium *Rickettsiales bacterium Ac37b* [64] and three Archaea species *Ignicoccus hospitalis kin4/i*, *Nanoarchaeum equitans* [29], and *Lokiarchaeum sp. GC14_75* [65]. Two giant DNA-viruses (giruses) were also analyzed, including the *Mimivirus* [44] and *Pandoravirus salinus* [46]. In addition, we downloaded several gene collections from the NCBI gene libraries containing the viral set (237,463 genes), plasmid set (95,214 genes), mitochondrial set (88,405 genes), and chloroplast set (80,807 genes). Table 1 gives a summary of the proteomes and gene sets.

The proteomes and gene sets listed above comprise 811,600 proteins, among which 2401 proteins (~0.3%) contain unknown "X" residues and were excluded for analysis in this work.

It should be pointed out that in the present analysis, only the primary protein at each gene locus is selected. The poplar (*P. trichocarpa*) proteome [55] was selected to test the potential influence of the versions of the proteomes and splicing alternatives. From the *P. trichocarpa* genome, there are three versions (v01, v02, and v03) of the proteomes, of which the v03 proteome has 41,434 primary proteins and 31,579 splicing alternatives (73,013 proteins in total). Using the primary proteins of all three versions and the full proteome of the v03 version as separated entries, a phylogenetic tree was constructed (Figure S3 in Supplementary Materials) and there is little difference with or without using alternative splicing proteins or by using different proteome versions.

5.2. Intrinsic Disorder (ID) Prediction. The PONDR-VSL2 algorithm [66] was applied to predict the ID content of all residues in a protein. This program had achieved ~81% accuracy for both short and long proteins. By default, a residue is in an ordered state if its PONDR score is less than 0.5, but in a disordered state when the PONDR score is larger than or equal to 0.5. PONDR scores of 0 and 1 corresponding to the fully ordered and fully disordered states, respectively. Here, this criterion was adopted and extended to calculate the ID content of a protein:

$$\mathrm{ID}_{\mathrm{pep}} = \frac{N_{\mathrm{D}}}{L}, \quad (1)$$

where N_{D} is the number of disordered residues and L is the total number of residues of the protein (i.e., protein length). $\mathrm{ID}_{\mathrm{pep}}$ is also termed as the "*rough definition*" of the disorder contents in [27] and ranges from 0 to 1, with 0 and 1 corresponding to the fully ordered and fully disordered proteins, respectively.

It had been suggested that the total proteome information content (PIC) could be defined as the total number of amino acids of the primary proteins (longest isoform at each gene locus) that the proteome carries [67]. In accordance with this definition, we also calculated the average intrinsic disorder content per residue as

$$\mathrm{ID}_{\mathrm{res}} = \sum_{i=1}^{X} \frac{D_i}{X}, \quad (2)$$

where X is the total number of amino acids and D_i is the PONDR score of the ith residue of the proteome or protein set. $\mathrm{ID}_{\mathrm{res}}$ corresponds to the definition adapted in [27]. Both $\mathrm{ID}_{\mathrm{pep}}$ and $\mathrm{ID}_{\mathrm{res}}$ are listed in Table 1. Because in present work distributions of genes (or proteins) are used to discuss the evolutionary dynamics, $\mathrm{ID}_{\mathrm{pep}}$ (simplified as ID in the main text) had been chosen to act as one of the attributes of the LD space.

5.3. Generation of the Fingerprints and Phylogenetic Analysis. To generate the fingerprints, the LD space of species X was first divided into $M \times N$ blocks (e.g., Table 2), M for $\ln(L)$ and N for ID. This separation is reasonable because both $\ln(L)$ and ID exhibit linearity (Figure 2). Then, the protein density in the ijth block (i in $\ln(L)$ and j in ID%) is calculated as $X_{ij} = n_{ij}/n_{\text{tot}}$, where n_{ij} is the number of proteins in the ijth block and $n_{\text{tot}} = \sum_{l=1}^{M}\sum_{d=1}^{N} n_{ld}$ is the total number of proteins in the proteome of species X. Normalization of the protein density is realized by default since $\sum X_{ij} = 1$.

Using the protein densities, the distance between two organisms A and B can be calculated using the Euclidean equation:

$$r_{AB} = \sqrt{\sum_{l=1}^{M}\sum_{d=1}^{N} (A_{ld} - B_{ld})^2}, \quad (3)$$

where r_{AB} is the distance between A and B and X_{ij} ($X = A$ or B) is the protein density in the ijth block. The calculated distance matrix is converted to the phylogenetic tree using the neighbor-joining method by the T-REX web server [68]. M and N and detailed block separations may serve as variables to fine tune the final phylogenetic tree. As a proof of concept, the reconstructed phylogenetic tree using $M = N = 10$ is shown in Figure 5.

The overall working flow of phylogenetic tree reconstruction is as follows: selection of the proteomes and protein sets → calculations and statistics of the intrinsic disorder contents (ID) and protein length of primary proteins (logarithm, $\ln(L)$) → calculations of the protein densities in all blocks (Table 2) → calculations of the Euclidian distance between each pair of proteomes or protein sets (3) → reconstruction of the phylogenetic tree based on the distance matrix.

Disclosure

The College of Engineering & Computer Science, SimCenter, University of Tennessee Chattanooga, 701 East M. L. King Blvd., Chattanooga, TN 37403, USA, is the current address of Hao-Bo Guo. Oak Ridge National Laboratory is managed by UT-Battelle LLC for the US DOE under Contract no. DE-AC05-00OR22725. A presentation for a part of this work has been given at the Quantitative Biology 2017 Meeting in Beijing.

Conflicts of Interest

The authors declare that they have no conflicts of interest.

Acknowledgments

This work is supported by the U.S. Department of Energy (DOE), Office of Science, Genomic Science Program, under Award no. DE-SC0008834.

Supplementary Materials

Table S1. Correlation coefficients between $\ln(L)$ and ID%. Table S2. Intervals that partition the LD spaces into $M \times N$ blocks with $M = N = 2$ and 5. Table S3. IDPs in the mitochondrion and chloroplast of *A. thaliana*. Figure S1. Protein-density contour maps (see Figure 1(a) in main text for the scale bar). Figure S2. Phylogenetic trees reconstructed from the protein distributions in the LD space using A—($M = N = 2$) and B ($M = N = 5$). Eukaryotes are in red, prokaryotes (bacteria and Archaea) in blue, and giruses in pink branches. MEGA5 (1) was used to plot the trees. Compared to that of the $M = N = 10$ tree (Figure 5), the branch length of the $M = N = 10$ tree is larger. Figure S3. Phylogenetic tree reconstructed from gene densities on the LD space. Different versions (v01–v03) of the *P. trichocarpa* proteomes have been used. By default of the present work, only proteins from primary transcripts are chosen for all proteomes. Here, for *P. trichocarpa* proteome v03, we tested both the primary transcripts (41,434 proteins) and all transcripts (73,013 proteins). We show here that progressive improvements including the splicing variants did not make significant changes in the phylogeny. *(Supplementary Materials)*

References

[1] D. Thirumalai, E. P. O'Brien, G. Morrison, and C. Hyeon, "Theoretical perspectives on protein folding," *Annual Review of Biophysics*, vol. 39, no. 1, pp. 159–183, 2010.

[2] K. A. Dill, K. Ghosh, and J. D. Schmit, "Physical limits of cells and proteomes," *Proceedings of the National Academy of Sciences of the United States of America*, vol. 108, no. 44, pp. 17876–17882, 2011.

[3] J. Z. Zhang, "Protein-length distributions for the three domains of life," *Trends in Genetics*, vol. 16, no. 3, pp. 107–109, 2000.

[4] L. Brocchieri and S. Karlin, "Protein length in eukaryotic and prokaryotic proteomes," *Nucleic Acids Research*, vol. 33, no. 10, pp. 3390–3400, 2005.

[5] V. N. Uversky, "Natively unfolded proteins: a point where biology waits for physics," *Protein Science*, vol. 11, no. 4, pp. 739–756, 2002.

[6] V. N. Uversky and A. K. Dunker, "Understanding protein non-folding," *Biochimica et Biophysica Acta (BBA) - Proteins and Proteomics*, vol. 1804, no. 6, pp. 1231–1264, 2010.

[7] P. Tompa, "Intrinsically disordered proteins: a 10-year recap," *Trends in Biochemical Sciences*, vol. 37, no. 12, pp. 509–516, 2012.

[8] V. N. Uversky, "A decade and a half of protein intrinsic disorder: Biology still waits for physics," *Protein Science*, vol. 22, no. 6, pp. 693–724, 2013.

[9] R. B. Berlow, H. J. Dyson, and P. E. Wright, "Functional advantages of dynamic protein disorder," *FEBS Letters*, vol. 589, no. 19, Part A, pp. 2433–2440, 2015.

[10] A. K. Dunker, S. E. Bondos, F. Huang, and C. J. Oldfield, "Intrinsically disordered proteins and multicellular organisms," *Seminars in Cell & Developmental Biology*, vol. 37, pp. 44–55, 2015.

[11] V. N. Uversky, "The multifaceted roles of intrinsic disorder in protein complexes," *FEBS Letters*, vol. 589, no. 19, Part A, pp. 2498–2506, 2015.

[12] P. Tompa, E. Schad, A. Tantos, and L. Kalmar, "Intrinsically disordered proteins: emerging interaction specialists," *Current Opinion in Structural Biology*, vol. 35, pp. 49–59, 2015.

[13] P. E. Wright and H. J. Dyson, "Intrinsically disordered proteins in cellular signalling and regulation," *Nature Reviews Molecular Cell Biology*, vol. 16, no. 1, pp. 18–29, 2015.

[14] B. Xue, A. K. Dunker, and V. N. Uversky, "Orderly order in protein intrinsic disorder distribution: disorder in 3500 proteomes from viruses and the three domains of life," *Journal of Biomolecular Structure and Dynamics*, vol. 30, no. 2, pp. 137–149, 2012.

[15] I. Yruela and B. Contreras-Moreira, "Protein disorder in plants: a view from the chloroplast," *BMC Plant Biology*, vol. 12, no. 1, p. 165, 2012.

[16] V. N. Uversky, V. Dave, L. M. Iakoucheva et al., "Pathological unfoldomics of uncontrolled chaos: intrinsically disordered proteins and human diseases," *Chemical Reviews*, vol. 114, no. 13, pp. 6844–6879, 2014.

[17] L. M. Iakoucheva, C. J. Brown, J. D. Lawson, Z. Obradović, and A. K. Dunker, "Intrinsic disorder in cell-signaling and cancer-associated proteins," *Journal of Molecular Biology*, vol. 323, no. 3, pp. 573–584, 2002.

[18] A. C. Joerger and A. R. Fersht, "The p53 pathway: origins, inactivation in cancer, and emerging therapeutic approaches," *Annual Review of Biochemistry*, vol. 85, no. 1, pp. 375–404, 2016.

[19] V. N. Uversky, I. Na, K. S. Landau, and R. O. Schenck, "Highly disordered proteins in prostate cancer," *Current Protein & Peptide Science*, vol. 18, no. 5, pp. 453–481, 2017.

[20] V. N. Uversky, "Targeting intrinsically disordered proteins in neurodegenerative and protein dysfunction diseases: another illustration of the D^2 concept," *Expert Review of Proteomics*, vol. 7, no. 4, pp. 543–564, 2010.

[21] S. J. Metallo, "Intrinsically disordered proteins are potential drug targets," *Current Opinion in Chemical Biology*, vol. 14, no. 4, pp. 481–488, 2010.

[22] D. Marasco and P. L. Scognamiglio, "Identification of inhibitors of biological interactions involving intrinsically disordered proteins," *International Journal of Molecular Sciences*, vol. 16, no. 4, pp. 7394–7412, 2015.

[23] J. S. Lazo and E. R. Sharlow, "Drugging undruggable molecular cancer targets," *Annual Review of Pharmacology and Toxicology*, vol. 56, no. 1, pp. 23–40, 2016.

[24] D. Kumar, N. Sharma, and R. Giri, "Therapeutic interventions of cancers using intrinsically disordered proteins as drug targets: c-Myc as model system," *Cancer Informatics*, vol. 16, 2017.

[25] S. Ambadipudi and M. Zweckstetter, "Targeting intrinsically disordered proteins in rational drug discovery," *Expert Opinion on Drug Discovery*, vol. 11, no. 1, pp. 65–77, 2016.

[26] U. Midic, C. J. Oldfield, A. K. Dunker, Z. Obradovic, and V. N. Uversky, "Protein disorder in the human diseasome: unfoldomics of human genetic diseases," *BMC Genomics*, vol. 10, Supplement 1, p. S12, 2009.

[27] M. Y. Lobanov and O. V. Galzitskaya, "How common is disorder? Occurrence of disordered residues in four domains of life," *International Journal of Molecular Sciences*, vol. 16, no. 8, pp. 19490–19507, 2015.

[28] Z. Peng, J. Yan, X. Fan et al., "Exceptionally abundant exceptions: comprehensive characterization of intrinsic disorder in all domains of life," *Cellular and Molecular Life Sciences*, vol. 72, no. 1, pp. 137–151, 2015.

[29] M. Podar, I. Anderson, K. S. Makarova et al., "A genomic analysis of the archaeal system Ignicoccus hospitalis-Nanoarchaeum equitans," *Genome Biology*, vol. 9, no. 11, article R158, 2008.

[30] S. G. Ball, D. Bhattacharya, and A. P. Weber, "Pathogen to powerhouse," *Science*, vol. 351, no. 6274, pp. 659-660, 2016.

[31] W. Neupert, "Mitochondrial gene expression: a playground of evolutionary tinkering," *Annual Review of Biochemistry*, vol. 85, no. 1, pp. 65–76, 2016.

[32] I. G. Johnston and B. P. Williams, "Evolutionary inference across eukaryotes identifies specific pressures favoring mitochondrial gene retention," *Cell Systems*, vol. 2, no. 2, pp. 101–111, 2016.

[33] J. F. Allen, "Why chloroplasts and mitochondria retain their own genomes and genetic systems: colocation for redox regulation of gene expression," *Proceedings of the National Academy of Sciences of the United States of America*, vol. 112, no. 33, pp. 10231–10238, 2015.

[34] J. F. Allen, W. B. M. de Paula, S. Puthiyaveetil, and J. Nield, "A structural phylogenetic map for chloroplast photosynthesis," *Trends in Plant Science*, vol. 16, no. 12, pp. 645–655, 2011.

[35] N. Pietrosemoli, J. A. Garcia-Martin, R. Solano, and F. Pazos, "Genome-wide analysis of protein disorder in *Arabidopsis thaliana*: implications for plant environmental adaptation," *PLoS One*, vol. 8, no. 2, article e55524, 2013.

[36] N. Perdigao, J. Heinrich, C. Stolte et al., "Unexpected features of the dark proteome," *Proceedings of the National Academy of Sciences of the United States of America*, vol. 112, no. 52, pp. 15898–15903, 2015.

[37] S. Proost, M. Van Bel, D. Vaneechoutte et al., "PLAZA 3.0: an access point for plant comparative genomics," *Nucleic Acids Research*, vol. 43, no. D1, pp. D974–D981, 2015.

[38] P. O. Wainright, G. Hinkle, M. L. Sogin, and S. K. Stickel, "Monophyletic origins of the metazoa: an evolutionary link with fungi," *Science*, vol. 260, no. 5106, pp. 340–342, 1993.

[39] C. R. Woese, O. Kandler, and M. L. Wheelis, "Towards a natural system of organisms: proposal for the domains Archaea, bacteria, and Eucarya," *Proceedings of the National Academy of Sciences of the United States of America*, vol. 87, no. 12, pp. 4576–4579, 1990.

[40] D. Moreira and P. Lopez-Garcia, "Ten reasons to exclude viruses from the tree of life," *Nature Reviews Microbiology*, vol. 7, no. 4, pp. 306–311, 2009.

[41] J.-M. Claverie and H. Ogata, "Ten good reasons not to exclude giruses from the evolutionary picture," *Nature Reviews Microbiology*, vol. 7, no. 8, p. 615, 2009.

[42] E. V. Koonin and P. Starokadomskyy, "Are viruses alive? The replicator paradigm sheds decisive light on an old but misguided question," *Studies in History and Philosophy of Science Part C: Studies in History and Philosophy of Biological and Biomedical Sciences*, vol. 59, pp. 125–134, 2016.

[43] P. Forterre, "To be or not to be alive: how recent discoveries challenge the traditional definitions of viruses and life," *Studies in History and Philosophy of Science Part C: Studies in History and Philosophy of Biological and Biomedical Sciences*, vol. 59, pp. 100–108, 2016.

[44] D. Raoult, S. Audic, C. Robert et al., "The 1.2-megabase genome sequence of Mimivirus," *Science*, vol. 306, no. 5700, pp. 1344–1350, 2004.

[45] M. G. Fischer, "Giant viruses come of age," *Current Opinion in Microbiology*, vol. 31, pp. 50–57, 2016.

[46] N. Philippe, M. Legendre, G. Doutre et al., "Pandoraviruses: amoeba viruses with genomes up to 2.5 Mb reaching that of parasitic eukaryotes," *Science*, vol. 341, no. 6143, pp. 281–286, 2013.

[47] D. Moreira and P. Lopez-Garcia, "Evolution of viruses and cells: do we need a fourth domain of life to explain the origin of eukaryotes?," *Philosophical Transactions of the Royal Society B-Biological Sciences*, vol. 370, no. 1678, article 20140327, 2015.

[48] P. Forterre and M. Gaia, "Giant viruses and the origin of modern eukaryotes," *Current Opinion in Microbiology*, vol. 31, pp. 44–49, 2016.

[49] J. C. Venter, M. D. Adams, E. W. Myers et al., "The sequence of the human genome," *Science*, vol. 291, no. 5507, pp. 1304–1351, 2001.

[50] M. Olivier, A. Aggarwal, J. Allen et al., "A high-resolution radiation hybrid map of the human genome draft sequence," *Science*, vol. 291, no. 5507, pp. 1298–1302, 2001.

[51] M. D. Adams, S. E. Celniker, R. A. Holt et al., "The genome sequence of *Drosophila melanogaster*," *Science*, vol. 287, no. 5461, pp. 2185–2195, 2000.

[52] J. Yu, S. Hu, J. Wang et al., "A draft sequence of the rice genome (*Oryza sativa* L. ssp. *indica*)," *Science*, vol. 296, no. 5565, pp. 79–92, 2002.

[53] R. Ming, R. VanBuren, C. M. Wai et al., "The pineapple genome and the evolution of CAM photosynthesis," *Nature Genetics*, vol. 47, no. 12, pp. 1435–1442, 2015.

[54] I. Arabidopsis Genome, "Analysis of the genome sequence of the flowering plant *Arabidopsis thaliana*," *Nature*, vol. 408, no. 6814, pp. 796–815, 2000.

[55] G. A. Tuskan, S. DiFazio, S. Jansson et al., "The genome of black cottonwood, *Populus trichocarpa* (Torr. & Gray)," *Science*, vol. 313, no. 5793, pp. 1596–1604, 2006.

[56] P. Amborella Genome, "The *Amborella* genome and the evolution of flowering plants," *Science*, vol. 342, no. 6165, article 1241089, 2013.

[57] S. A. Rensing, D. Lang, A. D. Zimmer et al., "The *Physcomitrella* genome reveals evolutionary insights into the conquest of land by plants," *Science*, vol. 319, no. 5859, pp. 64–69, 2008.

[58] J. M. Cherry, E. L. Hong, C. Amundsen et al., "Saccharomyces genome database: the genomics resource of budding yeast," *Nucleic Acids Research*, vol. 40, no. D1, pp. D700–D705, 2012.

[59] S. S. Merchant, S. E. Prochnik, O. Vallon et al., "The *Chlamydomonas* genome reveals the evolution of key animal and plant functions," *Science*, vol. 318, no. 5848, pp. 245–250, 2007.

[60] C. Aurrecoechea, J. Brestelli, B. P. Brunk et al., "GiardiaDB and TrichDB: integrated genomic resources for the eukaryotic protist pathogens *Giardia lamblia* and *Trichomonas vaginalis*," *Nucleic Acids Research*, vol. 37, pp. D526–D530, 2009.

[61] A. Karnkowska, V. Vacek, Z. Zubacova et al., "A eukaryote without a mitochondrial organelle," *Current Biology*, vol. 26, no. 10, pp. 1274–1284, 2016.

[62] F. R. Blattner, G. Plunkett 3rd, C. A. Bloch et al., "The complete genome sequence of *Escherichia coli* K-12," *Science*, vol. 277, no. 5331, pp. 1453–1462, 1997.

[63] B. E. Rubin, K. M. Wetmore, M. N. Price et al., "The essential gene set of a photosynthetic organism," *Proceedings of the National Academy of Sciences of the United States of America*, vol. 112, no. 48, pp. E6634–E6643, 2015.

[64] Z. Wang and M. Wu, "An integrated phylogenomic approach toward pinpointing the origin of mitochondria," *Scientific Reports*, vol. 5, no. 1, article 7949, 2015.

[65] A. Spang, J. H. Saw, S. L. Jorgensen et al., "Complex Archaea that bridge the gap between prokaryotes and eukaryotes," *Nature*, vol. 521, no. 7551, pp. 173–179, 2015.

[66] K. Peng, P. Radivojac, S. Vucetic, A. K. Dunker, and Z. Obradovic, "Length-dependent prediction of protein intrinsic disorder," *BMC Bioinformatics*, vol. 7, no. 1, p. 208, 2006.

[67] E. Schad, P. Tompa, and H. Hegyi, "The relationship between proteome size, structural disorder and organism complexity," *Genome Biology*, vol. 12, no. 12, article R120, 2011.

[68] A. Boc, A. B. Diallo, and V. Makarenkov, "T-REX: a web server for inferring, validating and visualizing phylogenetic trees and networks," *Nucleic Acids Research*, vol. 40, no. W1, pp. W573–W579, 2012.

[69] K. Tamura, D. Peterson, N. Peterson, G. Stecher, M. Nei, and S. Kumar, "MEGA5: molecular evolutionary genetics analysis using maximum likelihood, evolutionary distance, and maximum parsimony methods," *Molecular Biology and Evolution*, vol. 28, no. 10, pp. 2731–2739, 2011.

19

ISOdb: A Comprehensive Database of Full-Length Isoforms Generated by Iso-Seq

Shang-Qian Xie,[1] Yue Han,[2] Xiao-Zhou Chen,[3] Tai-Yu Cao,[1] Kai-Kai Ji,[1] Jie Zhu,[1] Peng Ling,[1] and Chuan-Le Xiao[2]

[1]Research Center for Terrestrial Biodiversity of the South China Sea, Institute of Tropical Agriculture and Forestry, Hainan University, Haikou 570228, China
[2]State Key Laboratory of Ophthalmology, Zhongshan Ophthalmic Center, Sun Yat-sen University, Guangzhou 510060, China
[3]School of Mathematics and Computer Science, Yunnan Minzu University, Kunming 650031, China

Correspondence should be addressed to Peng Ling; 18389807612@163.com and Chuan-Le Xiao; xiaochuanle@126.com

Academic Editor: Antonio Ferrante

The accurate landscape of transcript isoforms plays an important role in the understanding of gene function and gene regulation. However, building complete transcripts is very challenging for short reads generated using next-generation sequencing. Fortunately, isoform sequencing (Iso-Seq) using single-molecule sequencing technologies, such as PacBio SMRT, provides long reads spanning entire transcript isoforms which do not require assembly. Therefore, we have developed ISOdb, a comprehensive resource database for hosting and carrying out an in-depth analysis of Iso-Seq datasets and visualising the full-length transcript isoforms. The current version of ISOdb has collected 93 publicly available Iso-Seq samples from eight species and presents the samples in two levels: (1) sample level, including metainformation, long read distribution, isoform numbers, and alternative splicing (AS) events of each sample; (2) gene level, including the total isoforms, novel isoform number, novel AS number, and isoform visualisation of each gene. In addition, ISOdb provides a user interface in the website for uploading sample information to facilitate the collection and analysis of researchers' datasets. Currently, ISOdb is the first repository that offers comprehensive resources and convenient public access for hosting, analysing, and visualising Iso-Seq data, which is freely available.

1. Introduction

The variability of the transcriptome in an organism accounts for the variations in the phenotype and biological processes [1–4]. The alternative processing of primary RNA transcripts yields diverse spliced forms of the transcripts and mRNA isoforms. These isoforms may differ in structure, function, localization, or other properties [5–7]. Thus, the accurate landscape of transcript isoforms plays an important role in the understanding of gene function and gene regulation. At present, RNA-seq based on next-generation sequencing technology is a widely used approach for transcriptome profiling [8, 9]. While RNA-seq is often challenging to identify full-length gene isoform because of short read assembly, single-molecule real-time sequencing developed by Pacific Biosciences, known as PacBio SMRT, offers an alternative approach to generate longer reads and overcome the disadvantages of RNA-seq. Isoform sequencing (Iso-Seq) developed by PacBio SMRT provides long reads spanning entire transcript isoforms without requirement of assembly [10–12]. Though the PacBio sequencing technology is limited by a lower throughput, higher error rate, and higher cost per base and complemented with RNA-seq to achieve better effects [13], the Iso-Seq still has obvious advantages in improving annotations in reference genomes and identifying gene isoforms, alternative splicing (AS), and gene fusion events. Additionally, it helps in complementing the short/incomplete transcripts for species without a reference genome [14, 15]. However, till date, there is no database that provides comprehensive resources for a complete transcript isoform obtained from Iso-Seq data.

To facilitate the exploration of full-length isoforms in a transcriptome and benefit a broad range of investigators to further understand gene annotations and regulation, we present ISOdb, a comprehensive resource for hosting and carrying out an in-depth analysis of Iso-Seq datasets and visualising the full-length transcript isoforms. The current version of the database has collected 93 publicly available samples from eight species, which were processed and analysed by a unified pipeline (Figure 1). The outputs of ISOdb are presented in two levels: (1) sample level, including metainformation, long read distribution, isoform numbers, and AS events of each sample; (2) gene level, including the total isoform, novel isoform number, novel AS number, and isoform visualisation for each gene. To facilitate further analysis of researcher's datasets and update the database, ISOdb provides a user interface to upload the new sample information and a genome browser to query and visualise the full-length transcript isoforms. ISOdb is publicly available at http://isodb.xieslab.org.

2. Methods

2.1. Data Collection and Processing. The Iso-Seq data were collected from high-throughput RNA sequence read archive (SRA) database in NCBI. The current version contains 93 samples from eight animals and plants species: *Homo sapiens*, *Mus musculus*, *Gallus gallus*, *Gadus morhua*, *Arabidopsis thaliana*, *Gossypium barbadense*, *Triticum aestivum*, and *Amborella trichopoda*. The analysis tools include SMRT Analysis package, Quiver, GMAP, TAPIS, and SpliceGrapher were used in the pipeline of data processing. The workflow is summarized in Figure 1. Each sample was run through the Iso-Seq pipeline included in the SMRT Analysis software package (https://www.pacb.com/products-and-services/analytical-software/smrt-analysis). First, the raw sequence files produced from PacBio (bax.h5) were extracted, and reads of the insert (known as circular consensus sequence, CCS) were generated using ConsensusTools.sh with the parameters as described in the literature [16]. Subsequently, the reads were classified into full-length and non-full-length reads using pbclassify.py. The full-length reads were fed into the isoform-level clustering (ICE), and all the results were polished using Quiver [17]. Finally, we aligned the quivered fasta sequences against each reference genome by using GMAP [18] and analysed the spliced isoforms with TAPIS and SpliceGrapher by using the annotation file [16]. Table 1 shows the reference genome and related annotation files of the eight species.

2.2. Database Implementation. The database was implemented by PHP, MySQL, and JavaScript. The sample and gene information were stored and queried using MySQL and PHP. The JavaScript jQuery and D3.js library were used for producing dynamic and interactive data visualisation in the web browser. In addition, we integrated JBrowse in our database for visualising the full-length isoforms intuitively and the information of alignment against the reference genome for all Iso-Seq sequences in each species, as well as their annotation details were hosted in the genome browser.

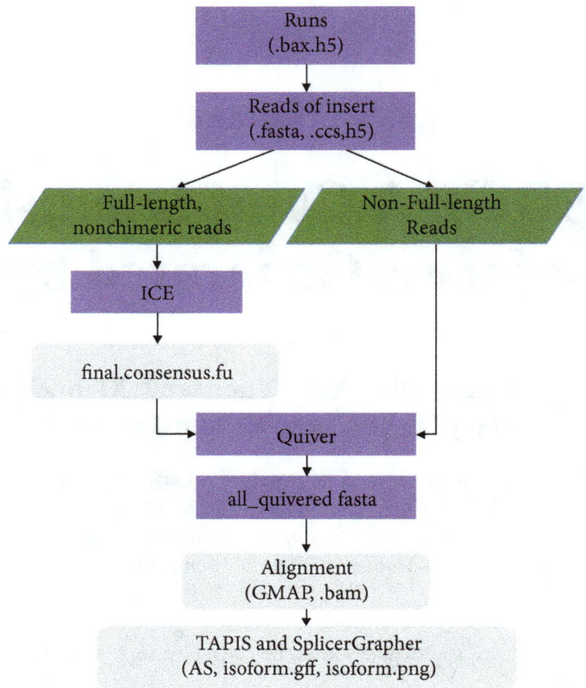

FIGURE 1: The analysis workflow of Iso-Seq data.

3. Usage and Features

The main function of ISOdb comprises home, browse, search, download, and help pages (Figure 2).

3.1. Search. This page provides a search option for the splice isoforms of genes in the database. Users can search genes by selecting a species and entering a gene symbol or NCBI gene ID in the search box of the search page (also appears in the home page). The output shows the information about the splice isoforms of the gene from all samples for the selected species, including the total isoform number, novel isoform number, and novel AS event number (Figure 3(a) and Figure 3(b)). Based on the transcript annotation file downloaded from NCBI (Table 1), the novel isoforms are identified by TAPIS and diagrammed by SpliceGrapher. In the detailed diagram of transcript isoforms, the grey block is annotated exon, purple block is the alternative $5'$ event, orange block is the alternative $3'$ event, and grey block with the blue border is intron retention (Figure 3(b)). Besides, the investigators can use a search box on the output page to filter the results. The JBrowse icon provides a hyperlink to a genome browser, which will be described in the next section.

3.2. Genome Browser. To explore the distribution of Iso-Seq reads for a given gene, ISOdb provides a genome browser to query and visualise the full-length read coverage and the transcript isoforms. A snapshot of an example of the "genome browser" is shown in Figure 3(c). The annotated gene track and reference are displayed on the top of the browser. Iso-Seq reads of the selected genes are shown at the bottom.

TABLE 1: The reference and annotation files for eight species.

Species	Reference	Link
Amborella trichopoda	AMTR1.0	https://www.ncbi.nlm.nih.gov/genome/12031
Arabidopsis thaliana	TAIR10	http://www.ncbi.nlm.nih.gov/genome/4
Gadus morhua	GadMor_May2010	https://www.ncbi.nlm.nih.gov/genome/2661
Gallus gallus	Gallus_gallus-5.0	https://www.ncbi.nlm.nih.gov/genome/111 ftp://ftp.ncbi.nih.gov/genomes/Gallus_gallus/GFF
Gossypium barbadense	GbV1.0	https://www.ncbi.nlm.nih.gov/genome/10770
Homo sapiens	GRCh38.p10	https://www.ncbi.nlm.nih.gov/genome/51
Mus musculus	GRCm38.p5	https://www.ncbi.nlm.nih.gov/genome/52
Triticum aestivum	CS42_TGAC_v1	http://www.ncbi.nlm.nih.gov/genome/11

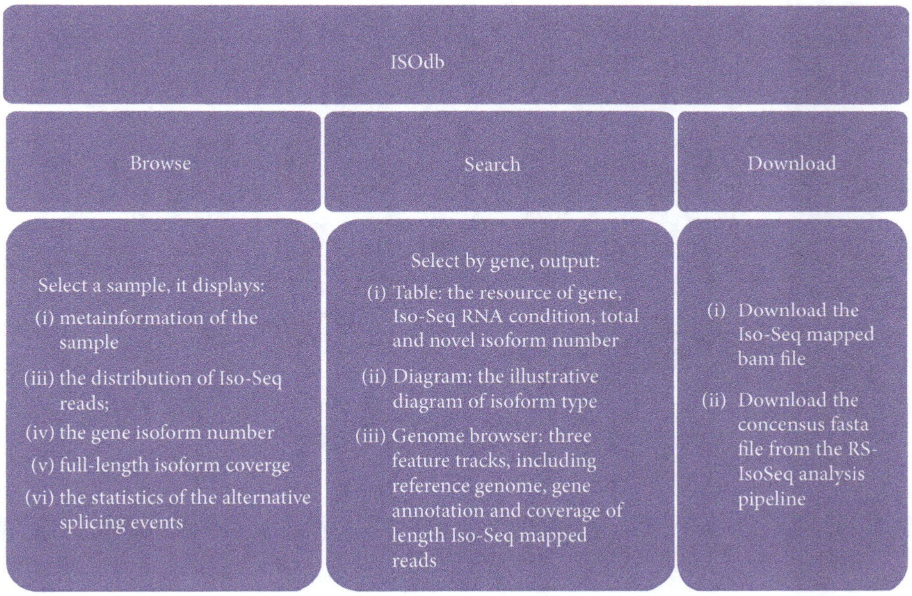

FIGURE 2: The main functions of ISOdb.

3.3. Browse. For each sample, this page displays (1) metainformation of the sample, including project/study/sample ID, experiment instrument, run number, release date, and experimental condition and PacBio sequencing chemistries (Figure 4); (2) plots showing overall statistics of each sample in three levels: reads, isoforms, and AS events. This section includes read distribution, isoform numbers against gene, full-length isoform numbers, and AS events of each sample (Figure 4).

3.4. Download. Investigators can download the bam file of the aligned Iso-Seq reads and consensus fasta sequences of full-length isoforms from each sample. The download page also has the search function so that the users can quickly find out the dataset of their interest for downloading.

4. Discussion

ISOdb is the first repository that offers comprehensive resources and convenient public access for hosting, analysing, and visualising Iso-Seq data. The accurate full-length spliced isoforms that are identified by Iso-Seq with no assembly are greatly beneficial for understanding gene annotations and gene regulation. As the numbers of studies using the Iso-Seq technique have been increasing significantly in the recent times, there is a great need for an integrated database that facilitates the exploration of data from Iso-Seq experiments. Thus, we developed the ISOdb by collecting 93 publicly available Iso-Seq samples from eight species and presented the samples along with the metainformation and the full-length splice isoform information of the genes.

Owing to the great advantage of Iso-Seq in identifying full-length transcript isoforms, we envision that Iso-Seq technology will be feasible to apply to a broader set of species and conditions and more such datasets will be released in future. To better collect and analyse datasets from investigators, we provide a user interface in the bottom of the website home page. While more samples are uploaded into ISOdb, some highly sample-dependent isoforms may be obtained. We will make efforts to continue improving

FIGURE 3: The gene search results from ISOdb. (a) The overview of search output, (b) the detailed diagram of total transcript isoforms (grey block: annotated exon; purple block: alternative 5′ event; orange block: alternative 3′ event; grey block with the blue border: intron retention), and (c) the genome browse of all isoforms.

the database in a timely manner, and the future updates will include more samples and integrate short reads from RNA-seq to calculate the abundance of transcript isoforms. We hope ISOdb will be a valuable resource for both experimental and computational biologists who are interested in transcriptomics.

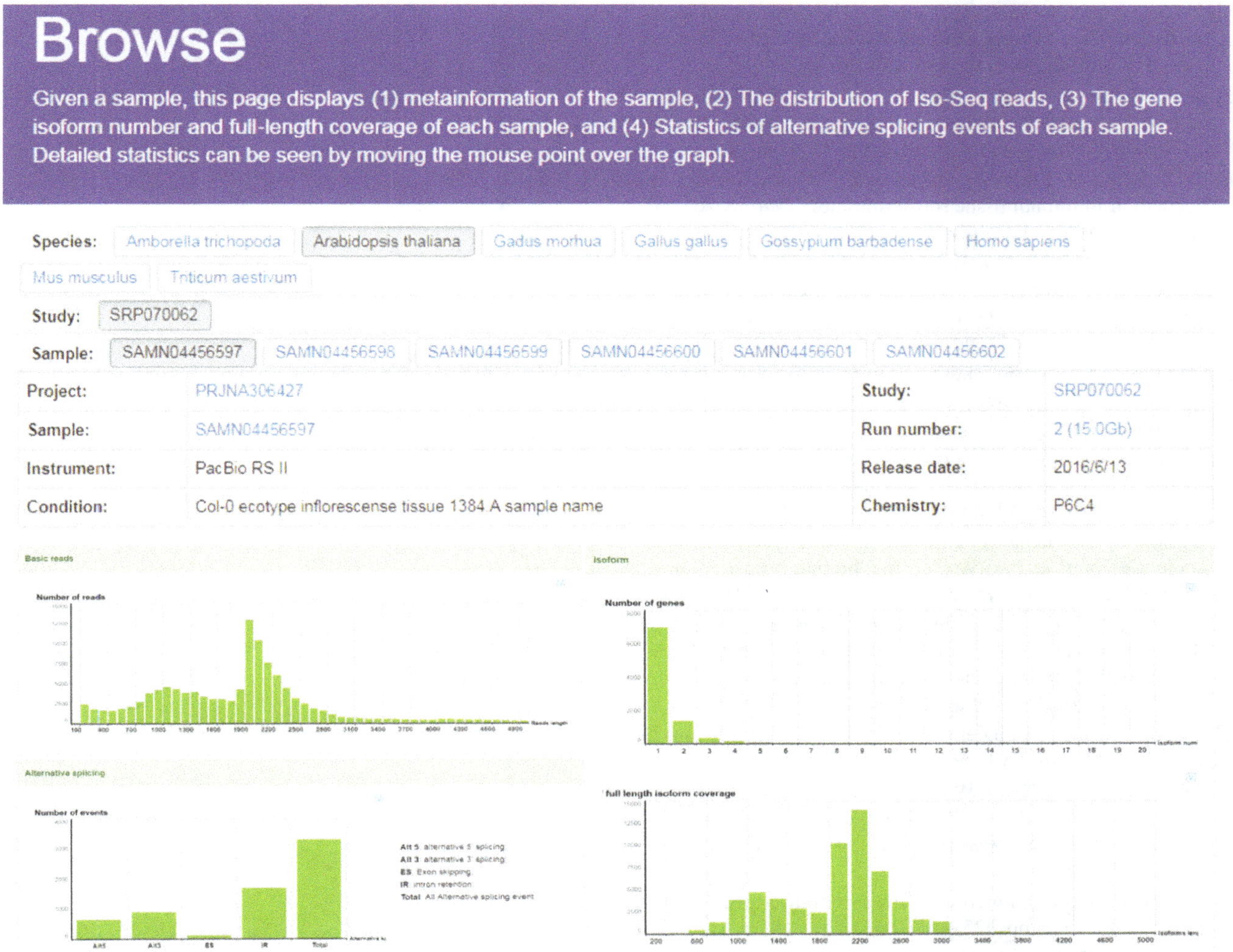

FIGURE 4: The sample browse results from ISOdb.

Conflicts of Interest

The authors declare that they have no competing interests.

Authors' Contributions

SQX and CLX designed the project, performed the analysis, wrote the website, and drafted the manuscript. YH performed the analysis and contributed back-end coding and constructed the database. XZC, TYC, KKJ, and JZ collected data and helped design the website. PL discussed the results and interpretation of final data. CLX conceived and directed the project. All authors read and approved the final manuscript. Shang-Qian Xie and Yue Han have contributed equally to this work.

Acknowledgments

We thank the anonymous reviewers for their constructive comments. This work was supported by grants from the National Natural Science Foundation of China (grant number 31760316, 31600667, and 31460297) and the Priming Scientific Research Foundation (grant number KYQD(ZR)1721) and Science Foundation for The Youth Teachers in 2017 (grant number hdkyxj201702) of Hainan University.

References

[1] D. J. Lockhart and E. A. Winzeler, "Genomics, gene expression and DNA arrays," *Nature*, vol. 405, no. 6788, pp. 827–836, 2000.

[2] A. Mortazavi, B. A. Williams, K. McCue, L. Schaeffer, and B. Wold, "Mapping and quantifying mammalian transcriptomes by RNA-Seq," *Nature Methods*, vol. 5, no. 7, pp. 621–628, 2008.

[3] B. Wang, E. Tseng, M. Regulski et al., "Unveiling the complexity of the maize transcriptome by single-molecule long-read sequencing," *Nature Communications*, vol. 7, p. 11708, 2016.

[4] Z. Kuang, J. D. Boeke, and S. Canzar, "The dynamic landscape of fission yeast meiosis alternative-splice isoforms," *Genome Research*, vol. 27, no. 1, pp. 145–156, 2017.

[5] A. J. Matlin, F. Clark, and C. W. J. Smith, "Understanding alternative splicing: towards a cellular code," *Nature Reviews Molecular Cell Biology*, vol. 6, no. 5, pp. 386–398, 2005.

[6] E. T. Wang, R. Sandberg, S. Luo et al., "Alternative isoform regulation in human tissue transcriptomes," *Nature*, vol. 456, no. 7221, pp. 470–476, 2008.

[7] Q. Pan, O. Shai, L. J. Lee, B. J. Frey, and B. J. Blencowe, "Deep surveying of alternative splicing complexity in the human transcriptome by high-throughput sequencing," *Nature Genetics*, vol. 40, no. 12, pp. 1413–1415, 2008.

[8] Z. Wang, M. Gerstein, and M. Snyder, "RNA-Seq: a revolutionary tool for transcriptomics," *Nature Reviews Genetics*, vol. 10, no. 1, pp. 57–63, 2009.

[9] A. E. Saliba, A. J. Westermann, S. A. Gorski, and J. Vogel, "Single-cell RNA-seq: advances and future challenges," *Nucleic Acids Research*, vol. 42, no. 14, pp. 8845–8860, 2014.

[10] D. Sharon, H. Tilgner, F. Grubert, and M. Snyder, "A single-molecule long-read survey of the human transcriptome," *Nature Biotechnology*, vol. 31, no. 11, pp. 1009–1014, 2013.

[11] E. Tseng, T. Clark, M. H. Ashby, and G. Shenykman, "Abstract 4898: full-length isoform sequencing of the human MCF-7 cell line using PacBio long reads," *Cancer Research*, vol. 75, Supplement 15, pp. 4898–4898, 2015.

[12] L. Shi, Y. Guo, C. Dong et al., "Long-read sequencing and de novo assembly of a Chinese genome," *Nature Communications*, vol. 7, p. 12065, 2016.

[13] A. Rhoads and K. F. Au, "PacBio sequencing and its applications," *Genomics, Proteomics & Bioinformatics*, vol. 13, no. 5, pp. 278–289, 2015.

[14] K. R. Chi, "Finding function in mystery transcripts," *Nature*, vol. 529, no. 7586, pp. 423–425, 2016.

[15] N. Singh, D. K. Sahu, R. Chowdhry et al., "IsoSeq analysis and functional annotation of the infratentorial ependymoma tumor tissue on PacBio RSII platform," *Meta Gene*, vol. 7, pp. 70–75, 2016.

[16] S. E. Abdel-Ghany, M. Hamilton, J. L. Jacobi et al., "A survey of the sorghum transcriptome using single-molecule long reads," *Nature Communications*, vol. 7, p. 11706, 2016.

[17] C. S. Chin, D. H. Alexander, P. Marks et al., "Nonhybrid, finished microbial genome assemblies from long-read SMRT sequencing data," *Nature Methods*, vol. 10, no. 6, pp. 563–569, 2013.

[18] T. D. Wu and C. K. Watanabe, "GMAP: a genomic mapping and alignment program for mRNA and EST sequences," *Bioinformatics*, vol. 21, no. 9, pp. 1859–1875, 2005.

20

Whole-Genome Sequencing and Comparative Genome Analysis Provided Insight into the Predatory Features and Genetic Diversity of Two *Bdellovibrio* Species Isolated from Soil

Omotayo Opemipo Oyedara,[1,2] **Aldo Segura-Cabrera**,[3] **Xianwu Guo**,[1] **Temidayo Oluyomi Elufisan**,[1,4] **Rafael Alejandro Cantú González**,[1] **and Mario A. Rodríguez Pérez**[1]

[1]*Instituto Politécnico Nacional, Centro de Biotecnología Genómica, 88710 Reynosa, TAMPS, Mexico*
[2]*Department of Biological Sciences, College of Science, Engineering and Technology, Faculty of Basic and Applied Science, Osun State University, PMB 4494, Osogbo, Osun State, Nigeria*
[3]*Red de Estudios Moleculares Avanzados, Instituto de Ecología, A.C., Xalapa Enriquez, VER, Mexico*
[4]*National Center for Technology Management, Agency of the Federal Ministry of Science and Technology (FMST), Obafemi Awolowo University, Ile-Ife, Nigeria*

Correspondence should be addressed to Omotayo Opemipo Oyedara; tayooyedara@yahoo.com

Academic Editor: Marco Gerdol

Bdellovibrio spp. are predatory bacteria with great potential as antimicrobial agents. Studies have shown that members of the genus *Bdellovibrio* exhibit peculiar characteristics that influence their ecological adaptations. In this study, whole genomes of two different *Bdellovibrio* spp. designated SKB1291214 and SSB218315 isolated from soil were sequenced. The core genes shared by all the *Bdellovibrio* spp. considered for the pangenome analysis including the epibiotic *B. exovorus* were 795. The number of unique genes identified in *Bdellovibrio* spp. SKB1291214, SSB218315, W, and *B. exovorus* JJS was 1343, 113, 857, and 1572, respectively. These unique genes encode hydrolytic, chemotaxis, and transporter proteins which might be useful for predation in the *Bdellovibrio* strains. Furthermore, the two *Bdellovibrio* strains exhibited differences based on the % GC content, amino acid identity, and 16S rRNA gene sequence. The 16S rRNA gene sequence of *Bdellovibrio* sp. SKB1291214 shared 99% identity with that of an uncultured *Bdellovibrio* sp. clone 12L 106 (a pairwise distance of 0.008) and 95–97% identity (a pairwise distance of 0.043) with that of other culturable terrestrial *Bdellovibrio* spp., including strain SSB218315. In *Bdellovibrio* sp. SKB1291214, 174 bp sequence was inserted at the host interaction *(hit)* locus region usually attributed to prey attachment, invasion, and development of host independent *Bdellovibrio* phenotypes. Also, a gene equivalent to *Bd0108* in *B. bacteriovorus* HD100 was not conserved in *Bdellovibrio* sp. SKB1291214. The results of this study provided information on the genetic characteristics and diversity of the genus *Bdellovibrio* that can contribute to their successful applications as a biocontrol agent.

1. Background

Studies on predatory bacteria have received much attention recently because of the possibility to harness their potentials for the biocontrol of pathogenic bacteria. *Bdellovibrio* spp. are versatile predatory bacteria that specialize in preying upon a wide range of Gram-negative bacteria, utilizing the resulting molecules from their attack for growth and reproduction [1]. Based on the mechanism of predation, there are two species of the genus *Bdellovibrio*, namely, *B. bacteriovorus* and *B. exovorus*. The former invade the periplasmic space of its prey while the latter attaches to the external surface (epibiotic) to derive its nutrients [2, 3]. Members of the genus *Bdellovibrio* are diverse with some of them exhibiting unique features that can influence their ecological adaptations. For instance, *B. bacteriovorus* strain W has the unique ability to develop a dormant structure called bdellocyst which can help them survive unfavorable conditions [4]. *B. bacteriovorus* is an obligate predatory bacterium. However, a strain isolated from Tiber River (*B. bacteriovorus* strain Tiberius) has shown the unique ability to grow simultaneously in the presence and absence of prey [5]. *B. bacteriovorus* strains that replicate and grow on nutrient-rich media without bacterial prey, usually called host-independent (HI) phenotypes often have mutations at a region of their genomes known as host interaction *(hit)* locus, tagged gene *Bd0108* in *B. bacteriovorus* HD100. The *hit* locus has been proposed to regulate the formation of type IV pilus needed for prey attachment and invasion [6].

Ancient and recent lateral gene transfers have been reported to occur in *Bdellovibrio* spp., and this may play a crucial role in their evolution probably leading to the development of unique features that can impact on their predatory lifestyle [5, 7, 8]. Thus, whole-genome sequence analysis can provide an in-depth understanding of variations in predation traits and evolution of *Bdellovibrio* spp. in turns helping in their successful application as biocontrol agents against bacterial pathogens. For instance, acquisition of pathogenic islands and alteration in their genomic structure via horizontal gene transfer may have an impact that can influence their application as biocontrol agents.

Bdellovibrio spp. found in soil are heterogeneous with different populations coexisting in the soil [9]. In our previous study, we isolated two different strains of *Bdellovibrio* spp. designated SKB1291214 and SSB218315 from soil samples in the same environment. The strains exhibited different phenotypes based on the time required to form plaque on Gram-negative bacteria prey lawns and prey range which was limited to some members of the family Enterobacteriaceae in *Bdellovibrio* sp. SKB1291214 [10]. Furthermore, the amplification of host interaction *(hit)* locus in *Bdellovibrio* sp. SKB1291214 using the PCR technique was unsuccessful. Therefore, we use whole-genome sequencing and comparative genomics as a tool to understand the genetic variations between these two strains and determine their relatedness with other reported genomes retrieved from the NCBI database.

2. Materials and Methods

2.1. Bacterial Strains and Genome Sequencing. Bdellovibrio spp. strains SKB1291214 and SSB218315 were isolated from soil samples obtained from different locations on a plot of land (26.069678N′, −98.313108W′ and 26.069446N′, −98.312902W′) within the Center for Genomic Biotechnology, National Polytechnic Institute (IPN as in Spanish) located in the city of Reynosa, Mexico. The *Bdellovibrio* spp. were cultured as described in our earlier report [10]. The genomic DNA (gDNA) was extracted using the Wizard® Genomic DNA Purification Kit (Madison, Wisconsin, USA) according to the manufacturer's instructions. The gDNA was subjected to optical density measurements in NanoDrop and Qubit (Thermo Fisher Scientific, Waltham, MA, USA). DNA migration in agarose gel electrophoresis was done to confirm the purity and concentration prior to fragmentation in Bioruptor (Diagenode Inc., Denville, NJ, USA). Fragmented gDNA was tested for size distribution and concentration using a 2200 Tapestation (Agilent Technologies Inc., Santa Clara, CA, USA) and subjected to Illumina library preparation using the Beckman SPRI-TE automated liquid handler and library prep reagents (Beckman Coulter, CA, USA). The resulting library was tested for size distribution and concentration by 2200, NanoDrop, and Qubit. The libraries were then loaded for Illumina NextSeq sequencing according to the standard operation. Paired-end 75 nucleotide (nt) reads were generated and checked for data quality using FASTQC (Babraham Institute, Cambridge, UK).

2.2. Genome Assembly and Annotation. The pair-end reads generated from the Illumina sequencing were trimmed using the Sickle tool 1.33 [11], assembled de novo using the SPAdes assembler version 3.10.0 [12], and then arranged into scaffolds using the MeDuSa scaffolder 1.3 [13]. The resulting contigs were then improved using Iterative Mapping and Assembly for Gap Elimination (IMAGE) [14]. Quast software was used to assess the quality of the generated scaffold based on the number of contigs and the N50 [15]. The genome sequences were automatically annotated using the NCBI Prokaryotic Genome Annotation Pipeline (PGAP) (https://www.ncbi.nlm.nih.gov/genome/annotation_prok/) and Rapid Annotation using Subsystem Technology (RAST) server [16]. Prophage sequences and genomic islands were predicted from the genomes using PHASTER [17] and IslandViewer4 [18] online application, respectively.

2.3. Phylogenetic Tree Construction and Estimation of Pairwise Evolutionary Divergence between 16S rRNA Gene Sequences. The 16S rRNA gene sequences were aligned using the MUSCLE alignment tool with default parameters, and a phylogenetic tree was constructed using the maximum likelihood method based on the Kimura 2-parameter model. Bootstrap values were calculated to test the robustness of interior node support and were obtained by conducting 1000 pseudoreplicates using MEGA© 6.0 software [19]. Pairwise evolutionary divergence (distance) was conducted in MEGA© 6.0 software using the Kimura 2-parameter model with 1000 bootstrap replications.

2.4. Comparative Genome Analysis. For the whole-genome comparative study, genomes of eight *Bdellovibrio* spp. were retrieved from the NCBI database and compared with the genomes of the study *Bdellovibrio* strains (*Bdellovibrio* sp. SKB1291214 and *B. bacteriovorus* SSB218315). The retrieved genomes include that of the epibiotic *B. exovorus* JSS (NC_020813), *B. bacteriovorus* strains HD100 (NC_005363), W (NZ_CP002190), Tiberius (NC_019567), 109J (NZ_CP007656), R0 (LUKE00000000), EC13 (LUKD00000000), and BER2 (LUKF00000000).

The similarity among the genomes based on average amino acid identity (AAI) was inferred using the ANI/AAI-Matrix Genome-based distance matrix calculator [20]. A pangenome analysis was carried out with the bacterial pangenome analysis (BPGA) tool [21] using the two study genomes and genomes of five reported *Bdellovibrio* spp. These include *B. exovorus* JSS (NC_020813) and *B. bacteriovorus* strains HD100 (NC_005363), W (NZ_CP002190), Tiberius (NC_019567), and 109J (NZ_CP007656). BLASTP search and functional annotation analysis of the core and unique genes were done with the BLAST2GO pipeline [22] using the default settings with the BLAST expectation value (*E* value) of 1.0*E* − 3. The *hit* locus regions of *Bdellovibrio* spp. HD100, SKB1291214, and SSB218315 were compared by constructing a genome map using the KBase online software (https://kbase.us/), followed by BLASTP analysis of the *hit* regions in ExPASy Bioinformatics Resource Portal (https://www.expasy.org/). Alignment of the regions corresponding to the *hit* locus in the different *Bdellovibrio* strains was done using the multiple sequence alignment tool, Clustal Omega [23].

2.5. Nucleotide Sequence Accession Numbers. The whole-genome shotgun project has been deposited at DDBJ/ENA/GenBank databases under the accession NELQ00000000 for *Bdellovibrio* sp. SKB1291214 (the version described in this paper is version NELQ01000000). The complete genome sequence of *B. bacteriovorus* SSB218315 was deposited in the same databases under accession number CP020946.

3. Result and Discussion

3.1. Genomic Features of Bdellovibrio spp. Strains SKB1291214 and SSB218315. The genomic features of *B. bacteriovorus* strains SSB218315 and *Bdellovibrio* sp. strain SKB1291214 are summarized in Table 1. The genome size of *B. bacteriovorus* SSB218315 and *Bdellovibrio* sp. SKB1291214 is 3,769,537 bp and 3,730,590 bp, respectively. *Bdellovibrio* spp. are small but have large genomes (approximately 3.7 Mb) that encode predation factors presumed important to seek and lyse prey cells [24]. The percentage GC content in *Bdellovibrio* sp. SKB1291214 (44.8%) is low compared to *B. bacteriovorus* SSB218315 (50.5%). Lambert et al. [25] reported some genes expressed during predation in *B. bacteriovorus* HD100. These include genes that are up- and down-regulated at the early stage of *B. bacteriovorus* HD100 (30 minutes) infection as it switches from the motile prey-seeking attack stage to the intraperiplasmic phase, when it establishes itself in the prey cell. And because of the phenotypic differences observed between *Bdellovibrio* spp. SKB1291214 and SSB218315 [10], genome analysis was done to identify and compare the gene equivalent described by Lambert et al. [25] in the study strains. From BLASTP analysis results, *B. bacteriovorus* SSB218315 have all the described gene equivalent (Additional file 1a). However, among the 75 described upregulated genes, *Bd1230 (lamb)*, *Bd0487*, and *Bd2298* equivalents in *B. bacteriovorus* HD100 were absent in the genome of *Bdellovibrio* sp. SKB1291214. The *Bd1230 (lamb)* gene encodes maltoporin, an outer membrane protein that is important for sugar transport in Gram-negative bacteria, and it is usually expressed when *Bdellovibrio* degrades its prey. The genes *Bd0487* and *Bd2298* are found only in the genome of *Bdellovibrio*, and they are significantly upregulated when

Table 1: Genomic features of *B. bacteriovorus* strains SSB218315 and *Bdellovibrio* sp. SKB1291214.

Genome information	*B. bacteriovorus* SSB218315	*Bdellovibrio* sp. SKB1291214
Chromosome size	3,769,537 bp	3,730,590 bp
Number of contigs	1	20
N_{50}	—	199,513
GC content	50.5%	44.80%
Total RNA	41	40
Complete rRNAs	3	3
tRNAs	34	33
Noncoding RNAs	4	4
Total genes	3620	3677
Total CDS	3579	3637
Coding CDS	3536	3588
Phage	1 (incomplete)	1 (incomplete)
Genomic island (integrated method)	69	146

Bdellovibrio enters the periplasmic phase of growth [25]. Furthermore, eight out of the forty-one reported downregulated gene equivalents implicated in the attack phase of *Bdellovibrio* were absent in the genome of SKB1291214. These genes include the equivalent of *Bd3260*, *Bd2608*, *Bd2400*, *Bd0737*, and *Bd0992 (cwlJ)* encoding putative membrane proteins and enzymes (putative lipase and cell wall hydrolase) that play a role in prey attachment and penetration. Two gene equivalents *Bd0880* and *Bd0931* encoding stress response proteins, a homologue of periplasmic adaptor protein CpxP and transcriptional regulator, and MerR family were also absent in the genome of *Bdellovibrio* sp. SKB1291214. *B. bacteriovorus* uses type IV pilus to attach and subsequently invade prey cells. The gene equivalent of *Bd0108* which encode proteins that function in regulating type IV pilus secretion in *Bdellovibrio* was also not present in the genome of SKB1291214. The missing gene equivalents described above might be playing important roles during *Bdellovibrio* predation. And thus, the absence of these genes in the genomes of *Bdellovibrio* sp. SKB1291214 can affect its rate of predation.

The RAST annotation server also predicted some genes presumed to enhance predation in *Bdellovibrio* spp. (Additional files 1b–f). These include genes encoding motility and chemotaxis factors, transport system including type IV pilus, stress response proteins, degradative proteins, and siderophores, and other defense factors.

The rapid motility of *Bdellovibrio* helps in prey location [26]. From RAST annotation and manual curation, about 75 genes encoding motility and chemotaxis factors were identified in the study *Bdellovibrio* strains. Among these factors are five adventurous gliding motility factors R, S, T, U, V, and *MglA* used by *Bdellovibrio* spp. to glide on solid surfaces and find prey in environments with a low water content such as biofilms [27, 28]. The RAST annotation server also predicted a sequence called diguanylate cyclase/phosphodiesterase (GGDEF and EAL domains) with PAS/PAC sensor(s) in *Bdellovibrio* spp. SKB1291214 (B9G69_13450, B9G69_

14735, B9G69_01735, and B9G69_08345) and SSB218315 (B9G79_16530, B9G79_14755, B9G79_11600, B9G79_00860, and B9G79_03750) as stress response proteins. Proteins that possess this GGDEF sequence secrete cyclic di-GMP, a signalling protein that controls *Bdellovibrio* to grow either as a predator that require prey for survival or a host-independent phenotype that can replicate on nutrient-rich medium. Four enzymatically competent GGDEF protein domains designated DgcA, DgcB, DgcC, and DgcD have been reported [29]. In the study of Hobley et al. [29], Δ*dgcA* mutants became a nonmotile host-independent strain that can grow axenically on nutrient-rich medium only. The ΔdgcA mutants can invade, replicate, and septate inside prey cells but cannot glide out of the prey cells to look for new prey. The Δ*dgcB* mutants became flagellated host-independent strains. The Δ*dgcC* mutants developed into predatory strains that are not capable of growing as host-independent (HI) strains. However, for ΔdgcC to grow axenically, they require additional or secondary mutation. Mutation of the *dgcD* gene did not result in any phenotypic alteration with mutants Δ*dgcD* growing both as host-dependent (HD) and HI. BLASTP analysis showed that GGDEF protein domains DgcA, DgcB, and DgcC are conserved in the two *Bdellovibrio* strains. However, the DgcD is not conserved in *Bdellovibrio* sp. SKB1291214, a similar result observed in *B. bacteriovorus* Tiberius [29].

Bdellovibrio spp. have been described to be nonpathogenic to human [30]. However, genomes of *Bdellovibrio* spp. SKB129124 and SSB218315 and other *Bdellovibrio* strains encode genes annotated as collagenase and hemolysin, virulence factors associated with some pathogens of human such as *Staphylococcus aureus* [31] and *Vibrio vulnificus* [32]. The genome of *B. bacteriovorus* SSB218315 also encodes genes annotated as RTX toxins, a factor that has different biological functions such as pore-forming leukotoxin, metalloprotease, and lipase activities [33]. BLAST2GO software was used to carry out a BLASTP search and assign gene ontology (GO) to the gene products of the sequences annotated as hemolysin, collagenase, and RTX toxins (Additional file 1g). The GO of the sequences annotated as hemolysin III is cytolysis, and the BLASTP analysis revealed a conserved hemolysin III-related protein domain. The BLASTP analysis of the annotated collagenase revealed a conserved U32 family peptidase protein domain. However, collagenase belongs to the U32 family peptidase [34]. Identification of the biological roles of hemolysin III and collagenase in *Bdellovibrio* spp. will aid in their successful application as biocontrol agents against human pathogens. Analysis using the Pfam database [35] revealed that the RTX-toxin sequences did not have toxin domain but rather a protein domain identified as a regulator of chromosome condensation (RCC1) repeat. Thus, the annotated RTX toxin might be performing a different role than being involved in toxin production.

Bacteria can acquire genomic islands (GEIs) via horizontal gene transfer (HGT). These GEIs can confer adaptive features such as antibiotic resistance, survival features, and metabolic activities, metabolism of complex compounds on the bacteria [36]. Some distinguishing features of GEIs include association with genes encoding tRNA, integrase, or transposase, and possession of the percentage G + C content that is different from another part of the genome [37]. Predicted GEIs of *Bdellovibrio* strains SKB1291214 and SSB218315 include hypothetical proteins, peptidase, septation protein spoVG (in *B. bacteriovorus* SSB218315), and survival protein *surA* which can aid the survival of bacteria at the stationary growth phase (Additional file 2).

3.2. Phylogeny and Amino Acid Identity of Bdellovibrio Species. The phylogenetic analysis was done to compare the 16S rRNA gene sequences of *Bdellovibrio* sp. SKB1291214 and *B. bacteriovorus* SSB218315 with sequences of other members of the genus *Bdellovibrio* and their relatives that belong to the genus *Bacteriovorax*, *Peredibacter*, and *Halobacteriovorax* (Figure 1). The 16S rRNA sequence of strains SKB1291214 and SSB218315 showed 96% similarity with a pairwise evolutionary distance of 0.043 (Additional file 3). The strain SKB1291214 shared 99% identity with an uncultured *Bdellovibrio* sp. clone12 L 106 (pairwise distance of 0.008) while strain SSB218315 shared 100% identity with other culturable terrestrial *B. bacteriovorus* which include *B. bacteriovorus* strain HD100 (pairwise distance 0.001) and Tiberius (pairwise distance 0.004). The phylogenetic tree showed that the two *Bdellovibrio* strains SKB1291214 and SSB218315 are phylogenetically different despite being isolated from soil samples in the same environment. Further species delineation was done to examine the AAI among the *Bdellovibrio* strains. For strains to belong to the same species, they must have ANI and AAI ≥ 95%, <10 Karlin genomic signature, and >70% in silico GGDH [38]. The AAI between strain SKB1291214 and other strains was very low (63.70–67.68%) while strain SSB218315 shared a high AAI value of 95% with *B. bacteriovorus* strains HD100, Tiberius, and 109J (Figure 2). The result showed that strain SSB218315 is closely related to HD100, Tiberius, and 109J and thus, they can conveniently be grouped as the same species. Meanwhile, considering the percentage GC content, phylogenetic tree clustering pattern, and AAI value, strain SKB1291214 could be grouped as a novel species; however, further analysis is needed.

3.3. Pangenome Analysis. A bacterial pangenome analysis (BPGA) tool was used to carry out pangenome analysis of eight *Bdellovibrio* spp. The pangenome is made up of 8134 genes, and the *Bdellovibrio* spp. shared 795 genes as core genomes (Figure 3, Additional file 4a). The BGPA predicted the pangenome of *Bdellovibrio* spp. as open based on the power law regression of the program (Additional file 4b). The total number of unique genes found in *Bdellovibrio* spp. SKB1291214 and SSB218315 is 1343 and 113, respectively (Table 2). The epibiotic *B. exovorus* JJS and bdellocyst-forming *B. bacteriovorus* W have a total of 1572 and 857 unique genes, respectively. The GO of the unique genes in *Bdellovibrio* sp. SKB1291214 revealed that they are rich in proteins involved in molecule transport, oxidation-reduction process, signal transduction, hydrolase activity phosphorylation, and nucleotide and ion binding (Additional files 4c–f). *B. exovorus* JJS has the highest number of unique genes (1572), and among these are three genes encoding type II

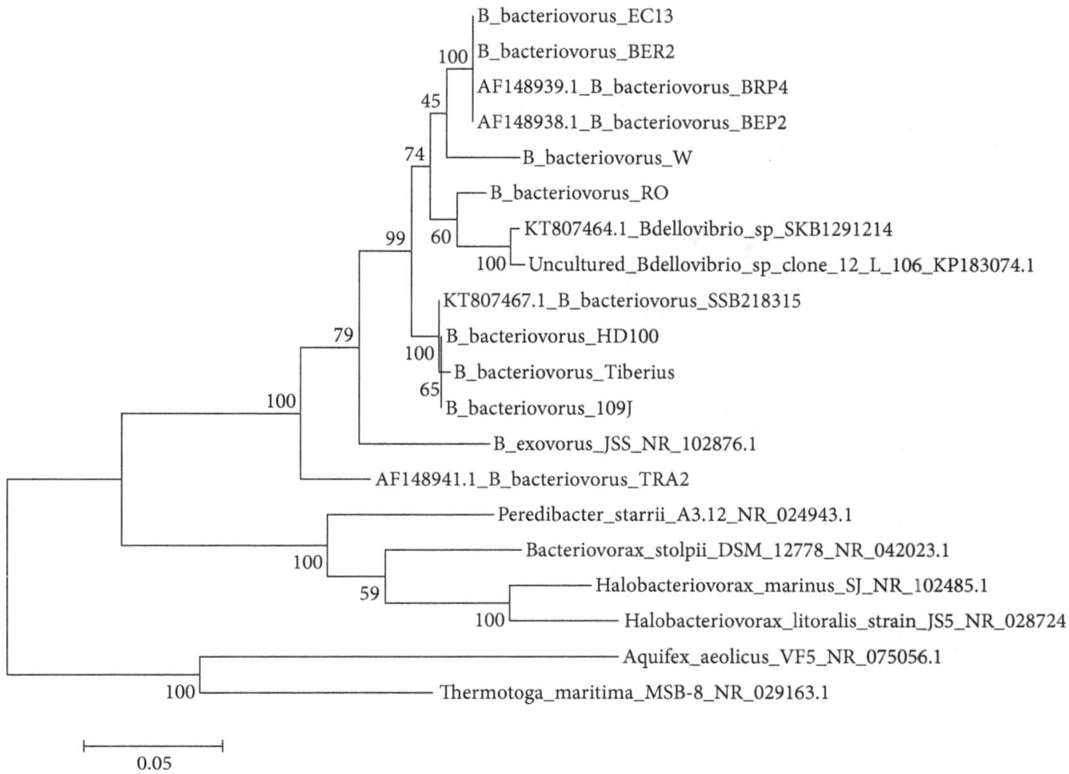

Figure 1: Molecular phylogenetic analysis by a maximum likelihood method using 16S rRNA gene sequences. The evolutionary history was inferred by using the maximum likelihood method based on the Kimura 2-parameter model with 1000 bootstrap replications. The percentage of trees in which the associated taxa clustered together is shown next to the branches. Initial tree for the heuristic search was obtained automatically by applying neighbor-joining and BioNJ algorithms to a matrix of pairwise distances estimated using the maximum composite likelihood (MCL) approach and then selecting the topology with the superior log likelihood value. The analysis involved 20 16S rRNA gene nucleotide sequences. Evolutionary analyses were conducted in MEGA6 [19].

CRISPR-associated endonuclease Cas1, CRISPR-associated Cas2, and type II CRISPR RNA-guided endonuclease Cas9 which usually act to defend prokaryotes against any invading foreign genetic material. These CRISPR genes are however absent in the genome of the periplasmic *Bdellovibrio* spp.

A comparative genomic study by Pasternak et al. [39] identified protein families that are specific to predatory bacteria and differentiate them from the nonpredatory bacteria. All the fifteen protein families reported to be specific to predatory bacteria were present in *Bdellovibrio* spp. SKB1291214 and SSB218315. Homologue of genes encoding two protein families, 4-diphosphocytidyl-2-C-methyl-D-erythritol kinase (B9G69_00295; mean similarity value = 57; *E* value = 1.33*E* − 134) and indole-3-glycerol phosphate synthase (B9G69_04640; mean similarity value = 74; *E* value = 0.0), reported to be specific to nonpredatory bacteria was found among the unique genes of *Bdellovibrio* sp. SKB1291214 (Additional file 4c). Predatory bacteria are different from the nonpredators based on the pathway utilized for the biosynthesis of isoprenoids. While the nonpredators use the deoxy-d-xylulose 5-phosphate (DOXP) or nonmevalonate pathway, the predators use the mevalonate pathway for the biosynthesis of isoprenoids [39]. The 4-diphosphocytidyl-2-C-methyl-D-erythritol kinase is an enzyme required for the nonmevalonate pathway synthesis of isoprenoid, and its presence in *Bdellovibrio* sp. SKB1291214 presumably is a result of horizontal gene transfer (HGT).

One of the differences between *B. exovorus* and *B. bacteriovorus* is the mechanism they use for prey attack [2, 3]. The latter is characterized by the invasion of the prey periplasm while the former are not capable of penetrating into their prey. During prey invasion in *B. bacteriovorus* HD100, three genes tagged *Bd0816*, *Bd3459*, and *Bd3460* play an important role. [26, 40, 41]. The *Bd0816* and *Bd3459* encode D-alanyl-D-alanine carboxypeptidase usually expressed at the point of prey entry while *Bd3460* encodes a protein called ankyrin which protects *Bdellovibrio* hydrolytic enzymes which it secretes during prey invasion.

Genes encoding D-alanyl-D-alanine carboxypeptidase were present among the unique genes of *Bdellovibrio* sp. SKB1291214 (B9G69_09970 and B9G69_09965) and *B. exovorus* JJS (A11Q_2041) (Additional files 4c and d). However, gene encoding ankyrin was absent in the genome of *B. exovorus* JJS but present in the genome *Bdellovibrio* spp. SKB1291214 and SSB218315 which are closer to the *B. bacteriovorus* HD100 compared to the epibiotic *B. exovorus*. While the presence of genes encoding D-alanyl-D-alanine carboxypeptidase is a general feature of the genus *Bdellovibrio* spp., the ankyrin-encoded genes are limited to the periplasmic members of the genus *Bdellovibrio*. The predation mechanism

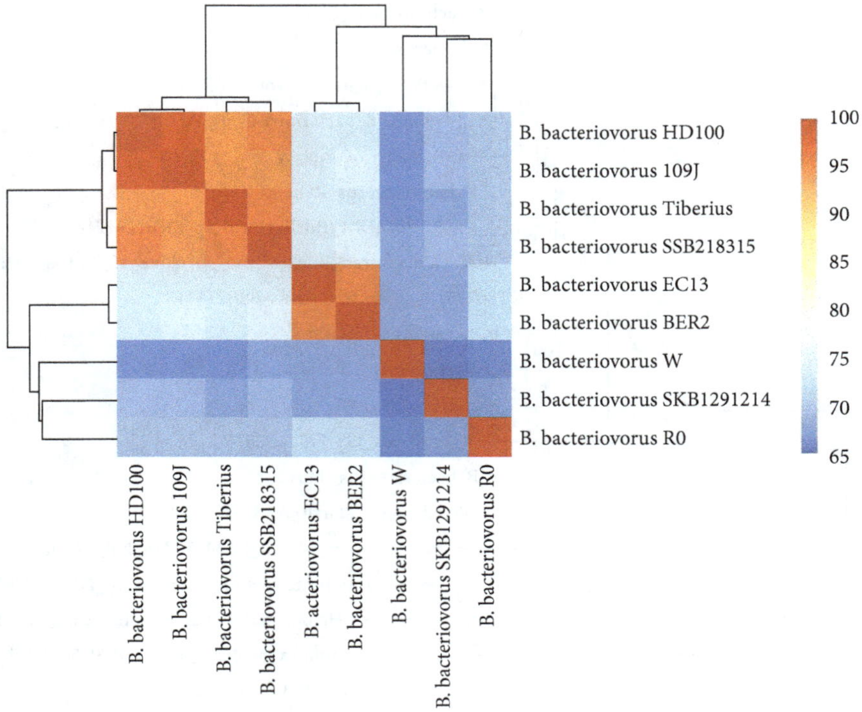

FIGURE 2: The average amino acid identity matrix clustering analysis of the whole genomes of nine *Bdellovibrio* strains. Colours represent bands of percent identity. The heatmap was generated in R package plots using heatmap.2 function.

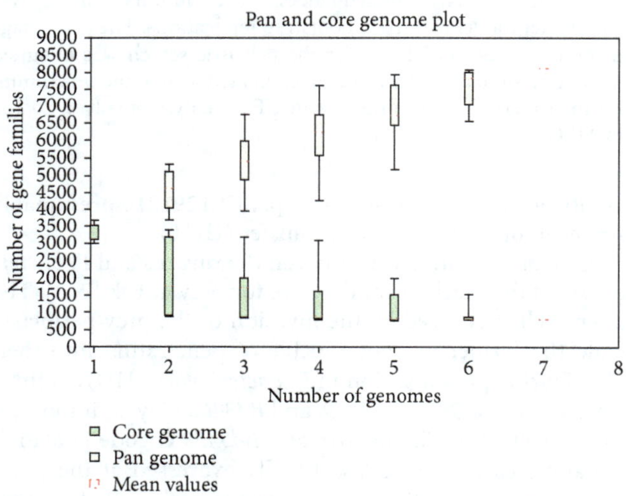

FIGURE 3: Plot showing the core and pangenomes of *Bdellovibrio* spp. The total number of genes or pangenome (yellow) and shared or core genome (green) for 7 *Bdellovibrio* strains are shown on the plot. The pangenome is made up of 8134 genes while the core genomes are made up of 795 genes.

of *B. exovorus* does not require prey invasion, hence, the possible reason why it does not have the gene equivalent of *Bd3460* in its genome. Furthermore, the unique genes of SKB1291214 also contain the Autographivirinae Erwinia phage-associated region coding for protein *AmsF* (B9G69_00395) which is involved in amylovoran biosynthesis. Amylovoran is an exopolysaccharide that plays a role in the pathogenesis of *Erwinia carotovora* [42].

The presence of genes encoding integrases (among predicted GEIs), transposases, phage-associated protein *AmsF*, and nonpredatory bacteria-associated 4-diphosphocytidyl-2-C-methyl-D-erythritol kinase (among the unique genes) suggests the occurrence of HGT in *Bdellovibrio* sp. SKB1291214. From the BLASTP analysis done with the BLAST2GO software using an E value threshold of $1E-6$, there is an indication that some of the unique genes are acquired horizontally from bacteria that belong to groups other than class Deltaproteobacteria. These groups include the Alphaproteobacteria (B9G69_00040, DUF4334 domain-containing and B9G69_13200, AraC family transcriptional regulator), Betaproteobacteria (B9G69_13210, Com family DNA-binding transcriptional regulator and B9G69_11675, FAD: FMN transferase), Gammaproteobacteria (B9G69_13180, glyoxalase bleomycin resistance dioxygenase; B9G69_13235, terminase small subunit; B9G69_13225, AlpA family transcriptional regulator; and B9G69_13230, bacteriophage), Epsilonproteobacteria (B9G69_11670, nuclease), Bacteroidetes (B9G69_18290, molybdopterin-binding oxidoreductase), and Cyanobacteria (B9G69_17905, HAMP domain-containing and B9G69_00045, 3 and saliva-related transmembrane). This result corroborates the earlier findings of Gophna et al. [7]. A study on the extent and frequency of HGT in *Bdellovibrio* spp. will provide useful information that can aid their successful application as biocontrol agents.

3.4. Analysis of the Host Interaction (hit) Locus. B. bacteriovorus has been described to have the ability to switch from being predatory usually referred to as host dependent (HD) to growing on nutrient-rich medium axenically, sometimes

Table 2: Number of core genes, accessory genes, unique genes, and exclusively absent genes obtained from the pangenome analysis of 7 strains of *Bdellovibrio* spp.

Genome number	Strain	Number of core genes	Number of accessory genes	Number of unique genes	Number of exclusively absent genes
1	*B. bacteriovorus* HD100	795	2641	65	1
2	*B. bacteriovorus* strain Tiberius	795	2579	301	1
3	*B. bacteriovorus* W	795	1063	857	54
4	*B. bacteriovorus* 109J	795	2635	132	1
5	*B. bacteriovorus* SSB218315	795	2575	113	6
6	*B. exovorus* JSS	795	200	1572	725
7	*Bdellovibrio* sp. SKB1291214	795	1366	1343	27

referred to as host independent (HI). Mutation at a region identified as host interaction *(hit)* locus has been reported to be responsible for the conversion from the HD to HI phenotypes. The *hit* locus has been described to be made up of an open reading frame (ORF) tagged *Bd0108* and part of ORF tagged *Bd0109* encoding a putative cell wall-associated protein in *B. bacteriovorus* HD100 [6]. There are *pil* genes located upstream of the *hit* locus (Figure 4(a)). These *pil* genes encode structural proteins for the formation of the type IV pilus system needed for prey adherence and colonization. The genes *Bd0113* and *Bd0114* are responsible for the pilus assembly while the TadA *(Bd0110)* and TadB *(Bd0111)* encode ATPase that provides energy for the type IV pilus secretion [43]. And downstream of the *hit* locus are genes tagged BD_RS00505 (new locus tag for *B. bacteriovorus* HD100 genes) and *Bd0103* in *B. bacteriovorus* HD100; both genes encode hypothetical proteins of unknown function. The above-described genes (the *Bd0108*, *Bd0109*, pil genes, BD_RS00505, and Bd0103) are inserted in between two genes *Bd0102* and *Bd0121* encoding chemotaxis factors. In our previous study, *hit* locus was successfully amplified in *B. bacteriovorus* SSB218315 using the PCR technique. The negative result obtained from the PCR amplification of the *hit* locus in *Bdellovibrio* sp. SKB1291214 made us construct genomic maps to compare the *hit* locus region between *Bdellovibrio* spp. HD100, SSB218315, and SKB1291214 (Figures 4(a)–4(c)). From the result of the BLASTP and multiple sequence alignment analysis, the region corresponding to the *Bd0108* (*hit* locus) is not conserved in *Bdellovibrio* sp. SKB1291214. (Additional file 5). Furthermore, a fragment of 174 bp absent in HD100 and SSB218315 was found inserted between gene equivalent *Bd0102* and *BD_RS00505* in SKB1291214 (Figure 4(b)). This fragment produces an insignificant *E* value with BLASTP analysis. Comparative analysis revealed that the gene equivalent *Bd0109* is conserved among the *Bdellovibrio* spp. including *Bdellovibrio* spp. SKB1291214 and *B. exovorus* JJS. Thus, this suggests that *Bd0109* gene may have an important role in the predatory activities of *Bdellovibrio* spp. Also, variations in the sequence of *Bd0108* may not be sufficient to hinder prey predation *Bdellovibrio* spp. Because *Bdellovibrio* spp. that have a mutation at the *hit* locus can be cultured axenically [6], we attempted to culture *Bdellovibrio* spp. SKB1291214 and SSB218315 on nutrient-rich medium in the absence of prey using three different techniques described by Ferguson et al. [44], Lambert and Sockett [45], and Seidler and Starr [46]. However, we could not successfully isolate the HI phenotypes using the three approaches, though all the yellow bacterial colonies obtained from the method exhibited the phenotypic characteristics described in the previous research.

4. Conclusion

Members of the genus *Bdellovibrio* have been reported to have potential applications as biocontrol agents against pathogens. This study focused on the whole-genome sequencing and comparative analysis of two *Bdellovibrio* spp. that showed phenotypic differences. The comparative analysis showed that *B. bacteriovorus* SSB218315 is genetically related to the soil-derived *B. bacteriovorus* HD100. We also observed that the *Bdellovibrio* sp. SKB1291214 is distinctively different from the epibiotic *B. exovorus*; although SKB1291214 showed traits associated with the intraperiplasmic predatory lifestyle, it is still different from SSB218315 and HD100 based on the 16S rRNA gene sequencing analysis, GC content, and AAI. The diversity was observed among the members of the genus *Bdellovibrio* thus suggesting the need to review the taxonomy of the genus *Bdellovibrio* in the nearest future. The pangenome analysis revealed that genomes of *Bdellovibrio* spp. have genes encoding different predation factors including signal transduction, hydrolytic, proteolytic, transport, and transport proteins that can help them survive as a bacterial predator. However, some factors such as hemolysin III and collagenase observed in the genomes need to be studied and characterized so that they will not have counterproductive effects when *Bdellovibrio* spp. are considered for applications as a biocontrol agent of pathogens in humans. Finally, *Bdellovibrio* sp. SKB1291214 have GEIs with atypical percent GC, *AmsF* protein, and a homologue of 4-diphosphocytidyl-2-C-methyl-D-erythritol kinase among its unique genes and an insertion of a 174 bp fragment in its *hit* locus region. These occurrences are presumptive indications of HGT in *Bdellovibrio* sp. SKB1291214.

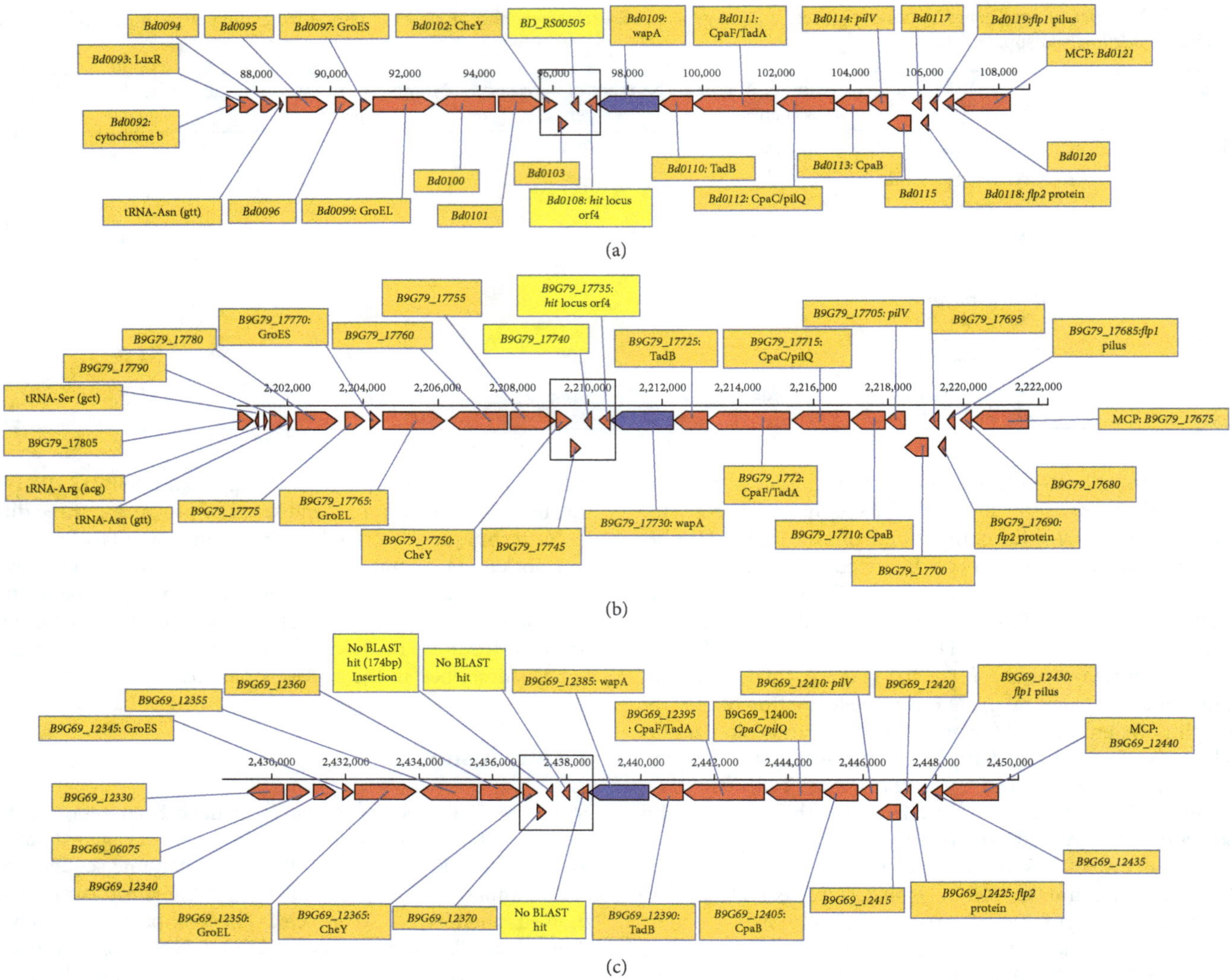

FIGURE 4: Diagrammatic comparison of the *hit* locus and the adjoining regions between *Bdellovibrio* spp. HD100 (a), SSB218315 (b), and SKB1291214 (c). The major differences can be observed at the region after the *wapA* gene. There was no BLAST hits for gene equivalent of *hit* locus orf4 *(Bd0108)*, *BD_RS00505* (new locus tag assigned to the fragment in *B. bacteriovorus* HD100) and another uncharacterized gene in *Bdellovibrio* sp. SKB1291214. The maps were generated using KBase software (http://biorxiv.org/content/early/2016/12/22/096354), and BLAST analysis was done in ExPASy Bioinformatics Resource Portal (http://www.expasy.org/). *CpaB*, *CpaF/TadA*, *TadB*, *pilQ/Cpac*, *pilV*, *flp1*, *flp2* (genes associated with type IV pilus secretion), chemotaxis protein (MCP: methyl accepting chemotaxis protein, chemotaxis protein *CheY*), heat-shock protein (*GroES* and *GroEL*), cell wall-associated protein *(wapA)*, host interaction *(hit)* locus orf.

Conflicts of Interest

The authors declare that they have no conflicts of interest.

Authors' Contributions

Omotayo Opemipo Oyedara and Mario A. Rodríguez Pérez contributed equally to this manuscript.

Acknowledgments

This work was funded by Secretaría de Investigación y Posgrado of Instituto Politecnico Nacional (Grant nos. 20161059 and 20170752). Consejo Nacional de Ciencia Y Tecnología (CONACYT), Mexico is acknowledged for providing Omotayo Opemipo Oyedara a doctoral scholarship (no. 595082/326342). Xianwu Guo and Mario A. Rodríguez Pérez hold scholarships from Comisión de Operación y Fomento de Actividades Académicas, Instituto Politécnico Nacional (COFAA-IPN). The publication fee for the manuscript was funded by COFAA-IPN.

Supplementary Materials

Supplementary 1. Additional file 1a: genome analysis of *Bdellovibrio* spp. SKB1291214 and SSB218315 for the genes expressed by *B. bacteriovorus* HD100 during predation as reported by Lambert et al. [25]. Additional files 1b–f: presumed predation-enhancing factors in *Bdellovibrio* spp. SKB1291214 and SSB218315 predicted by RAST annotation server. (1b) Flagellar and chemotaxis factor. (1c) Transport and type IV pilus proteins. (1d) Stress response proteins. (1e) Factors associated with the production of degradative enzymes for the metabolism of molecules. (1f) Siderophores and defense factors. Additional file 1g: BLASTp analysis of genes annotated as hemolysin, collagenase, and RTX toxin in the genomes of *Bdellovibrio* spp.

Supplementary 2. Additional files 2a and b: predicted genomic islands in *Bdellovibrio* spp. SKB1291214 and SSB218315, respectively.

Supplementary 3. Additional file 3: pairwise evolutionary distance among *Bdellovibrio* spp.

Supplementary 4. Additional files 4a–f: pangenome analysis of *Bdellovibrio* spp. using the BPGA pipeline. (4a) The core genes identified in the *Bdellovibrio* spp. (4b) The power and exponential fit law to predict the nature of *Bdellovibrio* spp. pangenome as open or close. (4c–f) The unique genes identified in *Bdellovibrio* spp. SKB1291214, JSS, W, and SSB218315, respectively.

Supplementary 5. Additional 5a and b: multiple sequence alignment of the *Bd0108* and *Bd0109* genes of *hit* locus, respectively.

References

[1] R. E. Sockett, "Predatory lifestyle of *Bdellovibrio bacteriovorus*," *Annual Review of Microbiology*, vol. 63, no. 1, pp. 523–539, 2009.

[2] H. Stolp and M. P. Starr, "*Bdellovibrio bacteriovorus* gen. et sp. n., a predatory, ectoparasitic, and bacteriolytic microorganism," *Antonie Van Leeuwenhoek*, vol. 29, no. 1, pp. 217–248, 1963.

[3] S. F. Koval, S. H. Hynes, R. S. Flannagan, Z. Pasternak, Y. Davidov, and E. Jurkevitch, "*Bdellovibrio exovorus* sp. nov., a novel predator of *Caulobacter crescentus*," *International Journal of Systematic and Evolutionary Microbiology*, vol. 63, Part 1, pp. 146–151, 2013.

[4] J. J. Tudor and S. F. Conti, "Characterization of germination and activation of *Bdellovibrio* bdellocysts," *Journal of Bacteriology*, vol. 133, no. 1, pp. 130–138, 1978.

[5] L. Hobley, T. R. Lerner, L. E. Williams et al., "Genome analysis of a simultaneously predatory and prey-independent, novel *Bdellovibrio bacteriovorus* from the River Tiber, supports *in silico* predictions of both ancient and recent lateral gene transfer from diverse bacteria," *BMC Genomics*, vol. 13, no. 1, p. 670, 2012.

[6] M. J. Capeness, C. Lambert, A. L. Lovering et al., "Activity of *Bdellovibrio hit* locus proteins, Bd0108 and Bd0109, links Type IVa pilus extrusion/retraction status to prey-independent growth signalling," *PLoS One*, vol. 8, no. 11, article e79759, 2013.

[7] U. Gophna, R. L. Charlebois, and W. F. Doolittle, "Ancient lateral gene transfer in the evolution of *Bdellovibrio bacteriovorus*," *Trends in Microbiology*, vol. 14, no. 2, pp. 64–69, 2006.

[8] A. Pan, I. Chanda, and J. Chakrabarti, "Analysis of the genome and proteome composition of *Bdellovibrio bacteriovorus*: indication for recent prey-derived horizontal gene transfer," *Genomics*, vol. 98, no. 3, pp. 213–222, 2011.

[9] E. Jurkevitch, D. Minz, B. Ramati, and G. Barel, "Prey range characterization, ribotyping, and diversity of soil and rhizosphere *Bdellovibrio* spp. isolated on phytopathogenic bacteria," *Applied and Environmental Microbiology*, vol. 66, no. 6, pp. 2365–2371, 2000.

[10] O. O. Oyedara, E. J. de Luna-Santillana, O. Olguin-Rodriguez et al., "Isolation of *Bdellovibrio* sp. from soil samples in Mexico and their potential applications in control of pathogens," *Microbiology Open*, vol. 5, no. 6, pp. 992–1002, 2016.

[11] N. A. Joshi and J. N. Fass, "Sickle: A Sliding-Window, Adaptive, Quality-Based Trimming Tool for FastQ Files," 2011, (Version 1.33) [Software]. https://github.com/najoshi/sickle.

[12] A. Bankevich, S. Nurk, D. Antipov et al., "SPAdes: a new genome assembly algorithm and its applications to single-cell sequencing," *Journal of Computational Biology*, vol. 19, no. 5, pp. 455–477, 2012.

[13] E. Bosi, B. Donati, M. Galardini et al., "MeDuSa: a multi-draft based scaffolder," *Bioinformatics*, vol. 31, no. 15, pp. 2443–2451, 2015.

[14] I. J. Tsai, T. D. Otto, and M. Berriman, "Improving draft assemblies by iterative mapping and assembly of short reads to eliminate gaps," *Genome Biology*, vol. 11, no. 4, article R41, 2010.

[15] A. Gurevich, V. Saveliev, N. Vyahhi, and G. Tesler, "QUAST: quality assessment tool for genome assemblies," *Bioinformatics*, vol. 29, no. 8, pp. 1072–1075, 2013.

[16] R. K. Aziz, D. Bartels, A. A. Best et al., "The RAST server: rapid annotations using subsystems technology," *BMC Genomics*, vol. 9, no. 1, p. 75, 2008.

[17] D. Arndt, J. R. Grant, A. Marcu et al., "PHASTER: a better, faster version of the PHAST phage search tool," *Nucleic Acids Research*, vol. 44, no. W1, pp. W16–W21, 2016.

[18] C. Bertelli, M. R. Laird, K. P. Williams et al., "IslandViewer 4: expanded prediction of genomic islands for larger-scale datasets," *Nucleic Acids Research*, vol. 45, no. W1, pp. W30–W35, 2017.

[19] K. Tamura, G. Stecher, D. Peterson, A. Filipski, and S. Kumar, "MEGA6: molecular evolutionary genetics analysis version 6.0," *Molecular Biology and Evolution*, vol. 30, no. 12, pp. 2725–2729, 2013.

[20] K. T. Konstantinidis and J. M. Tiedje, "Genomic insights that advance the species definition for prokaryotes," *Proceedings of the National Academy of Sciences of the United States of America*, vol. 102, no. 7, pp. 2567–2572, 2005.

[21] N. M. Chaudhari, V. K. Gupta, and C. Dutta, "BPGA - an ultra-fast pan-genome analysis pipeline," *Scientific Reports*, vol. 6, no. 1, article 24373, 2016.

[22] S. Götz, J. M. García-Gómez, J. Terol et al., "High-throughput functional annotation and data mining with the Blast2GO suite," *Nucleic Acids Research*, vol. 36, no. 10, pp. 3420–3435, 2008.

[23] F. Sievers, A. Wilm, D. Dineen et al., "Fast, scalable generation of high-quality protein multiple sequence alignments using Clustal Omega," *Molecular Systems Biology*, vol. 7, no. 1, p. 539, 2011.

[24] S. Rendulic, P. Jagtap, A. Rosinus et al., "A predator unmasked: life cycle of *Bdellovibrio bacteriovorus* from a genomic perspective," *Science*, vol. 303, no. 5658, pp. 689–692, 2004.

[25] C. Lambert, C. Y. Chang, M. J. Capeness, and R. E. Sockett, "The first bite—profiling the predatosome in the bacterial pathogen *Bdellovibrio*," *PLoS One*, vol. 5, no. 1, article e8599, 2010.

[26] C. Lambert, L. Hobley, C. Y. Chang, A. Fenton, M. Capeness, and L. Sockett, "A predatory patchwork: membrane and surface structures of *Bdellovibrio bacteriovorus*," *Advances in Microbial Physiology*, vol. 54, pp. 313–361, 2008.

[27] A. A. Medina, R. M. Shanks, and D. E. Kadouri, "Development of a novel system for isolating genes involved in predator-prey interactions using host independent derivatives of *Bdellovibrio bacteriovorus* 109J," *BMC Microbiology*, vol. 8, no. 1, p. 33, 2008.

[28] D. S. Milner, R. Till, I. Cadby et al., "Ras GTPase-like protein MglA, a controller of bacterial social-motility in myxobacteria, has evolved to control bacterial predation by *Bdellovibrio*," *PLoS Genetics*, vol. 10, no. 4, article e1004253, 2014.

[29] L. Hobley, R. K. Y. Fung, C. Lambert et al., "Discrete cyclic di-GMP-dependent control of bacterial predation versus axenic growth in *Bdellovibrio bacteriovorus*," *PLoS Pathogens*, vol. 8, no. 2, article e1002493, 2012.

[30] S. Gupta, C. Tang, M. Tran, and D. E. Kadouri, "Effect of predatory bacteria on human cell lines," *PLoS One*, vol. 11, no. 8, article e0161242, 2016.

[31] M. Otto, "*Staphylococcus aureus* toxins," *Current Opinion in Microbiology*, vol. 17, pp. 32–37, 2014.

[32] Y. C. Chen, M. C. Chang, Y. C. Chuang, and C. L. Jeang, "Characterization and virulence of hemolysin III from *Vibrio vulnificus*," *Current Microbiology*, vol. 49, no. 3, pp. 175–179, 2004.

[33] I. Linhartová, L. Bumba, J. Mašín et al., "RTX proteins: a highly diverse family secreted by a common mechanism," *FEMS Microbiology Reviews*, vol. 34, no. 6, pp. 1076–1112, 2010.

[34] H. J. Han, T. Taki, H. Kondo, I. Hirono, and T. Aoki, "Pathogenic potential of a collagenase gene from *Aeromonas veronii*," *Canadian Journal of Microbiology*, vol. 54, no. 1, pp. 1–10, 2007.

[35] R. D. Finn, J. Mistry, J. Tate et al., "The Pfam protein families database," *Nucleic Acids Research*, vol. 38, supplement 1, pp. D211–D222, 2009.

[36] M. Juhas, J. R. van der Meer, M. Gaillard, R. M. Harding, D. W. Hood, and D. W. Crook, "Genomic islands: tools of bacterial horizontal gene transfer and evolution," *FEMS Microbiology Reviews*, vol. 33, no. 2, pp. 376–393, 2009.

[37] G. S. Vernikos and J. Parkhill, "Resolving the structural features of genomic islands: a machine learning approach," *Genome Research*, vol. 18, no. 2, pp. 331–342, 2008.

[38] K. T. Konstantinidis and J. M. Tiedje, "Prokaryotic taxonomy and phylogeny in the genomic era: advancements and challenges ahead," *Current Opinion in Microbiology*, vol. 10, no. 5, pp. 504–509, 2007.

[39] Z. Pasternak, S. Pietrokovski, O. Rotem, U. Gophna, M. N. Lurie-Weinberger, and E. Jurkevitch, "By their genes ye shall know them: genomic signatures of predatory bacteria," *The ISME Journal*, vol. 7, no. 4, pp. 756–769, 2013.

[40] T. R. Lerner, A. L. Lovering, N. K. Bui et al., "Specialized peptidoglycan hydrolases sculpt the intra-bacterial niche of predatory *Bdellovibrio* and increase population fitness," *PLoS Pathogens*, vol. 8, no. 2, article e1002524, 2012.

[41] C. Lambert, I. T. Cadby, R. Till et al., "Ankyrin-mediated self-protection during cell invasion by the bacterial predator *Bdellovibrio bacteriovorus*," *Nature Communications*, vol. 6, no. 1, p. 8884, 2015.

[42] J. M. Koczan, M. J. McGrath, Y. Zhao, and G. W. Sundin, "Contribution of *Erwinia amylovora* exopolysaccharides amylovoran and levan to biofilm formation: implications in pathogenicity," *Phytopathology*, vol. 99, no. 11, pp. 1237–1244, 2009.

[43] G. Prehna, B. E. Ramirez, and A. L. Lovering, "The lifestyle switch protein *Bd0108* of *Bdellovibrio bacteriovorus* is an intrinsically disordered protein," *PLoS One*, vol. 9, no. 12, article e115390, 2014.

[44] M. A. Ferguson, J. L. Schmitt, A. R. Sindhurakar, C. B. Volle, M. E. Nuñez, and E. M. Spain, "Rapid isolation of host-independent *Bdellovibrio bacteriovorus*," *Journal of Microbiological Methods*, vol. 73, no. 3, pp. 279–281, 2008.

[45] C. Lambert and R. E. Sockett, "Laboratory maintenance of *Bdellovibrio*," *Current Protocols in Microbiology*, vol. 9, pp. 7B.2.1–7B.2.13, 2008.

[46] R. J. Seidler and M. P. Starr, "Isolation and characterization of host-independent *Bdellovibrios*," *Journal of Bacteriology*, vol. 100, no. 2, pp. 769–785, 1969.

21

Nonredundant, Highly Connected MicroRNAs Control Functionality in Breast Cancer Networks

Guillermo de Anda-Jáuregui,[1] Jesús Espinal-Enríquez,[1,2] Diana Drago-García,[3] and Enrique Hernández-Lemus[1,2]

[1]*Computational Genomics Division, National Institute of Genomic Medicine, 14610 Mexico City, Mexico*
[2]*Centro de Ciencias de la Complejidad, Universidad Nacional Autónoma de México, 04510 Mexico City, Mexico*
[3]*Department of Biological Regulation, Weizmann Institute of Science, 7610001 Rehovot, Israel*

Correspondence should be addressed to Enrique Hernández-Lemus; ehernandez@inmegen.gob.mx

Academic Editor: Changwon Park

Alterations to transcriptional regulation are an important factor in breast cancer. Noncoding RNA, such as microRNA (miR), have very influential roles in the transcriptional regulation of genes. Transcriptional regulation can be successfully modeled and analyzed using complex network theory. Particularly, interactions between two distinct classes of biological elements, such as miR and genes, can be approached through the bipartite network formalism. Based on bipartite network properties, it is possible to identify highly influential miRs in the network, such as those that have a large number of connections indicating regulation of a large set of genes. Some miRs in a network are nonredundant, which indicates that they are solely responsible of the regulation of a particular set of genes, which in turn may be associated to a particular biological process. We hypothesize that highly influential, nonredundant miRs, which we call *Commodore miRs* (Cdre-miRs), have an important role on the control of biological functions through transcriptional networks. In this work, we analyze the regulation of gene expression by miRs in healthy and cancerous breast tissue using bipartite miR-gene networks inferred from the Cancer Genome Atlas (TCGA) expression data. We observe differences in the degree, clustering coefficient and redundancy distributions for miRs and genes in the network, indicating differences in the way that these elements interact with each other. Furthermore, we identify a small set of five Cdre-miRs in the breast cancer network: miR-190b, miR-let7i, miR-292-b, miR-511, and miR-141. The neighborhood of genes controlled by each of these miRs is involved in particular biological functions such as dynein structure-associated processes, immune response, angiogenesis, cytokine activity, and cell motility. We propose that these Cdre-miRs are important control elements of biological functions deregulated in breast cancer.

1. Introduction

Breast cancer is a highly frequent cancer and one leading cause of death for women worldwide [1]. Aside from these important epidemiological aspects (and likely behind them) is the fact that breast cancer is a highly heterogeneous disease, both in its molecular origins and in its clinical manifestations; a fact that calls loudly for an improved understanding of the molecular mechanisms behind breast cancer development. Gene expression studies have provided with unprecedented information to characterize biomolecular activities leading to (or at least associated with) tumorigenesis. Recently, next generation sequencing has allowed us to accurately measure not only mRNA transcripts, but also regulatory noncoding RNA molecules such as microRNAs (miRs) and long noncoding RNAs (lncRNAs). miRs in particular have received a growing deal of attention due to their regulatory effects that seem to be fundamental for breast tumor establishment and progression [2].

In particular, the regulatory role of miRs seems to be central to processes involving cellular homoeostasis through processes such as apoptosis, proliferation, and migration that when deregulated give rise to well-known hallmarks of cancer [3–5]. Specific families of miRs playing either the role of "oncogenes" or "tumor suppressors" are commonly referred to as "oncomiRs" [6]. Specific miRNA regulation and coregulation patterns have been linked to oncogenic processes [7], specifically regarding breast cancer onset and evolution [2].

miR regulation at the transcriptional and posttranscriptional levels often occurs by inducing processes leading to mRNA destabilization [8]; transcriptome profiling is thus increasingly useful for the analysis of miR regulation in genome-wide settings. On the other hand, miRs have also been associated with other mechanisms of regulatory activity [9], in particular by widening target upregulation or downregulation [10]. These regulatory interactions are involved in mechanisms that may ensure biological robustness [11]. Regulatory relationships between miRs and genes (including transcription factors and other regulatory elements in the genome) are able to coexpress profiles that may be phenotype-inducing [12].

In order to improve our current knowledge on these matters, in particular regarding the transcriptional relation between miRNAs and their target genes (mRNAs), we propose the use of regulatory networks, using expression data from primary breast cancer tissue and matched control tissue. In our previous effort [13], we identified differences in the regulatory interactions of genes by other genes and by miRs; therefore, we decided to further explore these relationships using the bipartite graph formalism.

Bipartite networks are graphs composed by two disjoint sets of nodes and a set of edges such that each edge is surrounded by a node of each class. The two disjoint sets of nodes can be thought of as a *top* and *bottom* layers of nodes. The nature of bipartite graphs allows for the identification of topological parameters that describe both sets of nodes that are uniquely defined in this context. [14]. In this work, we will focus on three parameters: degree distributions, clustering coefficient, and redundancy coefficient.

1.1. Degree. The degree of a node is the number of edges that connect a node to its neighbors; it is thus the simplest measure of node connectivity. The degree distribution of a network is the most basic descriptor of a network [15]. In a bipartite network, the degree distribution is more informative if calculated for each set of nodes separately, as the connectivity of nodes in each set may be different. Thus, the degree distribution for a given set can be defined as follows:

$$P(K_{\text{NodeSet}}) = \frac{nk}{n}. \quad (1)$$

1.2. Clustering Coefficient. The clustering coefficient of a node is a measure of the local density of connections in which a given node participates. The clustering coefficient of a network is thus a measure of the cohesiveness of the network [16]. The clustering coefficient in a bipartite node is defined as follows:

$$\text{CC}(u) = \frac{\sum_{v \in N(N(v))} (|N(u) \cap N(v)| / |N(u) \cup N(v)|)}{|N(N(u))|}. \quad (2)$$

1.3. Redundancy Coefficient. Nodes of the same type can only be connected by being connected through nodes of another type. The redundancy coefficient of a node is the fraction of pairs of neighbors of the said node that are both linked to other nodes [16]; it is a measure of the importance of a given node in a layer for the connection of the nodes in the other layer. Redundant nodes can be removed from their layer without causing a disconnection of the nodes in the other layer, as seen in Figure 1. The formal definition for the redundancy coefficient is as follows [17]:

$$\text{RC}(u) = \frac{\left|\left\{\{u, w\} \subseteq N(v), \exists v' \neq v, \left(v', u\right) \in E, \left(v', w\right) \in E\right\}\right|}{|N(v)|(|N(v)| - 1)/2}. \quad (3)$$

Noncoding RNAs, such as miRs, have a regulatory biological role. Meanwhile, genes may be involved in several biological activities, as they are translated into proteins which can have various functional roles. Usually, biological functions involve the interaction of several biomolecules to produce physiologically observable phenomena [18]. By regulating a set of genes, a given miR can in fact control a number of biological functions in a particular phenotype. We should take into account the following two considerations based on topological properties of a bipartite network:

(i) miRs with small neighborhoods (i.e., with low degree) are influencing few genes, being less likely to have effect on a biological function.

(ii) miRs with high redundancy are less likely to be necessary for a given group of genes to be coregulated, as there are other miRs that will keep the group connected.

Therefore, we consider that highly connected, nonredundant miRs may have a major role in the control of biological functions through transcriptional networks: removal of these miRs would lead to the disconnection of a group of genes, which would involve a loss of the concerted regulation of this gene set, and in turn, of biological functionality. We have termed these miRs, *Commodore miRs* (Cdre-miRs).

2. Materials and Methods

2.1. miR-mRNA Bipartite Network Model. We constructed a bipartite network representing transcriptional regulation of genes by miRs in breast tissue, in the physiological and cancer states, hereafter referred to as the *healthy* and *cancer* groups.

We used gene and miR transcription profiles from samples from the Cancer Genome Atlas [19] breast cancer dataset [20]. Data processing and handling was performed following the pipeline previously implemented by our group [13] (Supplementary Material 1). The data processing pipeline included a differential expression analysis using DESeq2 [21] in order to identify the log fold change (LFC) of gene expression between the cancer and healthy groups.

Regulatory interactions between miRs and genes were inferred in terms of information theory. Mutual information (MI) [22] was calculated for each miR-gene pair, using the

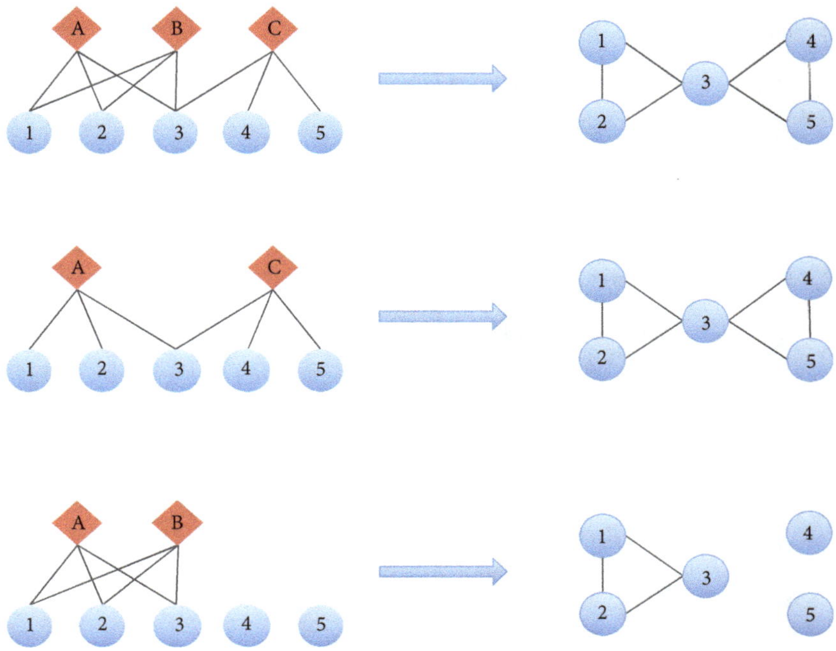

Figure 1: In this figure, we show a simple bipartite graph with 3 top nodes (A to C) and 5 bottom nodes (1 to 5). The projection of bottom nodes is shown on the right side. If node B is removed, no link in the projection is lost, as all bottom nodes connected through B are also connected through A; therefore, B is *redundant*. Meanwhile, removing node C causes the loss of connections of bottom nodes 4 and 5; therefore, C is *nonredundant*.

transcription profiles for either the healthy or cancer groups. Then, these interactions were filtered to conserve the 99.741%. MI calculation was performed using the engine of our in-house parallelized implementation of ARACNE [23].

Acknowledging that the canonical mode of action is that miRs regulate gene expression, we take this into account to interpret that mutual information interactions represent a directed relationship, from miR to gene; hence, higher values of mutual information indicate that the involved miR regulates directly the expression of the target gene.

2.2. Bipartite Network Analysis. Each network, healthy and cancer, was analyzed to identify general topological features, including number of miR and gene nodes, number of edges, and number of connected components. For each set of nodes, miR or genes, distributions were calculated for degree, clustering coefficient, and redundancy. Network analysis was performed using a combination of NetworkX for Python [16], igraph for R [24], and Cytoscape [25], as implemented previously [26].

2.3. Cdre-miR Identification. To identify highly influential, nonredundant Cdre-miRs, we separated all miRs in each network using two thresholds: a degree threshold of 100 and a redundancy threshold of 0.5.

2.4. Enrichment Analysis. For each Cdre-miR identified, we performed an over-representation analysis (ORA) of Gene Ontology (GO) biological processes [27, 28]. We performed the analysis using the WebGestalt [29] portal, with default parameters (number of genes in category between 5 and 2000, Benjamini-Hochberg multiple test adjustment, top 10 most significant processes).

3. Results

3.1. miR-Gene Bipartite Networks in Health and Cancer. We constructed two comparable networks representing the two states: healthy and cancer. Figure 2 shows a visualization of these networks. Table 1 contains the basic descriptors for each network; these networks are also provided, as GML files, in Supplementary Material 2. Figure 3 shows the cumulative frequency distribution for redundancy values for miRs and genes in healthy and cancer networks. Other distribution plots including degree and clustering coefficient of gene nodes are provided in Supplementary Material 2.

3.2. Identification of Cdre-miRs. In Figure 4, we show a scatter plot of miRs, with the y-axis representing degree and the x-axis representing redundancy. We divide this plot in four sectors delimited by the thresholds selected to define high connectivity (degree larger than 100) and low redundancy (redundancy lower than 0.5). Five miRs were identified to be highly connected and nonredundant in the cancer setting: miR-190b, miR-let7i, miR-292-b, miR-511, and miR-141. Importantly, no Cdre-miRs were identified in healthy breast tissue.

Figure 4(a) shows the five Cdre-miRs and their first gene neighbors, colored by LFC. As it is expected, there is little overlap between the neighborhoods of each Cdre-miR; however, they are not completely isolated from each other, as there is a small subset of genes that are regulated by more than one Cdre-miR. Figure 4(b) shows the 3

(a)

(b)

FIGURE 2: Network visualizations. (a) Healthy breast tissue. (b) Breast cancer tissue. Node color intensity in both networks is proportional to expression levels.

Table 1: Bipartite network parameters.

Parameter	Healthy	Cancer
Nodes, miR	97	414
Nodes, gene	2967	3240
Edges	16,589	14,063
Average degree, miR	171.02	33.97
Average degree, gene	5.59	4.34
Average clustering coefficient, miR	0.28	0.18
Average clustering coefficient, gene	0.24	0.27

Gene Ontology biological processes more enriched in the gene neighborhoods of each Cdre-miR. Complete enrichment analysis results may be found in Supplementary Material 3. In this panel, we also include the enrichment of the neighborhood intersection of miR-511 and let-7i: the *innate immunity* process.

4. Discussion

Understanding the role of miR-gene regulation is crucial to unveil the mechanisms behind tumorigenesis and progression in breast cancer. Bipartite networks offer a powerful tool to analyze the behavior of these interactions and at the same time the structural and functional relevance of specific miRs in the transcriptional regulatory program.

In this work, we have shown that the bipartite networks representing miR-gene regulation in health and breast cancer exhibit notable structural differences. This can be observed in terms of structural parameters, including degree, clustering, and redundancy coefficients (as seen in Table 1, Figures 3 and 5, and Supplementary Material 4); the overall distribution of these parameters is qualitatively different in each network, which indicates that they have essentially different topologies. This result is in agreement with the widely understood notion that miR regulation is altered in cancer.

miRs are, on average, more connected in the health context than in cancer. They also happen to be more redundant, as seen in Figure 4. Biologically, this represents an abundance of joint miR regulation over sets of genes in health, which provides a mechanism that confers a structural robustness and a stronger maintenance of the transcriptional regulatory program in a normal tissue. Meanwhile, miRs in cancer tend to be less connected on average and also less redundant.

4.1. Cdre-miRs and Their Role in Breast Cancer. A major finding in this work is that the presence of highly connected, nonredundant miRs, which we define as *Commodores*, is a phenomenon only observed in the breast cancer network. Cdre-miRs emerge as important gene expression regulators, as each of them alone is responsible for the regulation of large groups of genes. Perhaps, more interesting is the fact that, through the regulation of these gene sets, each Cdre-miR is an important control element of specific biological processes.

For instance, miR-141 is a widely studied oncomiR [30–38]. Our group has previously observed that this miR is one of the most relevant mediators of structural processes in breast cancer and particularly involved in the epithelial-to-mesenchymal transition (EMT), as well as in the opposite mesenchymal-to-epithelial transition (MET) [13]. Here, we show that its functional importance is reflected by its crucial role in the miR-gene network. The processes that mir-141 controls through the regulation of its gene neighborhood, which includes genes such as VIM and ZEB1-2, are directly related to EMT-MET: motility, cell migration, and extracellular matrix (ECM) organization.

Another miR that has been previously studied in the context of breast cancer is miR-190b. Previously, it has been shown that this miR is associated with estrogen-positive breast cancer [39] and hormone therapy resistance [40]. However, its specific role in the pathogenesis of the disease has not been described. We identified mir-190b as a Cdre-miR that is associated to the regulation of dynein assembly, vitamin metabolism, and cell proliferation of mammary epithelial cells, a well-known hallmark of cancer. As a Cdre-miR, it is possible that deregulation of this miR is mechanistically involved in the acquisition of these distinctive cancer features.

Other important cancer features are controlled by miR-29b. This miR has been found in serum of patients with breast cancer [41] and overexpressed in cervical cancer [42]. In this work, we identify it as an important regulator of transport, angiogenesis, and epithelial cell migration, processes that are predominantly controlled by this miR.

4.2. Concurrent Regulation of Innate Immune Response by Cdre-miRs. So far, we have identified processes that are uniquely controlled by a single Cdre-miR. However, we also found processes that are controlled by two different Cdre-miRs, through the regulation of different gene neighborhoods. For instance, let-7i controls genes related to leukocyte cell-cell adhesion, adaptive immune response, and cell activation. Meanwhile, miR-511 regulates genes associated to cytokine production, cell activation, and adaptive immune response; that is, they each control related, but nonoverlapping, processes.

Importantly, the overlap between these two miRs is the largest observed between any pair of Cdre-miRs, but is less than 25% of the size of either miRs' neighborhood (53 out of 238 and 213, resp.). Perhaps more interestingly, by evaluating the enrichment of the intersection of these two Cdre-miRs, we identify that their jointly regulated genes highly overrepresent the innate immune response process, which is not significantly enriched by neither of neighborhood alone. This can be interpreted as an instance of a coregulated process that requires combined action of two different Cdre-miRs.

5. Conclusion

Using bipartite networks, we identified differences in the miR-gene regulation between healthy breast tissue and breast cancer. We christened those miRs that are highly connected and nonredundant as *Commodore miRs*. The emergence of Cdre-miRs is a network structural property that is only found in the context of breast cancer.

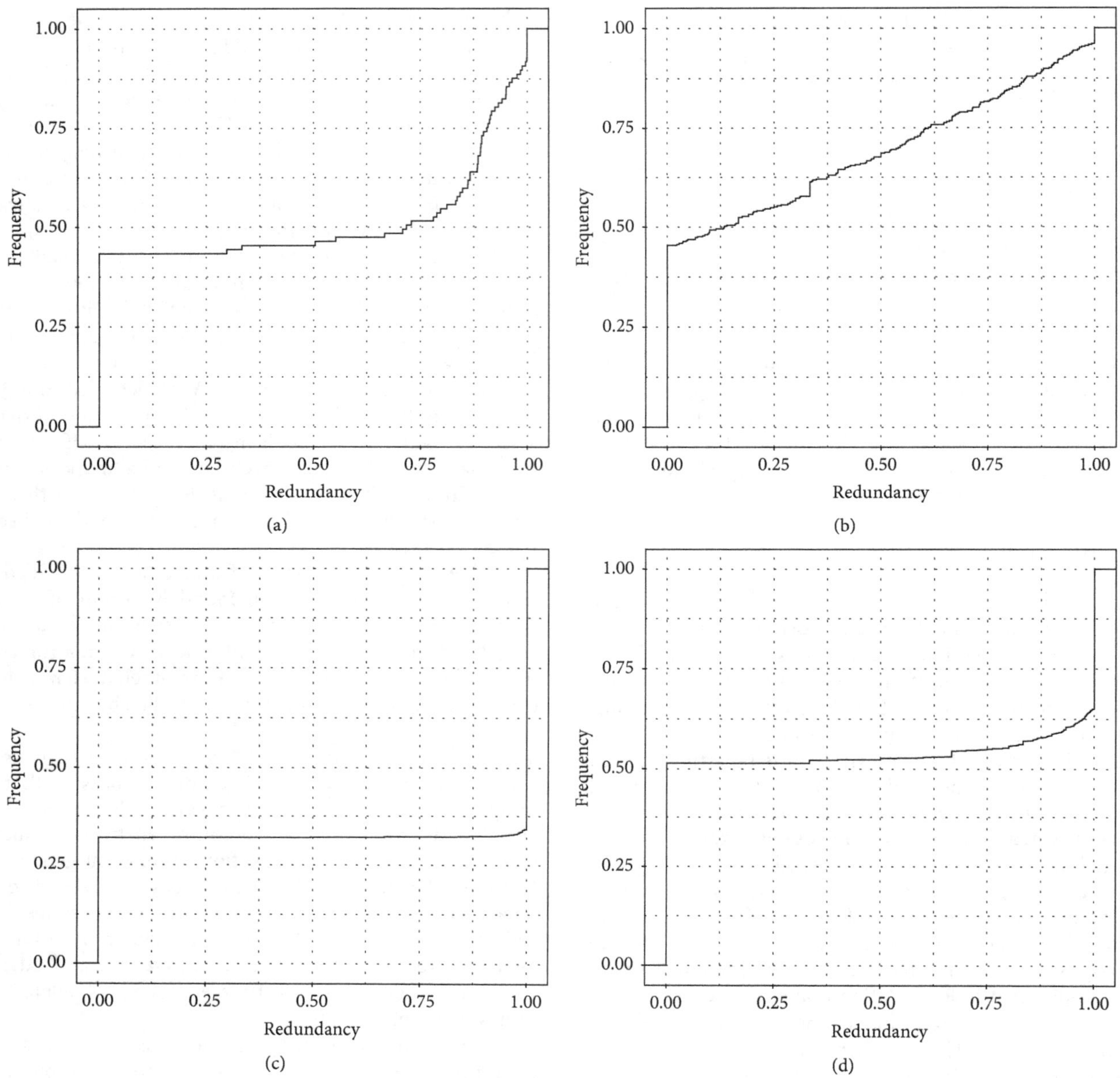

FIGURE 3: Redundancy coefficient cumulative frequency distribution plots. (a) miR, healthy. (b) miR, cancer. (c) gene, healthy. (d) gene, cancer. In each panel, x-axis represents redundancy coefficient values, and y-axis represents the normalized, accumulated frequency.

We identified five Cdre-miRs: miR-190, miR-29b, miR-141, miR-511, and let-7i that regulate specific biological processes. miR-190 is involved in cell structure, proliferation, and metabolism. miR-29b regulates cell transport, migration, and angiogenesis. miR-141 controls cell motility and extracellular matrix organization. miR-511 and let-7i independently regulate genes associated to cell adhesion and adaptive immune response through different gene sets. The latter shows that nonredundancy does not interfere with network robustness at a functional level.

Furthermore, innate immune response is a function which is not controlled by a single Cdre-miR, but it emerges in the context of the neighborhood overlap of miR-511 and let-7i. This is an example of how cooperation allows the emergence of new features that provide biological robustness to breast cancer. Identifying these elements is a step forward to identify actionable elements that provide controllability over the transcriptional networks.

Cdre-miRs may serve as novel biomarkers in breast cancer, which may be used to identify perturbation of biological functions in pathological phenotypes. Currently, therapeutic targeting of miRs in clinical oncology is limited; however, as technologies advance, the use of systems such as antagomiRs, microRNA mimics, and reporter systems will be available. In this context, we propose that Cdre-miRs may be attractive targets for the disruption of processes that are favorable for tumor growth and survival. With this in mind, Cdre-miRs may be an important

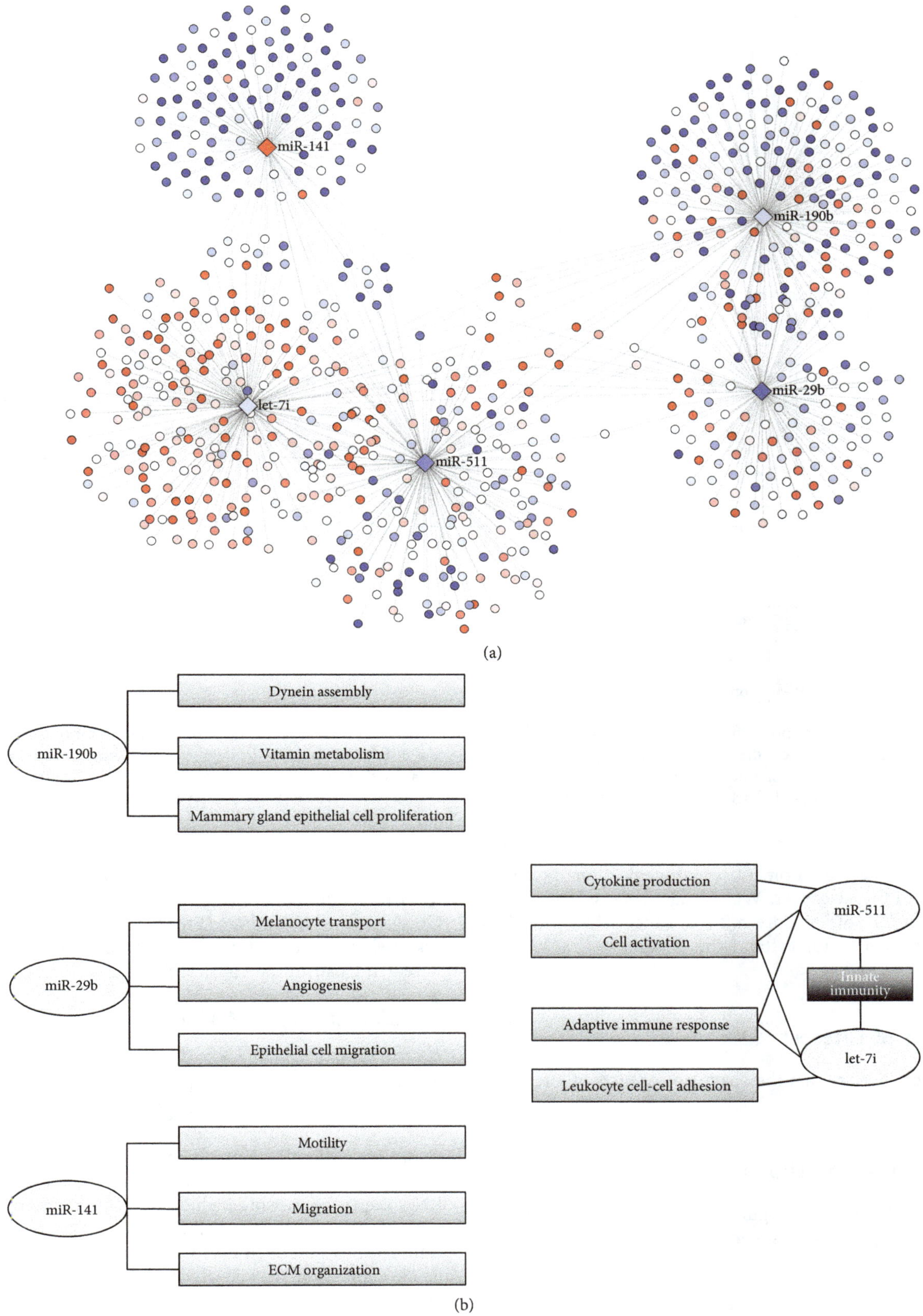

FIGURE 4: (a) shows the five Cdre-miRs and their gene neighbors. Nodes are colored in blue if subexpressed or red if overexpressed. (b) shows the GO biological processes enriched in each Cdre-miR neighborhood, in light grey. Additionally, the process enriched in the neighborhood intersection of mir-511 and let-7i, *innate immunity*, is shown in dark grey.

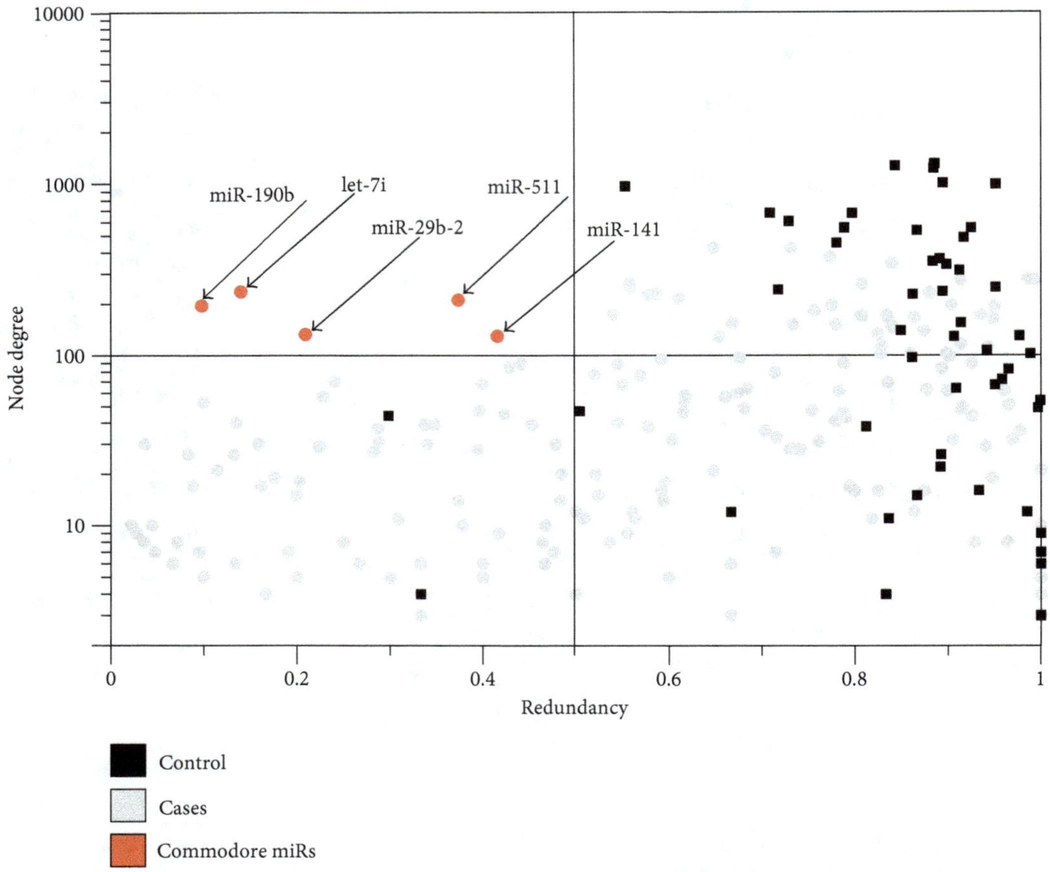

FIGURE 5: This scatterplot shows miR nodes in the healthy and cancer networks: grey dots for cancer and black squares for healthy. The x-axis represents the redundancy coefficient, and y-axis represents the degree of a miR. Only nodes with degrees larger than 2 are represented, as redundancy is not defined for nodes with a lower degree value. The plot is divided by lines representing thresholds for redundancy (0.5) and degree (100). The upper left sector of the plot is populated by five Cdre-miR nodes (red dots) with high connectivity and low redundancy.

element in the development of strategies for precision medicine in breast cancer. We strongly believe that further research in the role of Cdre-miRs in cancer shall provide a novel approach to understand the disease.

Conflicts of Interest

The authors declare that they have no conflicts of interest.

Authors' Contributions

Guillermo de Anda-Jáuregui and Jesús Espinal-Enríquez contributed equally to this work.

Acknowledgments

The research leading to these results has received funding from Grant no. 285544/2016 from the Consejo Nacional de Ciencia y Tecnología (CONACYT) as well as federal funding from the National Institute of Genomic Medicine (INMEGEN). One of the authors (Enrique Hernández-Lemus) also acknowledges support from the 2016 Marcos Moshinsky Research Chair in the Physical Sciences.

Supplementary Materials

Supplementary Material 1. Compressed (Zip) file containing the graphic pipeline followed in this work. We use raw counts from RNA-seq and miRNA-seq for breast tumors and adjacent control ("healthy") breast tissue from Genome Data Commons. MicroRNA preprocessing included mature count annotation, low-count filtering, and TMM normalization; RNA-seq preprocessing included mRNA annotation, RSEM normalization, and low-count filtering. This preprocessing is thoroughly described in [13]. These counts were integrated in an expression matrix which was used for a differential expression analysis (between cases and controls), and the inference of mutual information miR-gene networks. The final step is the identification of Cdre-miRs. Supplementary Material 2. Compressed (Zip) file containing GML representations of networks. Supplementary Material 3. Compressed

(Zip) file containing ORA results for each Cdre-miR gene neighborhood. Supplementary Material 4. Compressed (Zip) file containing degree and clustering coefficient distribution plots. *(Supplementary Materials)*

References

[1] J. Ferlay, I. Soerjomataram, R. Dikshit et al., "Cancer incidence and mortality worldwide: sources, methods and major patterns in GLOBOCAN 2012," *International Journal of Cancer*, vol. 136, no. 5, pp. E359–E386, 2015.

[2] E. O'Day and A. Lal, "MicroRNAs and their target gene networks in breast cancer," *Breast Cancer Research*, vol. 12, no. 2, p. 201, 2010.

[3] D. Hanahan and R. A. Weinberg, "The hallmarks of cancer," *Cell*, vol. 100, no. 1, pp. 57–70, 2000.

[4] D. Hanahan and R. A. Weinberg, "Hallmarks of cancer: the next generation," *Cell*, vol. 144, no. 5, pp. 646–674, 2011.

[5] M. Garofalo and C. M. Croce, "microRNAs: master regulators as potential therapeutics in cancer," *Annual Review of Pharmacology and Toxicology*, vol. 51, no. 1, pp. 25–43, 2011.

[6] W. C. S. Cho, "OncomiRs: the discovery and progress of microRNAs in cancers," *Molecular Cancer*, vol. 6, no. 1, p. 60, 2007.

[7] Y. Cai, X. Yu, S. Hu, and J. Yu, "A brief review on the mechanisms of miRNA regulation," *Genomics, Proteomics & Bioinformatics*, vol. 7, no. 4, pp. 147–154, 2009.

[8] D. Baek, J. Villén, C. Shin, F. D. Camargo, S. P. Gygi, and D. P. Bartel, "The impact of microRNAs on protein output," *Nature*, vol. 455, no. 7209, pp. 64–71, 2008.

[9] M. Selbach, B. Schwanhäusser, N. Thierfelder, Z. Fang, R. Khanin, and N. Rajewsky, "Widespread changes in protein synthesis induced by microRNAs," *Nature*, vol. 455, no. 7209, pp. 58–63, 2008.

[10] S. Vasudevan, Y. Tong, and J. A. Steitz, "Switching from repression to activation: microRNAs can up-regulate translation," *Science*, vol. 318, no. 5858, pp. 1931–1934, 2007.

[11] J. A. Vidigal and A. Ventura, "The biological functions of miRNAs: lessons from *in vivo* studies," *Trends in Cell Biology*, vol. 25, no. 3, pp. 137–147, 2015.

[12] G. Diaz, F. Zamboni, A. Tice, and P. Farci, "Integrated ordination of miRNA and mRNA expression profiles," *BMC Genomics*, vol. 16, no. 1, p. 767, 2015.

[13] D. Drago-García, J. Espinal-Enríquez, and E. Hernández-Lemus, "Network analysis of EMT and MET micro-RNA regulation in breast cancer," *Scientific Reports*, vol. 7, no. 1, p. 13534, 2017.

[14] J.-L. Guillaume and M. Latapy, "Bipartite graphs as models of complex networks," *Physica A: Statistical Mechanics and its Applications*, vol. 371, no. 2, pp. 795–813, 2006.

[15] M. Newman, *Networks: An Introduction*, Oxford University Press, 2010.

[16] A. A. Hagberg, D. A. Schult, and P. J. Swart, "Exploring network structure, dynamics, and function using NetworkX," in *Proceedings of the 7th Python in Science Conference*, pp. 11–15, Pasadena, CA USA, 2008.

[17] M. Latapy, C. Magnien, and N. Del Vecchio, "Basic notions for the analysis of large two-mode networks," *Social Networks*, vol. 30, no. 1, pp. 31–48, 2008.

[18] M. A. García-Campos, J. Espinal-Enríquez, and E. Hernández-Lemus, "Pathway analysis: state of the art," *Frontiers in Physiology*, vol. 6, 2015.

[19] The Cancer Genome Atlas Research Network, J. N. Weinstein, E. A. Collisson et al., "The cancer genome atlas pan-cancer analysis project," *Nature Genetics*, vol. 45, no. 10, pp. 1113–1120, 2013.

[20] C. X. Ma and M. J. Ellis, "The cancer genome atlas: clinical applications for breast cancer," *Oncology*, vol. 27, no. 12, pp. 1263–1269, 2013, 1274–9.

[21] M. I. Love, W. Huber, and S. Anders, "Moderated estimation of fold change and dispersion for RNA-seq data with DESeq2," *Genome Biology*, vol. 15, no. 12, 2014.

[22] E. Hernández-Lemus and J. M. Siqueiros-García, "Information theoretical methods for complex network structure reconstruction," *Complex Adaptive Systems Modeling*, vol. 1, no. 1, p. 8, 2013.

[23] A. A. Margolin, I. Nemenman, K. Basso et al., "ARACNE: an algorithm for the reconstruction of gene regulatory networks in a mammalian cellular context," *BMC Bioinformatics*, vol. 7, Supplement 1, p. S7, 2006.

[24] G. Csardi and T. Nepusz, "The igraph software package for complex network research," *InterJournal, Complex Systems*, vol. 1695, 2006.

[25] P. Shannon, A. Markiel, O. Ozier et al., "Cytoscape: a software environment for integrated models of biomolecular interaction networks," *Genome Research*, vol. 13, no. 11, pp. 2498–2504, 2003.

[26] G. de Anda-Jáuregui, K. Guo, B. A. McGregor, and J. Hur, "Exploration of the anti-inflammatory drug space through network pharmacology: applications for drug repurposing," *Frontiers in Physiology*, vol. 9, p. 151, 2018.

[27] M. Ashburner, C. A. Ball, J. A. Blake et al., "Gene ontology: tool for the unification of biology," *Nature Genetics*, vol. 25, no. 1, pp. 25–29, 2000.

[28] The Gene Ontology Consortium, "Expansion of the gene ontology knowledgebase and resources," *Nucleic Acids Research*, vol. 45, D1, pp. D331–D338, 2017.

[29] J. Wang, D. Duncan, Z. Shi, and B. Zhang, "WEB-based GEne SeT AnaLysis Toolkit (WebGestalt): update 2013," *Nucleic Acids Research*, vol. 41, no. W1, pp. W77–W83, 2013.

[30] Y.-S. Yao, W.-S. Qiu, R.-Y. Yao et al., "miR-141 confers docetaxel chemoresistance of breast cancer cells via regulation of EIF4E expression," *Oncology Reports*, vol. 33, no. 5, pp. 2504–2512, 2015.

[31] H.-P. Tsai, S.-F. Huang, C.-F. Li, H.-T. Chien, and S.-C. Chen, "Differential microRNA expression in breast cancer with different onset age," *PLoS One*, vol. 13, no. 1, article e0191195, 2018.

[32] X. Li, S. Roslan, C. N. Johnstone et al., "MiR-200 can repress breast cancer metastasis through ZEB1-independent but moesin-dependent pathways," *Oncogene*, vol. 33, no. 31, pp. 4077–4088, 2014.

[33] P. Li, T. Xu, X. Zhou et al., "Downregulation of miRNA-141 in breast cancer cells is associated with cell migration and invasion: involvement of ANP32E targeting," *Cancer Medicine*, vol. 6, no. 3, pp. 662–672, 2017.

[34] J. Finlay-Schultz, D. M. Cittelly, P. Hendricks et al., "Progesterone downregulation of miR-141 contributes to expansion of stem-like breast cancer cells through maintenance of progesterone receptor and Stat5a," *Oncogene*, vol. 34, no. 28, pp. 3676–3687, 2015.

[35] B. G. Debeb, L. Lacerda, S. Anfossi et al., "miR-141-mediated regulation of brain metastasis from breast cancer," *Journal*

of the National Cancer Institute, vol. 108, no. 8, article djw026, 2016.

[36] V. Damiano, G. Brisotto, S. Borgna et al., "Epigenetic silencing of miR-200c in breast cancer is associated with aggressiveness and is modulated by ZEB1," *Genes, Chromosomes & Cancer*, vol. 56, no. 2, pp. 147–158, 2017.

[37] M. Alunni-Fabbroni, L. Majunke, E. K. Trapp et al., "Whole blood microRNAs as potential biomarkers in post-operative early breast cancer patients," *BMC Cancer*, vol. 18, no. 1, p. 141, 2018.

[38] N. Abedi, S. Mohammadi-Yeganeh, A. Koochaki, F. Karami, and M. Paryan, "miR-141 as potential suppressor of β-catenin in breast cancer," *Tumour Biology*, vol. 36, no. 12, pp. 9895–9901, 2015.

[39] G. Cizeron-Clairac, F. Lallemand, S. Vacher, R. Lidereau, I. Bieche, and C. Callens, "M*iR-190b*, the highest upregulated miRNA in ERα-positive compared to ERα-negative breast tumors, a new biomarker in breast cancers?," *BMC Cancer*, vol. 15, no. 1, p. 499, 2015.

[40] T. Joshi, D. Elias, J. Stenvang et al., "Integrative analysis of miRNA and gene expression reveals regulatory networks in tamoxifen-resistant breast cancer," *Oncotarget*, vol. 7, no. 35, pp. 57239–57253, 2016.

[41] O. Shaker, M. Maher, Y. Nassar, G. Morcos, and Z. Gad, "Role of microRNAs -29b-2, -155, -197 and -205 as diagnostic biomarkers in serum of breast cancer females," *Gene*, vol. 560, no. 1, pp. 77–82, 2015.

[42] L. S. Servín-González, A. J. Granados-López, and J. A. López, "Families of microRNAs expressed in clusters regulate cell signaling in cervical cancer," *International Journal of Molecular Sciences*, vol. 16, no. 6, pp. 12773–12790, 2015.

Permissions

All chapters in this book were first published in IJG, by Hindawi Publishing Corporation; hereby published with permission under the Creative Commons Attribution License or equivalent. Every chapter published in this book has been scrutinized by our experts. Their significance has been extensively debated. The topics covered herein carry significant findings which will fuel the growth of the discipline. They may even be implemented as practical applications or may be referred to as a beginning point for another development.

The contributors of this book come from diverse backgrounds, making this book a truly international effort. This book will bring forth new frontiers with its revolutionizing research information and detailed analysis of the nascent developments around the world.

We would like to thank all the contributing authors for lending their expertise to make the book truly unique. They have played a crucial role in the development of this book. Without their invaluable contributions this book wouldn't have been possible. They have made vital efforts to compile up to date information on the varied aspects of this subject to make this book a valuable addition to the collection of many professionals and students.

This book was conceptualized with the vision of imparting up-to-date information and advanced data in this field. To ensure the same, a matchless editorial board was set up. Every individual on the board went through rigorous rounds of assessment to prove their worth. After which they invested a large part of their time researching and compiling the most relevant data for our readers.

The editorial board has been involved in producing this book since its inception. They have spent rigorous hours researching and exploring the diverse topics which have resulted in the successful publishing of this book. They have passed on their knowledge of decades through this book. To expedite this challenging task, the publisher supported the team at every step. A small team of assistant editors was also appointed to further simplify the editing procedure and attain best results for the readers.

Apart from the editorial board, the designing team has also invested a significant amount of their time in understanding the subject and creating the most relevant covers. They scrutinized every image to scout for the most suitable representation of the subject and create an appropriate cover for the book.

The publishing team has been an ardent support to the editorial, designing and production team. Their endless efforts to recruit the best for this project, has resulted in the accomplishment of this book. They are a veteran in the field of academics and their pool of knowledge is as vast as their experience in printing. Their expertise and guidance has proved useful at every step. Their uncompromising quality standards have made this book an exceptional effort. Their encouragement from time to time has been an inspiration for everyone.

The publisher and the editorial board hope that this book will prove to be a valuable piece of knowledge for researchers, students, practitioners and scholars across the globe.

List of Contributors

Renato Graciano de Paula, Amanda Cristina Campos Antoniêto, Liliane Fraga Costa Ribeiro, Cláudia Batista Carraro, Karoline Maria Vieira Nogueira, Douglas Christian Borges Lopes , Alinne Costa Silva, Mariana Taíse Zerbini, Wellington Ramos Pedersoli, Mariana do Nascimento Costa and Roberto Nascimento Silva
Molecular Biotechnology Laboratory, Department of Biochemistry and Immunology, Ribeirao Preto Medical School (FMRP), University of Sao Paulo, Ribeirao Preto, SP, Brazil

Hesham A. M. Gomaa
Pharmacology Department, College of Pharmacy, Jouf University, Sakakah, Saudi Arabia
Assistant Professor at Biochemistry Department, Faculty of Pharmacy, Nahda University, Beni-Suef, Egypt

Asmaa T. Ali
Assistant Lecturer at Biochemistry Department, Faculty of Pharmacy, Nahda University, Beni-Suef, Egypt

M. Abdel Gabbar
Lecturer at Biochemistry Division, Chemistry Department, Beni-Suef University, Ph.D. in Biochemistry, Beni-Suef, Egypt

M. A. Kandeil
Professor and Head of Biochemistry Department Faculty of Veterinary Medicine, Beni-Suef University, Beni-Suef, Egypt

Artur Gurgul, Klaudia Pawlina-Tyszko, Tomasz Szmatoła and Igor Jasielczuk
Laboratory of Genomics, Department of Animal Genetics and Molecular Biology, National Research Institute of Animal Production, Krakowska 1, 32-083 Balice, Poland

Monika Bugno-Poniewierska
Institute of Veterinary Sciences, University of Agriculture in Krakow, Mickiewicza 24/28 Av, 30-059 Krakow, Poland

Stefan Dobosz
Department of Salmonid Research, Inland Fisheries Institute in Olsztyn, Rutki, 83-330 Żukowo, Poland

Konrad Ocalewicz
Department of Marine Biology and Ecology, Institute of Oceanography, Faculty of Oceanography and Geography, University of Gdansk, M. Piłsudskiego 46 Av, 81-378 Gdynia, Poland

Ningyang Li, Zhichang Qiu, Xiaoming Lu, Bingchao Shi, Xiaozhen Tang and Xuguang Qiao
Key Laboratory of Food Processing Technology and Quality Control in Shandong Province, College of Food Science and Engineering, Shandong Agricultural University, No. 61, Daizong Road, Tai'an, Shandong Province 271018, China

Xiudong Sun
State Key Laboratory of Crop Biology, College of Horticulture Science and Engineering, Shandong Agricultural University, No. 61, Daizong Road, Tai'an, Shandong Province 271018, China

Anna V. Shchennikova, Marya A. Slugina, Alexey V. Beletsky, Mikhail A. Filyushin, Andrey A. Mardanov, Olga A. Shulga, Elena Z. Kochieva, Nikolay V. Ravin and Konstantin G. Skryabin
Federal State Institution "Federal Research Centre "Fundamentals of Biotechnology" of the Russian Academy of Sciences", Moscow 119071, Russia

Marya A. Slugina, Elena Z. Kochieva and Konstantin G. Skryabin
Lomonosov Moscow State University, Moscow 119991, Russia

Li Chen, Huifang Yan, Binbin Cao, Ye Wu, Qiang Gu, Yanling Yang, Zhen Shi, Zhixian Yang, Xingzhi Chang, Yuehua Zhang, Xiru Wu, Yuwu Jiang and Jingmin Wang
Department of Pediatrics, Peking University First Hospital, Beijing 100034, China

Li Chen
Department of Pediatric Cardiology, Beijing Anzhen Hospital, Capital Medical University, Beijing 100029, China

Jiangxi Xiao
Department of Radiology, Peking University First Hospital, Beijing 100034, China

Huixia Yang, Junya Chen and Yu Sun
Department of Obstetrics and Gynecology, Peking University First Hospital, Beijing 100034, China

Hong Pan
Department of Central Laboratory, Peking University First Hospital, Beijing 100034, China

Lei Li, Xi Wang, Huimin Lu, Xinping Xi, Mei Zhou, Tianbao Chen and Lei Wang
Natural Drug Discovery Group, School of Pharmacy, Queen's University, Belfast BT9 7BL, UK

Qing Wu
School of Chinese Materia Medica, Beijing University of Chinese Medicine, Beijing 102488, China

Chris J. Watson
Centre for Experimental Medicine, School of Medicine, Dentistry and Biomedical Sciences, Queen's University, Belfast BT9 7BL, UK

Túlio De Lima Campos
Bioinformatics Core Facility, Aggeu Magalhães Institute (IAM), Oswaldo Cruz Foundation (FIOCRUZ), Recife PE, Brazil

Ricardo Durães-Carvalho, Otávio Valério de Carvalho and Lindomar José Pena
Department of Virology, Aggeu Magalhães Institute (IAM), Oswaldo Cruz Foundation (FIOCRUZ), Recife PE, Brazil

Antonio Mauro Rezende
Department of Microbiology, Aggeu Magalhães Institute (IAM), Oswaldo Cruz Foundation (FIOCRUZ), Recife PE, Brazil

Alain Kohl
MRC-University of Glasgow Centre for Virus Research, Glasgow G61 1QH, UK

Gabriel Luz Wallau
Department of Entomology, Aggeu Magalhães Institute (IAM), Oswaldo Cruz Foundation (FIOCRUZ), Recife PE, Brazil

Chuphong Thongnak, Apichaya Puangpetch and Chonlaphat Sukasem
Division of Pharmacogenomics and Personalized Medicine, Department of Pathology, Faculty of Medicine Ramathibodi Hospital, Mahidol University, Bangkok, Thailand

Chuphong Thongnak, Apichaya Puangpetch and Chonlaphat Sukasem
Laboratory for Pharmacogenomics, Somdech Phra Debaratana Medical Center (SDMC), Ramathibodi Hospital, Bangkok, Thailand

Chuphong Thongnak
Department of Clinical Pathology, Faculty of Medicine Vajira Hospital, Navamindradhiraj University, Bangkok, Thailand

Areerat Hnoonual
Graduate Program in Biomedical Sciences, Faculty of Medicine, Prince of Songkla University, Songkhla, Thailand

Duangkamol Tangviriyapaiboon
Rajanagarindra Institute of Child Development, Chiang Mai, Thailand

Suchaya Silvilairat
Division of Pediatric Cardiology, Department of Pediatrics, Faculty of Medicine, Chiang Mai University, Chiang Mai, Thailand

Ekawat Pasomsub and Wasun Chantratita
Excellence Center for Genomic Medicine, Faculty of Medicine Ramathibodi Hospital, Mahidol University, Bangkok, Thailand
Virology and Molecular Microbiology Unit, Department of Pathology, Faculty of Medicine Ramathibodi Hospital, Mahidol University, Bangkok, Thailand

Pornprot Limprasert
Human Genetics Unit, Department of Pathology, Faculty of Medicine, Prince of Songkla University, Hat Yai, Songkhla, Thailand
Faculty of Medicine, King Mongkut's Institute of Technology Ladkrabang, Bangkok, Thailand

Zhe Long, Tianjiao Li, Zhao Chen, Yun Peng, Chunrong Wang, Xiaocan Hou, Hongyu Yuan, Puzhi Wang, Yue Xie, Lang He, Xin Zhou, Huirong Peng, Beisha Tang and Hong Jiang
Department of Neurology, Xiangya Hospital, Central South University, Changsha, Hunan 410008, China

Zhe Long
Sydney Medical School and the Brain & Mind Institute, The University of Sydney, 94 Mallett St, Camperdown, NSW 2050, Australia

Rong Qiu
School of Information Science and Engineering, Central South University, Changsha, Hunan 410083, China

Kun Xia
Laboratory of Medical Genetics, Central South University, Changsha, Hunan 410078, China

Beisha Tang
National Clinical Research Center for Geriatric Diseases, Xiangya Hospital, Central South University, Changsha, Hunan 410008, China

Beisha Tang and Hong Jiang
Key Laboratory of Hunan Province in Neurodegenerative Disorders, Central South University, Changsha, Hunan 410008, China

Beisha Tang
Parkinson's Disease Center of Beijing Institute for Brain Disorders, Beijing 100069, China
Collaborative Innovation Center for Brain Science, Shanghai 200032, China
Collaborative Innovation Center for Genetics and Development, Shanghai 200433, China

Beisha Tang and Hong Jiang
Xinjiang Medical University, Xinjiang 830011, China

Luana de Fátima Alves, Gabriel Lencioni Lovate and Guilherme Marcelino Viana de Siqueira
Department of Biochemistry, Faculdade de Medicina de Ribeirão Preto, University of São Paulo, Ribeirão Preto, SP, Brazil

Cauã Antunes Westmann
Department of Cell Biology, Faculdade de Medicina de Ribeirão Preto, University of São Paulo, Ribeirão Preto, SP, Brazil

Tiago Cabral Borelli and María-Eugenia Guazzaroni
Department of Biology, Faculdade de Filosofia, Ciências e Letras de Ribeirão Preto, University of São Paulo, Ribeirão Preto, SP, Brazil

Suliman Khan and Hongwei Hou
The Key Laboratory of Aquatic Biodiversity and Conservation of Chinese Academy of Sciences, Institute of Hydrobiology, Chinese Academy of Sciences, Wuhan, Hubei 430072, China

Pengfei Duan and Lunguang Yao
Collaborative Innovation Center of Water and Security for Water Source Region of Mid-Line of South-to-North Diversion Project, College of Agricultural Engineering, Nanyang Normal University, Nanyang, Henan, China

Xiaofang Jin, Dandan Liu, Yeyun Li and Changjun Jiang
State Key Laboratory of Tea Plant Biology and Utilization, Anhui Agricultural University, Hefei, Anhui 230036, China

Xiaofang Jin, Linlong Ma, Ziming Gong and Dan Cao
Fruit and Tea Research Institute, Hubei Academy of Agricultural Sciences, Wuhan, Hubei 430064, China

Lili Yang, Shijun Hao, Haiyan Song, Tao Liu and Peiyong Zheng
Institute of Digestive Diseases, Longhua Hospital, China-Canada Center of Research for Digestive Diseases (ccCRDD), Shanghai University of Traditional Chinese Medicine, Shanghai 200032, China

Weili Lin
Key Laboratory of Computational Biology, CAS-MPG Partner Institute for Computational Biology, Shanghai Institutes for Biological Sciences, Chinese Academy of Sciences, Shanghai 200032, China
University of Chinese Academy of Sciences, Beijing 100049, China

Colleen A. Nugent
Digestive Diseases and Nutrition Center, Women and Children's Hospital of Buffalo, Department of Pediatrics, The State University of New York at Buffalo, Buffalo, NY 14214, USA

Cintia Mareque, Martín Beracochea and Federico Battistoni
Departamento de Bioquímica y Genómica Microbianas, Instituto de Investigaciones Biológica Clemente Estable, Avenida Italia 3318, Montevideo 11600, Uruguay

Thais Freitas da Silva, Renata Estebanez Vollú and Lucy Seldin
Laboratório de Genética Microbiana, Instituto de Microbiologia Paulo de Góes, Universidade Federal do Rio de Janeiro (UFRJ), Rio de Janeiro, Brazil

Olga Raskina
Institute of Evolution University of Haifa, Aba-Hushi Avenue 199, Mount Carmel, Haifa 498838, Israel

Heng Lu and Yi Chen
Department of Cardiovascular Surgery, Union Hospital, Fujian Medical University, Fuzhou, China

Linlin Li
Department of Nephrology, Union Hospital, Fujian Medical University, Fuzhou, China

Hao-Bo Guo, Yue Ma and Hong Guo
Department of Biochemistry and Cellular and Molecular Biology, University of Tennessee, Knoxville, TN 37996, USA

Gerald A. Tuskan and Xiaohan Yang
Biosciences Division, Oak Ridge National Laboratory, Oak Ridge, TN 3783, USA

Shang-Qian Xie, Tai-Yu Cao, Kai-Kai Ji, Jie Zhu and Peng Ling
Research Center for Terrestrial Biodiversity of the South China Sea, Institute of Tropical Agriculture and Forestry, Hainan University, Haikou 570228, China

Yue Han and Chuan-Le Xiao
State Key Laboratory of Ophthalmology, Zhongshan Ophthalmic Center, Sun Yat-sen University, Guangzhou 510060, China

Xiao-Zhou Chen
School of Mathematics and Computer Science, Yunnan Minzu University, Kunming 650031, China

Omotayo Opemipo Oyedara, Xianwu Guo, Temidayo Oluyomi Elufisan, Rafael Alejandro Cantú González and Mario A. Rodríguez Pérez
Instituto Politécnico Nacional, Centro de Biotecnología Genómica, 88710 Reynosa, TAMPS, Mexico

Omotayo Opemipo Oyedara
Department of Biological Sciences, College of Science, Engineering and Technology, Faculty of Basic and Applied Science, Osun State University, PMB 4494, Osogbo, Osun State, Nigeria

Aldo Segura-Cabrera
Red de Estudios Moleculares Avanzados, Instituto de Ecología, A.C., Xalapa Enriquez, VER, Mexico

Temidayo Oluyomi Elufisan
National Center for Technology Management, Agency of the Federal Ministry of Science and Technology (FMST), Obafemi Awolowo University, Ile-Ife, Nigeria

Guillermo de Anda-Jáuregui, Jesús Espinal-Enríquez and Enrique Hernández-Lemus
Computational Genomics Division, National Institute of Genomic Medicine, 14610 Mexico City, Mexico

Jesús Espinal-Enríquez and Enrique Hernández-Lemus
Centro de Ciencias de la Complejidad, Universidad Nacional Autónoma de México, 04510 Mexico City, Mexico

Diana Drago-García
Department of Biological Regulation, Weizmann Institute of Science, 7610001 Rehovot, Israel

Index